TRANSLATIONS SERIES IN MATHEMATICS AND ENGINEERING

TRANSLATIONS SERIES IN MATHEMATICS AND ENGINEERING

M.I. Yadrenko
Spectral Theory of Random Fields

1983, viii + 259 pp.
ISBN 0-911575-00-6 Optimization Software, Inc.
ISBN 0-387-90823-4 Springer-Verlag New York Berlin Heidelberg Tokyo
ISBN 3-540-90823-4 Springer-Verlag Berlin Heidelberg New York Tokyo

G.I. Marchuk
Mathematical Models In Immunology

1983, xxv + 353 pp.
ISBN 0-911575-01-4 Optimization Software, Inc.
ISBN 0-387-90901-X Springer-Verlag New York Berlin Heidelberg Tokyo
ISBN 3-540-90901-X Springer-Verlag Berlin Heidelberg New York Tokyo

A.A. Borovkov, Ed.
Advances In Probability Theory: Limit Theorems and Related Problems

1984, xiv + 378 pp.
ISBN 0-911575-03-0 Optimization Software, Inc.
ISBN 0-387-90945-1 Springer-Verlag New York Berlin Heidelberg Tokyo
ISBN 3-540-90945-1 Springer-Verlag Berlin Heidelberg New York Tokyo

V.A. Dubovitskij
The Ulam Problem of Optimal Motion of Line Segments

1985, xiv + 114 pp.
ISBN 0-911575-04-9 Optimization Software, Inc.
ISBN 0-387-90946-X Springer-Verlag New York Berlin Heidelberg Tokyo
ISBN 3-540-90946-X Springer-Verlag Berlin Heidelberg New York Tokyo

N.V Krylov, R.S. Liptser, and A.A. Novikov, Eds.
Statistics and Control of Stochastic Processes

1985, xiv + 507 pp.
ISBN 0-911575-18-9 Optimization Software, Inc.
ISBN 0-387-96101-1 Springer-Verlag New York Berlin Heidelberg Tokyo
ISBN 3-540-96101-1 Springer-Verlag Berlin Heidelberg New York Tokyo

Yu. G. Evtushenko
Numerical Optimization Techniques

1985, xiv + 561 pp.
ISBN 0-911575-07-3 Optimization Software, Inc.
ISBN 0-387-90949-4 Springer-Verlag New York Berlin Heidelberg Tokyo
ISBN 3-540-90949-4 Springer-Verlag Berlin Heidelberg New York Tokyo

Continued on page 453

Vladimir F. Dem'yanov
and Leonid V. Vasil'ev

NONDIFFERENTIABLE OPTIMIZATION

Optimization Software, Inc.
Publications Division, New York

Authors
Vladimir F. Dem'yanov
Department of Applied Mathematics
and Control Processes
Leningrad State University
Leningrad
USSR

Leonid V. Vasil'ev
Department of Mathematics
Leningrad Polytechnic Institute
Leningrad
USSR

Series Editor
A.V. Balakrishnan
School of Engineering
University of California
Los Angeles, CA 90024
USA

Translator
Prof. Tetsushi Sasagawa
Department of Mechanical
Engineering
Sophia University
7-1, Kioi-cho, Chiyoda-ku
Tokyo
Japan

Library of Congress Cataloging-in-Publication Data

Dem'ianov, V.F. (Vladimir Fedorovich), 1938-
Nondifferentiable optimization.

(Translations series in mathematics and engineering)
Includes bibliographies and index.
1. Operations research. 2. Mathematical
optimization. I. Vasil'ev, L.V. (Leonid Vasil'evich)
II. Title. III. Series.
T57.6.D4613 1985 519.8 85-18736
ISBN 0-911575-09-X

Printed in the United States of America.

Worldwide Distribution Rights by Springer-Verlag New York Inc.,
175 Fifth Avenue, New York, New York 10010, USA and
Springer-Verlag Berlin Heidelberg New York Tokyo,
Heidelberg Platz 3, Berlin-Wilmersdorf-33, The Federal
Republic of Germany

ISBN 0-911575-09-X Optimization Software, Inc.
ISBN 0-387-90951-6 Springer-Verlag New York Berlin Heidelberg Tokyo
ISBN 3-540-90951-6 Springer-Verlag Berlin Heidelberg New York Tokyo

About the Authors

Vladimir F. Dem'yanov received his Ph.D. from Leningrad State University in 1964 and his D.Sci. degree from the Computing Center of the USSR Academy of Sciences in 1972. He was appointed Professor at the Applied Mathematics Department of Leningrad State University in 1970. His main interest is Nondifferentiable Optimization.

Leonid V. Vasil'ev received his Ph.D. from Leningrad State University in 1977. In 1980 he was appointed Professor at the Mathematics Department of the Leningrad Polytechnic Institute. His main interest is also Nondifferentiable Optimization.

Contents

Preface

Of recent coinage, the term "nondifferentiable optimization" (NDO) covers a spectrum of problems related to finding extremal values of nondifferentiable functions. Problems of minimizing nonsmooth functions arise in engineering applications as well as in mathematics proper. The Chebyshev approximation problem is an ample illustration of this. Without loss of generality, we shall consider only *minimization* problems.

Among nonsmooth minimization problems, minimax problems and convex problems have been studied extensively ([31], [36], [57], [110], [120]). Interest in NDO has been constantly growing in recent years (monographs: [30], [81], [127] and articles and papers: [14], [20], [87]-[89], [98], [130], [135], [140]-[142], [152], [153], [160], all dealing with various aspects of nonsmooth optimization).

For solving an arbitrary minimization problem, it is necessary to:

1. Study properties of the objective function, in particular, its differentiability and directional differentiability.
2. Establish necessary (and, if possible, sufficient) conditions for a global or local minimum.
3. Find the direction of descent (steepest or, simply, feasible--in appropriate sense).
4. Construct methods of successive approximation.

In this book, the minimization problems for nonsmooth functions of a finite number of variables are considered. Of fundamental importance are necessary conditions for an extremum (for example, [24], [45], [57], [73], [74], [103], [159], [163], [167], [168].

In the case of smooth functions, the importance of the concept of a gradient is well known. However, for nonsmooth functions, gradients do not exist. For a maximum function and a convex function, the subgradient plays a role similar to that of the gradient: with every point x_0 we associate a compact set $\partial f(x_0)$, which is called the subdifferential of the function $f(x)$ at the point x_0. Using the subdifferential, it is possible to:

1. Find the directional derivative of the function at the point x_0:

$$\frac{\partial f(x_0)}{\partial g} \equiv \lim_{\alpha \to +0} \alpha^{-1}[f(x_0 + \alpha g) - f(x_0)] = \max_{v \in \partial f(x_0)} (v,g) .$$

2. Verify necessary conditions for a minimum: for a point x^* to be a minimum point of the function $f(x)$ on E_n, it is necessary that

$$0 \in \partial f(x^*) .$$

3. Find the direction of steepest descent: if $0 \notin \partial f(x_0)$, then the direction

$$g(x_0) = -v(x_0) \, \|v(x_0)\|^{-1} ,$$

where $\|v(x_0)\| = \min_{v \in \partial f(x_0)} \|v\|$, $v(x_0) \in \partial f(x_0)$, is the direction of steepest descent of the function $f(x)$ at the point x_0.

Such an important role of the subdifferential has prompted an attempt to extend the concept of a subdifferential to Lipschitzian functions: F.H. Clarke [133], [134]; J. Warga [9], [168]; B.H. Pshenichnyj [104]; N.Z. Shor [126], [127]; A. Gol'dshtejn [139], [140], among others.

Using subdifferentials and subgradients, it is possible to construct several methods of successive approximation for minimizing convex functions, maximum functions, as well as other classes of functions ([30], [36], [91], [127], [149]-[151], [156], [170], [171]).

The problem of minimizing a smooth function $f(x)$ on the set

$$\Omega = \{x \in E_n \mid h_i(x) \leqslant 0 \quad \forall i \in 1:N\} ,$$

where $h_i(x)$ is a smooth function on E_n, is in fact a problem of NDO, because the set Ω can be represented as follows:

$$\Omega = \{x \in E_n \mid h(x) \leqslant 0\} ,$$

where $h(x) = \max_{i \in 1:N} h_i(x)$ is no longer a smooth function.

The objective of this book is a systematic exposition of the theory of optimization of nondifferentiable functions. In Chapter 1, the basic results from the theory of convex functions, convex sets, and point-to-set mappings are introduced. Much attention is paid to ε-subdifferentials and properties of ε-subdifferential mappings. Convex functions are essential not only because they constitute a large class of nonsmooth functions, but also because the tools of the theory of convex functions can be extended to more general classes of nonsmooth functions.

This concept of a convex function and of a maximum function is tied in with that of a directional derivative. Quite a few authors, among those cited above, do not use directional derivatives in their generalizations of the subdifferential. However, in optimization problems, the directional derivative is more natural, as well as more useful.

In Chapter 2, a new class of nondifferentiable functions, that is, the class of quasidifferentiable functions, is described. For such functions the concept of a quasidifferential, which is closely related to that of a directional derivative, plays a significant role. It appears that for each point there exists a pair of convex sets (quasidifferential). The quasidifferential is a generalization of the concept of a derivative (for smooth functions) and of a subdifferential (for convex functions).

The notion of quasidifferentials simplifies considerably the statement of necessary conditions for an extremum and the problem of finding the directions of steepest descent and ascent. The principal formulas of quasidifferential calculus, which is indeed a generalization of the classical quasidifferential calculus, are established next. The class of quasidifferentiable functions is a linear space closed with respect to all "differentiable" operations as well as operations of taking pointwise maxima and minima (while the class of convex functions is not a linear space but a convex cone). The concept of quasidifferentiable sets is a natural extension. A necessary condition for an extremum of a quasidifferentiable function on a quasidifferentiable set is established in terms of quasidifferentials, which essentially extends the class of problems which can be investigated analytically. For a large class of quasidifferentiable functions, it is possible to algorithmize the process of verifying necessary conditions, as well as the process of finding steepest descent or ascent directions. However, numerical techniques still need to be developed.

Chapters 3 and 4 are devoted to numerical methods for solving NDO problems, including minimization of convex functions and maximum functions. Successive approximation methods are classified as relaxation and non-relaxation methods. A method is called the relaxation method if the value of a function at each step is smaller than that at the preceding step. We discuss both classes of methods, but not the advantages of one versus the other, because the "dragon" of optimization is multiheaded and it takes a special sword to cut-off each head. Thus, the method of subgradient descent is simple to instrument but converges very slowly. Many methods depend on the aims and available means. Sometimes, it is possible to make a rough but quick approximation; in other cases, high accuracy may be needed and computational complexity is not a problem.

Most of these methods are "first-order" methods, since the first-order approximations (derivative, subgradient, subdifferential) are used. One might expect that a further development of the NDO theory will involve higher-order methods.

Some material is relegated to exercises. We do not consider stochastic procedures ([51], [83], [107], [117]); nor problems of game theory ([21], [58], [63], [64]) and those of multicriteria optimization ([195]), where NDO is needed.

Included in this book are the results of recent research in nonsmooth optimization, obtained at the Department of Applied Mathematics/Control Processes and at the Institute of Computational Mathematics of Leningrad State University.

Some results were reported at the Seventh and Eighth All-Union Summer Schools on Optimization at Shchukino (1977) and Shushenskoe (1979).

The authors are indebted to N.N. Moiseev for his encouragement and support throughout; and to A.B. Pevnyj and A.M. Rubinov for a careful reading of the manuscipt and constructive suggestions. Thanks are also due to E.F. Vojton, M.K. Gavurin, Yu.M. Ermol'ev, S.S. Kukateladze, V.N. Malozemov, B.N. Pshenichnyj, and V.M. Tikhomirov for their many helpful comments for improving the exposition.

Notation

$\inf\{f(x) \mid x \in A\}$ is shortened to $\inf_{x \in A} f(x)$.

$$\Omega = \{v \in E_n \mid \exists\, \alpha_0 > 0: x_0 + \alpha v \in A \;\forall\, \alpha \in [0, \alpha_0]\}$$

is interpreted as follows: Ω is a set of points $v \in E_n$, for which there exists an $\alpha_0 > 0$ such that $x_0 + \alpha v \in A$ for all $\alpha \in [0, \alpha_0]$.

The set of integers from p to q is denoted by p:q.

The lower and upper limits are denoted by $\underline{\lim}$ and $\overline{\lim}$, respectively.

The number of elements of a set A is denoted by $|A|$.

The symbol $\equiv$ implies "equal by definition."

The symbol ■ indicates the end of a proof.

Material which is used in the sequel is delineated by an asterisk.

Chapter 1

FUNDAMENTALS OF CONVEX ANALYSIS AND RELATED PROBLEMS

1. CONVEX SETS. CONVEX HULLS. SEPARATION THEOREM

1. In what follows we shall consider the n-dimensional Euclidean space E_n of vectors $x = (x^{(1)},\dots,x^{(n)})$. The space E_n is assumed to be linear. Let us introduce some notation:

$$0_n = 0 = (0,\dots,0) \in E_n ,$$

$$x_k = (x_k^{(1)},\dots,x_k^{(n)}) ,$$

$$(x_1,x_2) = \sum_{i=1}^{n} x_1^{(i)} x_2^{(i)} ,$$

$$||x|| = \sqrt{(x,x)} ,$$

$$x^2 = (x,x) .$$

The Cauchy-Buniakowski inequality

$$|(x_1,x_2)| \le ||x_1|| \; ||x_2||$$

is valid for all vectors $x_1,x_2 \in E_n$.

The vectors $x_1,\dots,x_r$ are said to be *linearly independent* if the equality $\sum_{k=1}^{r} \alpha_k x_k = 0$ implies that all coefficients α_k, $k \in 1:r$ are equal to zero.

If $r \ge n+1$, then the vectors $x_1,\dots x_r$ are *linearly dependent*, i.e., there exist scalars $\beta_1,\dots,\beta_r$ such that $\sum_{k=1}^{r} \beta_k^2 > 0$ (i.e., the β_k are not all equal to zero) and

$$\sum_{k=1}^{r} \beta_k x_k = 0 \tag{1.1}$$

If $r \geq n+2$, then we have the equality

$$\sum_{k=1}^{r} \beta_k = 0 \tag{1.2}$$

in addition to (1.1).

To prove this, we introduce the vectors

$$\bar{x}_k = (1, x_k^{(1)}, \ldots, x_k^{(n)}) \in E_{n+1}, \quad k \in 1:r, \quad r \geq n+2 .$$

Since any $n+2$ vectors in E_{n+1} are linearly dependent, there exist scalars β_k such that $\sum_{k=1}^{r} \beta_k^2 > 0$ and

$$\sum_{k=1}^{r} \beta_k \bar{x}_k = 0_{n+1} . \tag{1.3}$$

It follows from (1.3) that

$$\sum_{k=1}^{r} \beta_k x_k = 0_n , \qquad \sum_{k=1}^{r} \beta_k = 0$$

(here we have set the first component and each of the n remaining components equal to zero).

The set which contains no elements is said to be *empty* and is denoted by $\emptyset$.

Let

$$S_\delta(x_0) = \{x \in E_n \mid \|x-x_0\| \leq \delta\} , \qquad \delta > 0 .$$

The set $S_\delta(x_0)$ is said to be a δ-*neighborhood* of the point x_0. A point x_0 is said to be an *interior point* of a set G if there exists a $\delta > 0$ such that $S_\delta(x_0) \subset G$. We shall denote the set of interior points of a set G by $\operatorname{int} G$ (this set may be empty).

A set $G \subset E_n$ is said to be *open* if for any x_0 there exists a

$\delta > 0$ such that $S_\delta(x_0) \subset G$. It is obvious that $G = \text{int}\, G$ for any open set G.

A set of points x which may be represented in the form $x = \lim_{k\to\infty} x_k$, where $x_k \in G$ $\forall k \in 1:\infty$, is said to be the *closure* of a set $G \subset E_n$. We shall denote the closure of a set G by $\bar{G}$.

A set $G \subset E_n$ is said to be *closed* if $x_0 \in G$ follows from the relation $x_k \xrightarrow[k\to\infty]{} x_0$, $x_0 \in G$ $\forall k \in 1:\infty$. It is obvious that $G = \bar{G}$ for any closed set G.

A point x_0 is said to be a *boundary point* of a set $G \subset E_n$ if, for any $\delta > 0$, its δ-neighborhood $S_\delta(x_0)$ includes at least one point which does not belong to G and at least one point which does belong to G (here x_0 may not belong to G). We shall denote the set of boundary points of a set G by G_{fr}.

A set G is said to be *bounded* if there exists a real number $K < +\infty$ such that $\|x\| \le K$ $\forall x \in G$.

A set G is said to be *unbounded* if for any $K > 0$ there exists an $x \in G$ such that $\|x\| > K$.

It is obvious that the union, intersection, sum and difference of two bounded sets are again bounded sets.

The intersection of two sets of which at least one is bounded is a bounded set.

If A and B are closed sets, then their union and intersection are again closed sets. However, this property no longer holds for the sum, difference and algebraic difference.

EXAMPLE 1. Let

$$A = \{x = (x^{(1)}, x^{(2)}) \in E_2 \mid x^{(2)} \ge \frac{1}{x^{(1)}},\ x^{(1)} > 0\},$$

$$B = \{x = (x^{(1)}, x^{(2)}) \in E_2 \mid x^{(1)} = 0,\ x^{(2)} \le 0\}.$$

It is obvious that the sets A and B are closed but not bounded.

The set

$$C = A+B = \{x=(x^{(1)},x^{(2)}) \mid x^{(1)} > 0,\ x^{(2)} \in (-\infty,\infty)\}$$

is not closed because we have

$$\bar{C} = \{x=(x^{(1)},x^{(2)}) \mid x^{(1)} \geq 0,\ x^{(2)} \in (-\infty,\infty)\} \neq C .$$

However, if the sets A and B are closed and at least one of them is bounded, then their sum (and algebraic difference) is also closed.

A set which has the property that, for every sequence constructed from its elements, we can select a convergent subsequence the limit of which belongs to the original set is said to be *compact*. It is well known that a set in E_n is compact iff it is closed and bounded.

<u>DEFINITION 1</u>. A set $\Omega \subset E_n$ is said to be *convex* if, in addition to two arbitrary points $x_1, x_2 \in \Omega$, the set contains the line segment connecting these points, i.e., $[x_1,x_2] \subset \Omega$, where

$$[x_1,x_2] = \{x \in E_n \mid x = \alpha x_1 + (1-\alpha)x_2,\ \alpha \in [0,1]\} .$$

A convex set Ω is said to be *strictly convex* if for any $x_1, x_2 \in \Omega$, $x_1 \neq x_2$, and any $\alpha \in (0,1)$ we have $x_\alpha = \alpha x_1 + (1-\alpha)x_2 \in \operatorname{int}\Omega$.

There exists another definition of a convex set.

<u>DEFINITION 1*</u>. A set $\Omega \subset E_n$ is said to be *convex* if, in addition to two arbitrary points x_1, x_2, the set includes the point $\frac{1}{2}(x_1+x_2)$, i.e., if, for any $x_1, x_2 \in \Omega$, the center point of the line segment connecting the points x_1, x_2 also belongs to Ω.

Clearly, if a set Ω is convex in the sense of Definition 1, then it is also convex in the sense of Definition 1*.

The converse is not true.

EXAMPLE 2. Let Ω be the set of rational numbers of $[0,1]$.

Obviously, this set is convex in the sense of Definition 1*, but it is not convex in the sense of Definition 1.

PROBLEM 1.1. Prove that if a set Ω is closed and convex in the sense of Definition 1*, then it is convex in the sense of Definition 1 as well.

REMARK 1. Throughout this book, when we refer to convex sets, we have in mind the convexity described by Definition 1.

LEMMA 1.1. Let $\Omega \subset E_n$ be a closed set and assume that, for any $x_1, x_2 \in \Omega$, there exists an $\alpha \in (0,1)$ (where α depends on x_1, x_2) such that $x_\alpha = \alpha x_1 + (1-\alpha)x_2 \in \Omega$. Then the set Ω is convex.

Proof. Suppose that this is not true, i.e., that the set Ω is not convex. Then there exist $x_1, x_2 \in \Omega$, $x_1 \neq x_2$ and an $\alpha_0 \in (0,1)$ such that

$$x_{\alpha_0} = \alpha_0 x_1 + (1-\alpha_0)x_2 \notin \Omega .$$

Let

$$G_1 = \Omega \cap [x_1, x_{\alpha_0}] , \qquad G_2 = \Omega \cap [x_{\alpha_0}, x_2] .$$

These sets are closed and, moreover, $x_{\alpha_0} \notin G_1$, $x_{\alpha_0} \notin G_2$. Hence, we can find

$$\min_{x \in G_1} \|x - x_{\alpha_0}\| = \|\bar{x} - x_{\alpha_0}\| \neq 0 ,$$

$$\min_{x \in G_2} \|x - x_{\alpha_0}\| = \|\bar{\bar{x}} - x_{\alpha_0}\| \neq 0 ,$$

where $\bar{x} \in G_1$, $\bar{\bar{x}} \in G_2$, i.e., $\bar{x} \in \Omega$, $\bar{\bar{x}} \in \Omega$, and there should be no point of Ω on $(\bar{x}, x_{\alpha_0}]$ and $[x_{\alpha_0}, \bar{\bar{x}})$. This contradicts the assumptions of the lemma (here $(x,y] = [x,y]\setminus\{x\}$, $[x,y) = [x,y]\setminus\{y\}$). ∎

2. Let $G \subset E_n$. The set

$$\text{co } G = \left\{ x = \sum_{k=1}^{r} \alpha_k x_k \mid x_k \in G,\ \alpha_k \ge 0,\ \sum_{k=1}^{r} \alpha_k = 1,\ r \text{ an arbitrary natural number} \right\}$$

is said to be the *convex hull* of the set G.

THEOREM 1.1 (Carathéodory's Theorem). Any vector $x \in \text{co } G$ can be represented as a convex combination of $n+1$ or fewer vectors from the set G.

Proof. Let $x \in \text{co } G$, i.e.,

$$x = \sum_{k=1}^{r} \alpha_k x_k, \qquad \alpha_k \ge 0, \qquad \sum_{k=1}^{r} \alpha_k = 1.$$

Without loss of generality, let all α_k be positive. We assume that $r \ge n+2$. The proof shown above (see (1.2)) implies that there exist scalars β_k, $\sum_{k=1}^{r} \beta_k^2 > 0$ such that

$$\sum_{k=1}^{r} \beta_k x_k = 0, \qquad \sum_{k=1}^{r} \beta_k = 0. \tag{1.4}$$

Let $\varepsilon = \min_{\{k \mid \beta_k > 0\}} \beta_k^{-1} \alpha_k > 0$ and $\bar{\alpha}_k = \alpha_k - \varepsilon \beta_k$, $k \in 1:r$.

Here, $\varepsilon > 0$ because, by virtue of (1.4), there exist both positive and negative coefficients β_k (at least one of each).

Now, clearly, we have

$$x = \sum_{k=1}^{r} \bar{\alpha}_k x_k, \qquad \sum_{k=1}^{r} \bar{\alpha}_k = 1, \qquad \bar{\alpha}_k \ge 0 \quad \forall k.$$

It is also clear that at least one of the scalars $\bar{\alpha}_k$ is reduced to zero, i.e., x has already been represented as a convex combination of no more than $r - 1$ vectors from G.

In a similar way, we can obtain a representation with $r \le n+1$. ∎

LEMMA 1.2. If $G \subset E_n$ is a bounded closed set, then $\text{co } G$ is a convex bounded closed set.

Proof. Let $x \in \operatorname{co} G$, i.e.,

$$x = \sum_{k=1}^{r} \alpha_k x_k = (x^{(1)},\dots,x^{(n)}) , \qquad \alpha_k \geq 0 ,$$
$$\sum_{k=1}^{r} \alpha_k = 1 , \qquad x_k \in G . \tag{1.5}$$

If G is bounded, then the components of any vector in G are bounded in absolute values by the same number M.

Hence, for the vector x, we have

$$|x^{(i)}| = \left| \sum_{k=1}^{r} \alpha_k x_k^{(i)} \right| \leq \sum_{k=1}^{r} \alpha_k |x_k^{(i)}| \leq M \sum_{k=1}^{r} \alpha_k = M ,$$

i.e., the components of the vector x (see (1.5)) in $\operatorname{co} G$ are bounded by the same number M. Therefore, $\operatorname{co} G$ is a bounded set.

Now let us prove the convexity of $\operatorname{co} G$.

Let $x \in \operatorname{co} G$, $y \in \operatorname{co} G$. Without loss of generality (if necessary, by adding terms with null coefficients), we can let

$$x = \sum_{k=1}^{r} \alpha_k x_k , \qquad y = \sum_{k=1}^{r} \alpha'_k x_k ,$$

$$x_k \in G , \qquad \alpha_k \geq 0 , \qquad \alpha'_k \geq 0 , \qquad \sum_{k=1}^{r} \alpha_k = \sum_{k=1}^{r} \alpha'_k = 1 .$$

Then the convex combination of x and y:

$$x_\alpha = \alpha x + (1-\alpha) y , \qquad \alpha \in [0,1]$$

becomes

$$x_\alpha = \sum_{k=1}^{r} (\alpha\alpha_k + (1-\alpha)\alpha'_k) x_k = \sum_{k=1}^{r} \bar{\alpha}_k x_k , \tag{1.6}$$

where $\bar{\alpha}_k = \alpha\alpha_k + (1-\alpha)\alpha'_k$. It is obvious that

$$\bar{\alpha}_k \geq 0 , \qquad \sum_{k=1}^{r} \bar{\alpha}_k = \alpha \sum_{k=1}^{r} \alpha_k + (1-\alpha) \sum_{k=1}^{r} \alpha'_k = 1 ,$$

i.e., $x_\alpha \in \operatorname{co} G$ for any $\alpha \in [0,1]$. This implies that the set $\operatorname{co} G$ is convex.

Let us show that the set $\text{co}\, G$ is closed. Let $x_s \to x^*$, $x_s \in \text{co}\, G$. Then

$$x_s = \sum_{k=1}^{r_s} \alpha_{sk} x_{sk}, \qquad \alpha_{sk} \geq 0,$$
$$\sum_{k=1}^{r_s} \alpha_{sk} = 1, \qquad x_{sk} \in G, \qquad k \in 1:r_s, \qquad s \in 1:\infty. \tag{1.7}$$

According to Carathéodory's theorem, we can choose a representation x_s such that $r_s \leq n+1$. Let us suppose that $r_s = n+1$ (if necessary, using arbitrary vectors from G with null coefficients in representation (1.7)).

Since G is a bounded set and $\alpha_{sk} \in [0,1]$, we can choose a sequence $\{s_i\}$ such that

$$x_{s_i k} \xrightarrow[s_i \to \infty]{} x_{0k}, \qquad \alpha_{s_i k} \xrightarrow[s_i \to \infty]{} \alpha_{0k}.$$

It is clear from (1.7) that

$$x^* = \sum_{k=1}^{n+1} \alpha_{0k} x_{0k}, \qquad x_{0k} \in G, \qquad \alpha_{0k} \geq 0, \qquad \sum_{k=1}^{n+1} \alpha_{0k} = 1,$$

i.e., $x^* \in \text{co}\, G$. This completes the proof of the lemma. ∎

PROBLEM 1.2. Prove that if G is a convex set, then $\text{co}\, G = G$.

REMARK 2. The following property is obvious: if $A_i \subset E_n$, $i \in I = 1:N$, are convex sets, then

$$\text{co} \bigcup_{i \in I} A_i = \left\{ x = \sum_{i \in I} \alpha_i x_i \;\middle|\; x_i \in A_i,\ \alpha_i \geq 0,\ \sum_{i \in I} \alpha_i = 1 \right\}.$$

THEOREM 1.2 (Separation Theorem). Let $\Omega \subset E_n$ be a closed convex set and let $x_0 \notin \Omega$. Then there exist a vector $g_0 \in E_n$, $\|g_0\| = 1$, and a number $a > 0$ such that, for any $x \in \Omega$,

$$(x - x_0, g_0) \leq -a. \tag{1.8}$$

Proof. Let us find

$$\min_{z\in\Omega}\|x_0-z\|^2 = \|x_0-z_0\|^2 \equiv \rho_0^2 ,$$

where $\rho_0 = \|x_0-z_0\|$. Here, as is easily seen, the minimal value can be achieved. Since $x_0 \notin \Omega$, then $\rho_0 > 0$. Hence we let

$$g_0 = -\rho_0^{-1}(z_0 - x_0) . \tag{1.9}$$

We shall show that for all $x \in \Omega$ the inequality

$$(x-x_0, g_0) \leq -\rho_0 \tag{1.10}$$

is valid. Let us assume the opposite. Then there exists an $\bar{x} \in \Omega$ such that

$$(\bar{x}-x_0, g_0) = -b_0 > -\rho_0 . \tag{1.11}$$

Next, let

$$x_\alpha = \alpha\bar{x} + (1-\alpha)z_0 = z_0 + \alpha(\bar{x} - z_0) , \qquad \alpha \in [0,1] .$$

Then, clearly, $x_\alpha \in [z_0, \bar{x}]$ and, moreover, we have

$$\begin{aligned}(x_0 - x_\alpha)^2 &= (x_0 - z_0)^2 - 2\alpha(\bar{x} - z_0, x_0 - z_0) + \alpha^2(\bar{x} - z_0)^2\\ &= \rho_0^2 - 2\alpha[(\bar{x}-x_0, x_0-z_0) - (z_0-x_0, x_0-z_0)] + \alpha^2(\bar{x}-z_0)^2.\end{aligned}$$

Considering (1.9) and (1.11), we obtain

$$(x_0 - x_\alpha)^2 = \rho_0^2 - 2\alpha[-b_0\rho_0 + \rho_0^2] + \alpha^2(\bar{x} - z_0)^2 . \tag{1.12}$$

Since $\rho_0^2 - b_0\rho_0 > 0$ from (1.11), it follows from (1.12) that, for sufficiently small $\alpha \in (0,1)$, $(x_0-x_\alpha)^2 < \rho_0^2$. This contradicts the definition of the point z_0.

Thus, since (1.10) is equivalent to (1.8), the theorem is proved. ∎

<u>REMARK 3</u>. From the proof of this theorem it follows that if $\Omega \subset E_n$ is a closed convex set and $x_0 \in E_n$, then there exists a unique point $z_0 \in \Omega$ such that

$$(z-x_0, z_0-x_0) \geq \|z_0 - x_0\|^2 \qquad \forall x \in \Omega .$$

COROLLARY 1. If $\Omega \subset E_n$ is a closed convex set and x_0 is a boundary point of the set Ω, then there exists a vector $g_0 \in E_n$, $\|g_0\| = 1$, such that

$$(x-x_0, g_0) \le 0 \qquad \forall x \in \Omega . \tag{1.13}$$

Proof. Let x_0 be a boundary point of Ω. Then there exists a sequence $\{x_k\}$ such that $x_k \notin \Omega$, $x_k \to x_0$. Since $x_k \notin \Omega$, then by Theorem 1.2 we have for all $x \in \Omega$

$$(x-x_k, g_k) \le -a_k , \tag{1.14}$$

where

$$a_k = \min_{z \in \Omega} \|x_k - z\| = \|x_k - z_k\| > 0 ,$$

$$g_k = -\|z_k - x_k\|^{-1}(z_k - x_k) , \qquad \|g_k\| = 1 .$$

It is clear that $a_k > 0$ and $a_k = \min_{z \in \Omega} \|x_k - z\| \le \|x_0 - x_k\|$, i.e. $a_k \to 0$.

Without loss of generality, we can assume that $g_k \to g_0$, $\|g_0\| = 1$.

Taking the limit in (1.14), we get

$$(x-x_0, g_0) \le 0 \qquad \forall x \in \Omega . \qquad \blacksquare$$

REMARK 4. Condition (1.8) implies that the plane

$$(x-x_0, g_0) = 0 \tag{1.15}$$

which passes through the point x_0 has the property that the set Ω lies completely on one side of this plane and, moreover, that the distance from the set Ω to this plane is not less than α.

Condition (1.13) implies that the set Ω lies completely on one side of the plane (1.15).

COROLLARY 2. Let $\Omega \subset E_n$ be a convex set and let $x_0 \in \Omega$. Then there exists a vector $g_0 \in E_n$, $||g|| = 1$, such that

$$(x-x_0, g_0) \leq 0 \qquad \forall\, x \in \bar{\Omega} .$$

PROBLEM 1.3. Prove the following theorem and corollary.

THEOREM 1.3. Let Ω_1 and Ω_2 be closed convex sets and let at least one of them be bounded. If the sets Ω_1 and Ω_2 have no common points, then there exist a vector g_0, a point $x_0 \in \Omega_1$ and a number $a > 0$ such that

$$(x-x_0, g_0) \leq 0 \qquad \forall x \in \Omega_1 ,$$

$$(x-x_0, g_0) \geq a \qquad \forall x \in \Omega_2 .$$

Hint. Find

$$\min_{x\in\Omega_1,\, y\in\Omega_2} ||x-y||^2 = ||x_0 - y_0||^2 .$$

COROLLARY. Let Ω_1 and Ω_2 be closed convex sets and, moreover, let at least one of them be solid. If these sets have no common interior points, i.e., if

$$\text{int } \Omega_1 \cap \text{int } \Omega_2 = \emptyset , \tag{1.16}$$

then there exist a vector g_0 and a point $x_0 \in \Omega_1$ such that

$$(x-x_0, g_0) \leq 0 \qquad \forall x \in \Omega_1 ,$$

$$(x-x_0, g_0) \geq 0 \qquad \forall x \in \Omega_2 .$$

REMARK 5. The assertion in this corollary can be strengthened by replacing (1.16) by one of the conditions:

$$\text{int } \Omega_1 \cap \Omega_2 = \emptyset , \qquad \Omega_1 \cap \text{int } \Omega_2 = \emptyset . \tag{1.17}$$

2. POINT-TO-SET MAPPINGS

1. Let Ω be some arbitrary set. We denote by $\Pi(\Omega)$ the collection of all nonempty subsets of the set Ω. Let Ω_1 and Ω_2 be arbitrary sets in finite-dimensional spaces. A mapping $A(x)$ which transforms each point $x \in \Omega_1$ to some subset of the set Ω_2 is said to be a *point-to-set mapping* (p.s.m.) from Ω_1 to Ω_2 or a mapping of Ω_1 to $\Pi(\Omega_2)$. We shall denote this by $A:\Omega_1 \to \Pi(\Omega_2)$.

Let $A \subset E_n$, $B \subset E_n$. Let us define

$$\rho(A,B) = \sup_{w \in A} \inf_{v \in B} \|v-w\| + \sup_{v \in B} \inf_{w \in A} \|v-w\| . \tag{2.1}$$

The number $\rho(A,B)$ is called the *distance* between the sets A and B in the *Hausdorff metric* (or the *Hausdorff distance*).

The value $\sup_{w \in A} \inf_{v \in B} \|v-w\|$ is called the *distance* of A from the set B (in the Hausdorff sense), and the value $\sup_{v \in B} \inf_{w \in A} \|v-w\|$ is called the *distance* of B from the set A (in the Hausdorff sense).

PROBLEM 2.1. Prove that if A and B are closed sets in the space E_n, then the equality $\rho(A,B) = 0$ is equivalent to the relation $A = B$.

REMARK 1. The assumption that A and B are closed in Problem 2.1 is essential, as is seen from the following example.

Let

$$A = \{x = (x^{(1)}, x^{(2)}) \in E_2 \mid x^{(i)} \text{ rational numbers}, \; |x^{(i)}| \le 1, \; i \in 1:2\} ,$$

$$B = \{x = (x^{(1)}, x^{(2)}) \in E_2 \mid |x^{(i)}| \le 1, \; i \in 1:2\} .$$

The set A is not closed. Obviously $\rho(A,B) = 0$, but $A \neq B$.

A mapping $A(x)$ is said to be *continuous in the Hausdorff sense (H-continuous)* at a point x_0 if

$$\rho(A(x), A(x_0)) \xrightarrow[x \to x_0]{} 0 . \tag{2.2}$$

Point-to-set mapping $A:\Omega_1 \to \Pi(\Omega_2)$, where $\Omega_1 \subset E_n$, $\Omega_2 \subset E_m$, is said to be *upper semicontinuous* (u.s.c.) at a point $x_0 \in \Omega_1$ if it follows from $x_k \to x_0$, $v_k \in A(x_k)$, $v_k \to v_0$ that $v_0 \in A(x_0)$.

We notice that there are various definitions of upper semicontinuity [86], which coincide if Ω_2 is bounded.

The set

$$Z = \{[x,y] \in \Omega_1 \times \Omega_2 \mid x \in \Omega_1, \quad y \in A(x)\} \tag{2.3}$$

is called the *graph* of mapping $A(x)$.

PROBLEM 2.2. Prove that a mapping $A(x)$ is upper semicontinuous on Ω_1 iff the graph Z (see (2.3)) is closed in $\Omega_1 \times \Omega_2$.

PROBLEM 2.3. Let p.s.m. $A:\Omega_1 \to \Pi(\Omega_2)$ be upper semicontinuous on Ω_1. Then, prove that for an arbitrary $x \in \Omega_1$ the set $A(x)$ is closed.

The converse is not true, as is seen from the following example.

Let $\Omega_1 = \Omega_2 = E_1$,

$$A(x) = \begin{cases} [0,1] , & x \neq 0 , \\ [0,\frac{1}{2}] , & x = 0 . \end{cases}$$

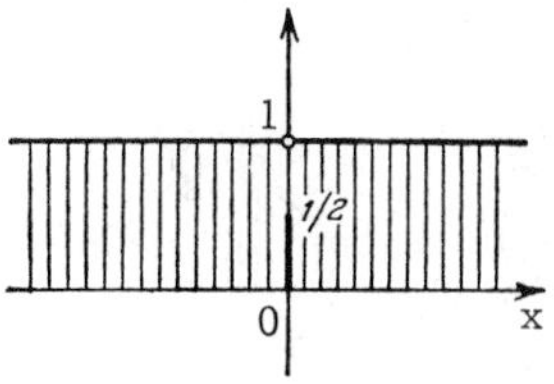

Figure 1

The set $A(x)$ is closed for an arbitrary x but the mapping $A(x)$ is not u.s.c. at the point $x_0 = 0$ (Figure 1).

A mapping $A:\Omega_1 \to \Pi(\Omega_2)$ is said to be *lower semicontinuous* (l.s.c.) at a point $x_0 \in \Omega_1$ if for all $v_0 \in A(x_0)$ and an arbitrary sequence $\{x_k\}$, $x_k \to x_0$, $x_k \in \Omega_1$, there exist $v_k \in A(x_k)$ such that that $v_k \to v_0$.

A mapping $A(x)$ is said to be *continuous in the Kakutani sense (K-continuous)* at a point x_0 if it is both upper and lower semicontinuous at this point.

The following example illustrates the concepts of upper and lower semicontinuity.

EXAMPLE 1. Let $\Omega_1 = \Omega_1 = E_1$. Let us consider the mappings

$$A_1(x) = \begin{cases} [x,1+x] , & x \neq 0 , \\ [0,\frac{1}{2}] , & x = 0 , \end{cases} \qquad A_2(x) = \begin{cases} [x,1+x] , & x \neq 0 , \\ [0,2] , & x = 0 . \end{cases}$$

At the point $x_0 = 0$, the mapping $A_1(x)$ is lower semicontinuous but not upper semicontinuous. On the other hand, the mapping $A_2(x)$ is upper semicontinuous but not lower semicontinuous (Figures 2,3).

A mapping $A(x)$ is said to be *bounded* on Ω_1 if for any bounded subset $G \subset \Omega_1$ there exists a bounded subset $B \subset \Omega_2$ such that $A(x) \subset B \quad \forall x \in G$.

A mapping $A(x)$ is said to be *closed* at a point x_0 if the set $A(x_0)$ is closed.

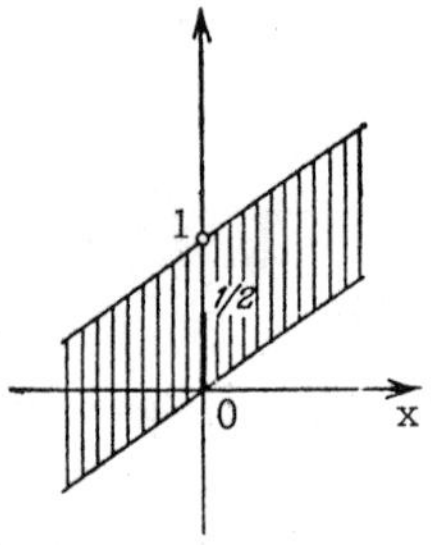

Figure 2

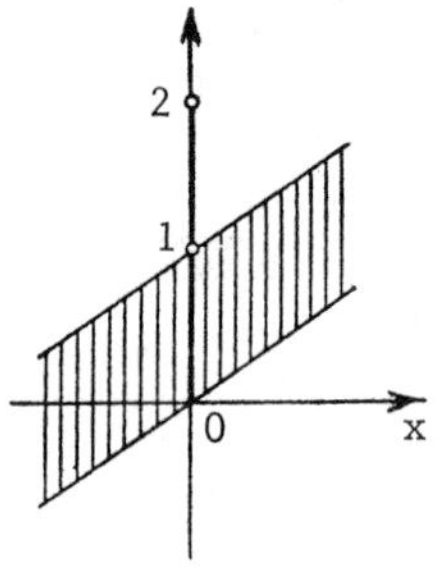

Figure 3

THEOREM 2.1 [78]. If a mapping $A(x)$ is bounded on some neighborhood of a point x_0 and closed at the point x_0, then the continuity of $A(x)$ in the Kakutani sense at the point x_0 follows from the continuity of $A(x)$ in the Hausdorff sense at the point x_0 and vice versa, i.e., the notions of H-continuity and K-continuity are equivalent in this case.

Proof. Let p.s.m. $A(x)$ be continuous in the Hausdorff sense at the point x_0, i.e., we have (2.2). Let us show first that $A(x)$ is u.s.c.

Let $x_k \to x_0$, $v_k \in A(x_k)$, $v_k \to v_0$.

Assume that $v_0 \notin A(x_0)$. Due to the fact that $A(x_0)$ is closed, there exists an $a > 0$ such that for sufficiently large k

$$\inf_{v\in A(x_0)} \|v-v_k\| = \min_{v\in A(x_0)} \|v-v_k\| \geq a > 0 .$$

Then we have (see (2.1))

$$\rho(A(x_k),A(x_0)) \geq \inf_{v\in A(x_0)} \|v-v_k\| \geq a > 0 ,$$

which contradicts (2.2). Hence p.s.m. $A(x)$ is u.s.c.

Now we shall establish that p.s.m. $A(x)$ is lower semicontinuous. Let $v_0 \in A(x_0)$ and $x_k \to x_0$. Define

$$b_k = \inf_{v\in A(x_k)} \|v-v_0\| .$$

Since $b_k \leq \rho(A(x_0),A(x_k))$ and $A(x)$ is continuous in the Hausdorff sense, then $\rho(A(x_0),A(x_k)) \to 0$, and hence $b_k \to 0$.

Then, for any sequence $\{\gamma_k\}$ such that $\gamma_k \to +0$, there exist $\gamma_k \in A(x_k)$ such that $\|v_k-v_0\| \leq b_k + \gamma_k$, i.e., $v_k \to v_0$, and this implies (by virtue of the arbitrariness of $v_0 \in A(x_0)$ and the sequence $\{x_k\}$) that $A(x)$ is lower semicontinuous. K-continuity follows from this lower semicontinuity and the upper semicontinuity established above.

Now let p.s.m. $A(x)$ be K-continuous at a point x_0. Suppose that it is not H-continuous at this point. Then there exist a number $a > 0$ and a sequence $\{x_k\}$ such that

$$x_k \to x_0 , \qquad \rho(A(x_k),A(x_0)) \geq 2a > 0 . \tag{2.4}$$

In this case (see (2.1)) we have for every fixed k either

$$\sup_{w\in A(x_k)} \inf_{v\in A(x_0)} \|v-w\| \geq a \tag{2.5}$$

or

$$\sup_{v\in A(x_0)} \inf_{w\in A(x_k)} \|v-w\| \geq a . \tag{2.6}$$

Without loss of generality, we may assume that either (2.5) or

(2.6) is valid for all of the sequences $\{x_k\}$. We shall investigate each of these cases in detail.

In case (2.5), there exist $w_k \in A(x_k)$ such that

$$\inf_{v\in A(x_0)} \|v-w_k\| \geq \frac{a}{2} . \tag{2.7}$$

By assumption, p.s.m. $A(x)$ is bounded in a neighborhood of the point x_0, and hence the sequence $\{w_k\}$ is also bounded. Let us consider a convergent subsequence $\{w_{k_s}\}$, $w_{k_s} \to w_0$. From (2.7) it is obvious that $w_0 \notin A(x_0)$, and this contradicts the upper semicontinuity of $A(x)$ at the point x_0. Hence case (2.5) is not possible.

In case (2.6), there exist $v_k \in A(x_0)$ for each k such that

$$\inf_{w\in A(x_k)} \|v_k-w\| \geq \frac{a}{2} . \tag{2.8}$$

Since the set $A(x_0)$ is closed and bounded, there exists a subsequence $\{k_s\}$ such that $v_{k_s} \to v_0 \in A(x_0)$. From (2.8) we have for sufficiently large k_s

$$\inf_{w\in A(x_{k_s})} \|v_0-w\| \geq \frac{a}{4} ,$$

which contradicts the lower semicontinuity of the mapping $A(x)$ at the point x_0. ∎

COROLLARY. If $A(x)$ is H-continuous and closed at a point x_0, then it is also K-continuous.

If $A(x)$ is K-continuous at a point x_0 and bounded in some neighborhood of the point x_0, then it is also H-continuous.

REMARK 2. The requirements of the conditions in Theorem 2.1 are minimal. Indeed, let $A(x)$ be both H-continuous and K-continuous. Since $A(x)$ is K-continuous, it is upper semicontinuous and therefore (see Problem 2.3) $A(x_0)$ is closed. We shall now give an example which shows that it is essential that $A(x)$ be bounded in a neighborhood of the point x_0 (and also at the point x_0 itself, naturally). Let $\Omega_1 = \Omega_2 = E_1$,

$$A(x) = \begin{cases} [-1+x, 1+x] \cup \{x^{-1}\}, & x \neq 0 \\ [-1,1], & x = 0. \end{cases}$$

The mapping $A(x)$ is not bounded in a neighborhood of the point $x_0 = 0$.

Clearly, the mapping $A(x)$ is lower semicontinuous at the point $x_0 = 0$, because, for any $v_0 \in [-1,1]$ and any sequence $\{x_k\}$ such that $x_k \to 0$, there exists a sequence $\{v_k\}$ such that $v_k = (v_0 + x_k) \in A(x_k)$, $v_k \to v_0$.

On the other hand, if $x_k \to 0$, $v_k \in [-1+x_k, 1+x_k] \subset A(x_k)$, $v_k \to v_0$, then $v_0 \in [-1,1]$, i.e., $v_0 \in A(x_0)$; in other words, the mapping $A(x)$ is upper semicontinuous. Therefore $A(x)$ is both u.s.c. and l.s.c., that is, it is K-continuous at the point $x_0 = 0$.

However, for $|x^{-1}| - |x| > 1$, we have

$$\rho(A(x), A(x_0)) > |x^{-1}| - 1 - |x| + |x| = |x^{-1}| - 1 \to +\infty,$$

i.e., p.s.m. $A(x)$ is not H-continuous at the point $x_0 = 0$.

2. Let $A \subset E_n$, $B \subset E_n$.

LEMMA 2.1. If

$$\sup_{v \in A} (v, g) = \sup_{v \in B} (v, g) \tag{2.9}$$

for all $g \in E_n$, then

$$\overline{\text{co } A} = \overline{\text{co } B}. \tag{2.10}$$

Proof. Assume the opposite, i.e., that $\overline{\text{co } A} \neq \overline{\text{co } B}$. Then there exists either a point $v_0 \in \overline{\text{co } A}$ such that $v_0 \notin \overline{\text{co } B}$ or a point $w_0 \in \overline{\text{co } B}$ such that $w_0 \notin \overline{\text{co } A}$. Let us assume, for the sake of arguments, that we have a point $v_0 \in \overline{\text{co } A}$, $v_0 \notin \overline{\text{co } B}$. Then, by the separation theorem, there exist a number $a > 0$ and a vector g_0, $\|g_0\| = 1$, such that

$$(w-v_0, g_0) \le -a \qquad \forall w \in \overline{\text{co } B} . \tag{2.11}$$

Since $v_0 \in \overline{\text{co } A}$, we have (see Theorem 1.1)

$$v_0 = \lim_{k\to\infty} v_k , \qquad v_k = \sum_{i=1}^{n+1} \alpha_{ki} v_{ki} ,$$

$$v_{ki} \in A , \quad \alpha_{ki} \ge 0 , \quad \sum_{i=1}^{n+1} \alpha_{ki} = 1 .$$

Then

$$(v_k, g_0) = \sum \alpha_{ki}(v_{ki}, g_0) \le \sum \alpha_{ki} \sup_{v\in A} (v, g_0) = \sup_{v\in A} (v, g_0) .$$

Taking the limit, we obtain

$$(v_0, g_0) \le \sup_{v\in A} (v, g_0) . \tag{2.12}$$

Since $B \subset \overline{\text{co } B}$, it follows from (2.11) and (2.12) that

$$\sup_{w\in B} (w, g_0) \le (v_0, g_0) - a \le \sup_{v\in A} (v, g_0) - a ,$$

which contradicts (2.9). This proves the lemma. ∎

COROLLARY. If A and B are closed convex sets, then (2.9) is equivalent to the relation $A = B$.

It is easy to prove that

$$\sup_{v\in A} (v, g) = \sup_{v\in \text{co } A} (v, g) = \sup_{v\in \overline{\text{co } A}} (v, g) . \tag{2.13}$$

REMARK 3. By an argument similar to that used in the proof of Lemma 2.1, we can show that if

$$\sup_{v\in A} (v, g) \ge \sup_{v\in B} (v, g) \qquad \forall g \in E_n ,$$

then $\overline{\text{co } A} \supset \overline{\text{co } B}$.

3. Let a set $A(x)$ be closed, convex and bounded for all

$$x \in \text{int } S_\delta(x_0) ,$$

where

$$\delta > 0 , \qquad S_\delta(x_0) = \{x \in E_n \mid \|x-x_0\| \le \delta\} .$$

Define

$$h(x,g) = \max_{v \in A(x)} (v,g) .$$

<u>LEMMA 2.2</u>. If the function $h(x,g)$ is continuous on $S_\delta(x_0) \times E_n$, then the mapping $A(x)$ is H-continuous at the point x_0.

<u>Proof</u>. Let us first establish that the mapping $A(x)$ is upper semi-continuous at the point x_0. Let $x_k \to x_0$, $v_k \in A(x_k)$, $v_k \to v_0$. We have to show that

$$v_0 \in A(x_0) . \tag{2.14}$$

Suppose that this is not the case, i.e., that $v_0 \notin A(x_0)$. By the separation theorem there exist a number $a > 0$ and a vector $g_0 \in E_n$, $||g_0|| = 1$, such that

$$(g_0, v-v_0) \leq -a < 0 \qquad \forall v \in A(x_0) ,$$

yielding

$$h(x_0,g_0) = \max_{v \in A(x_0)} (v,g_0) \leq (v_0,g_0) - a . \tag{2.15}$$

On the other hand, we have $h(x_k,g_0) = \max_{v \in A(x_k)} (v,g_0) \geq (v_k,g_0)$.

Here, taking the limit as $k \to \infty$ and using the continuity of $h(x,g_0)$ with respect to x and inequality (2.15), we obtain

$$h(x_0,g_0) \geq (v_0,g_0) \geq \max_{v \in A(x_0)} (v,g_0) + a = h(x_0,g_0) + a ,$$

which is impossible since $a > 0$. This contradiction proves the upper semicontinuity of $A(x)$ at the point x_0.

Now we establish the lower semicontinuity of $A(x)$. Let $v_0 \in A(x_0)$.

Take an arbitrary sequence $\{x_k\}$, $x_k \to x_0$. We have to show that there exist $v_k \in A(x_k)$ such that $v_k \to v_0$. Suppose that this is not the case.

Without loss of generality, we can assume that

$$\rho(A(x_k),v_0) = \min_{v\in A(x_k)} ||v-v_0|| = ||v_k-v_0|| \geq a > 0 .$$

From the necessary condition for the minimum of the function $\phi(v) = ||v-v_0||^2$ on the set $A(x_k)$ (see Remark 3 in Section 1) we have

$$(||v_k-v_0||^{-1}(v_k - v_0), v - v_0) \geq a \qquad \forall v \in A(x_k) . \qquad (2.16)$$

Let

$$g_k = -||v_k-v_0||^{-1}(v_k - v_0) .$$

Without loss of generality, we can assume that $g_k \to g_0$. Then, from (2.16), we obtain

$$h(x_k,g_k) = \max_{v\in A(x_k)} (v,g_k) \leq (v_0,g_k) - a$$

$$\leq \max_{v\in A(x_0)} (v,g_k) - a = h(x_0,g_k) - a .$$

Here, taking the limit as $k\to\infty$ and making use of the continuity of h with respect to x and g, we get

$$h(x_0,g_0) \leq h(x_0,g_0) - a ,$$

which is impossible since $a > 0$. This contradiction proves the lower semicontinuity of the p.s.m. $A(x)$. Upper semicontinuity has already been established above. Hence, $A(x)$ is closed at the point x_0 and bounded in a neighborhood of the point x_0. Thus, from Theorem 2.1, $A(x)$ is H-continuous.

The lemma is proved. ■

REMARK 4. It may not be necessary to require that $A(x)$ be bounded because this will follow from the continuity of $h(x,g)$ with respect to g. However, in this case we must define $h(x,g)$ by

$$h(x,g) = \sup_{v\in A(x)} (v,g) . \qquad (2.17)$$

LEMMA 2.3. Let an upper semicontinuous p.s.m. $A:\Omega_1 \to \Pi(\Omega_2)$ be given on Ω_1, where Ω_2 is a compact set in E_m, and an upper semicontinuous p.s.m. $B:\Omega_1 \times \Omega_2 \to \Pi(E_p)$ be given on $\Omega_1 \times \Omega_2$. Then the point-to-set mapping $C(x) = \bigcup_{z \in A(x)} B(x,z)$ is upper semicontinuous on Ω_1.

Proof. Let $x^* \in \Omega_1$, $x_k \to x^*$, $v_k \to v^*$, $v_k \in C(x_k)$. Since $v_k \in C(x_k)$, then for each k there exists a point $z_k \in A(x_k)$ such that $v_k \in B(x_k, z_k)$.

Let us select from the sequence $\{z_k\}$ a convergent subsequence $\{z_{k_s}\}$, $z_{k_s} \to z^*$. Because $A(x)$ is upper semicontinuous, we have $z^* \in A(x^*)$.

We also have

$$(x_{k_s}, z_{k_s}) \to (x^*, z^*) , \qquad v_{k_s} \to v^* , \qquad v_{k_s} \in B(x_{k_s}, z_{k_s}) .$$

Since $B(x,z)$ is upper semicontinuous, we obtain $v^* \in B(x^*, z^*)$, i.e., $v^* \in C(x^*)$. The lemma is proved. ∎

LEMMA 2.4. Let a point-to-set mapping

$$G: \Omega_1 \to \Pi(\Omega_2) , \qquad \Omega_1 \subset E_n , \qquad \Omega_2 \subset E_m ,$$

be upper semicontinuous and bounded on Ω_1. Then $\operatorname{co} G(x)$ is upper semicontinuous on Ω_1.

Proof. Let

$$x^* \in \Omega_1, \qquad x_s \in \Omega_1, \qquad x_s \to x^*, \qquad v_s \to v^*, \qquad v_s \in \operatorname{co} G(x_s) .$$

We need to establish that $v^* \in \operatorname{co} G(x^*)$.

By Carathéodory's theorem (Theorem 1.1) the representation

$$v_s = \sum_{k=1}^{m+1} \alpha_{sk} v_{sk}$$

is valid for each s, where $\alpha_{sk} \geq 0$, $\sum_{k=1}^{m+1} \alpha_{sk} = 1$, $v_{sk} \in G(x_s)$ $\forall k \in 1:(m+1)$.

Since $G(x)$ is bounded, we can select a sequence $\{s_i\}$ such that

$$v_{s_i k} \xrightarrow[i\to\infty]{} v_k^* , \qquad \alpha_{s_i k} \to \alpha_k^* , \qquad k \in 1:(m+1) .$$

Then we have

$$v^* = \sum_{k=1}^{m+1} \alpha_k^* v_k^* , \qquad \alpha_k^* \geq 0 , \qquad \sum_{k=1}^{m+1} \alpha_k^* = 1$$

and, furthermore, since

$$x_{s_i} \xrightarrow[i\to\infty]{} x^* , \qquad v_{s_i k} \to v_k^* , \qquad v_{s_i k} \in G(x_{s_i})$$

and $G(x)$ is upper semicontinuous, we obtain $v_k^* \in G(x^*)$ for each $k \in 1:(m+1)$, which completes the proof. ■

PROBLEM 2.4. Prove that if $G_1(x)$ and $G_2(x)$ are upper semicontinuous and bounded on Ω_1, then the mapping $co\{G_1(x) \cup G_2(x)\}$ is upper semicontinuous on Ω_1.

3. CONVEX CONE. CONE OF FEASIBLE DIRECTIONS. CONJUGATE CONE

1. DEFINITION 1. A set $\Gamma \subset E_n$ is said to be a *cone* if in addition to x it contains points $v = \lambda x$ for any $\lambda > 0$.

Let $\Omega \subset E_n$ be a closed convex set and let $x_0 \in \Omega$. We consider the set

$$\gamma(x_0) = \{v = \lambda(x - x_0) \mid \lambda > 0, \ x \in \Omega\} .$$

Clearly, $\gamma(x_0)$ is a cone and $0 \in \gamma(x_0)$.

The closure of the cone $\gamma(x_0)$ is the *cone of feasible directions* of the set Ω at the point x_0 and is denoted by $\Gamma(x_0)$.

REMARK 1. If $x_0 \in \operatorname{int} \Omega$, then, clearly,

$$\gamma(x_0) = \Gamma(x_0) = E_n .$$

LEMMA 3.1. If Ω is a closed convex set and $x_0 \in \Omega$, then $\Gamma(x_0)$ is a closed convex cone.

Proof. It is clear from the definition of a closed set that $\Gamma(x_0)$

is a closed cone. To prove the convexity of $\Gamma(x_0)$, it is sufficient to show that $\gamma(x_0)$ is convex. Let $v_1, v_2 \in \gamma(x_0)$, i.e., $v_1 = \lambda_1(x_1 - x_0)$, $v_2 = \lambda_2(x_2 - x_0)$, where $\lambda_1 > 0$, $\lambda_2 > 0$, $x_1, x_2 \in \Omega$.

We need to show that

$$v_\alpha = \alpha v_1 + (1-\alpha)v_2 \in \gamma(x_0) \qquad \text{for } \alpha \in [0,1] .$$

Let

$$\bar{\lambda} = \alpha\lambda_1 + (1-\alpha)\lambda_2 > 0 , \qquad \alpha_0 = \frac{\alpha\lambda_1}{\bar{\lambda}} , \qquad 0 \le \alpha_0 \le 1 .$$

From the convexity of Ω, the point $\bar{x} = \alpha_0 x_1 + (1-\alpha_0)x_2$ belongs to Ω. Then we have

$$\begin{aligned} v_\alpha &= \alpha\lambda_1(x_1 - x_0) + (1-\alpha)\lambda_2(x_2 - x_0) \\ &= \bar{\lambda}[\alpha_0(x_1 - x_0) + (1-\alpha_0)(x_2 - x_0)] = \bar{\lambda}(\bar{x} - x_0) \in \gamma(x_0) , \end{aligned}$$

and this completes the proof. ∎

DEFINITION 2. Let $\Gamma \subset E_n$ be a cone. The set

$$\Gamma^+ = \{w \in E_n \mid (w,v) \ge 0 \quad \forall v \in \Gamma\}$$

is called the *conjugate cone* of Γ.

EXAMPLES

1. If $\Gamma = E_n$, then $\Gamma^+ = \{0\}$.
2. If $\Gamma = \{x = (x^{(1)}, \dots, x^{(n)}) \mid x^{(i)} \ge 0 \quad \forall i \in 1:n\}$, then $\Gamma^+ = \Gamma$.
3. If $\Gamma = \{x \in E_n \mid (A,x) \le 0\}$, where $A \in E_n$, then $\Gamma^+ = \{x = -\lambda A \mid \lambda \ge 0\}$.
4. If $\Gamma = \{0\}$, then $\Gamma^+ = E_n$.
5. If $\Gamma(x_0)$ is a cone of feasible directions of a set Ω at the point x_0 and $x_0 \in \operatorname{int}\Omega$, then, as we have already observed, $\Gamma(x_0) = E_n$ and, hence, $\Gamma^+(x_0) = \{0\}$.

PROBLEM 3.1. Let $\Gamma = \{x \in E_n \mid (A,x) = 0\}$. Prove that

$$\Gamma^+ = \{v = \lambda A \mid \lambda \in (-\infty, \infty)\} .$$

LEMMA 3.2. If $\Gamma \subset E_n$ is a cone, then $\Gamma^+ \subset E_n$ is a closed convex cone.

Proof. Follows from the definition of the conjugate cone.

PROBLEM 3.2. Prove that if $\Gamma(x_0)$ is the cone of feasible directions of a set Ω at a point x_0, then the following relation holds:

$$\Gamma^+(x_0) = \{w \in E_n \mid (x-x_0,w) \geq 0 \quad \forall x \in \Omega\} . \tag{3.1}$$

THEOREM 3.1. Let $\Gamma \subset E_n$ be a closed convex cone and let $G \subset E_n$ be a convex compact set. In order that Γ and G have no common points i.e., that

$$\Gamma \cap G = \emptyset , \tag{3.2}$$

it is necessary and sufficient that there exists a vector $w_0 \in \Gamma^+$ such that

$$\max_{x \in G} (w_0,x) < 0 . \tag{3.3}$$

Proof. Sufficiency. Since $w_0 \in \Gamma^+$, and from the definition of the conjugate cone, we have

$$(w_0,v) \geq 0 \qquad \forall v \in \Gamma . \tag{3.4}$$

On the other hand, by assumption (see (3.3)) we have

$$(w_0,x) < 0 \qquad \forall x \in G . \tag{3.5}$$

Inequalities (3.4) and (3.5) imply that Γ and G have no common points.

Necessity. Let Γ and G have no common points. Find (Figure 4)

$$\rho = \min_{v \in \Gamma, x \in G} ||v-x|| = ||v_0 - x_0|| . \tag{3.6}$$

From condition (3.2) we have $||v_0-x_0|| > 0$.

Let $g_0 = \| v_0-x_0 \|^{-1}(v_0-x_0)$. We show first that

$$(v_0,g_0) = 0 . \tag{3.7}$$

Assume the opposite, i.e., $(v_0, g_0) = a \neq 0$. Then $v_0 \neq 0$.

Consider an arbitrary point $v_\lambda = v_0\lambda$ $(\lambda \neq 1)$. For $\lambda > 0$, $v_\lambda \in \Gamma$ (because Γ is a cone) and

$$(v_\lambda - x_0)^2 = (v_0 - x_0 + (\lambda-1)v_0)^2$$
$$= \|v_0 - x_0\|^2 + 2(\lambda-1)a\|v_0-x_0\| + (\lambda-1)^2 v_0^2 \quad .$$

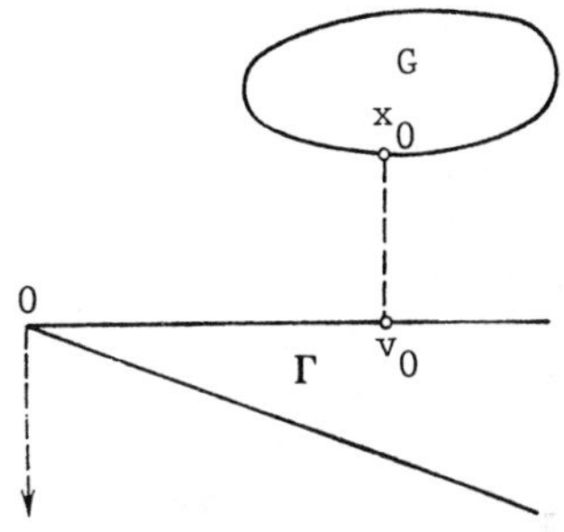

Figure 4

Hence, for a λ which is sufficiently close to 1 and satisfies

$$\operatorname{sign}(\lambda-1) = -\operatorname{sign} a \quad ,$$

we obtain $\|v_\lambda - x_0\|^2 < \|v_0 - x_0\|^2$, which contradicts (3.6).

Using the same argument as in the proof of Theorem 1.2 (inequality (1.10)), we can prove that

$$(x, g_0) \leq -\|v_0 - x_0\| \qquad \forall x \in G \quad . \tag{3.8}$$

We shall now establish that

$$(v, g_0) \geq 0 \qquad \forall v \in \Gamma \quad . \tag{3.9}$$

In fact, considering (3.7), this is clear from

$$(v, g_0) = (v - x_0, g_0) + (x_0 - v_0, g_0) + (v_0, g_0) \geq \rho - \rho = 0 \quad . \tag{3.10}$$

Let $w_0 = g_0$. Then (3.3) follows from (3.8), and we can conclude from (3.9) that $w_0 = g_0 \in \Gamma^+$. The theorem is proved. ■

<u>THEOREM 3.2</u>. Let $\Gamma \subset E_n$ be a closed convex cone and let $G \subset E_n$ be a convex compact set. In order that

$$\operatorname{int} \Gamma \cap G = \emptyset \quad , \qquad \Gamma \cap \operatorname{int} G = \emptyset \quad , \tag{3.11}$$

it is necessary and sufficient that there exists a vector $w_0 \in \Gamma^+$ such that

$$\max_{x \in G} (w_0, x) \leq 0 . \tag{3.12}$$

PROBLEM 3.3. Prove Theorem 3.2.

2. LEMMA 3.3. If $\Gamma \subset E_n$ is a closed convex cone, then

$$\Gamma^{++} = \Gamma . \tag{3.13}$$

Proof. Let $v_0 \in \Gamma$. Then by the definition of the conjugate cone, we have $(w, v_0) \geq 0$ for any $w \in \Gamma^+$, i.e., $v_0 \in \Gamma^{++}$ and hence $\Gamma \subset \Gamma^{++}$.

Let us show the converse inclusion. Assume that $v_0 \in \Gamma^{++}$ but $v_0 \notin \Gamma$.

Let $G = \{v_0\}$. From Theorem 3.1, there exists a $w_0 \in \Gamma^+$ such that $(w_0, v_0) < 0$, i.e., $v_0 \notin \Gamma^{++}$, which contradicts the assumption. ∎

COROLLARY 1. Let $\Gamma \subset E_n$ be a convex cone and $\bar{\Gamma}$ be its closure. Then

$$\Gamma^{++} = \bar{\Gamma} . \tag{3.14}$$

Proof. $\bar{\Gamma}$ is a closed cone. According to (3.13), $\bar{\Gamma}^{++} = \bar{\Gamma}$.

On the other hand, $\bar{\Gamma}^+ = \Gamma^+$. ∎

COROLLARY 2. If $\Gamma_i \subset E_n$, $i \in 1:s$, are convex cones and

$$\sum_{i=1}^{s} \Gamma_i = \left\{ v = \sum_{i=1}^{s} v_i \;\middle|\; v_i \in \Gamma_i, \quad i \in 1:s \right\} ,$$

then

$$\left(\sum_{i=1}^{s} \Gamma_i \right)^+ = \bigcap_{i=1}^{s} \Gamma_i^+ . \tag{3.15}$$

This follows from the definition of the conjugate cone and the fact that $0 \subset \bar{\Gamma}_i$, $i \in 1:s$. ∎

COROLLARY 3. If $\Gamma_i \subset E_n$, $i \in 1:s$, are closed convex cones, then

$$\left(\bigcap_{i=1}^{s} \Gamma_i \right)^+ = \overline{\sum_{i=1}^{s} \Gamma_i^+} . \tag{3.16}$$

Proof.

$$\left(\bigcap_{i=1}^{s} \Gamma_i \right)^+ = \left(\bigcap_{i=1}^{s} (\Gamma_i^+)^+ \right)^+ = \left(\sum_{i=1}^{s} \Gamma_i^+ \right)^{++} = \overline{\sum_{i=1}^{s} \Gamma_i^+} . \qquad ∎$$

PROBLEM 3.4. Prove that if $K_1,\ldots,K_m$ are closed cones in E_n, then either $K_1+\cdots+K_m$ is a closed cone or there exists a nontrivial representation of zero. The latter implies that there exist $x_i \in K_i$ such that $x_1+\cdots+x_m = 0$ and not all of the x_i are equal to zero.

REMARK 2. In (3.16), closure is essential. Let us consider one example which shows that the sum of two closed convex cones is not necessarily a closed set.

EXAMPLE 1. Consider two cones in E_3:

$$K_1 = \{x = (x^{(1)},x^{(2)},x^{(3)}) \mid x^{(1)} = x^{(3)} = 0,\quad x^{(2)} \le 0\},$$

$$K_2 = \{x = (x^{(1)},x^{(2)},x^{(3)}) \mid (x^{(1)})^2 + (x^{(3)})^2 \le 2x^{(2)}x^{(3)},\ x^{(2)} \ge 0\}$$

(the circle $(x^{(1)})^2 + (x^{(3)})^2 = 2x^{(3)}$ is a generatrix of the cone K_2 on the plane $x^{(2)} = 1$). Clearly, K_1 and K_2 are closed cones.

However, their sum

$$K_1 + K_2 = \{x \in E_3 \mid x^{(3)} > 0\} \cup \{x \mid x^{(1)} = x^{(3)} = 0\}$$

is not a closed cone.

PROBLEM 3.5. Prove that if sets $\Omega_1,\ldots,\Omega_s$ are polyhedral (i.e., each of them is given by a finite number of inequalities of the form $(x,A_i) \le b_i \quad \forall i \in 1:m$), then their sum (the set $\Omega_1+\cdots+\Omega_s$) is a closed set.

3. Let G be an arbitrary set in E_n. Denote by $K(G)$ the convex cone spanned by G:

$$K(G) = \left\{v = \sum_{k=1}^{r} \alpha_k v_k \mid v_k \in G,\quad \alpha_k \le 0,\quad k \in 1:r,\quad r \text{ an arbitrary integer}\right\}.$$

The cone $K(G)$ is called the *convex conical hull* of the set G.

PROBLEM 3.6. Prove that if Γ is a convex cone, then $\Gamma + \Gamma = \Gamma$.

Why is the convexity important?

LEMMA 3.4. Any point $v \in K(G)$ can be represented in the form

$$v = \sum_{k=1}^{r} \alpha_k v_k , \qquad v_k \in G , \qquad \alpha_k > 0 , \qquad k \in 1:r ,$$

where $1 \le r \le n$ and the vectors v_k, $k \in 1:r$, are linearly independent.

Proof. Let $v \in K(G)$. Then

$$v = \sum_{k=1}^{r} \alpha_k v_k , \qquad v_k \in G , \qquad \alpha_k \ge 0 .$$

We can assume that $\alpha_k > 0$ $\forall k \in 1:r$. If the v_k are linearly dependent, then there exist β_k, $k \in 1:r$, such that

$$\sum_{k=1}^{r} \beta_k^2 > 0 , \qquad \sum_{k=1}^{r} \beta_k v_k = 0 . \tag{3.17}$$

We can also assume that there exist positive numbers among the $\{\beta_k\}$ (because it follows from (3.17) that there exists at least one non-null coefficient among the $\{\beta_k\}$ and we can regard it as positive, if necessary by changing the sign of all β_k, when (3.17) holds. Let

$$\varepsilon = \min_{\{k \mid \beta_k > 0\}} \beta_k^{-1} \alpha_k , \qquad \bar{\alpha}_k = \alpha_k - \varepsilon \beta_k , \qquad k \in 1:r .$$

It is clear that $v = \sum_{k=1}^{r} \bar{\alpha}_k v_k$ and there is at least one hull coefficient among the $\bar{\alpha}_k$ (the rest being nonnegative).

Thus, if there exist linearly dependent v_k in the representation of a vector $v \in K(G)$, then we can get a new representation involving fewer vectors. Proceeding further in this manner, we finally get a representation in which all the v_k are linearly independent. Here, it is clear that there are no more than n linearly independent vectors v_k. The lemma is proved. ∎

LEMMA 3.5. Let $G \subset E_n$ be a bounded closed set and $0 \notin \text{co}\, G$. Then $K(G)$ is a closed convex cone.

Proof. It is clear that $K(G)$ is a convex cone. Let us prove that it is also closed. First of all we note that

$$K(G) = \{v = \lambda v' \mid \lambda \geq 0,\ v' \in \text{co } G\}, \qquad (3.18)$$

as may easily be verified.

Now let $v_k \to v^*$, $v_k \in K(G)$. We need to prove that $v^* \in K(G)$. From (3.18) we have

$$v_k = \lambda_k v_k', \qquad \lambda_k \geq 0, \qquad v_k' \in \text{co } G. \qquad (3.19)$$

Since G is a bounded set, co G is also bounded. Furthermore, by assumption $0 \notin \text{co } G$. Hence

$$0 < m \leq \|v'\| \leq M < \infty \qquad \forall\, v' \in \text{co } G.$$

The sequence $\{v_k'\}$ in (3.19) is bounded because $v_k' \in \text{co } G$ and satisfies the above inequalities, i.e., $0 < m \leq \|v_k'\| \leq M < \infty$. From this fact and (3.19) we can conclude that $\{\lambda_k\}$ is also a bounded sequence. Therefore, we can choose a subsequence $\{k_s\}$ such that

$$\lambda_{k_s} \to \lambda^*, \qquad v_{k_s}' \to v'.$$

Clearly, $v^* = \lambda^* v'$, $\lambda^* \geq 0$, $v' \in \text{co } G$. This means that $v^* \in K(G)$, which completes the proof. ∎

REMARK 3. The assumption $0 \notin \text{co } G$ in Lemma 3.5 is essential, as is clear from the following example.

Let

$$G = \{x = (x^{(1)}, x^{(2)}) \in E_2 \mid (x^{(1)} - 1)^2 + (x^{(2)})^2 \leq 1\}.$$

Here $0 \in \text{co } G = G$, $K(G) = \{0\} \cup \{x = (x^{(1)}, x^{(2)}) \mid x^{(1)} > 0\}$, i.e., $K(G)$ is not a closed set.

However, if G consists of a finite number of points, then the assumption $0 \notin \text{co } G$ in Lemma 3.5 can be dropped.

LEMMA 3.6. If a set G consists of a finite number of points, then $K(G)$ is a closed convex cone.

Proof. Let $v^* \to v^*$, $v_k \in K(G)$. We need to prove that $v^* \in K(G)$. From Lemma 3.4,

$$v_k = \sum_{i=1}^{r_k} \lambda_{ki} v_{ki} , \qquad \lambda_{ki} > 0 , \qquad v_{ki} \in G .$$

Since G consists of a finite number of points, it is possible to find a subsequence of $\{v_{k_s i}\}$ which contains the same vectors, i.e.,

$$v_{k_s} = \sum_{i=1}^{r} \lambda_{k_s i} v_i , \qquad \lambda_{k_s i} > 0 , \qquad v_i \in G , \qquad r \leq n,$$

where the v_i are linearly independent.

Let us prove that the sequence $\{\lambda_{k_s} = (\lambda_{k_s 1},\ldots,\lambda_{k_s r}) \in E_r\}$ is bounded. Suppose that this is not the case. Without loss of generality, we can assume that

$$\gamma_{k_s} = \sum_{i=1}^{r} \lambda_{k_s i} \to \infty$$

(recall that all $\lambda_{k_s i} > 0$). Then

$$v_{k_s} = \gamma_{k_s} \sum_{i=1}^{r} \lambda'_{k_s i} v_i = \gamma_{k_s} v'_{k_s} ,$$

where

$$v'_{k_s} = \sum_{i=1}^{r} \lambda'_{k_s i} v_i , \qquad \lambda'_{k_s i} = \gamma_{k_s}^{-1} \lambda_{k_s i} > 0 .$$

Since

$$\sum_{k=1}^{r} \lambda'_{k_s i} = 1 , \qquad \text{then} \qquad v'_{k_s} \in \text{co } G .$$

We can assume that

$$\lambda'_{k_s i} \to \lambda_i , \qquad \lambda_i \geq 0 , \qquad \sum_{i=1}^{r} \lambda_i = \Lambda .$$

But

$$v'_{k_s} = \frac{v_{k_s}}{\gamma_{k_s}} , \qquad v_{k_s} \to v^* , \qquad \gamma_{k_s} \to +\infty ,$$

and therefore $v'_{k_s} \to 0$. Since

$$v'_{k_s} = \sum \lambda'_{k_s i} v_i , \qquad \lambda'_{k_s i} \to \lambda_i , \qquad \lambda_i \geq 0 , \qquad \sum \lambda_i = 1 ,$$

then $\sum_{i=1}^{r} \lambda_i v_i = 0$, which contradicts the assumed linear independence of $\{v_i\}$. Thus, $\{\lambda_{k_s}\}$ is a bounded sequence. Choosing any convergent subsequence, we get

$$v^* = \sum_{i=1}^{r} \lambda_i^* v_i , \qquad \text{where} \quad v_i \in G, \quad \lambda_i^* \geq 0,$$

i.e., $v^* \in K(G)$. This completes the proof of the Lemma. ∎

4. We consider the point-to-set mappings $\gamma(x)$, $\Gamma(x) \equiv \bar{\gamma}(x)$ and $\Gamma^+(x)$ defined above for $x \in \Omega$.

LEMMA 3.7. The mapping $\gamma(x)$ is lower semicontinuous on Ω.

Proof. Let $x_0 \in \Omega$. We have to prove that if $v_0 \in \gamma(x_0)$ and $x_i \to x_0$, $x_i \in \Omega$, then there exists a sequence of points $\{v_i\}$, $v_i \in \gamma(x_i)$, such that $v_i \to v_0$. If $v_0 = 0$, then the assertion obviously holds. Assume that $v_0 \neq 0$. Since $v_0 \in \gamma(x_0)$, then we can use the representation $v_0 = \lambda_0(\bar{x} - x_0)$, where $\lambda_0 > 0$, $\bar{x} \in \Omega$, $\bar{x} \neq x_0$. Let us define $v_i = \lambda_0(\bar{x} - x_i)$.

Then, clearly, $v_i \in \gamma(x_i)$ and $v_i \to v_0$. ∎

LEMMA 3.8. The mapping $\Gamma^+(x)$ is upper semicontinuous on Ω.

Proof. Let $x_0 \in \Omega$ and $x_k \to x_0$, $y_k \to y_0$, $x_k \in \Omega$, $y_k \in \Gamma^+(x_k)$.

It is necessary to prove that $y_0 \in \Gamma^+(x_0)$. Since $y_k \in \Gamma^+(x_k)$, then (see Problem 3.2)

$$(x - x_k, y_k) \geq 0 \qquad \forall x \in \Omega .$$

Taking the limit here for a fixed $x \in \Omega$, we have $(x - x_0, y_0) \geq 0$. This inequality holds for all $x \in \Omega$, and this implies that $y_0 \in \Gamma^+(x_0)$. The lemma is proved. ∎

5. Let $A \subset E_n$ be a compact set and let

$$\Gamma = \{v \in E_n \mid (z,v) \le 0 \quad \forall z \in A\} .$$

It is easily verified that Γ is a convex cone.

LEMMA 3.9. The relation

$$\Gamma^+ = -\bar{K}(\text{co } A) \tag{3.20}$$

holds, where $K(G)$ is the convex conical hull of a set G.

Proof. Let us denote by B the right-hand side of (3.20). Let $w \in B$, i.e.,

$$w = \lim_{k\to\infty} w_k , \quad w_k = -\lambda_k z_k , \quad \lambda_k \ge 0 , \quad z_k = \sum_{i=1}^{r_k} \alpha_{ki} z_{ki} ,$$

$$\alpha_{ki} \ge 0 , \quad \sum_{i=1}^{r_k} \alpha_{ki} = 1 , \quad z_{ki} \in A \quad \forall i \in 1:r_k .$$

By virtue of Lemma 3.4, and without loss of generality, we can assume that

$$r_k = r \le n \qquad \forall k .$$

Let us take an arbitrary $v \in \Gamma$. By the definition of the set Γ we have

$$(v, z_{ki}) \le 0 \qquad \forall i \in 1:r .$$

Therefore

$$(v,w_k) = -\lambda_k \sum_{i=1}^{r} \alpha_{ki} (z_{ki}, v) \ge 0 \qquad \forall v \in \Gamma ,$$

and this implies that $w_k \in \Gamma^+$. Since Γ^+ is a closed set, then $w \in \Gamma^+$, i.e.,

$$B \subset \Gamma^+ . \tag{3.21}$$

Now let us prove that

$$\Gamma^+ \subset B . \tag{3.22}$$

Assume that this is not the case. Then there exists a $w_0 \in \Gamma^+$ such that $w_0 \notin B$. By the separation theorem, there exists a $g \notin E_n$, $||g|| = 1$, such that

$$(w_0, g) < 0 , \tag{3.23}$$

$$(v, g) \geq 0 \qquad \forall v \in B . \tag{3.24}$$

Since for any $z \in A$ we have $-z \in B$, it follows from (3.24) that

$$(z, g) \leq 0 \qquad \forall z \in A .$$

In other words, $g \in \Gamma$. However, $w_0 \in \Gamma^+$. Hence $(g, w_0) \geq 0$, which contradicts (3.23). Therefore, (3.22) is proved. Relation (3.20) now follows from (3.21) and (3.22). ■

4. CONVEX FUNCTIONS. CONTINUITY AND DIRECTIONAL DIFFERENTIABILITY

1. Let a function $f(x)$ be defined on a convex set $S \subset E_n$. In what follows, unless we state otherwise, we assume that the function $f(x)$ is finite on its domain of definition, i.e., it takes finite values at any point in the domain of definition.

The set

$$\{[\beta, x] \in E_{n+1} \mid x \in S,\ \beta \in E_1,\ \beta = f(x)\}$$

is called the *graph* of the function $f(x)$. The set

$$\text{epi } f = \{[\beta, x] \in E_{n+1} \mid x \in S,\ \beta \in E_1,\ \beta \geq f(x)\}$$

is called the *epigraph* of the function $f(x)$ defined on the set S.

A function is said to be *convex* on S if

$$f(\alpha x_1 + (1-\alpha)x_2) \leq \alpha f(x_1) + (1-\alpha) f(x_2) \quad \forall x_1, x_2 \in S, \quad \forall \alpha \in [0,1] . \tag{4.1}$$

The epigraph of a convex function is a convex set.

A function $f(x)$ is said to be *strictly convex* on S if

$$f(\alpha x_1 + (1-\alpha)x_2) < \alpha f(x_1) + (1-\alpha) f(x_2)$$

$$\forall x_1, x_2 \in S, \quad x_1 \neq x_2, \quad \forall \alpha \in (0,1) . \tag{4.2}$$

A function $f(x)$ is called *strongly concave* on S if there exists an $m > 0$ such that

$$f(\alpha x_1 + (1-\alpha)x_2) \le \alpha f(x_1) + (1-\alpha)f(x_2) - \alpha(1-\alpha)m\|x_1-x_2\|^2$$

$$\forall x_1, x_2 \in S, \quad \forall \alpha \in [0,1] . \tag{4.3}$$

A function $f(x)$ is called *concave* on S if

$$f(\alpha x_1 + (1-\alpha)x_2) \ge \alpha f(x_1) + (1-\alpha)f(x_2)$$

$$\forall x_1, x_2 \in S, \quad \forall \alpha \in [0,1] .$$

Strictly concave and *strongly concave* functions are defined analogously.

Note that a *linear* function $f(x) = (A,x) + b$, where $A \in E_n$, $b \in (-\infty,\infty)$, is both convex and concave at the same time. From the definition it is also seen that if $f_i(x)$, $i \in 1:N$, are convex functions on S, then the function

$$f(x) = \sum_{i=1}^{N} \alpha_i f_i(x) , \quad \alpha_i \ge 0 , \quad i \in 1:N ,$$

is also convex. In particular, the function $F(x) = Af(x)$ is convex if $A > 0$ and $f(x)$ is a convex function.

PROBLEM 4.1. Prove that if a function $f(x)$ is convex and $x_i \in S$, $\alpha_i \ge 0$, $i \in 1:p$, $\sum_{i=1}^{p} \alpha_i = 1$, then

$$f\left(\sum_{i=1}^{p} \alpha_i x_i\right) \le \sum_{i=1}^{p} \alpha_i f(x_i) . \tag{4.4}$$

This is called *Jensen's inequality*.

PROBLEM 4.2. Prove that if $f(x)$ is a continuous function on S and

$$f(\tfrac{1}{2}(x_1 + x_2)) \le \tfrac{1}{2}[f(x_1) + f(x_2)] \quad \forall x_1, x_2 \in S , \tag{4.5}$$

then the function $f(x)$ is convex on S.

This property does not hold if $f(x)$ is not continuous.

REMARK 1. It is obvious that if $f_i(x)$, $i \in 1:N$, are convex functions and at least one of them is strongly (strictly) convex, then the function $f(x) = \sum_{i=1}^{N} f_i(x)$ is also strongly (strictly) convex.

PROBLEM 4.3. Prove that if $f(x)$ is a convex function and $f(x) \geq 0$ $\forall x \in S$, then $f_1(x) = f^2(x)$ is also convex. Note that in this problem the condition $f(x) \geq 0$ is essential, as shown by the following example.

EXAMPLE 1. Let $x \in E_1$, $f(x) = x^2 - 1$, $f_1(x) = f^2(x)$.

Clearly, $f_1(x)$ is not a convex function.

LEMMA 4.1. If $f(x)$ is a convex function on S, then, for all $\alpha \notin [0,1]$ and $x_1, x_2 \in S$ such that $x_\alpha = \alpha x_1 + (1-\alpha)x_2 \in S$, the inequality

$$f(\alpha x_1 + (1-\alpha)x_2) \geq \alpha f(x_1) + (1-\alpha)f(x_2) \tag{4.6}$$

holds.

Proof. Suppose that this lemma does not hold. Then there exist $x_1, x_2 \in S$ and a point $\bar{x} = \alpha x_1 + (1-\alpha)x_2 \in S$, where $\alpha \notin [0,1]$, such that

$$f(\bar{x}) < \alpha f(x_1) + (1-\alpha)f(x_2). \tag{4.7}$$

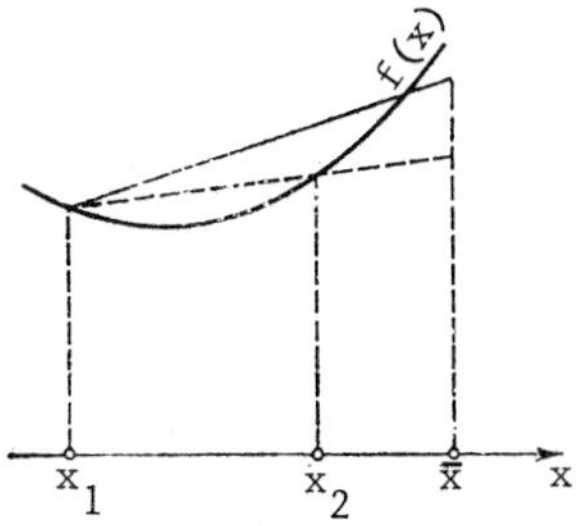

Figure 5

Let $\alpha < 0$ for the sake of discussion. Then, as may easily be verified (Figure 5), we have

$$x_2 = \beta x_1 + (1-\beta)\bar{x},$$

where $\beta = -\alpha(1-\alpha)^{-1} \in (0,1)$. Because the $f(x)$ is convex and from (4.7) we obtain

$$\begin{aligned} f(x_2) &\leq \beta f(x_1) + (1-\beta)f(\bar{x}) \\ &< \beta f(x_1) + (1-\beta)(\alpha f(x_1) + (1-\alpha)f(x_2)) \\ &= (\beta + (1-\beta)\alpha)f(x_1) + (1-\beta)(1-\alpha)f(x_2) = f(x_2), \end{aligned}$$

which is impossible. Similarly, we get a contradiction when $\alpha > 1$ (in this case we have to exchange the positions of the points x_1 and x_2 in Figure 5). ■

LEMMA 4.2. If $f(x)$ is a strongly convex function on S (see (4.3)) then

$$f(\alpha x_1 + (1-\alpha)x_2) \geq f(x_1) + (1-\alpha)f(x_2) - \alpha(1-\alpha)m\|x_1-x_2\|^2 \tag{4.8}$$

for all $x_1, x_2 \in S$ and $\alpha \notin [0,1]$ such that $[\alpha x_1 + (1-\alpha)x_2] \in S$. The proof of this lemma is analogous to that of Lemma 4.1.

Assume that the lemma does not hold. Then there exist $x_1, x_2 \in S$, $x_1 \neq x_2$, $\alpha \notin [0,1]$ such that

$$f(\bar{x}) < \alpha f(x_1) + (1-\alpha)f(x_2) - \alpha(1-\alpha)m\|x_1-x_2\|^2 , \tag{4.9}$$

where $\bar{x} = \alpha x_1 + (1-\alpha)x_2 \in S$.

Let $\alpha < 0$ for the sake of discussion. Then we have again

$$x_2 = \beta x_1 + (1-\beta)\bar{x} , \qquad \beta = -\alpha(1-\alpha)^{-1} \in (0,1) .$$

From (4.3) and (4.9) we obtain

$$\begin{aligned} f(x_2) &\leq \beta f(x_1) + (1-\beta)f(\bar{x}) - \beta(1-\beta)m\|x_1-\bar{x}_2\|^2 \\ &< \beta f(x_1) + (1-\beta)[\alpha f(x_1) + (1-\alpha)f(x_2) - \alpha(1-\alpha)m\|x_1-\bar{x}\|^2] \\ &\qquad - \beta(1-\beta)m\|x_1-\bar{x}\|^2 \\ &= f(x_2) + m(1-\beta)[-\alpha(1-\alpha)\|x_1-x_2\|^2 - \beta\|x_1-\bar{x}\|^2] . \end{aligned} \tag{4.10}$$

Here, since $x_1 - \bar{x} = (1-\alpha)(x_1-x_2)$, the quantity in the last square brackets of (4.10) is equal to zero. Hence, from (4.10) it follows that $f(x_2) < f(x_2)$, which is impossible. Similarly, we also get a contradiction for $\alpha > 1$. ■

PROBLEM 4.4. Prove that if $f(x)$ is a strictly convex function on S, then

$$f(\alpha x_1 + (1-\alpha)x_2) \;>\; \alpha f(x_1) + (1-\alpha)f(x_2) \tag{4.11}$$

for $x_1 \neq x_2$, $\alpha \notin [0,1]$ such that $\alpha x_1 + (1-\alpha)x_2 \in S$.

2. Let a function $f(x)$ be convex on a convex set $S \subset E_n$.

<u>THEOREM 4.1</u>. The function $f(x)$ is continuous at any interior point of the set S.

<u>Proof</u>. Let $x_0 \in \text{int}\, S$. Denote by e_i, $i \in 1{:}n$, the basis vectors of the coordinate system: $e_i = (0,\dots,0,1,0,\dots,0)$. Since $x_0 \in \text{int}\, S$, then there exist $\beta_i > 0$ such that $x_0 \pm g_i \in S$, where $g_i = \beta_i e_i$.

Define $g_{n+i} = -g_i$ and let $x \to x_0$. Then we can represent a point x as $x = x_0 + \sum_{i=1}^{2n} \alpha_i g_i$, where $\alpha_i \geq 0$ and, moreover, $\min\{\alpha_i, \alpha_{i+n}\} = 0$ for all $i \in 1{:}n$. Since $\sum_{i=1}^{2} \alpha_i \to 0$ as $x \to x_0$ (clearly, the α_i depend on x), let us assume that we already have $\sum_{i=1}^{2} \alpha_i < 1$. Then, from

$$x = x_0 + \sum_{i=1}^{2n} \alpha_i g_i = \left(1 - \sum_{i=1}^{2n} \alpha_i\right) x_0 + \sum_{i=1}^{2n} \alpha_i (x_0 + g_i)$$

and the convexity of $f(x)$, we have

$$f(x) \leq \left(1 - \sum_{i=1}^{2n} \alpha_i\right) f(x_0) + \sum_{i=1}^{2n} \alpha_i f(x_0 + g_i)$$

or

$$f(x) - f(x_0) \leq \sum_{i=1}^{2n} \alpha_i (f(x_0+g_i) - f(x_0)) \quad . \tag{4.12}$$

We now notice that the point x_0 can be represented in the form $x_0 = \frac{1}{2}(x + x')$, where

$$x' = x_0 + \sum_{i=1}^{2n} \alpha_i' g_i \,, \qquad \alpha_i' = \begin{cases} \alpha_{i+n}\,, & i \leq n\,, \\ \alpha_{i-n}\,, & i > n\,. \end{cases}$$

It is clear that $\alpha_i' \geq 0$. Using the relation

$$f(x_0) \leq \tfrac{1}{2}[f(x) + f(x')]$$

and an argument analogous to that applied above to get (4.12), we obtain

$$f(x_0) - f(x) \le f(x') - f(x_0) \le \sum_{i=1}^{2n} \alpha_i' [f(x_0+g_i) - f(x_0)] . \tag{4.13}$$

The continuity of the function $f(x)$ at the point x_0 now follows from (4.12) and (4.13) because

$$\alpha_i \ge 0 , \quad \alpha_i' \ge 0 , \quad \sum_{i=1}^{2n} \alpha_i = \sum_{i=1}^{2n} \alpha_i' \xrightarrow[x \to x_0]{} 0 . \quad ■$$

3. Let $g \in E_n$. In addition, let S be a convex set and let $x_0 \in S$.

Assume that there exists an $\alpha_0 > 0$ such that $x_0 + \alpha_0 g \in S$. Let a function $f(x)$ be defined on S.

A function $f(x)$ is said to be *differentiable* at a point x_0 in a direction g if the following finite limit exists:

$$\frac{\partial f(x_0)}{\partial g} \equiv \lim_{\alpha \to +0} \alpha^{-1}[f(x_0+\alpha g) - f(x_0)] .$$

In this case, the number $\frac{\partial f(x_0)}{\partial g}$ is called the *derivative* of a function $f(x)$ at the point x_0 *in the direction* g. Clearly,

$$f(x_0 + \alpha g) = f(x_0) + \alpha \frac{\partial f(x_0)}{\partial g} + o(x_0, g, \alpha) ,$$

where $\frac{o(x_0,g,\alpha)}{\alpha} \xrightarrow[\alpha \to +0]{} 0$.

If a function $f(x)$ is differentiable at a point x_0 in any direction $g \in E_n$, then we shall say that it is *directionally differentiable at the point* x_0.

By definition, it is clear that if $g_1 = \beta g$, where $\beta > 0$, then

$$\frac{\partial f(x_0)}{\partial g_1} = \beta \frac{\partial f(x_0)}{\partial g} .$$

We notice that directional differentiability at a point does not imply continuity at that point.

EXAMPLE 2. Let $x = (x^{(1)}, x^{(2)}) \in E_2$. Denote by G (Figure 6) the set bounded by the axis $L = \{(x^{(1)}, x^{(2)}) \mid x^{(1)} \geq 0,\ x^{(2)} = 0\}$ and the curve

$$K = \left\{(x^{(1)}, x^{(2)}) \mid x^{(1)} = \cos\phi \cos\frac{\phi}{4};\ x^{(2)} = \sin\phi\cos\frac{\phi}{4};\ \phi \in [0, 2\pi]\right\}.$$

Let

$$f(x) = \begin{cases} 1, & x \in G, \\ 0, & x \notin G \end{cases}$$

and let $x_0 = (0,0) = 0$. It is clear that, for any $g = (g^{(1)}, g^{(2)})$, the derivative $\frac{\partial f(x_0)}{\partial g}$ exists and is equal to zero, although the function $f(x)$ is discontinuous at the point x_0.

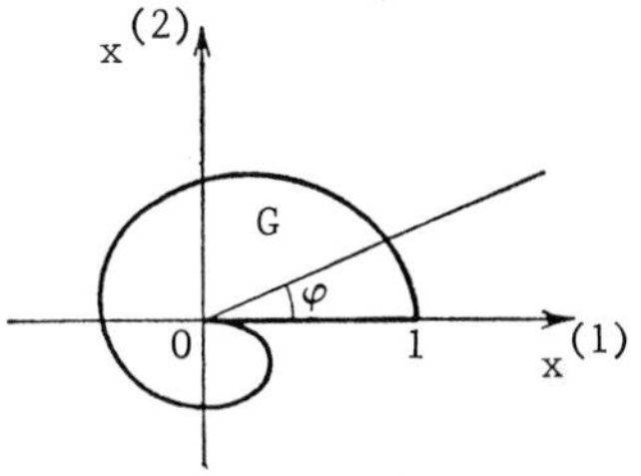

Figure 6

A function $f(x)$ is *differentiable* at a point $x_0 \in \text{int}\, S$ if there exists a vector $v(x_0)$ such that

$$f(x_0 + \alpha g) = f(x_0) + \alpha(g, v(x_0)) + o(x_0, g, \alpha) \qquad \forall g \in E_n,\ \|g\| = 1,$$

where $\frac{o(x_0, g, \alpha)}{\alpha} \xrightarrow[\alpha \to 0]{} 0$ uniformly with respect to g. If a function $f(x)$ is differentiable at a point x_0, then it is differentiable at that point in any direction g and $\frac{\partial f(x_0)}{\partial g} = (g, v(x_0))$. Moreover, in this case, the function $f(x)$ has first-order partial derivatives at the point x_0 and

$$v(x_0) = \frac{\partial f(x_0)}{\partial x} \equiv \left(\frac{\partial f(x_0)}{\partial x^{(1)}}, \ldots, \frac{\partial f(x_0)}{\partial x^{(n)}}\right).$$

The vector $v(x_0) = \frac{\partial f(x_0)}{\partial x}$ is called the *gradient* of the function $f(x)$ at the point x_0. For the function $f(x)$ in Example 2, we have the representation

$$f(x_0 + \alpha g) = f(x_0) + o(x_0, g, \alpha),$$

but the function $f(x)$ is not differentiable at the point $x_0 = (0,0)$ because $\frac{o(x_0,g,\alpha)}{\alpha} \to 0$ nonuniformly with respect to g, $\|g\| = 1$.

THEOREM 4.2. Let $f(x)$ be a convex function on S and let $x_0 \in \text{int } S$. Then the function $f(x)$ is differentiable at the point x_0 in any direction $g \in E_n$.

Proof. We need to prove the existence of a finite limit

$$\frac{\partial f(x_0)}{\partial g} \equiv \lim_{\alpha\to+0} h(\alpha) \quad ,$$

where $h(\alpha) = \alpha^{-1}[f(x_0+\alpha g) - f(x_0)]$. Since $x_0 \in \text{int } S$, there exists a $\delta > 0$ such that the interval $(x_0-\delta g, x_0+\delta g) \subset S$. Let $\alpha_0 \in (0,\delta)$.

For $\alpha \in (0,\alpha_0]$ we have

$$f(x_0+\alpha g) = f(\beta x_0 + (1-\beta)f(x_0+\alpha_0 g)) \quad ,$$

where $\beta = \alpha_0^{-1}(\alpha_0 - \alpha) \in (0,1]$. By virtue of the convexity of $f(x)$,

$$f(x_0+\alpha g) \leq \beta f(x_0) + (1-\beta)f(x_0+\alpha_0 g) \quad .$$

Hence

$$\begin{aligned} h(\alpha) &= \alpha^{-1}[f(x_0+\alpha g) - f(x_0)] \\ &\leq \alpha^{-1}(1-\beta)(f(x_0+\alpha_0 g) - f(x_0)) \\ &= \alpha_0^{-1}[f(x_0+\alpha_0 g) - f(x_0)] \equiv h(\alpha_0) \quad . \end{aligned}$$

Thus, $h(\alpha)$ does not increase as $\alpha \to +0$. On the other hand, we have

$$\begin{aligned} f(x_0) &= f((\alpha+\alpha_0)^{-1}\alpha_0(x_0+\alpha g) + (\alpha+\alpha_0)^{-1}\alpha(x_0-\alpha_0 g)) \\ &\leq (\alpha+\alpha_0)^{-1}\alpha_0 f(x_0+\alpha g) + (\alpha+\alpha_0)^{-1}\alpha f(x_0-\alpha_0 g) \quad . \end{aligned}$$

Therefore,

$$f(x_0+\alpha g) \geq \alpha_0^{-1}(\alpha+\alpha_0)f(x_0) - \alpha_0^{-1}\alpha f(x_0-\alpha_0 g) \quad .$$

Hence

$$h(\alpha) \geq \alpha_0^{-1}[f(x_0) - f(x_0-\alpha_0 g)] \quad ,$$

i.e., the function $h(\alpha)$ is bounded from below as $\alpha \to +0$. The existence of the finite limit $\frac{\partial f(x_0)}{\partial g} \equiv \lim_{\alpha \to +0} h(\alpha)$ now follows from the monotonicity and boundedness of the function $h(\alpha)$. The theorem is proved. ∎

<u>COROLLARY</u>. From the proof of the above theorem, it is clear that

$$\frac{\partial f(x_0)}{\partial g} = \inf_{\alpha>0} \alpha^{-1}[f(x_0+\alpha g_0 - f(x_0)] \ .$$

<u>LEMMA 4.3</u>. If $f(x)$ is both convex and differentiable on an open convex set S, then

$$f(x_0+g) \geq f(x_0) + (f'(x_0),g)$$

for all $x_0 \in S$ and $g \in E_n$ such that $x_0+g \in S$. Here

$$f'(x_0) = \frac{\partial f(x_0)}{\partial x} \equiv \left\{\frac{\partial f(x_0)}{\partial x^{(1)}}, \ldots, \frac{\partial f(x_0)}{\partial x^{(n)}}\right\} \ .$$

<u>Proof</u>. Let $x_1 = x_0+g$. By virtue of the convexity of $f(x)$ we have

$$f(x_0+\alpha g) = f(\alpha x_1 + (1-\alpha)x_0) \leq \alpha f(x_1) + (1-\alpha)f(x_0) \qquad \forall \alpha \in [0,1].$$

Hence

$$\alpha^{-1}[f(x_0+\alpha g) - f(x_0)] \leq f(x_1) - f(x_0) \ .$$

Taking the limit as $\alpha \to +0$, we obtain

$$(f'(x_0),g) \leq f(x_1) - f(x_0) = f(x_0+g) - f(x_0) \ ,$$

which completes the proof. ∎

4. Let

$$f(x) = \sup_{y \in G} \phi(x,y) \ , \qquad (4.14)$$

where $x \in \Omega \subset E_n$, Ω is a convex set, G is an arbitrary set in any space, and the function $\phi(x,y)$ is convex with respect to x on Ω for every fixed $y \in G$.

Let us also assume that the function $f(x)$ given by relation

(4.14) is finite on Ω. It is easy to show that in this case the function $f(x)$ is also convex with respect to x on Ω. To demonstrate this, let $x_\alpha = \alpha x_1 + (1-\alpha)x_2$, where $x_1, x_2 \in \Omega$, $\alpha \in [0,1]$.

From the convexity of $\Omega(x,y)$ with respect to x, we have

$$\phi(x_\alpha,y) \leq \alpha\phi(x_1,y) + (1-\alpha)\phi(x_2,y)$$

and hence we obtain

$$\begin{aligned} f(x_\alpha) = \sup_{y \in G} \phi(x_\alpha,y) &\leq \sup_{y\in G} [\alpha\phi(x_1,y) + (1-\alpha)\phi(x_2,y)] \\ &\leq \alpha \sup_{y\in G} \phi(x_1,y) + (1-\alpha) \sup_{y\in G} \phi(x_2,y) \\ &= \alpha f(x_1) + (1-\alpha)f(x_2) \ . \end{aligned}$$

This implies the convexity of $f(x)$ on Ω. ■

5. Now, let

$$f(x) = \inf_{y\in G} \phi(x,y) \ , \tag{4.15}$$

where $x \in \Omega \subset E_n$, Ω is a convex set, G is a convex set in an arbitrary space, and the function $\phi(x,y)$ is convex jointly with respect to variables $[x,y]$ on $\Omega \times G$. Assume also that the function $f(x)$ given by (4.15) is finite on Ω.

LEMMA 4.4. The function $f(x)$ defined by (4.15) is convex on Ω.

Proof. Let us take arbitrary $x_1, x_2 \in \Omega$ and let sequences $\{y_{1k}\}$ and $\{y_{2k}\}$ be such that

$$y_{ik} \in G \quad \forall\, i \in 1:2 \ , \quad \forall\, k \in 1:\infty \ ,$$

$$\phi(x_i,y_{ik}) \xrightarrow[k\to\infty]{} f(x_i) = \inf_{y\in G} \phi(x_i,y) \ . \tag{4.16}$$

Let

$$x_\alpha = \alpha x_1 + (1-\alpha)x_2 \ , \qquad y_{\alpha k} = \alpha y_{1k} + (1-\alpha)y_{2k} \ ,$$
$$\alpha \in [0,1] \ .$$

Since the function $\phi(x,y)$ is convex with respect to the set of variables, then we have

$$\begin{aligned}\phi(x_\alpha, y_{\alpha k}) &= \phi(\alpha x_1 + (1-\alpha)x_2,\ \alpha y_{1k} + (1-\alpha)y_{2k}) \\ &\le \alpha\phi(x_1, y_{1k}) + (1-\alpha)\phi(x_2, y_{2k}) \ .\end{aligned}$$

Hence

$$\begin{aligned}f(x_\alpha) &= \inf_{y\in G} \phi(x_\alpha, y) \le \phi(x_\alpha, y_{\alpha k}) \\ &\le \alpha\phi(x_1, y_{1k}) + (1-\alpha)\phi(x_2, y_{2k}) \ .\end{aligned}$$

From this inequality and (4.16), taking the limit, we obtain

$$f(x_\alpha) \le \alpha f(x_1) + (1-\alpha)f(x_2) \qquad \forall \alpha \in [0,1] \ ,$$

and this implies that the function $f(x)$ is convex with respect to x on Ω. ∎

REMARK 2. For the function (4.14) to be convex we did not require that the set G be convex. However, convexity of G is essential for convexity of function (4.15). In the above lemma, we assume that the function $\phi(x,y)$ is convex with respect to the set of variables x and y. The following example shows that convexity with respect to each of variables x and y is not sufficient for convexity with respect to the set of variables.

EXAMPLE 3. Let $x \in E_1$, $y \in E_1$, $\phi(x,y) = xy$. The function $\phi(x,y)$ is convex with respect to x for every fixed y and convex with respect to y for every fixed x. Let $x_\alpha = \alpha x_1 + (1-\alpha)x_2$, $y_\alpha = \alpha y_1 + (1-\alpha)y_2$, where $\alpha \in [0,1]$. For $\phi(x,y)$ to be convex with respect to the set of variables, it is necessary that, for all x_1, x_2, y_1, y_2 and $\alpha \in [0,1]$, the inequality

$$\phi(x_\alpha, y_\alpha) \le \alpha\phi(x_1, y_1) + (1-\alpha)\phi(x_2, y_2) \tag{4.17}$$

be satisfied. Here we note that

$$\begin{aligned} F(\alpha) &\equiv \phi(x_\alpha, y_\alpha) - \alpha\phi(x_1,y_1) - (1-\alpha)\phi(x_2,y_2) \\ &= (\alpha x_1 + (1-\alpha)x_2)(\alpha y_1 + (1-\alpha)y_2) - \alpha x_1 y_1 - (1-\alpha)x_2 y_2 \\ &= \alpha(1-\alpha)(y_1 - y_2)(x_2 - x_1) \quad . \end{aligned}$$

It is obvious that if $\alpha \in (0,1)$, $x_1 \neq x_2$, $y_1 \neq y_2$ and $\text{sign}\,(y_1 - y_2) = \text{sign}\,(x_1 - x_2)$, then $F(\alpha) > 0$ and, in this case inequality (4.17) is not satisfied, i.e., the function $\phi(x,y)$ is not convex with respect to the set of variables.

REMARK 3. In general, the inf operation does not conserve convexity property, i.e., even if a function $\phi(x,y)$ is convex with respect to x for every fixed $y \in G$, the function $f(x)$ given by (4.15) is not necessarily convex.

EXAMPLE 4. Let $x \in E_1$, $y \in 1:2$, $\phi(x,1) = x$, $\phi(x,2) = -x$.

Then we have

$$f(x) = \inf_{y \in 1:2} \phi(x,y) = \begin{cases} -x\ , & x \geq 0\ , \\ x\ , & x < 0\ . \end{cases}$$

Obviously, the function $f(x)$ is not convex. Thus, although the sup operation always conserves the convexity property (see Subsection 4), the inf operation, generally speaking, does not.

PROBLEM 4.5. Let a function $f(x)$ be given on E_n and be continuously differentiable. Show that the following statements are equivalent:

1. $f(x)$ is a convex function on E_n.
2. $f(x_2) - f(x_1) \geq (f'(x_1),\ x_2 - x_1) \quad \forall x_1, x_2 \in E_n$.
3. For every fixed x_0 and g, the function $(f'(x_0 + \alpha g, g)$ is nondecreasing as α increases.
4. Let $f(x)$ be a twice continuously differentiable function on E_n. The matrix of second-order derivatives $f''(x)$ is nonnegative definite, i.e.,

$$(f''(x)g,g) \geq 0 \qquad \forall\, x,g \in E_n.$$

Here

$$f''(x) = \begin{pmatrix} \frac{\partial^2 f}{\partial x^{(1)}\partial x^{(1)}} & \frac{\partial^2 f}{\partial x^{(1)}\partial x^{(2)}} & \cdots & \frac{\partial^2 f}{\partial x^{(1)}\partial x^{(n)}} \\ \cdots & \cdots & \cdots & \cdots \\ \frac{\partial^2 f}{\partial x^{(n)}\partial x^{(1)}} & \frac{\partial^2 f}{\partial x^{(n)}\partial x^{(2)}} & \cdots & \frac{\partial^2 f}{\partial x^{(n)}\partial x^{(n)}} \end{pmatrix}.$$

PROBLEM 4.6. Let $f(x)$ be a continuously differentiable function on E_n. Show that the following statements are equivalent:

1. $f(x)$ is strictly convex function.
2. $f(x_2) - f(x_1) > (f'(x_1),\ x_2-x_1) \quad \forall x_1,x_2 \in E_n,\ x_1 \neq x_2.$
3. $(f'(x+\alpha g),g)$ is an increasing function of α as α increases.

Is the following statement equivalent to 1-3?

Let $f(x)$ be a twice continuously differentiable function on E_n. The matrix of second-order derivatives $f''(x)$ is positive definite, i.e., $(f''(x)g,g) > 0 \quad \forall x,g \in E_n,\ g \neq 0$.

PROBLEM 4.7. Prove that if $f(x)$ is a strongly convex and twice continuously differentiable function on E_n, then there exists an $m > 0$ such that

$$(f''(x)g,g) \geq m\|g\|^2 \qquad \forall x,g \in E_n .$$

PROBLEM 4.8. Prove that for the quadratic function

$$f(x) = \tfrac{1}{2}(Ax,x) + (b,x) \quad ,$$

where A is an $(n\times n)$-matrix, $b \in E_n$, the concepts of strict convexity and strong convexity coincide (i.e., any strongly convex function is strictly convex, and vice versa).

PROBLEM 4.9. Prove that if $f(x)$ is a continuous function on E_n and

$$f(\tfrac{1}{2}(x_1+x_2)) \le \tfrac{1}{2}[f(x_1)+f(x_2)] - \gamma\|x_1-x_2\|^2 \qquad \forall x_1,x_2 \in E_n ,$$

where $\gamma > 0$ is the same for all $x_1, x_2 \in E_n$, then the function $f(x)$ is strongly convex on E_n, and vice versa.

6. Quasiconvex Functions. Let a function $f(x)$ be defined on a convex set $S \subset E_n$. Assume that $f(x)$ is finite. A function $f(x)$ is said to be *quasiconvex* on S if

$$f(\alpha x + (1-\alpha)y) \le \max\{f(x), f(y)\}$$
$$\forall\ \alpha \in [0,1],\ \forall\ x,y \in S . \qquad (4.18)$$

If inequality (4.18) holds strictly for $\alpha \in (0,1)$, $x \neq y$, then the function $f(x)$ is said to be *strictly quasiconvex*. Clearly, any convex function is quasiconvex.

If functions $f_i(x)$, $i \in 1:m$, are quasiconvex on S, then the function $f(x) = \max\limits_{i\in 1:m} f_i(x)$ is also quasiconvex on S.

From (4.18) it follows that if $x_i \in S$, $\alpha_i \ge 0$, $\sum\limits_{i=1}^{m} \alpha_i = 1$, then $f\left(\sum\limits_{i\in 1:m} \alpha_i x_i\right) \le \max\limits_{i\in 1:m} f(x_i)$.

<u>LEMMA 4.5</u>. A function $f(x)$ is quasiconvex on S iff, for any $c \in E_1$, the set $D_c = \{x\in S \mid f(x)\le c\}$ is convex.

<u>Proof</u>. Necessity. Let us set $c \in E_1$.

If $x \in D_c$ and $y \in D_c$, then from (4.18) we have

$$f(\alpha x + (1-\alpha)y) \le \max\{f(x), f(y)\} \le c \qquad \forall\,\alpha \in [0,1] ,$$

and this implies that the set D_c is convex.

Sufficiency. Let the set D_c be convex for any $c \in E_1$.

Let us take arbitrary points $x \in S$ and $y \in S$ and let $c_0 = \max\{f(x), f(y)\}$. Then $x \in D_{c_0}$, $y \in D_{c_0}$ and therefore

$$\alpha x + (1-\alpha)y \in D_{c_0} \qquad \forall\,\alpha \in [0,1] ,$$

i.e., (4.18) holds.

We shall call a point $\bar{x} \in S$ a *virtual local minimum* point of a function on S if it follows from $\bar{\bar{x}} \in S$ and $f(\bar{\bar{x}}) < f(\bar{x})$ that there exist a point y on the segment $[\bar{x},\bar{\bar{x}}]$ such that $f(y) > f(\bar{x})$, and a $\delta > 0$ such that $f(\bar{x}) \le f(x)$ $\forall x \in S_\delta(\bar{x})$.

Let $M^* = \{x \in S \mid f(x) = \inf_{y \in S} f(y)\}$.

COROLLARY. If a function f(x) is quasiconvex on S, then the set M^* is convex and any virtual local minimum point of the function f(x) on S is a global minimum point of f(x) on S.

Proof. The convexity of the set M^* follows from Lemma 4.5.

Let $\bar{x} \in S$ be a virtual local minimum point of the function f(x) on S and let $\bar{\bar{x}} \in M^*$. Assume that $\bar{x} \notin M^*$, i.e., $f(\bar{\bar{x}}) < f(\bar{x})$. From the definition of a virtual local minimum point, there exists an $\alpha \in [0,1]$ such that

$$f(\alpha\bar{x} + (1-\alpha)\bar{\bar{x}}) > f(\bar{x}) \quad . \qquad (4.19)$$

By virtue of the quasiconvexity of the function f(x) on the set S, we have

$$f(\alpha\bar{x} + (1-\alpha)\bar{\bar{x}}) \le \max \{f(\bar{x}), f(\bar{\bar{x}})\} = f(\bar{x}) ,$$

and this contradicts (4.19). ■

REMARK 4. A point $x_0 \in S$ is called a *local minimum point* of a function f(x) on S if there exists $r > 0$ such that

$$f(x) \ge f(x_0) \qquad \forall x \in \{x \in S \mid \|x-x_0\| \le r\} .$$

Note that a local minimum point of f(x) on S may happen to be a global maximum point of f(x) on S, which cannot occur in the case of a virtual local minimum point of f(x) on S unless f(x) is a constant function on S.

LEMMA 4.6. Let a function f(x) be convex on S and a function g(x) be linear on S. Then the function $h(x) = f(x)(g(x))^{-1}$ is quasiconvex on the set $\Omega = \{x \in S \mid g(x) > 0\}$.

Proof. Let us set $x \in \Omega$, $y \in \Omega$ and $\alpha \in [0,1]$. It is obvious that $\alpha x + (1-\alpha)y \in \Omega$ because the set Ω is convex. Then we have

$$\begin{aligned} h(\alpha x + (1-\alpha)y) &= \frac{f(\alpha x + (1-\alpha)y)}{g(\alpha x + (1-\alpha)y)} \\ &\le [\alpha f(x) + (1-\alpha)f(y)][\alpha g(x) + (1-\alpha)g(y)]^{-1} \\ &= \left[\alpha g(x)\frac{f(x)}{g(x)} + (1-\alpha)g(y)\frac{f(y)}{g(y)}\right] \\ &\qquad \times [\alpha g(x) + (1-\alpha)g(y)]^{-1} \\ &\le \max\left\{\frac{f(x)}{g(x)}, \frac{f(y)}{g(y)}\right\} = \max\{h(x), h(y)\} \ . \end{aligned}$$ ∎

COROLLARY. The set of minimum points of the function $h(x)$ on S is convex.

Now let a quasiconvex function $f(x)$ be directionally differentiable on E_n. The the function $\frac{\partial f(x)}{\partial g}$ is quasiconvex with respect to $g \in E_n$. In fact, letting $x_0 \in E_n$, $g = \beta g_1 + (1-\beta)g_2$, $g_1, g_2 \in E_n$, $\beta \in [0,1]$, we have

$$\begin{aligned} \frac{\partial f(x_0)}{\partial g} &= \lim_{\alpha\to+0} \alpha^{-1}[f(x_0+\alpha g) - f(x_0)] \\ &= \lim_{\alpha\to+0} \alpha^{-1}[f(\beta(x_0+\alpha g_1) + (1-\beta)(x_0+\alpha g_2)) - f(x_0)] \\ &\le \lim_{\alpha\to 0} \alpha^{-1}[\max\{f(x_0+\alpha g_1), f(x_0+\alpha g_2)\} - f(x_0)] \\ &= \max\{\lim_{\alpha\to+0} \alpha^{-1}[f(x_0+\alpha g_1) - f(x_0)], \lim_{\alpha\to 0} \alpha^{-1}[f(x_0+\alpha g_2) - f(x_0)]\} \end{aligned}$$ ∎

Let

$$\tilde{f}(x) = \inf\{\mu \in E_1 \mid (\mu, x) \in \text{co epi } f\} \ .$$

It is clear that the function $\tilde{f}(x)$ is convex on E_n and that $\inf_{x\in E_n} \tilde{f}(x) = \inf_{x \in E_n} f(x)$. This implies that, in principle, the problem of finding a minimum of a quasiconvex function $f(x)$ is reduced to that of finding the minimum of the convex function $\tilde{f}(x)$.

An important example of a quasiconvex (but not convex) problem is that of best uniform approximation of a function by a linear fractional function

$$\max_{t\in[a,b]} \left| a_0(t) - \frac{\sum_{i=1}^{n} x_i a_i(t)}{\sum_{i=1}^{n} x_i b_i(t)} \right| \to \min_{x\in E_n}$$

$$\sum x_i b_i(t) > 0 ,$$

where $a_0(t)$, $a_i(t)$, $b_i(t)$, $i \in 1:n$, are continuous functions on $[a,b]$.

5. SUBGRADIENTS AND SUBDIFFERENTIALS OF CONVEX FUNCTIONS

1. Let a function $f(x)$ be convex and finite on E_n and let $x_0 \in E_n$.

The set

$$\partial f(x_0) = \{v \in E_n \mid f(x)-f(x_0) \geq (v,x-x_0) \ \forall x \in E_n\} \tag{5.1}$$

is called a *subdifferential* of the function $f(x)$ at the point x_0. An element $v \in \partial f(x_0)$ is called a *subgradient* (or *generalized gradient*) of the function $f(x)$ at the point x_0.

THEOREM 5.1. The set $\partial f(x_0)$ is nonempty, convex, closed and bounded.

Proof. Let us introduce the set

$$\bar{Z} = \{[\beta,x] \in E_{n+1} \mid \beta \in E_1,\ x \in E_n,\ \beta \geq f(x)\} .$$

The set $\bar{Z}$ is convex, closed and nonempty (it is the epigraph of the function $f(x)$). The point $[f(x_0),x_0]$ is a boundary point of the set $\bar{Z}$.

From Corollary 1 of Theorem 1.2, there exist a number c and a vector v, which are not simultaneously equal to zero, i.e.,

$$c^2 + v^2 > 0 \ , \tag{5.2}$$

such that

$$c\beta + (v,x) \geq cf(x_0) + (v,x_0) \qquad \forall [\beta, x] \in \bar{Z} \ . \tag{5.3}$$

Letting $x = x_0$, we have $c(\beta - f(x_0)) \geq 0$ $\forall \beta \geq f(x_0)$. Hence $c > 0$. However, for the case $c = 0$, it follows from (5.3) that

$$(v, x-x_0) \geq 0 \qquad \forall x \in E_n \ . \tag{5.4}$$

This implies that $v = 0$, which contradicts (5.2). Therefore, c must be positive. Now, assuming that $\beta = f(x)$ in (5.3), we obtain

$$f(x) - f(x_0) \geq (-c^{-1}v, \ x-x_0) \qquad \forall x \in E_n \ ,$$

i.e., $v_0 = -c^{-1}v \in \partial f(x_0)$. Thus, the set $\partial f(x_0)$ is nonempty. The convexity and closedness of $\partial f(x_0)$ are obvious. It remains to show the boundedness. We assume that $\partial f(x_0)$ is not bounded. Then there exists a sequence $\{v_k\}$ such that $v_k \in \partial f(x_0)$, $||v_k|| \to \infty$. Let $x_k = x_0 + z_k$, where $z_k = \delta_0 ||v_k||^{-1} v_k$. From (5.1) we have

$$f(x_k) - f(x_0) \geq (v_k, x_k - x_0) = (v_k, z_k) = \delta_0 ||v_k|| \to +\infty \ . \tag{5.5}$$

Note that $||x_k - x_0|| = ||z_k|| = \delta_0$ for all k. Since the function f(x) is convex on E_n it is continuous, f(x) must be bounded on the set $S_{\delta_0}(x_0)$, which contradicts (5.5). This contradiction proves the theorem. ∎

<u>COROLLARY</u>. The point-to-set mapping $\partial f(x)$ is bounded on any bounded set.

<u>Proof</u>. Is analogous to that of the above theorem.

<u>REMARK 1</u>. From the definition (see (5.1)) it is clear that the point-to-set mapping $\partial f(x)$ is upper semicontinuous at any point $x_0 \in E_n$.

However, as will be shown below, it is, generally speaking, not lower semicontinuous.

LEMMA 5.1. Let $x_0 \in E_n$ and $f(x)$ be a convex function on E_n. For any $\varepsilon > 0$ there exists a $\delta > 0$ such that

$$\partial f(x) \subset \partial f(x_0) + S_\varepsilon(0) \qquad \forall x \in S_\delta(x_0) . \tag{5.6}$$

Proof. Assume that this is not the case. Then there exist a number $\varepsilon_0 > 0$ and sequences $\{x_k\}, \{v_k\}$ such that

$$x_k \to x_0 , \qquad v_k \in \partial f(x_k) ,$$

$$\rho(v_k, \partial f(x_0)) \equiv \min_{v\in\delta f(x_0)} \| v_k - v \| \geq \varepsilon_0 . \tag{5.7}$$

Since the mapping $\partial f(x)$ is bounded on $S_\delta(x_0)$, the sequence $\{v_k\}$ is bounded.

Without loss of generality, we can assume that $v_k \to v_0$. Then it follows from (5.7) that $v_0 \notin \partial f(x_0)$, and this contradicts the upper semicontinuity of the mapping $\partial f(x)$ at the point x_0. ∎

THEOREM 5.2. The relation

$$\frac{\partial f(x_0)}{\partial g} = \max_{v\in\partial f(x_0)} (v,g)$$

holds.

Proof. The directional differentiability of $f(x)$ is established in Theorem 4.2. Let $h(\alpha) = \alpha^{-1}[f(x_0+\alpha g) - f(x_0)]$ and let us find $\lim_{\alpha\to+0} h(\alpha)$. By virtue of (5.1) we have for any $v \in \partial f(x_0)$

$$f(x_0+\alpha g) - f(x_0) \geq (v,\alpha g) .$$

Hence

$$\frac{\partial f(x_0)}{\partial g} \equiv \lim_{\alpha\to+0} h(\alpha) \geq \max_{v\in\partial f(x_0)} (v,g) . \tag{5.8}$$

Let us introduce the sets

$$Z = \{[\beta,x] \in E_{n+1} \mid x \in E_n,\ \beta > f(x)\} ,$$

$$L = \{[\beta,x] \mid \beta = f(x_0) + \alpha\frac{\partial f(x_0)}{\partial g},\ x = x_0 + \alpha g,\ \alpha > 0\} .$$

We shall show that the ray L and the set Z have no common points. First of all, we note that it is shown in the proof of the Theorem 4.2 that $h(\alpha)$ does not increase as $\alpha \to +0$. Therefore

$$\frac{\partial f(x_0)}{\partial g} \equiv \lim_{\alpha'\to+0} h(\alpha') \le h(\alpha) \qquad \forall\alpha > 0 ,$$

i.e.,

$$f(x_0+\alpha g) \ge f(x_0) + \alpha\frac{\partial f(x_0)}{\partial g} . \tag{5.9}$$

Here we assume that L and Z have no common points. Then there exists at least one point $[\beta_1,x_1] \in Z \cap L$; i.e.,

$$\beta_1 > f(x_1) , \qquad \beta_1 = f(x_0) + \alpha_1\frac{\partial f(x_0)}{\partial g} , \qquad x_1 = x_0 + \alpha_1 g ,$$

where $\alpha_1 > 0$. Hence we have

$$f(x_0) + \alpha_1\frac{\partial f(x_0)}{\partial g} > f(x_0+\alpha_1 g) ,$$

which contradicts (5.9). Therefore, L and Z have no common points

Let us construct the following set

$$P = Z - L = \{w = v-v' \mid v \in Z,\ v' \in L\} .$$

Clearly, P is a convex set and $0 \notin P$.

Due to Corollary 2 of the separation theorem, there exists a point $[c,v] \in E_{n+1}$ such that

$$c^2 + v^2 > 0 ,$$

$$c\beta + (v,x) \ge c\left(f(x_0) + \alpha\frac{\partial f(x_0)}{\partial g}\right) + (v,\ x_0+\alpha g) \tag{5.10}$$

for all α, β, x such that $\beta \ge f(x)$, $\alpha \ge 0$.

Using the same argument as in the proof of Theorem 5.1, we can conclude that $c > 0$. For $\beta = f(x)$, $\alpha = 0$, we have

$$f(x) - f(x_0) \ge -c^{-1}(v,\ x-x_0) \qquad \forall\, x \in E_n ,$$

i.e., $v_0 = -c^{-1}v \in \partial f(x_0)$. Hence, for $\beta = f(x)$, $\alpha > 0$, it follows from (5.10) that

$$f(x) - f(x_0) \geq (v_0, x-x_0) + \alpha\left(\frac{\partial f(x_0)}{\partial g} - (v_0,g)\right) .$$

For $x = x_0$ and $\alpha > 0$ we obtain

$$\frac{\partial f(x_0)}{\partial g} \leq (v_0,g) \leq \max_{v\in\partial f(x)} (v,g) .$$

This and (5.8) prove the theorem. ∎

COROLLARY. Let $\frac{\partial f(x_0)}{\partial g} = -\alpha < 0$ and $v_1 \in \partial f(x_0)$ be such that

$$(v_1,g) = \max_{v\in\partial f(x)} (v,g) = \frac{\partial f(x_0)}{\partial g} .$$

Then, for $g_1 = -g$, we have

$$\frac{\partial f(x_0)}{\partial g_1} = \max_{v\in\partial f(x)} (v,g_1) \geq (v_1,g_1) = -(v_1,g) = -\frac{\partial f(x_0)}{\partial g} = \alpha > 0 ,$$

i.e., if the function $f(x)$ decreases in some direction, then it must inevitably increase in the opposite direction.

REMARK 2. In the sequel we shall use the following representation of a convex function:

$$f(x) = \max_{z\in E_n} [f(z) + (v(z), x-z)] , \tag{5.11}$$

where $v(z)$ is an arbitrary vector in $\partial f(x)$.

Let us prove (5.11). If $v(z) \in \partial f(z)$, then

$$f(x) \geq f(z) + (v(z), x-z) .$$

Since this inequality holds for any $z \in E_n$, we have

$$f(x) \geq \sup_{z\in E_n} [f(z) + (v(z), x-z)] . \tag{5.12}$$

The relation

$$\sup_{z\in E_n} [f(z) + (v(z), x-z)] \geq f(x)$$

is obvious. From this and (5.12) we obtain (5.11) (we can write the maximum instead of the supremum because the supremum is attained at $z = x$). Also,

$$f(x) = \max_{z \in S} [f(z) + (v(z), x-z)] \quad , \qquad (5.13)$$

where $S \subset E_n$ is an arbitrary set which contains x as an interior point.

Hence, a convex function is the maximum of functions which are linear with respect to x.

REMARK 3. From Theorem 5.1 it follows that if $f(x)$ is a convex function, then for any $x_0 \in E_n$ there exists a $v_0 = v(x_0) \in E_n$ (possibly not unique) such that

$$f(x) - f(x_0) \geq (v(x_0), x-x_0) \qquad \forall x \in E_n \quad .$$

Let us prove that the converse is also true. That is, if for any $x_0 \in E_n$ there exists a $v(x_0) \in E_n$ such that the above inequality is valid, then the function $f(x)$ is convex.

Let us take arbitrary $x_1, x_2 \in E_n$, $\alpha \in [0,1]$ and let $x_0 = \alpha x_1 + (1-\alpha)x_2$. Due to the assumption, there exists a $v_0 = v(x_0)$ such that

$$f(x) - f(x_0) \geq (v_0, x-x_0) \qquad \forall x \in E_n \quad ,$$

implying that for $x = x_1$

$$f(x_1) \geq f(x_0) + (1-\alpha)(v_0, x_1-x_2) \quad .$$

For $x = x_2$, analogously, we have

$$f(x_2) \geq f(x_0) - \alpha(v_0, x_1-x_2) \quad .$$

Multiplying the first inequality of the last two inequalities by α and the second one by $(1-\alpha)$ and summing them, we obtain

$$\alpha f(x_1) + (1-\alpha)f(x_2) \geq f(x_0) = f(\alpha x_1 + (1-\alpha)x_2) \quad .$$

Since $x_1, x_2 \in E_n$ and $\alpha \in [0,1]$ are arbitrary, this implies that $f(x)$ is a convex function. ■

This property can be viewed as a definition of a convex function.

2. Subdifferentials of a Sum and a Maximum Function.

Let $f_1(x)$ and $f_2(x)$ be convex functions on E_n. The function

$$f(x) = f_1(x) + f_2(x) \tag{5.14}$$

is also convex.

LEMMA 5.2. The relation

$$\partial f(x) = \partial f_1(x) + \partial f_2(x) \tag{5.15}$$

holds.

Proof. For any $g \in E_n$ we have

$$\frac{\partial f(x)}{\partial g} = \max_{v \in \partial f(x)} (v,g) \quad . \tag{5.16}$$

On the other hand, it is seen from (5.14) that

$$\begin{aligned}
\frac{\partial f(x)}{\partial g} &\equiv \lim_{\alpha \to +0} \alpha^{-1}[f(x+\alpha g) - f(x)] \\
&= \lim_{\alpha \to +0} \alpha^{-1}[f_1(x+\alpha g) - f_1(x)] + \lim_{\alpha \to +0} \alpha^{-1}[f_2(x+\alpha g) - f_2(x)] \\
&= \frac{\partial f_1(x)}{\partial g} + \frac{\partial f_2(x)}{\partial g} \quad .
\end{aligned} \tag{5.17}$$

Note that

$$\frac{\partial f_1(x)}{\partial g} = \max_{v \in \partial f_1(x)} (v,g) \ , \qquad \frac{\partial f_2(x)}{\partial g} = \max_{v \in \partial f_2(x)} (v,g) \ .$$

Then, from (5.16) and (5.17) we have

$$\begin{aligned}
\max_{v \in \partial f(x)} (v,g) &= \max_{v \in \partial f_1(x)} (v,g) + \max_{v \in \partial f_2(x)} (v,g) \\
&= \max_{v \in [\partial f_1(x) + \partial f_2(x)]} (v,g) \quad .
\end{aligned}$$

Since this equality holds for all $g \in E_n$ and the sets $\partial f(x)$

and $\partial f_1(x) + \partial f_2(x)$ are closed and convex, then, due to Corollary of Lemma 2.1, we get

$$f(x) = \partial f_1(x) + \partial f_2(x) ,$$

which completes the proof. ■

PROBLEM 5.1. Prove that if $f(x)$ is a convex function on E_n and $f_1(x) = Af(x)$, $A > 0$, then

$$\partial f_1(x) = A\partial f(x) . \tag{5.18}$$

LEMMA 5.3. Let $f(x)$ be a continuously differentiable convex function on E_n. Then

$$\partial f(x_0) = \{f'(x_0)\} . \tag{5.19}$$

Proof. Let us fix $x_0 \in E_n$. It is proved in Lemma 4.3 that in this case we have

$$f(x) - f(x_0) \geq (f'(x_0), x-x_0) \qquad \forall x \in E_n .$$

This means that $v = f'(x_0) \in \partial f(x_0)$.

Next, let $v \in \partial f(x_0)$, i.e.,

$$f(x) - f(x_0) \geq (v, x-x_0) \qquad \forall x \in E_n . \tag{5.20}$$

Then, $h(x) \equiv f(x) - f(x_0) + (v, x_0-x)$ is a continuously differentiable function on E_n and, moreover, (5.20) implies $h(x) \geq 0$ $\forall x \in E_n$. Since $h(x_0) = 0$, the point x_0 is a minimum point of the function $h(x)$ on E_n. According to the necessary condition for a minimum, we have $h'(x_0) = 0$, i.e., $f'(x_0) - v = 0$ or $v = f'(x_0)$. This proves the lemma. ■

LEMMA 5.4. Let $f_i(x)$, $i \in I \equiv 1:N$, be convex functions on E_n. Also, let $f(x) = \max_{i \in I} f_i(x)$. Then the function $f(x)$ is convex and, moreover,

$$\partial f(x) = \text{co} \{\partial f_i(x) \mid i \in R(x)\} , \tag{5.21}$$

where

$$R(x) = \{i \in I \mid f_i(x) = f(x)\} .$$

Proof. Denote by $D(x)$ the right-hand side of (5.21). The convexity of the function $f(x)$ is obvious. By Theorem 5.2 we have

$$\frac{\partial f(x)}{\partial g} = \max_{v \in \partial f(x)} (v,g) . \tag{5.22}$$

On the other hand, for fixed x and g we obtain

$$f(x+\alpha g) = f(x) + \alpha\frac{\partial f(x)}{\partial g} + o(\alpha) = \max_{i \in I}\left[f_i(x) + \alpha\frac{\partial f_i(x)}{\partial g} + o_i(\alpha)\right],$$

where $\alpha^{-1}o(\alpha) \xrightarrow[\alpha\to+0]{} 0$, $\alpha^{-1}o_i(\alpha) \xrightarrow[\alpha\to+0]{} 0 \quad \forall i \in I$.

For sufficiently small $\alpha > 0$, it follows that

$$f(x+\alpha g) = \max_{i\in R(x)} f_i(x+\alpha g) = \max_{i \in R(x)}\left[f_i(x) + \alpha\frac{\partial f_i(x)}{\partial g} + o_i(\alpha)\right]$$

$$= f(x) + \alpha \max_{i\in R(x)} \frac{\partial f_i(x)}{\partial g} + o(\alpha) ,$$

where

$$\alpha^{-1}o(\alpha) \xrightarrow[\alpha\to+0]{} 0 , \qquad o(\alpha) \in [\bar{o}(\alpha),\bar{\bar{o}}(\alpha)] ,$$

$$\bar{o}(\alpha) = \min_{i\in R(x)} o_i(\alpha) , \qquad \bar{\bar{o}}(\alpha) = \max_{i\in R(x)} o_i(\alpha) .$$

Hence

$$\frac{\partial f(x)}{\partial g} = \max_{i\in R(x)} \frac{\partial f_i(x)}{\partial g} = \max_{i\in R(x)} \max_{v_i\in\partial f_i(x)} (v_i,g) = \max_{v\in D(x)} (v,g) .$$

From this and (5.22), it follows that

$$\max_{v\in\partial f(x)} (v,g) = \max_{v\in D(x)} (v,g) .$$

Since this relation holds for all $g \in E_n$, we have, due to Corollary of Lemma 3.2, that $\partial f(x) = D(x)$. ∎

EXAMPLE 1. Let $x \in E_1$, $f(x) = |x| = \max\{f_1(x), f_2(x)\}$, where $f_1(x) = -x$, $f_2(x) = x$. Here, $I = 1:2$. Since the functions $f_1(x)$ and $f_2(x)$ are linear (therefore, also convex), the function $f(x)$

is also convex. Noting that $f_1'(x) = -1$, $f_2'(x) = 1$ and

$$R(x) = \begin{cases} \{1\}, & x < 0, \\ \{2\}, & x > 0, \\ \{1,2\}, & x = 0, \end{cases}$$

we have from formula (5.2) and Lemma 5.3

$$\partial f(x) = \text{co}\,\{f_i'(x) \mid i \in R(x)\},$$

i.e.,

$$\partial f(x) = \begin{cases} -1, & x < 0, \\ +1, & x > 0, \\ [-1,1], & x = 0. \end{cases}$$

Now, just for this function, we shall show that the mapping $\partial f(x)$ is not lower semicontinuous. Let us take the points $x_0 = 0$, $v_0 = 0.5$ and choose a sequence $\{x_i\}$ such that $x_i \to -0$. Since the functions $f_1(x)$ and $f_2(x)$ are differentiable, we have

$$\partial f(0) = \text{co}\,\{f_1'(0), f_2'(0)\} = \text{co}\,\{-1,1\} = [-1,1].$$

Hence, clearly, $v_0 \in \partial f(x_0)$. However, for the sequence $\{x_i\}$, we have

$$\partial f(x_i) = \{f_1'(x_i)\} = \{-1\} \qquad \forall i.$$

Therefore, there does not exist a sequence $\{v_i\}$, $v_i \in \partial f(x_i)$ which converges to $v_0 = 0.5$. This implies that the mapping $\partial f(x)$ is not lower semicontinuous at the point $x_0 = 0$ and, therefore, it is not continuous in the Kakutani sense.

<u>EXAMPLE 2</u>. Let

$$x = (x^{(1)}, \ldots, x^{(n)}) \in E_n,$$

$$f(x) = \max_{i \in 1:n} |x^{(i)}| \equiv \max_{i \in 1:n} f_i(x),$$

where $f_i(x) = |x^{(i)}|$. It follows from Example 1 that

$$\partial f_i(x) = \begin{cases} -e_i & \text{if } x^{(i)} < 0 , \\ e_i & \text{if } x^{(i)} > 0 , \\ \text{co } \{e_i, -e_i\} & \text{if } x^{(i)} = 0 , \end{cases}$$

where e_i are unit vectors, $e_i = (0,\dots,0,\underset{i}{1},0,\dots,0)$.

By (5.21)

$$\partial f(x) = \text{co } \{w_i \mid i \in R(x)\} ,$$

$$R(x) = \{i \in 1:n \mid |x^{(i)}| = f(x)\} ,$$

$$w_i = \begin{cases} e_i & \text{if } x^{(i)} \geq 0 , \\ -e_i & \text{if } x^{(i)} \leq 0 . \end{cases}$$

Hence $\partial f(0) = \text{co } \{\pm e_i \mid i \in 1:n\}$. Observe that $f(x) = (w,x)$, where $w \in \partial f(x)$.

PROBLEM 5.2. Prove the formula analogous to (5.21) for the case

$$f(x) = \max_{y \in G} \phi(x,y) , \tag{5.23}$$

where $G \subset E_m$ is a compact set, the function $\phi(x,y)$ is continuous in y and convex with respect to x for any fixed $y \in G$.

3. LEMMA 5.5. A convex function is Lipschitzian on any bounded convex set, i.e., for any bounded $G \subset E_n$, there exists an $L < \infty$ such that

$$|f(x) - f(z)| \leq L\|x-z\| \qquad \forall x \in G, \ \forall z \in G .$$

Proof. By the definition of $\partial f(z)$ we have

$$f(x) - f(z) \geq (v, x-z) , \qquad \text{where } v \in \partial f(z) .$$

By virtue of the corollary of Theorem 5.1, there exists an $L < \infty$ such that

$$\|v\| \leq L \qquad \forall v \in \partial f(z), \ \forall z \in G .$$

Hence

$$f(x) - f(z) \geq -L\|x-z\| .$$

By exchanging the roles of z and x and repeating the above procedure, we obtain

$$f(z) - f(x) \geq -L\|z-x\| .$$

The last two inequalities yield

$$|f(x) - f(z)| \leq L\|x-z\| \qquad \forall x \in G, \quad \forall z \in G ,$$

and this implies that the $f(x)$ is Lipschitzian on G. The lemma is proved. ■

Let us fix $x_0 \in E_n$ and $g \in E_n$. Let $O(\alpha)$ be an arbitrary vector function such that

$$O(\alpha) \in E_n , \qquad \alpha^{-1}\|O(\alpha)\| \xrightarrow[\alpha\to+0]{} 0 .$$

<u>LEMMA 5.6</u>. If a function $f(x)$ is Lipschitzian at a neighborhood of the point x_0 and directionally differentiable at the point x_0, then

$$\lim_{\alpha\to+0} \alpha^{-1}[f(x_0 + \alpha g + O(\alpha)) - f(x_0)] = \frac{\partial f(x_0)}{\partial g}$$

$$\equiv \lim_{\alpha\to+0} \alpha^{-1}[f(x_0+\alpha g) - f(x_0)].$$

<u>Proof</u>. Since the function $f(x)$ is Lipschitzian, we have

$$|f(x_0+\alpha g+O(\alpha)) - f(x_0+\alpha g)| \leq L\|O(\alpha)\| = L|o(\alpha)| ,$$

i.e.,

$$f(x_0 + \alpha g + O(\alpha)) - f(x_0 + \alpha g) = o(\alpha) ,$$

where

$$o(\alpha) = \|O(\alpha)\| , \qquad \alpha^{-1}o(\alpha) \xrightarrow[\alpha\to+0]{} 0 .$$

Therefore,

$$\lim_{\alpha\to+0} \alpha^{-1}[f(x_0 + \alpha g + O(\alpha)) - f(x_0)]$$

$$= \lim_{\alpha\to+0} [\alpha^{-1}[f(x_0+\alpha g) - f(x_0)] + \alpha^{-1}o(\alpha)]$$

$$= \lim_{\alpha\to+0} \alpha^{-1}[f(x_0+\alpha g) - f(x_0)] + \lim_{\alpha\to+0} \alpha^{-1}o(\alpha) = \frac{\partial f(x_0)}{\partial g} .$$

■

Note that Lemma 5.6 is also valid for any convex function because a convex function is directionally differentiable and Lipschitzian in a neighborhood of any point x_0 (see Theorem 4.2 and Lemma 5.5).

DEFINITION. A function of one variable $\psi(t)$ given on $[0,T]$, $T < \infty$, is said to be *absolutely continuous* if, for any $\varepsilon > 0$, there exists a $\delta > 0$ such that

$$\sum_{k=1}^{N} |\psi(t_k'') - \psi(t_k')| < \varepsilon$$

if

$$\sum_{k=1}^{N} (t_k'' - t_k') < \delta ,$$

where $0 \le t_1' < t_1'' \le t_2' < t_2'' \le \cdots \le t_k' < t_k'' \le \cdots \le t_N' < t_N'' \le T$ and N is an arbitrary natural number.

Let $f(x)$ be a convex function on E_n and let $x(t)$ be an n-dimensional continuous vector-valued function given on $[0,T]$ and satisfying a Lipschitz condition, i.e., there exists a $K_1 > 0$ such that

$$\|x(t') - x(t'')\| \le K_1|t'-t''| \qquad \forall t',t'' \in [0,T] .$$

Let $\psi(t) = f(x(t))$. By virtue of Lemma 5.5, the convex function $f(x)$ is Lipschitzian with respect to x on an arbitrary bounded set $G \subset E_n$.

The function $\psi(t)$ is Lipschitzian on $[0,T]$ and since $f(x)$ and $x(t)$ are Lipschitzian with respect to x and t, respectively,

$$|\psi(t') - \psi(t)| = |f(x(t')) - f(x(t))| \le K\|x(t') - x(t)\|$$
$$\le KK_1|t'-t| \qquad \forall t',t \in [0,T] .$$

From this, it is easy to see that $\psi(t)$ is an absolutely continuous function on $[0,T]$.

It is well known (see, for example, [115]) that there exists an integrable function $\omega(t)$ on $[0,T]$ such that

$$\psi(t) = \psi(0) + \int_0^t \omega(\tau)\, d\tau . \tag{5.24}$$

This integral is interpreted in the Lebesgue sense. In this case, for almost all $t \in [0,T]$,

$$\omega(t) = \lim_{\Delta\to 0} \Delta^{-1}(\psi(t+\Delta) - \psi(t)) \equiv \psi'(t) , \tag{5.25}$$

i.e., the function $\psi(t)$ is differentiable almost everywhere. If in addition, $x(t)$ is a differentiable vector-valued function, i.e., $x(t+\Delta) = x(t) + \Delta\dot{x}(t) + o(\Delta)$, where $\Delta^{-1}\|o(\Delta)\| \xrightarrow[\Delta\to 0]{} 0$, then it follows from (5.24), (5.25) and Lemma 5.6 that

$$f(x(t)) = f(x(0)) + \int_0^t \frac{\partial f(x(\tau))}{\partial \dot{x}(\tau)}\, d\tau . \tag{5.26}$$

LEMMA 5.7. Let $S \subset E_n$ be a convex open set and let $f_1(x)$ and $f_2(x)$ be convex functions on S. If

$$\partial f_1(x) = \partial f_2(x) \qquad \forall x \in S ,$$

then $f_1(x) - f_2(x) = C$, where C is a constant.

Proof. Let us take an arbitrary $x_0 \in S$.

For any $x \in S$ we have from (5.26)

$$f_1(x) = f_1(x_0) + \int_0^1 \frac{\partial f_1(x_0+\tau g)}{\partial g}\, d\tau ,$$

$$f_2(x) = f_2(x_0) + \int_0^1 \frac{\partial f_2(x_0+\tau g)}{\partial g}\, d\tau ,$$

where $g = x - x_0$. However,

$$\frac{\partial f_1(x_0+\tau g)}{\partial g} = \max_{v\in\partial f_1(x_0+\tau g)} (v,g) ,$$

$$\frac{\partial f_2(x_0+\tau g)}{\partial g} = \max_{v\in\partial f_2(x_0+\tau g)} (v,g) .$$

Since $\partial f_1(x) = \partial f_2(x)$ $\forall x \in S$, we obtain

$$f_1(x) - f_2(x) = f_1(x_0) = f_2(x_0) \equiv C ,$$

which proves the lemma. ∎

Thus, a convex function is defined by its subdifferential mapping with the accuracy up to a constant summand.

LEMMA 5.8. Let $f(x)$ be a convex function on E_n. Also, let $x_1 \neq x_2$. In order that

$$\partial f(x_1) \cap \partial f(x_2) \neq \emptyset , \tag{5.27}$$

it is necessary for all $\alpha \in (0,1)$ and sufficient for at least one $\alpha \in (0,1)$ that the equality

$$f(\alpha x_1 + (1-\alpha)x_2) = \alpha f(x_1) + (1-\alpha)f(x_2) \tag{5.28}$$

be valid.

Proof. Necessity. Let $v \in \partial f(x_1)$, $v \in \partial f(x_2)$. By the definition of a subgradient

$$f(z) - f(x_1) \geq (v,z-x_1) \qquad \forall z \in E_n , \tag{5.29}$$

$$f(z) - f(x_2) \geq (v,z-x_2) \qquad \forall z \in E_n . \tag{5.30}$$

Letting $z = x_2$ in (5.29) and $z = x_1$ in (5.30), we get

$$f(x_2) - f(x_1) = (v, x_2-x_1) . \tag{5.31}$$

Take any $\alpha \in (0,1)$ and let $x_\alpha = \alpha x_1 + (1-\alpha)x_2$. From (5.29) and (5.31),

$$f(x_\alpha) - f(x_1) \geq (1-\alpha)(v, x_2-x_1) = (1-\alpha)(f(x_2) - f(x_1)) ,$$

i.e.,

$$f(x_\alpha) \geq \alpha f(x_1) + (1-\alpha)f(x_2) . \tag{5.32}$$

On the other hand, since the function $f(x)$ is convex, then

$$f(x_\alpha) \le \alpha f(x_1) + (1-\alpha)f(x_2) \qquad \forall\, \alpha \in [0,1].$$

Hence, (5.28) follows from this and (5.32).

Sufficiency. Let (5.28) be valid for some $\alpha \in (0,1)$. Take an arbitrary $v \in \partial f(x_\alpha)$. Then

$$f(z) \ge f(x_\alpha) + (v,\ z-x_\alpha) \qquad \forall\, z \in E_n ,$$

which plus (5.28) yield

$$\begin{aligned} f(z) &\ge \alpha f(x_1) + (1-\alpha)f(x_2) + (v,\ z - \alpha x_1 - (1-\alpha)x_2) \qquad (5.33)\\ &= \alpha[f(x_1) + (z-x_1,\ v)] + (1-\alpha)[f(x_2) + (z-x_2,\ v)] \\ &\qquad\qquad \forall\, z \in E_n . \end{aligned}$$

For $z = x_1$,

$$f(x_1) \ge \alpha f(x_1) + (1-\alpha)[f(x_2) + (x_1-x_2,\ v)] ,$$

i.e.,

$$(1-\alpha)[f(x_1) - f(x_2)] \ge (1-\alpha)(x_1-x_2,\ v) .$$

Since $1 - \alpha > 0$, then

$$f(x_1) - f(x_2) \ge (x_1-x_2,\ v) .$$

Analogously, for $z = x_2$ we obtain from (5.33)

$$f(x_2) - f(x_1) \ge (x_2-x_1,\ v) ,$$

which plus the preceding inequality yield

$$f(x_1) - f(x_2) = (x_1-x_2,\ v) . \qquad (5.34)$$

On the other hand, we have from (5.33) that

$$f(z) - f(x_1) \ge (z-x_1,\ v) + (1-\alpha)[f(x_2) - f(x_1) + (x_1-x_2,\ v)] \qquad \forall\, z \in E_n . \qquad (5.35)$$

By virtue of (5.34), the term in the square brackets in (5.35) is equal to zero. Hence

$$f(z) - f(x_1) \geq (z-x_1, v) \qquad \forall z \in E_n ,$$

i.e.,

$$v \in \partial f(x_1) . \tag{5.36}$$

Analogously, we have from (5.33)

$$f(z) - f(x_2) \geq (z-x_2, v) + \alpha[f(x_1) - f(x_2) - (x_1-x_2, v)] \qquad \forall z \in E_n,$$

and we get again from (5.34)

$$f(z) - f(x_2) \geq (z-x_2, v) \qquad \forall z \in E_n,$$

i.e., $v \in \partial f(x_2)$. It follows from this and (5.36) that

$$\partial f(x_\alpha) \subset [\partial f(x_1) \cap \partial f(x_2)] .$$

Therefore, this establishes (5.27) and the lemma has been proved. ■

4. Let us fix $x_0 \in E_n$. Then, for any g, $\|g\| = 1$, we have

$$f(x_0+\alpha g) = f(x_0) + \alpha \frac{\partial f(x_0)}{\partial g} + o(\alpha, g) , \tag{5.37}$$

where

$$\alpha^{-1} o(\alpha, g) \xrightarrow[\alpha \to +0]{} 0 . \tag{5.38}$$

Let us show that the convergence to zero in (5.38) is uniform with respect to g, $\|g\| = 1$. From (5.26) we have

$$f(x_0+\alpha g) = f(x_0) + \int_0^\alpha \frac{\partial f(x_0+\tau g)}{\partial g} d\tau . \tag{5.39}$$

By Theorem 5.2,

$$\frac{\partial f(x)}{\partial g} = \max_{v \in \partial f(x)} (v, g) .$$

From this and (5.39),

$$f(x_0+\alpha g) - f(x_0) \leq \alpha \max_{v \in B(\alpha)} (v, g) , \tag{5.40}$$

where $B(\alpha) = \bigcup_{\|x-x_0\| \leq \alpha} \partial f(x)$. Since the mapping $\partial f(x)$ is upper semicontinuous (see Remark 1), then

$$\max_{v\in B(\alpha)} (v,g) \le \max_{v\in\partial f(x_0)} (v,g) + q(\alpha) = \frac{\partial f(x_0)}{\partial g} + q(\alpha) , \tag{5.41}$$

where $q(\alpha) \xrightarrow[\alpha\to+0]{} 0$, $q(\alpha)$ not depending on g, $\|g\|=1$, any longer.

By virtue of (5.1),

$$f(x_0+\alpha g) - f(x_0) \ge \alpha \frac{\partial f(x_0)}{\partial g} ,$$

which plus (5.40), (5.41) yield

$$0 \le f(x_0+\alpha g) - f(x_0) - \alpha \frac{\partial f(x_0)}{\partial g} \le \alpha q(\alpha) .$$

Since $q(\alpha)$ does not depend on g, $\|g\|=1$, the uniformity of (5.38) with respect to g, $\|g\|= 1$, follows from (5.37).

We note that, generally speaking, the function $o(\alpha,g) \equiv o(\alpha,g,x_0)$ is not even continuous in x_0.

5. Let $f(x)$ be a finite concave function on E_n, i.e.,

$$f(\alpha x_1 + (1-\alpha)x_2) \ge \alpha f(x_1) + (1-\alpha)f(x_2) \tag{5.42}$$

$$\forall x_1,x_2 \in E_n, \quad \forall \alpha \in [0,1].$$

The set

$$\bar\partial f(x_0) = \{v \in E_n \mid f(x)-f(x_0) \le (v,x-x_0) \ \forall x \in E_n\} \tag{5.43}$$

is said to be the *superdifferential* of the function $f(x)$ at the point x_0 and an element $v \in \bar\partial f(x_0)$ is said to be a *supergradient* of $f(x)$ at the point x_0. Since the function $f_1(x) = -f(x)$ is convex, it is obvious that

$$\bar\partial f(x_0) = -\partial f_1(x_0) .$$

Therefore, the propositions analogous to those proved for convex functions are valid for superdifferentials and supergradients as well:

The set $\bar\partial f(x_0)$ is nonempty, convex, closed and bounded.

A concave function $f(x)$ is differentiable in any direction and, moreover,

$$\frac{\partial f(x_0)}{\partial g} = \max_{v \in \partial f(x_0)} (v,g) . \tag{5.44}$$

PROBLEM 5.3. Prove the formula analogous to (5.15) for the sum of concave functions.

PROBLEM 5.4. Find the superdifferential of the function

$$f(x) = \min_{i \in I} f_i(x) ,$$

where $I = 1:N$; $f_i(x)$ is a concave function on E_n.

PROBLEM 5.5. Prove that for a concave function $f(x)$ and a continuously differentiable vector-valued function $x(t)$ we have the representation

$$f(x(t)) = f(x(0)) + \int_0^t \frac{\partial f(x(\tau))}{\partial \dot{x}(\tau)} d\tau . \tag{5.45}$$

The integral in this equality is interpreted in the Lebesgue sense.

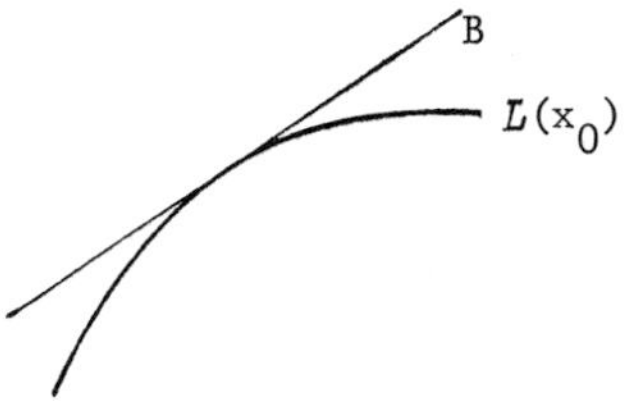

Figure 7

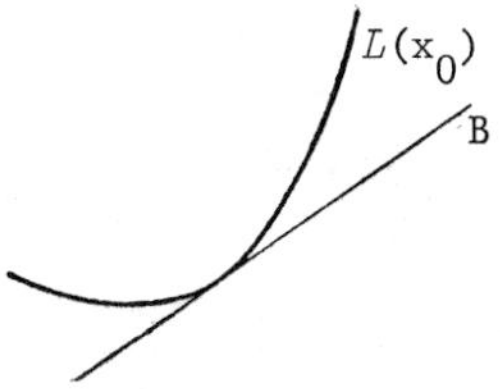

Figure 8

REMARK 4. The meaning of the term "subgradient" (and "subdifferential") becomes clear if we observe that the plane

$$L(x_0) = \{[\beta,x] \in E_{n+1} \mid x \in E_n, \ \beta = f(x_0)+(v,x-x_0)\} ,$$

where $v \in \partial f(x_0)$, passes through the point $[f(x_0),x_0]$ and lies *on or below* the graph of the function $f(x)$ (Figure 7):

$$B = \{[\beta,x] \mid x \in E_n, \ \beta = f(x)\}$$

(see Definition (5.1)).

Analogously, for the supergradient of concave functions, the plane $L(x_0)$, where $v \in \bar{\partial} f(x_0)$, lies *on or above* the graph of the function $f(x)$ (Figure 8) and also passes through the point $[f(x_0), x_0]$.

REMARK 5. Let $f(x)$ be a convex function on E_1. The following geometric interpretation of the subdifferential at a point x_0 is possible.

Through the point $(f(x_0), x_0) \in E_2$ we draw a support line (or a line of support) for the epigraph of the function $f(x)$. Let α be the slope tangent of this line to the axis Ox. Then, $\alpha \in \partial f(x_0)$.

Conversely, if $\alpha \in \partial f(x_0)$, then there exists a line passing through the point $(f(x_0), x_0)$, being a support line for the epigraph of the function $f(x)$ and such that its slope tangent to the axis Ox is equal to α.

6. DISTANCE FROM A SET TO A CONE. CONDITIONS FOR A MINIMUM

1. Let a finite convex function $f(x)$ be defined on a convex set $S \subset E_n$.

We established in Section 4 that the function $f(x)$ is differentiable with respect to any direction at each interior point of the set S. Let $\Omega \subset \operatorname{int} S$ be a convex closed set.

LEMMA 6.1. For the function $f(x)$ to attain the minimum value on Ω at a point $x^* \in \Omega$, it is necessary and sufficient that

$$\min_{g \in \Gamma(x^*),\, \|g\|=1} \frac{\partial f(x^*)}{\partial g} \geq 0 , \tag{6.1}$$

where $\Gamma(x^*)$ is the cone of feasible directions of the set Ω at the point x^*.

Proof. Necessity. Let $x^* \in \Omega$ be a minimum point of $f(x)$ on Ω. Suppose that (6.1) is not satisfied. Then, since $\Gamma(x^*) = \overline{\gamma(x^*)}$, there exists a vector

$$g_0 \in \gamma(x^*) \equiv \{v = \lambda(z-x^*) \mid \lambda > 0,\ z \in \Omega\}, \qquad \|g_0\| = 1,$$

such that

$$\frac{\partial f(x^*)}{\partial g_0} = -\alpha < 0 . \tag{6.2}$$

Since

$$f(x^*+\alpha g_0) = f(x^*) + \alpha\frac{\partial f(x^*)}{\partial g_0} + o(\alpha), \tag{6.3}$$

$$o(\alpha)\alpha^{-1} \xrightarrow[\alpha\to+0]{} 0$$

and, for sufficiently small $\alpha > 0$, we have $x^* + \alpha g_0 \in \Omega$, then the existence of an $\alpha_0 > 0$ such that

$$f(x^*+\alpha g_0) < f(x^*), \qquad x^*+\alpha g_0 \in \Omega \quad \forall \alpha \in (0,\alpha_0]$$

follows from (6.2) and (6.3). This contradicts the fact that x^* is a minimum point of $f(x)$ on Ω.

Sufficiency. Let (6.1) be valid. By the definition of $\partial f(x^*)$ we have for $x \neq x^*$

$$f(x) \geq f(x^*) + \max_{v\in\partial f(x^*)} (v,x-x^*) = f(x^*) + \|x-x^*\| \frac{\partial f(x^*)}{\partial g(x)},$$

where $g(x) = \|x-x^*\|^{-1}(x-x^*)$. It follows from this and (6.1) that

$$f(x) \geq f(x^*) + \|x-x^*\| \min_{g\in\Gamma(x^*),\|g\|=1} \frac{\partial f(x^*)}{\partial g} \geq f(x^*) \quad \forall x \in \Omega,$$

which proves the lemma. ∎

<u>COROLLARY</u>. Condition (6.1) is equivalent to the following condition:

$$\min_{g\in\Gamma(x^*),\|g\|=1} \frac{\partial f(x^*)}{\partial g} = 0 .$$

<u>REMARK 1</u>. A necessary condition for a minimum of any directionally differentiable function is

$$\inf_{g\in\gamma(x^*),\|g\|=1} \frac{\partial f(x^*)}{\partial g} \geq 0,$$

where

$$\gamma(x^*) = \{g = \lambda(x-x^*) \mid \lambda > 0,\ x \in \Omega\} .$$

PROBLEM 6.1. Prove this proposition.

REMARK 2. We note that the minimum in (6.1) is attained because the function $\max_{v\in\partial f(x^*)}(v,g)$ is convex with respect to g (see Subsection 4.4). Obviously, it is finite and, hence, continuous in g on E_n and attains its minimum on any compact set in E_n, in particular on the set $\{g\in\Gamma(x^*) \mid \|g\|=1\}$.

2. We give now some lemmas concerning geometric properties, which will often be used in the sequel.

Let $B\subset E_n$ be a convex compact set, let $\Gamma\subset E_n$ be a convex closed cone and let Γ^+ be the conjugate cone of Γ. Also, let

$$C = B-\Gamma^+ , \qquad C_{fr} = C\setminus \operatorname{int} C ,$$

$$\psi = \min_{\|g\|=1,g\in\Gamma} \max_{v\in B}(v,g) , \qquad \phi = \min_{\|g\|=1} \sup_{v\in C}(v,g) ,$$

$$\rho = \inf_{v\in C_{fr}}\|v\| , \qquad d = \inf_{v\in C}\|v\| .$$

Since the sets B and Γ^+ are closed and B is bounded, the infima in the definitions of ρ and d are attained, i.e.,

$$\rho = \min_{v\in C_{fr}}\|v\| , \qquad d = \min_{v\in C}\|v\| .$$

If $d>0$, then $\rho=d$, i.e., ρ is the distance from the origin to the set C (or, what is just the same, the distance from the set B to the cone Γ^+). On the other hand, if $d=0$, then ρ is the radius of the maximum sphere which is centered on O and contained in C.

LEMMA 6.2. The following equality holds:

$$\psi = \phi . \tag{6.4}$$

Proof. Let us show that, for any $g\in E_n$, we have

$$\sup_{v\in C}(v,g) = \begin{cases} \max\limits_{v\in B}(v,g) & \text{if } g\in\Gamma , \\ +\infty & \text{if } g\notin\Gamma . \end{cases} \qquad (6.5)$$

To do this, we first note that

$$\sup_{v\in C}(v,g) = \sup_{v_1\in B, v_1\in[-\Gamma^+]}(v_1+v_2, g) = \max_{v_1\in B}(v_1,g) + \sup_{v_1\in[-\Gamma^+]}(v_2,g) .$$

If $g\in\Gamma = (\Gamma^+)^+$, then $(-v_2,g)\geq 0 \ \ \forall v_2\in[-\Gamma^+]$. Since $0\in\Gamma^+$, we have $\sup\limits_{v\in[-\Gamma^+]}(v,g) = 0$. If $g\notin\Gamma$, then we can find $v_2'\in[-\Gamma^+]$ such that $(-v_2',g)<0$. Since Γ^+ is a cone, hence $\sup\limits_{v\in[-\Gamma^+]}(v,g) = +\infty$. Thus,

$$\sup_{v\in[-\Gamma^+]}(v,g) = \begin{cases} 0 & \text{if } g\in\Gamma , \\ +\infty & \text{if } g\notin\Gamma . \end{cases} \qquad (6.6)$$

Then (6.5) and (6.4) follow from (6.6) because (6.5) holds for any $g\in E_n$. The lemma is proved. ■

As the above proof shows, the infimum in the definition of ϕ is attained, i.e., $\phi = \min\limits_{||g||=1}\sup\limits_{v\in C}(v,g)$.

<u>LEMMA 6.3</u>. The condition $\psi\geq 0$ is equivalent to the condition $d=0$.

<u>Proof</u>. Let $\psi\geq 0$. Then, by virtue of (6.4), $\phi\geq 0$. Suppose that $d>0$. Let $v_0\in C$ and let $||v_0|| = d$. It was established in the proof of Theorem 1.2 (see inequality (1.10)) that

$$(v,v_0) \geq ||v_0||^2 . \qquad (6.7)$$

Let $g_0 = -||v_0||^{-1}v_0$. From (6.7) we have

$$\phi = \min_{||g||=1}\sup_{v\in C}(v,g) \leq \sup_{v\in C}(v,g_0) = -||v_0|| = -d < 0 . \qquad (6.8)$$

Thus, we get a contradiction, i.e., the equality $d=0$ follows from the condition $\psi\geq 0$.

Now assume that $d=0$. This implies that $0\in C$. Then

$$\psi = \phi = \min_{||g||=1}\sup_{v\in C}(v,g) \geq \min_{||g||=1}(0,g) = 0 ,$$

i.e., the inequality $\psi \geq 0$ follows from the condition $d = 0$. This completes the proof. ∎

COROLLARY. The condition $\psi < 0$ is equivalent to $d > 0$.

LEMMA 6.4. The following relation holds:

$$\psi = \begin{cases} -\rho = -d & \text{if } d > 0 , \\ \rho & \text{if } d = 0 . \end{cases} \tag{6.9}$$

Proof. First we consider the case where $d > 0$. From (6.8) and the fact that $\rho = d$ in this case, we have $\psi \leq -\rho$. Next we establish the opposite inequality. Let $v_0 \in C$ and let $\|v_0\| = d = \rho$. We define $g_0 = -\|v_0\|^{-1} v_0$ and show that for any $g \in E_n$, $\|g\| = 1$, $g \neq g_0$,

$$\sup_{v \in C} (v,g) > -\|v_0\| = -d . \tag{6.10}$$

To do this, let $g \in E_n$, $\|g\| = 1$, $g \neq g_0$. Then

$$\|v_0\| > (-v_0, g) \geq -\sup_{v \in C} (v,g) .$$

From (6.4) and (6.10) we can conclude that

$$\psi = \min_{\|g\|=1} \sup_{v \in C} (v,g) \geq -\rho$$

Note that, by virtue of (6.5), $g_0 \in \Gamma$. Thus, if $d > 0$, then $\psi = -\rho = -d$.

Now let $d = 0$. Then $\psi \geq 0$ by Lemma 6.3. In this case, as was already noticed, ρ is the radius of the maximum sphere centered at the origin and contained in C, i.e., $S_\rho(0) \subset C$. For any $\delta > 0$ such that $S_\delta(0) \subset C$, we have

$$\delta = \max_{v \in S_\delta(0)} (v,g) \leq \sup_{v \in C} (v,g) \qquad \forall g \in E_n, \ \|g\| = 1 .$$

Hence $\delta \leq \phi$. In particular, $\rho \leq \phi$. From (6.4)

$$\rho \leq \phi = \psi . \tag{6.11}$$

Next let us prove

$$S_\psi(0) \subset C . \tag{6.12}$$

Assume the opposite. Then we can find vector $v_0 \in E_n$ such that $v_0 \in S_\psi(0)$ but $v_0 \notin C$. Clearly

$$\|v_0\| \le \psi . \tag{6.13}$$

By the separation theorem, we can find a vector $g_0 \in E_n$, $\|g_0\| = 1$, and a number $\alpha > 0$ such that $(v,g_0) \le (v_0,g_0) - \alpha \quad \forall v \in C$.

Therefore

$$\psi = \phi = \min_{\|g\|=1} \sup_{v \in C} (v,g) \le \sup_{v \in C} (v,g_0) \le (v_0,g_0) - \alpha .$$

From this and (6.13) we obtain $\psi \le \|v_0\| - \alpha \le \psi - \alpha < \psi$. This is a contradiction. Thus inclusion (6.12) as well as the inequality

$$\psi \le \rho \tag{6.14}$$

are established. Considering both (6.11) and (6.14), we conclude that if $d = 0$, then $\psi = \rho$. This completes the proof. ■

COROLLARY 1. Let $\psi < 0$. If a vector $g_0 \in \Gamma$, $\|g_0\| = 1$, is such that

$$\max_{v \in B} (v,g_0) = \psi ,$$

then

$$\begin{aligned} (v,g_0) &\le -\rho \qquad \forall v \in B , \\ (w,g_0) &\ge 0 \qquad \forall w \in \Gamma^+ . \end{aligned} \tag{6.15}$$

REMARK 3. A vector $v_0 \in C$, for which $\|v_0\| = d$, is unique.

This follows from the fact that the strongly convex function (v,v) attains its minimum on any closed convex set at a unique point.

COROLLARY 2. Let C be a closed convex set (not necessarily bounded) and let

$$\psi = \min_{\|g\|=1} \sup_{v \in C} (v,g) , \qquad d = \min_{v \in C} \|v\| , \qquad \rho = \min_{v \in C_{fr}} \|v\| .$$

Then, the condition $\psi \geq 0$ is equivalent to the condition $d = 0$ and we have the relation

$$\psi = \begin{cases} -\rho = -d & \text{if} \quad d > 0 , \\ \rho & \text{if} \quad d = 0 . \end{cases}$$

Moreover, a vector $v_0 \in C$, for which $||v_0|| = d$, is unique.

COROLLARY 3. Let the set B consist of a unique point b.

If $b \notin \Gamma^+$, then

$$\min_{v \in \Gamma^+} ||b-v|| = \max_{g \in [-\Gamma], ||g||=1} (b,g) = \max_{g \in [-\Gamma], ||g|| \leq 1} (b,g) .$$

This obviously follows from (6.9). It is also clear that

$$\min_{v \in \Gamma^+} ||b-v|| = \max_{g \in [-\Gamma], ||g|| \geq 1} (b,g) \qquad \forall b \in E_n .$$

3. A Geometric Interpretation of the Conditions for a Minimum.

THEOREM 6.1. Condition (6.1) is equivalent to the condition

$$\partial f(x^*) \cap \Gamma^+(x^*) \neq \emptyset . \tag{6.16}$$

Proof. We note that (6.16) is equivalent to

$$0 \in [\partial f(x^*) - \Gamma^+(x^*)] . \tag{6.17}$$

Let (6.1) be satisfied, i.e.,

$$\min_{g \in \Gamma(x^*), ||g||=1} \max_{v \in \partial f(x^*)} (v,g) \geq 0 . \tag{6.18}$$

Let $B = \partial f(X^*)$, $\Gamma = \Gamma(x^*)$. Then from (6.18) we obtain $\psi \geq 0$.

By Lemma 6.3 we can conclude that $d = \min_{v \in B - \Gamma^+} ||v|| = 0$, i.e., (6.17), and hence (6.16), holds.

Now, let (6.16) and (6.17) be valid. Then $d = 0$ and, by Lemma 6.3, $\psi \geq 0$. This proves the theorem. ∎

COROLLARY. If $\Omega = E_n$, the condition (6.16) is equivalent to the condition

$$0 \in \partial f(x^*) . \tag{6.19}$$

If, in addition, $f(x)$ is a continuously differentiable function, then (6.16) becomes the classical condition

$$f'(x^*) = 0 . \tag{6.20}$$

4. The Steepest Descent Direction.

Let $x_0 \in \Omega$. Define

$$d(x_0) = \min_{\substack{v \in \partial f(x_0) \\ w \in \Gamma^+(x_0)}} \|v-w\|_* . \tag{6.21}$$

Then, there exist points $v(x_0) \in \partial f(x_0)$ and $w(x_0) \in \Gamma^+(x_0)$ such that

$$d(x_0) = \|v(x_0) - w(x_0)\| . \tag{6.22}$$

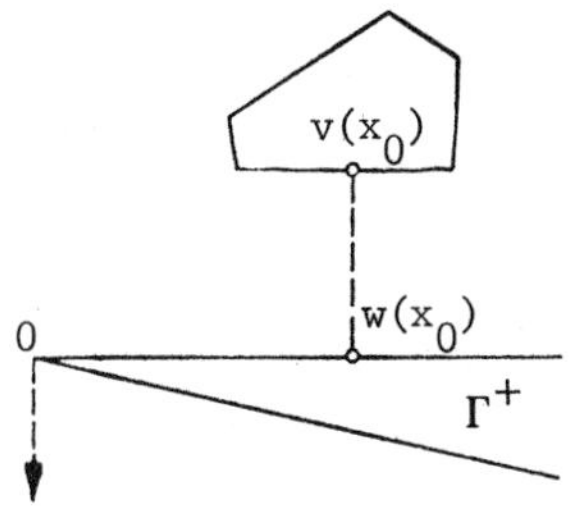

Figure 9

The vector $v(x_0) - w(x_0)$ satisfying (6.22) is unique (see Remark 3) though the pair of points $v(x_0)$ and $w(x_0)$ can be nonunique (Figure 9).

If $d(x_0) = 0$, then, according to Theorem 6.1, the point x_0 is a minimum point of the function $f(x)$ on Ω. Consider the case $d(x_0) > 0$.

Define

$$g(x_0) = \|w(x_0) - v(x_0)\|^{-1} (w(x_0) - v(x_0)) . \tag{6.23}$$

DEFINITION. Let $d(x_0) > 0$. A vector $g_0 \in \Gamma$, $\|g_0\| = 1$, such that

$$\frac{\partial f(x_0)}{\partial g} = \inf_{g \in \Gamma(x_0), \|g\|=1} \frac{\partial f(x_0)}{\partial g} , \tag{6.24}$$

is said to be the steepest descent direction of the function $f(x)$ at the point x_0 with respect to the set Ω.

THEOREM 6.2. If $d(x_0) < 0$, then the vector $g(x_0)$ is the steepest descent direction of the function $f(x)$ at the point x_0 with respect to the set Ω and

$$\frac{\partial f(x_0)}{\partial g(x_0)} = -d(x_0) \quad . \tag{6.25}$$

This direction is unique.

Proof. Let $B = \partial f(x)$, $\Gamma = \Gamma(x_0)$. Then from (6.22), (6.23) and (6.8) we have

$$\sup_{v \in C} (v, g(x_0)) = -\|v(x_0) - w(x_0)\| = -d(x_0) \ .$$

Since $d(x_0) < +\infty$, we then conclude from (6.5) that

$$g(x_0) \in \Gamma \ , \quad -d(x_0) = \max_{v \in \partial f(x_0)} (v, g(x_0)) = \frac{\partial f(x_0)}{\partial g(x_0)} \ . \tag{6.26}$$

Hence, from (6.10) and (6.5) we obtain for all $g \in \Gamma(x_0)$, $\|g\| = 1$,

$$\frac{\partial f(x_0)}{\partial g} = \max_{v \in \partial f(x_0)} (v, g) \geq -d(x_0) = \frac{\partial f(x_0)}{\partial g(x_0)} \ ,$$

i.e., the vector $g(x_0)$ is the steepest descent direction of the function $f(x)$ at the point x_0 with respect to Ω. The uniqueness of the steepest descent direction also follows from (6.10). Equality (6.25) is clear from (6.26). ■

REMARK 4. The direction $g(x_0)$ belongs to $\Gamma(x_0)$, i.e., $g(x_0) \in \Gamma(x_0)$. However, it may happen that $g(x_0) \notin \gamma(x_0)$, i.e., this direction can lead out of the set Ω.

PROBLEM 6.2. Let $M^* = \{x \in E_n \mid f(x) = \inf_{z \in E_n} f(z)\}$. Show that if $M^* \neq \emptyset$ and $\operatorname{int} M^* \neq \emptyset$, then

$$\partial f(x) = \{0\} \quad \forall x \in \operatorname{int} M^* \ .$$

7. ε-SUBDIFFERENTIALS

1. Let $f(x)$ be a finite convex function on E_n. Fix $x_0 \in E_n$, $\varepsilon \geq 0$ and let

$$d_\varepsilon f(x_0) = \{v \in E_n \mid f(x) - f(x_0) \geq (v, x - x_0) - \varepsilon \quad \forall x \in E_n\}. \tag{7.1}$$

The set $\partial_\varepsilon f(x_0)$ is said to be the ε-*subdifferential* of the function $f(x)$ at the point x_0. An element $v \in \partial_\varepsilon f(x_0)$ is said to be an ε-*subgradient* of the function $f(x)$ at the point x_0.

THEOREM 7.1. For any fixed $x_0 \in E_n$ and $\varepsilon \geq 0$, the set $\partial_\varepsilon f(x_0)$ is nonempty, closed, convex and bounded.

Proof. Since $\partial f(x_0) \subset \partial_\varepsilon f(x_0)$ $\forall \varepsilon \geq 0$, the set $\partial_\varepsilon f(x_0)$ is not empty. The closedness and convexity are clear from the definition.

Now we show the boundedness. Fix $\delta > 0$. The continuity of the function $f(x)$ implies that there exists a $c < \infty$ such that

$$\delta^{-1}[f(x) - f(x_0) + \varepsilon] \leq c \tag{7.2}$$

$$\forall x \in S_\delta(x_0) \equiv \{x \in E_n \mid \|x - x_0\| \leq \delta\} .$$

Taking an arbitrary $v \in \partial_\varepsilon f(x_0)$ and the point $x_1 = x_0 + \delta v \|v\|^{-1}$, we get

$$f(x_1) \geq f(x_0) + (v, x_1 - x_0) - \varepsilon = f(x_0) + \delta\|v\| - \varepsilon. \tag{7.3}$$

Hence

$$\|v\| \leq \delta^{-1}[f(x_1) - f(x_0) + \varepsilon] \leq c \qquad \forall v \in \partial_\varepsilon f(x_0) , \tag{7.4}$$

which proves the theorem. ∎

COROLLARY. The mapping $\partial_\varepsilon f : [0, \infty) \times E_n \to \Pi(E_n)$ is bounded on any bounded set in $[0, \infty) \times E_n$.

Proof. Is analogous to that of the boundedness in Theorem 7.1.

EXAMPLE 1. Let $f(x) = \delta(x-a)^2$, where $a \in E_n$, $\delta > 0$ are fixed. Let us find $\partial_\varepsilon f(x)$. Let $v \in \partial_\varepsilon f(x)$ and let us represent v in the form $v = f'(x) + r = 2\delta(x-a) + r$. By definition,

$$f(z) - f(x) \geq (v, z-x) - \varepsilon \qquad \forall z \in E_n ,$$

i.e.,

$$\delta(z-a)^2 - \delta(x-a)^2 \geq (2\delta(x-a)+r,\ z-x) - \varepsilon \quad \forall z \in E_n . \tag{7.5}$$

Therefore

$$\delta(z-x)^2 - (r, z-x) + \varepsilon \geq 0 \qquad \forall z \in E_n . \quad (7.6)$$

Let us find the minimum of the left-hand side of this inequality.

The minimum is attained for $z^* = x + 0.5\delta^{-1}r$ and (7.6) is reduced to $-0.25\delta^{-1}r^2 + \varepsilon \geq 0$ for $z = z^*$, i.e., $\|r\| \leq 2\sqrt{\varepsilon\delta}$. This condition is necessary and sufficient for the fulfillment of (7.6). Thus,

$$\partial_\varepsilon f(x) = \{v = 2\delta(x-a) + r \mid r \in E_n, \ \|r\| \leq 2\sqrt{\delta\varepsilon}\} . \quad (7.7)$$

<u>PROBLEM 7.1</u>. Construct the set $\partial_\varepsilon f(x)$ for the function $f(x) = (Ax,x)$, where A is an $n\times n$ symmetric positive definite matrix.

<u>EXAMPLE 2</u>. Let $f(x) = (a,x) + b$, where $a \in E_n$, $b \in E_1$.

Let us take $v \in \partial_\varepsilon f(x)$. Then,

$$f(z) - f(x) \geq (v, z-x) - \varepsilon ,$$

or

$$(a-v, z-x) \geq -\varepsilon \qquad \forall z \in E_n . \quad (7.8)$$

It is clear that $v = a$ and, hence, $\partial_\varepsilon f(x) = \partial f(x) = \{a\}$.

2. Fix $x_0 \in E_n$ and $\varepsilon \geq 0$. Let

$$B_\varepsilon f(x_0) = \{v \in E_n \mid \exists x \in E_n : v \in \partial f(x), \ f(x) - f(x_0) \geq (v, x-x_0) - \varepsilon\} .$$

<u>LEMMA 7.1</u>. The following relation holds:

$$\partial_\varepsilon f(x_0) = \overline{B_\varepsilon f(x_0)} . \quad (7.10)$$

<u>Proof</u>. Is omitted because relation (7.10) is a special case for $\Omega = E_n$ of Theorem 10.1, which will be proved in the sequel.

From (7.1) we get the following.

<u>COROLLARY</u>. Let x_0 be fixed.

Then, for any $x \in E_n$ and $v \in \partial f(x)$ we can find an $\varepsilon > 0$ such that $v \in \partial_\varepsilon f(x_0)$.

REMARK 1. Using (7.10), we obtain the following geometric interpretation of the ε-subdifferential: a subgradient v of the function $f(x)$ at a point $\bar{x}$ belongs to $\partial_\varepsilon f(x_0)$ if the linear function $\ell(y) = (v, y-\bar{x}) + f(\bar{x})$, which describes a supporting hyperplane for the set epi f at the point $\bar{x}$, takes a value greater than or equal to $f(x_0) - \varepsilon$ at the point x_0 (Figure 10).

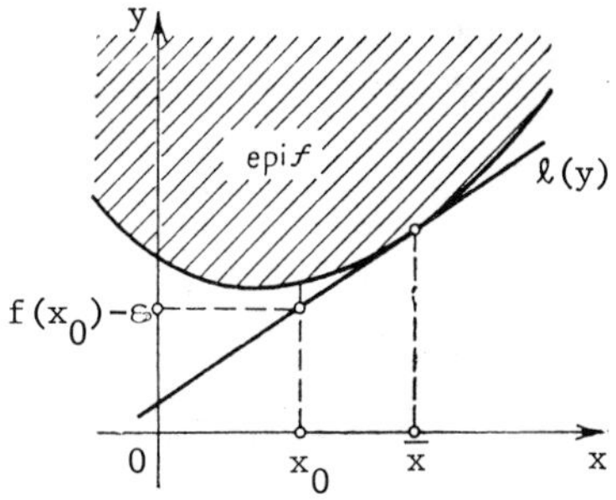

Figure 10

Let us introduce an example which shows that the closure in (7.10) is essential.

EXAMPLE 3. Let $x \in E_1$. Consider the function

$$f(x) = \begin{cases} 3-2x, & x \le 1, \\ x^{-1}, & x \ge 1. \end{cases}$$

It is not difficult to verify that $f(x)$ is a convex function. Let us take the point $x_0 = 2$ and $\varepsilon = 0.5$. We have

$$f(x) - f(2) = \begin{cases} 3-2x-0.5, & x \le 1, \\ x^{-1} - 0.5, & x \ge 1. \end{cases} \tag{7.11}$$

Clearly

$$f(x) - f(2) \ge -0.5 = (0, x-2) - 0.5 \qquad \forall x \in E_1 .$$

Therefore

$$0 \in \partial_{0.5} f(2) . \tag{7.12}$$

On the other hand,

$$\partial f(x) = \begin{cases} \{-2\}, & x < 1, \\ [-2,-1], & x = 1, \\ \{-x^{-2}\}, & x > 1 . \end{cases}$$

Therefore, there is no $x \in E_1$ such that $0 \in \partial f(x)$.

Recalling the definition of the set $B_\varepsilon f(x_0)$ (see (7.9)), we can conclude that $0 \notin B_{0.5} f(2)$ though, of course, $0 \in \overline{B_{0.5} f(2)}$. Thus, we cannot remove the sign of closure in (7.10). However, as will be shown below, this is possible in some cases.

LEMMA 7.2. If $f(x)$ be a strongly convex function (see Subsection 4.1), then

$$\partial_\varepsilon f(x_0) = B_\varepsilon f(x_0) . \tag{7.13}$$

To show this, we first note that the inclusion

$$B_\varepsilon f(x_0) \subset \partial_\varepsilon f(x_0) \tag{7.14}$$

follows from (7.10). Next we prove the converse inclusion.

Let $v \in \partial_\varepsilon f(x_0)$, i.e.,

$$f(z) - f(x_0) \geq (v, z-x_0) - \varepsilon \qquad \forall z \in E_n . \tag{7.15}$$

Let $h(z) = f(z) - f(x_0) + (v, x_0 - z)$. Since $f(x)$ is a strongly convex function, then $h(z)$ is a strongly convex function as well, and, hence, the function $h(z)$ attains its minimum value at some point $x' \in E_n$: $h(x') = \min_{z \in E_n} h(z)$. Then,

$$h(x') \leq h(z) \qquad \forall z \in E_n ,$$

i.e.,

$$f(x') - (v, x') \leq f(z) - (v, z)$$

or

$$f(z) - f(x') \geq (v, z-x') \qquad \forall z \in E_n ,$$

and this implies that $v \in \partial f(x')$. From (7.15) we have $(v, x_0 - x') + f(x') - f(x_0) \geq -\varepsilon$, i.e., $v \in B_\varepsilon f(x_0)$. Thus

$$\partial_\varepsilon f(x_0) \subset B_\varepsilon f(x_0) . \tag{7.16}$$

Hence (7.13) follows from (7.14) and (7.16). ∎

3. Let $f_1(x)$ and $f_2(x)$ be convex finite functions. Let $f(x) = f_1(x) + f_2(x)$. By Lemma 5.2,

$$\partial f(x) = \partial f_1(x) + \partial f_2(x) \quad . \tag{7.17}$$

For ε-subdifferentials, the analogous relation no longer holds.

THEOREM 7.2. The following representation holds:

$$\partial_\varepsilon f(x_0) = \bigcup_{\substack{\varepsilon_1+\varepsilon_2=\varepsilon \\ \varepsilon_1\ge 0,\varepsilon_2\ge 0}} [\partial_{\varepsilon_1} f_1(x_0) + \partial_{\varepsilon_2} f_2(x_0)] \ . \tag{7.18}$$

Proof. Denote by D the right-hand side of (7.18). The inclusion

$$D \subset \partial_\varepsilon f(x_0) \tag{7.19}$$

is obvious. Let us take $v \in \partial_\varepsilon f(x_0)$. By Lemma 7.1, we can find a sequence $\{v_k\}$ such that

$$v_k \in B_\varepsilon f(x_0) \ , \qquad v_k \xrightarrow[k\to\infty]{} v \ . \tag{7.20}$$

By the definition of $B_\varepsilon f(x_0)$, for each v_k we can find x_k such that

$$v_k \in \partial f(x_k) \ , \qquad (v_k, x_0-x_k) + f(x_k) - f(x_0) \ge -\varepsilon \ .$$

From (7.17) we have

$$v_k = v_k' + v_k'' \ , \tag{7.21}$$

where $v_k' \in \partial f_1(x_k)$, $v_k'' \in \partial f_2(x_k)$, and, in addition, from (7.20)

$$[(v_k', x_0-x_k) + f_1(x_k) - f_1(x_0)] + [(v_k'', x_0-x_k) + f_2(x_k) - f_2(x_0)] \ge -\varepsilon \ . \tag{7.22}$$

As the terms in the square brackets on the left-hand side of (7.22) are nonpositive, (7.22) can hold only in the case where $\varepsilon_{k_1} \ge 0$ and $\varepsilon_{k_2} \ge 0$ are such that

$$(v_k', x_0-x_k) + f_1(x_k) - f_1(x_0) \ge -\varepsilon_{k1} \ ,$$

$$(v_k'', x_0-x_k) + f_2(x_k) - f_2(x_0) \ge -\varepsilon_{k2} \ ,$$

and $\varepsilon_{k_1}+\varepsilon_{k_2}=\varepsilon$, i.e., $v'_k\in B_{\varepsilon_{k1}}f_1(x_0)$, $v''_k\in B_{\varepsilon_{k2}}f_2(x_0)$.

Since $f_1(x)$ and $f_2(x)$ are finite functions,

$$B_{\varepsilon_{k1}}f_1(x_0)\subset\partial_{\varepsilon_{k1}}f_1(x_0)\subset\partial_\varepsilon f_1(x_0)\quad,$$

$$B_{\varepsilon_{k2}}f_2(x_0)\subset\partial_{\varepsilon_{k2}}f_2(x_0)\subset\partial_\varepsilon f_2(x_0)\quad,$$

and the sets $\partial_\varepsilon f_1(x_0)$ and $\partial_\varepsilon f_2(x_0)$ are bounded, then, without loss of generality, we can assume that

$$v'_k\to v'\ ,\qquad v''_k\to v''\ ,\qquad \varepsilon_{k1}\to\varepsilon'\ \geq 0\ ,\qquad \varepsilon_{k2}\to\varepsilon''\ \geq 0\ .$$

Here $\varepsilon'+\varepsilon''=\varepsilon$. Because $v'_k\in\partial_{\varepsilon_{k_1}}f_1(x_0)$, we have

$$f_1(z)-f_1(x_0)\ \geq\ (v'_k,\ z-x_0)-\varepsilon_{k1}\qquad \forall z\in E_n\ .$$

Taking here the limit for every fixed $z\in E_n$ yields $f_1(z)-f_1(x_0)\geq(v',z-x_0)-\varepsilon'$, i.e., $v'\in\partial_{\varepsilon'}f_1(x_0)$. Similarly, we obtain $v''\in\partial_{\varepsilon''}f_2(x_0)$. Hence, from (7.20) and (7.21) we have

$$v\ =\ v'+v''\in[\partial_{\varepsilon'}f_1(x_0)+\partial_{\varepsilon''}f_2(x_0)]\ \subset\ D\quad.$$

It means that $\partial_\varepsilon f(x_0)\subset D$. From this and (7.19), we obtain (7.18). ∎

<u>COROLLARY 1</u>. If $f_i(x)$, $i\in I\equiv 1:N$, are convex finite functions, then

$$\partial_\varepsilon\left(\sum_{i=I}f_i(x)\right)\ =\ \bigcup_{\substack{\sum_{i\in I}\varepsilon_i=\varepsilon,\ \varepsilon_i\geq 0}}\left[\sum_{i\in I}\partial_{\varepsilon_i}f_i(x)\right]\quad.$$

<u>COROLLARY 2</u>. If $f_2(x)$ is a linear function, then

$$\partial_\varepsilon f(x)\ =\ \partial_\varepsilon f_1(x)+\partial f_2(x)\quad.$$

This follows from the fact that, for the linear function $f_2(x)=(a,x)+b$,

$$\partial_\varepsilon f_2(x)\ =\ \partial f_2(x)\ =\ \{a\}$$

(see Example 2).

PROBLEM 7.2. Construct the set $\partial_\varepsilon f(x)$ for the function

$$f(x) = (Ax,x) + (a,x) ,$$

where A is an $n \times n$ positive definite matrix and $a \in E_n$.

REMARK 2. Note that the obvious relation

$$\partial_\varepsilon [bf(x)] = b\partial_{\varepsilon b^{-1}} f(x)$$

holds. Here, $f(x)$ is a convex function and $b > 0$.

REMARK 3. Let

$$f_\varepsilon(x) = \sup_{z \in E_n} [f(z) + (v_\varepsilon(z), x-z)] , \tag{7.23}$$

where $v_\varepsilon(z) \in \partial_\varepsilon f(z)$ (for each z, one arbitrary vector $v_\varepsilon(z)$ in $\partial_\varepsilon f(z)$ is chosen). From the definition of $\partial_\varepsilon f(z)$, it is obvious that

$$0 \le f_\varepsilon(x) - f(x) \le \varepsilon \tag{7.24}$$

(compare with (5.11)).

4. Let us find the ε-subdifferential of a maximum function. Let $f_i(x)$, $i \in I \equiv 1:N$, be convex functions on E_n. Let

$$f(x) = \max_{i \in I} f_i(x) . \tag{7.25}$$

Taking $x_0 \in E_n$, $\varepsilon \ge 0$, we shall construct the following set:

$$K_\varepsilon(x_0) = \left\{ v \in E_n \;\middle|\; \begin{array}{l} v = \sum_{i \in I_1} \alpha_i v_i, \quad \alpha_i > 0, \quad \sum_{i \in I_1} \alpha_i = 1, \quad I_1 \subset I, \\ v_i \in \partial_{\varepsilon_i \alpha_i^{-1}} f_i(x_0) , \\ \sum_{i = I_1} \varepsilon_i + \varepsilon_0 = \varepsilon, \quad \varepsilon_i \ge 0, \quad \varepsilon_0 \ge 0, \\ \sum_{i \in I_1} \alpha_i f_i(x_0) \ge f(x_0) - \varepsilon_0 . \end{array} \right. \tag{7.26}$$

THEOREM 7.3. For the maximum function described by (7.25), the following representation holds:

$$\partial_\varepsilon f(x_0) = \overline{K_\varepsilon(x_0)} . \tag{7.27}$$

Proof. Take an arbitrary $v \in K_\varepsilon(x_0)$. Then,

$$v = \sum_{i \in I_1} \alpha_i v_i , \qquad I_1 \subset I , \qquad v_i \in \partial_{\varepsilon_i \alpha_i^{-1}} f_i(x_0) .$$

Hence

$$f_i(z) - f_i(x_0) \geq (v_i, z-x_0) - \varepsilon_i \alpha_i^{-1} \qquad \forall z \in E_n , \quad \forall i \in I_1 . \quad (7.28)$$

Multiplying both sides of (7.28) by α_i and summing them, we obtain

$$\sum_{i \in I_1} \alpha_i f_i(z) - \sum_{i \in I_1} \alpha_i f_i(x_0) \geq (v, z-x_0) - \sum_{i \in I_1} \varepsilon_i \qquad \forall z \in E_n . \quad (7.29)$$

Since $\sum_{i \in I_1} \alpha_i f_i(z) \leq f(z)$, we have from (7.26) and (7.29)

$$f(z) - f(x_0) \geq (v, z-x_0) - \sum_{i \in I_1} \varepsilon_i - \varepsilon_0 = (v, z-x_0) - \varepsilon \qquad \forall z \in E_n .$$

Hence $v \in \partial_\varepsilon f(x_0)$. Thus we have proved that $K_\varepsilon(x_0) \subset \partial_\varepsilon f(x_0)$. The closedness of $\partial_\varepsilon f(x_0)$ implies that

$$\overline{K_\varepsilon(x_0)} \subset \partial_\varepsilon f(x_0) . \quad (7.30)$$

Now let $v \in \partial_\varepsilon f(x_0)$. Then, by 7.1., $v \in \overline{B_\varepsilon f(x_0)}$, i.e.,

$$v = \lim_{k \to \infty} w_k , \qquad w_k \in B_\varepsilon f(x_0) . \quad (7.31)$$

By the definition of $B_\varepsilon f(x_0)$, for each k, there exists a point $x_k \in E_n$ such that

$$w_k \in \partial f(x_k) , \qquad (w_k, x_0 - x_k) + f(x_k) - f(x_0) \geq -\varepsilon . \quad (7.32)$$

By formula (5.21) we have

$$\partial f(x_k) = \text{co}\, \{\partial f_i(x_k) \mid i \in R(x_k)\} ,$$

where

$$R(x_k) = \{i \in I \mid f_i(x_k) - f(x_k)\} .$$

Therefore,

$$w_k = \sum_{i \in R(x_k)} \alpha_{ki} v_{ki} , \qquad v_{ki} \in \partial f_i(x_k) ,$$

$$\sum_{i \in R(x_k)} \alpha_{ki} = 1 , \qquad \alpha_{ki} \ge 0 ,$$

i.e.,

$$f_i(z) - f_i(x_k) \ge (v_{ki}, z-x_k) \qquad \forall z \in E_n . \tag{7.33}$$

Let

$$R_k = \{i \in R(x_k) \mid \alpha_{ki} > 0\} .$$

From (7.33) we have for each $i \in R_k$

$$f_i(z) - f_i(x_0) \ge (v_{ki}, z-x_0) + [f_i(x_k) - f_i(x_0) + (v_{ki}, x_0-x_k)] \quad \forall z . \tag{7.34}$$

Let

$$N_{ki} \equiv f_i(x_k) - f_i(x_0) + (v_{ki}, x_0-x_k) = -\alpha_{ki}^{-1}\varepsilon_{ki} . \tag{7.35}$$

Then, from (7.34) we have for each $i \in R_k$

$$f_i(z) - f_i(x_0) \ge (v_{ki}, z-x_0) - \alpha_{ki}^{-1}\varepsilon_{ki} \qquad \forall z , \tag{7.36}$$

i.e.,

$$v_{ki} \in \partial_{\alpha_{ki}^{-1}\varepsilon_{ki}} f_i(x_0) . \tag{7.37}$$

From (7.36) we obtain for $z = x_0$

$$N_{ki} = -\alpha_{ki}^{-1}\varepsilon_{ki} \le 0 . \tag{7.38}$$

Hence,

$$\begin{aligned}\sum_{i \in R_k} \alpha_{ki} N_{ki} &= -\sum_{i \in R_k} \varepsilon_{ki} \\ &= (w_k, x_0-x_k) + \sum_{i \in R_k} \alpha_{ki} f_i(x_k) - \sum_{i \in R_k} \alpha_{ki} f_i(x_0) .\end{aligned} \tag{7.39}$$

Since

$$\sum_{i \in R_k} \alpha_{ki} f_i(x_k) = f(x_k) , \qquad \sum_{i \in R_k} \alpha_{ki} f_i(x_0) \le f(x_0) ,$$

we have from (7.39) and (7.32)

$$\sum_{i \in R_k} \alpha_{ki} N_{ki} = -\sum_{i \in R_k} \varepsilon_{ki} \ge (w_k, x_0-x_k) + f(x_k) - f(x_0) \ge -\varepsilon .$$

Thus,

$$\sum_{i \in R_k} \varepsilon_{ki} \leq \varepsilon . \tag{7.40}$$

From (7.38) it follows also that

$$\varepsilon_{ki} \geq 0 . \tag{7.41}$$

Multiplying (7.35) by α_{ki} and summing the results yields

$$f(x_k) - \sum_{i \in R_k} \alpha_{ki} f_i(x_0) + (w_k, x_0 - x_k) = - \sum_{i \in R_k} \varepsilon_{ki} ,$$

i.e.,

$$\sum_{i \in R_k} \alpha_{ki} f_i(x_0) = f(x_k) + (w_k, x_0 - x_k) + \sum_{i \in R_k} \varepsilon_{ki} .$$

Define $\varepsilon_{k0} = \varepsilon - \sum_{i \in R_k} \varepsilon_{ki}$. It is seen from (7.40) that

$$\varepsilon_{k0} \geq 0 . \tag{7.42}$$

From (7.32) we get

$$\sum_{i \in R_k} \alpha_{ki} f_i(x_0) \geq f(x_0) - \varepsilon + \sum_{i \in R_k} \varepsilon_{ki} = f(x_0) - \varepsilon_{k0} , \tag{7.43}$$

where

$$\sum_{i \in R_k} \varepsilon_{ki} + \varepsilon_{k0} = \sum_{i \in R_k} \varepsilon_{ki} + \varepsilon - \sum_{i \in R_k} \varepsilon_{ki} = \varepsilon .$$

It follows from (7.37), (7.40), (7.43) and the definition (7.26) that

$$w_k \in K_\varepsilon(x_0) .$$

From this and (7.31) we can conclude that $v \in \overline{K_\varepsilon(x_0)}$, i.e.,

$$\partial_\varepsilon f(x_0) \subset \overline{K_\varepsilon(x_0)} . \tag{7.44}$$

The proposition of this theorem follows now from (7.30) and (7.44).

<u>PROBLEM 7.3</u>. Let G be a compact set. Find $\partial_\varepsilon f(x_0)$ for the function $f(x) = \max_{y \in G} \phi(x,y)$, where $\phi(x,y)$ is convex with respect to x for any $y \in G$ and continuous in x and y on $E_n \times G$.

EXAMPLE 4. Let us illustrate an application of relation (7.27) using the following example. Let $x \in E_1$, $f(x) = |x|$, $\varepsilon \ge 0$. Let us find $\partial_\varepsilon f(x)$.

Since $|x| = \max\{x,-x\}$, then $f(x) = \max_{i \in 1:2} f_i(x)$, where $f_1(x) = x$, $f_2(x) = -x$. It was shown in Example 2 that

$$\begin{aligned} \partial_\varepsilon f_1(x) &= \{+1\} \qquad \varepsilon \ge 0, \qquad \forall x \in E_1, \\ \partial_\varepsilon f_2(x) &= \{-1\} \qquad \varepsilon \ge 0, \qquad \forall x \in E_1. \end{aligned} \tag{7.45}$$

In (7.26), we can take as I_1 the following sets:

(1) $I_1 = \{1\}$; (2) $I_1 = \{2\}$; (3) $I_1 = \{1,2\}$.

Therefore, from (7.45) we obtain

$$K_\varepsilon(x) = K_1 \cup K_2 \cup K_3, \tag{7.46}$$

where

$$K_1 = \{v \in E_1 \mid v = 1,\ x > |x| - \varepsilon_0,\ \varepsilon_0 \in [0,\varepsilon]\},$$

$$K_2 = \{v \in E_1 \mid v = -1,\ -x \ge |x| - \varepsilon_0,\ \varepsilon_0 \in [0,\varepsilon]\},$$

$$\begin{aligned} K_3 = \{v \in E_1 \mid\ & v = \alpha + (1-\alpha)(-1) = 2\alpha - 1,\quad \alpha \in (0,1), \\ & \alpha x + (1-\alpha)(-x) \ge |x| - \varepsilon_0,\ \varepsilon_0 \in [0,\varepsilon]\}. \end{aligned}$$

If $x > 0$, then $|x| = x$ and hence

$$K_1 = \{1\}, \qquad K_2 = \begin{cases} \{-1\} & \text{if } x \le 0.5\varepsilon, \\ \emptyset & \text{if } x > 0.5\varepsilon, \end{cases}$$

$$K_3 = \left\{v \in E_1 \mid v = 2\alpha - 1,\ \alpha \in (0,1),\ \alpha \ge 1 - \frac{\varepsilon_0}{2x},\ \varepsilon_0 \in [0,\varepsilon]\right\}.$$

It is clear that

$$\partial_\varepsilon f(x) = \overline{K_\varepsilon(x)} = \begin{cases} [1 - \varepsilon x^{-1}, 1] & \text{if } x > 0.5\varepsilon, \\ [-1,1] & \text{if } 0 \le x \le 0.5\varepsilon. \end{cases} \tag{7.47}$$

Analogously we have $|x| = -x$ in the case $x < 0$ and hence

$$K_2 = \{-1\}, \qquad K_1 = \begin{cases} \{1\} & \text{if } x \ge -0.5\varepsilon, \\ \emptyset & \text{if } x < -0.5\varepsilon, \end{cases}$$

$$K_3 = \{v \in E_1 \mid v = 2\alpha - 1,\ \alpha \in (0,1),\ \alpha \le -\frac{\varepsilon_0}{2x},\ \varepsilon_0 \in [0,\varepsilon]\} .$$

Therefore

$$\partial_\varepsilon f(x) = \begin{cases} [-1,1] & \text{if} \quad 0 \ge x \ge -0.5\varepsilon , \\ [-1,-1-\varepsilon x^{-1}] & \text{if} \quad x < -0.5\varepsilon . \end{cases} \tag{7.48}$$

From (7.47) and (7.48) we finally obtain

$$\partial_\varepsilon f(x) = \begin{cases} [1-\varepsilon x^{-1},1] & \text{if} \quad x > 0.5\varepsilon , \\ [-1,1] & \text{if} \quad |x| \le 0.5\varepsilon , \\ [-1,-1-\varepsilon x^{-1}] & \text{if} \quad x < -0.5\varepsilon . \end{cases}$$

8. DIRECTIONAL ε-DERIVATIVES. CONTINUITY OF ε-SUBDIFFERENTIAL MAPPING

1. Let $f(x)$ be a convex finite function on E_n and let $\varepsilon \ge 0$, $x_0 \in E_n$, $g \in E_n$.

The quantity

$$\frac{\partial_\varepsilon f(x_0)}{\partial g} = \max_{v \in \partial_\varepsilon f(x_0)} (v,g) \tag{8.1}$$

is said to be the ε-*derivative of the function* $g(x)$ *at the point* x_0 *in the direction* g.

<u>THEOREM 8.1.</u> The following relation holds:

$$\frac{\partial_\varepsilon f(x_0)}{\partial g} = \inf_{\alpha>0} \alpha^{-1}[f(x_0+\alpha g) - f(x_0) + \varepsilon] . \tag{8.2}$$

<u>Proof.</u> Let

$$h(\alpha) = \alpha^{-1}[f(x_0+\alpha g) - f(x_0) + \varepsilon] , \qquad D = \inf_{\alpha>0} h(\alpha) . \tag{8.3}$$

Since we have from the definition

$$f(x_0+\alpha g) - f(x_0) \ge (v,\alpha g) - \varepsilon \qquad \forall v \in \partial_\varepsilon f(x_0), \quad \forall \alpha \ge 0 ,$$

then

$$(v,g) \le h(\alpha) \qquad \forall \alpha \ge 0, \quad \forall v \in \partial_\varepsilon f(x_0) .$$

Hence

$$\frac{\partial_\varepsilon f(x_0)}{\partial g} \equiv \max_{v \in \partial_\varepsilon f(x_0)} (v,g) \le \inf_{\alpha>0} h(\alpha) \equiv D . \quad (8.4)$$

It follows from (8.4) that $D > -\infty$.

Here we fix x_0 and g. Let us introduce the sets

$$Z = \{[\beta,x] \in E_{n+1} \mid x \in E_n,\ \beta > f(x)\} ,$$

$$L = \{[\beta,x] \in E_{n+1} \mid x = x_0+\alpha g,\ \beta = f(x_0)+\alpha D-\varepsilon,\ \alpha > 0\} .$$

Next we shall show that

$$L \cap Z = \emptyset . \quad (8.5)$$

Assume the opposite, i.e., that for some α_1, β_1 and x_1 the following relations hold:

$$\beta_1 > f(x_1) , \qquad \beta_1 = f(x_0) + \alpha_1 D - \varepsilon ,$$

$$x_1 = x_0 + \alpha_1 g , \quad \alpha_1 > 0 . \quad (8.6)$$

From (8.6), in particular, we have

$$f(x_0) + \alpha_1 D - \varepsilon > f(x_1) .$$

Therefore,

$$D > \frac{1}{\alpha_1} [f(x_1) - f(x_0) + \varepsilon]$$

$$= \frac{1}{\alpha_1} [f(x_0+\alpha_1 g) - f(x_0) + \varepsilon] = h(\alpha_1) . \quad (8.7)$$

However, (8.7) contradicts (8.3). Thus, (8.5) has been proved.

Define the set $P = Z - L$. Clearly, it is convex and, in addition, $0 \notin P$. Due to Corollary 2 of the separation theorem, there exists a point $[c,v] \in E_{n+1}$ such that

$$c \in E_1 , \qquad v \in E_n , \qquad c^2 + v^2 > 0 , \quad (8.8)$$

$$c\beta + (v,x) \ge c(f(x_0) + \alpha D - \varepsilon) + (v, x_0+\alpha g) \quad (8.9)$$

$$\forall \beta, x, \alpha : \beta \ge f(x) , \quad \alpha \ge 0 .$$

Now let us show that $c > 0$. For $\alpha = 0$, $x = x_0$ we have from (8.9)

$$c(\beta - f(x_0) + \varepsilon) \geq 0 \qquad \forall \beta \geq f(x_0) .$$

Hence, $c \geq 0$. If we had the equality

$$c = 0 , \tag{8.10}$$

then from (8.8) we would have the relation

$$\|v\| > 0 .$$

On the other hand, for $\alpha = 0$, we obtain from (8.9) and (8.10)

$$(v, x-x_0) \geq 0 , \qquad \forall x \in E_n ,$$

which is possible only for $v = 0$. This contradicts (8.11).

Thus, we have $c > 0$. Therefore, from (8.9) we get for $\beta = f(x)$, $\alpha = 0$

$$f(x) - f(x_0) \geq (-c^{-1}v, x-x_0) - \varepsilon \qquad \forall x \in E_n ,$$

i.e.,

$$v^* = (-c^{-1}v) \in \partial_\varepsilon f(x_0) .$$

From (8.9) with $\beta = f(x)$, it follows that

$$f(x) - f(x_0) \geq (v^*, x-x_0) - \varepsilon + \alpha(D - (v^*,g)) .$$

For $x = x_0$ we have

$$D \leq (v^*,g) + \alpha^{-1}\varepsilon \qquad \forall \alpha > 0 .$$

Taking the limit as $\alpha \to +\infty$ yields

$$D \leq (v^*,g) \leq \max_{v \in \partial_\varepsilon f(x_0)} (v,g) \equiv \frac{\partial_\varepsilon f(x_0)}{\partial g} .$$

Relation (8.2) follows from this and (8.4). The theorem has been proved. ∎

<u>REMARK 1</u>. For $\varepsilon = 0$ the function $h(\alpha) = \alpha^{-1}[f(x_0+\alpha g) - f(x_0)]$ is nonincreasing as $\alpha \to +0$ (it follows from the proof of Theorem 4.2); hence

$$\frac{\partial f(x_0)}{\partial g} \equiv \lim_{\alpha\to+0} \alpha^{-1}[f(x_0+\alpha g) - f(x_0)] = \inf_{\alpha>0} \alpha^{-1}[f(x_0+\alpha g) - f(x_0)] ,$$

which coincides with (8.2) for $\varepsilon = 0$.

2. Let $f(x)$ be a finite convex function. Consider the mapping $\partial_\varepsilon f\colon [0,\infty) \times E_n \to \Pi(E_n)$.

THEOREM 8.2. The mapping $\partial_\varepsilon f(x)$ is H-continuous in ε and x on $(0,\infty) \times E_n$.

Proof. Let us arbitrarily fix $\varepsilon_0 > 0$ and $x \in E_n$. First we consider the case where $f(x)$ is a strongly convex function, i.e., there exists an $m > 0$ such that

$$f(0.5(x+y)) \le 0.5[f(x)+f(y)] - 0.25m\|x-y\|^2 \qquad \forall x,y \in E_n. \tag{8.12}$$

Take $v \in \partial_\varepsilon f(x)$, i.e.,

$$f(z) \ge f(x) + (v,z-x) \qquad \forall z \in E_n. \tag{8.13}$$

Let $z = 0.5(x+y)$. It follows from (8.12) and (8.13) that

$$0.5(f(x)+f(y)) - 0.25m\|x-y\|^2 \ge f(0.5(x+y)) \equiv f(z)$$

$$\ge f(x) + (v,z-x) = f(x) + 0.5(v,y-x) \qquad \forall y \in E_n.$$

Hence we have

$$f(y) \ge f(x) + (v,y-x) + 0.5m\|x-y\|^2 \qquad \forall y \in E_n. \tag{8.14}$$

(More precisely, we can put here m instead of $0.5m$, as will be shown in Section 9 (see (9.17).)

In Theorem 8.1 we proved that

$$\max_{v\in\partial_\varepsilon f(x)} (v,g) = \inf_{\alpha>0} \alpha^{-1}[f(x+\alpha g) - f(x) + \varepsilon] \equiv \inf_{\alpha>0} h(x,g,\alpha,\varepsilon) ,$$

where

$$h(x,g,\alpha,\varepsilon) = \frac{1}{\alpha}[f(x+\alpha g) - f(x) + \varepsilon] .$$

Here we note that

$$h(x,g,\alpha,\varepsilon) \xrightarrow[\alpha\to+0]{} +\infty$$

for $\varepsilon > 0$. Since $\varepsilon_0 > 0$ and $f(x)$ is a continuous function, then there exist an $\alpha_0 > 0$ and a $\beta > 0$ such that

$$\inf_{\alpha>0} h(x,g,\alpha,\varepsilon) = \inf_{\alpha\ge\alpha_0} h(x,g,\alpha,\varepsilon) \qquad \forall x \in S_\beta(x_0), \quad \forall \varepsilon : |\varepsilon-\varepsilon_0| \le \beta, \quad \forall g : \|g\| = 1 .$$

On the other hand, it follows from (8.14) that $h(x,g,\alpha,\varepsilon) \xrightarrow[\alpha\to+\infty]{} +\infty$; hence there exists an $\alpha_1 > \alpha_0$ such that

$$\inf_{\alpha>0} h(x,g,\alpha,\varepsilon) = \inf_{\alpha_0\le\alpha\le\alpha_1} h(x,g,\alpha,\varepsilon) \qquad \forall x \in S_\beta(x_0), \quad \forall \varepsilon : |\varepsilon-\varepsilon_0| \le \beta, \quad \forall g : \|g\| = 1 .$$

Since h is a continuous function in all variables, then

$$\max_{v\in\partial_\varepsilon f(x)} (v,g) = \min_{\alpha_0\le\alpha\le\alpha_1} h(x,g,\alpha,\varepsilon) \equiv H(x,g,\varepsilon) .$$

The function $H(x,g,\varepsilon) \equiv \max_{v\in\partial_\varepsilon f(x)} (v,g)$ is also continuous in x, g and ε. Hence, by Lemma 2.2, the mapping $\partial_\varepsilon f(x)$ is H-continuous in ε and x at the point $[\varepsilon_0,x_0]$ if $f(x)$ is a strongly convex function.

Now, let $f(x)$ be a convex (but not necessarily strongly convex) function. Define the function

$$F_\delta(x) = f(x) + \delta(x - x_0)^2 = f(x) + f_\delta(x) ,$$

where $\delta > 0$. This function is strongly convex. Hence the mapping $\partial_\varepsilon F_\delta(x)$ is H-continuous in ε and x at the point $[\varepsilon_0,x_0]$. By Theorem 7.2,

$$\partial_\varepsilon F_\delta(x) = \bigcup_{\substack{\varepsilon_1+\varepsilon_2=\varepsilon \\ \varepsilon_1\ge0,\,\varepsilon_2\ge0}} [\partial_{\varepsilon_1} f(x) + \partial_{\varepsilon_2} f_\delta(x)] . \tag{8.15}$$

Since

$$\partial_\varepsilon f_\delta(x) = \{v = 2\delta(x-x_0)+r \mid \|r\| \le 2\sqrt{\varepsilon\delta}\}$$

(see (7.7)), then it is seen from this and (8.15) that

$$\rho(\partial_\varepsilon F_\delta(x),\ \partial_\varepsilon f(x)) \to 0 \tag{8.16}$$

as $\delta \to 0$, where $\rho(A,B)$ is the distance between A and B in the Hausdorff metric (see (2.1)). Since the mapping $\partial_\varepsilon F_\delta(x)$ is H-continuous in ε and x, we conclude from (8.16) that $\partial_\varepsilon f(x)$ is also H-continuous in ε and x at the point $[\varepsilon_0, x_0]$. The theorem has been proved. ∎

3. DEFINITION. Let $\varepsilon \geq 0$. A point $x_0 \in E_n$ is said to be *ε-stationary* if

$$0 \in \partial_\varepsilon f(x_0) \ . \tag{8.17}$$

LEMMA 8.1. Relation (8.17) is equivalent to the relation

$$0 \leq f(x_0) - f^* \leq \varepsilon \ , \tag{8.18}$$

where

$$f^* = \inf_{x \in E_n} f(x) \ .$$

Proof. By the definition of $\partial_\varepsilon f(x_0)$, (8.17) is equivalent to the inequality

$$f(x) \geq f(x_0) + (0, x - x_0) - \varepsilon = f(x_0) - \varepsilon \quad \forall\ x \in E_n \ ,$$

which proves the lemma. ∎

Now consider the case where x_0 is not ε-stationary. Then

$$\inf_{||g||=1} \frac{\partial_\varepsilon f(x_0)}{\partial g} = \min_{||g||=1} \max_{v \in \partial_\varepsilon f(x_0)} (v,g) = \max_{v \in \partial_\varepsilon f(x_0)} (v,g) < 0$$

(see the proof of Theorem 6.2).

Let

$$g_0 = -||v_0||^{-1} v_0 \qquad \text{where} \quad ||v_0|| = \min_{v \in \partial_\varepsilon f(x_0)} ||v|| \ .$$

The direction $g_0 = g(x_0)$ is called the *ε-steepest descent* direction of the function $f(x)$ at the point x_0. This direction is unique. This proof is similar to the proof of the uniqueness in Theorem 6.2 (see Remark 4 of Section 6).

From (8.2) we have

$$\inf_{\alpha>0} \alpha^{-1}[f(x_0+\alpha g_0) - f(x_0) + \varepsilon] < 0$$

implying the existence of an $\alpha_0 > 0$ such that

$$f(x_0 + \alpha_0 g_0) - f(x_0) + \varepsilon < 0 ,$$

i.e.,

$$\inf_{\alpha>0} f(x_0 + \alpha g_0) < f(x_0) - \varepsilon . \tag{8.19}$$

Hence the function $f(x)$ can be decreased by not less than ε in the direction g_0.

This property is valid for any $g \in E_n$ such that

$$\frac{\partial_\varepsilon f(x_0)}{\partial g} < 0 .$$

4. Let $\Omega \subset E_n$ be a closed convex set and $f(x)$ be a convex function on Ω. It was proved (in Theorem 6.1) that for the functio f(x) to attain its minimum value on Ω, it is necessary and sufficient that

$$\partial f(x^*) \cap \Gamma^+(x^*) \neq \emptyset , \tag{8.20}$$

where $\Gamma^+(x^*)$ is the cone which is conjugate to the cone of feasible directions.

<u>LEMMA 8.2</u>. If $x_0 \in \Omega$ and

$$\partial_\varepsilon f(x_0) \cap \Gamma^+(x_0) \neq \emptyset , \tag{8.21}$$

then

$$0 \leq f(x_0) - f^* \leq \varepsilon , \tag{8.22}$$

where

$$f^* = \inf_{x\in\Omega} f(x) .$$

Proof. It follows from (8.21) that there exists a $v_0 \in \partial_\varepsilon f(x_0)$ such that $v_0 \in \Gamma^+(x_0)$, i.e.,

$$(x-x_0, v_0) \geq 0 \quad \forall x \in \Omega. \tag{8.23}$$

From the definition of $\partial_\varepsilon f(x_0)$ we have

$$f(x) - f(x_0) \geq (v, x-x_0) - \varepsilon \quad \forall x \in E_n . \tag{8.24}$$

Hence, considering (8.23), the inequality

$$f(x) - f(x_0) \geq -\varepsilon \quad \forall x \in \Omega$$

holds. This inequality is equivalent to (8.22). ■

5. Let a linear function $g(x)$ be given on E_n and a convex function $f(x)$ be given on the set $\Omega = \{x \in E_n \mid g(x) > 0\}$. Let

$$h(x) = f(x) g^{-1}(x) \qquad (\text{where } g^{-1}(x) = \frac{1}{g(x)}) ,$$

$$h^* = \inf_{x \in \Omega} h(x) , \qquad M^* = \{x \in \Omega \mid h(x) = h^*\} .$$

LEMMA 8.3. Each local minimum point of the function $h(x)$ on Ω is a global minimum point of $h(x)$ on Ω. The set M^* is convex.

Proof. Let $x_0 \in \Omega$ be a local minimum point of the function $h(x)$ on Ω. This implies that there exists a $\delta > 0$ such that $S_\delta(x_0) \subset \Omega$, because Ω is open, and

$$h(x) \geq h(x_0) \qquad \forall x \in S_\delta(x_0) . \tag{8.25}$$

It is necessary to show that $h(x) \geq h(x_0)$ for all $x \in \Omega$. We take an arbitrary point $\bar{x} \in \Omega \backslash S_\delta(x_0)$. Let the segment connecting the points $\bar{x}$ and x_0 intersect the sphere $C_\delta(x_0) = \{x \mid \|x-x_0\| = \delta\}$ at a point x_1.

Then, it follows from (8.25) that

$$f(x_1) \geq g(x_1) h(x_0) , \qquad f(x_0) = g(x_0) h(x_0) . \tag{8.26}$$

Clearly, there exists a $\beta > 1$ such that $\bar{x} = \beta x_1 + (1-\beta)x_0$. Due to the convexity of the function $f(x)$ and the linearity of the function $g(x)$

$$f(\bar{x}) \geq \beta f(x_1) + (1-\beta)f(x_0) , \qquad g(\bar{x}) \not\leq \beta g(x_1) + (1-\beta)g(x_0)$$

(see Lemma 4.1.).

Considering this and (8.26), we obtain

$$\begin{aligned} f(\bar{x}) &\geq \beta g(x_1)h(x_0) + (1-\beta)g(x_0)h(x_0) \\ &= (\beta g(x_1) + (1-\beta)g(x_0))h(x_0) \geq g(\bar{x})h(x_0) , \end{aligned}$$

i.e., $h(\bar{x}) \geq h(x_0)$.

Therefore, since $\bar{x}$ is an arbitrary point in $\Omega\backslash S_\delta(x_0)$ and (8.25) is valid, then

$$h(x) \geq h(x_0) = h^* \quad \forall x \in \Omega . \qquad (8.27)$$

It follows from (8.27) that x_0 is a global minimum point of the function $h(x)$ on the set Ω. Since $h(x)$ is quasiconvex, the set M^* is convex (see Subsection 4.6). ∎

Let us take $x_0 \in E_n$, $v \in E_n$, $||v|| = 1$ and $\varepsilon \geq 0$.

<u>COROLLARY</u>. Any local minimum point of the function

$$h(\alpha) = \alpha^{-1}[f(x_0+\alpha v) - f(x_0) + \varepsilon]$$

on $(0,\infty)$ is a global minimum point of the function $h(\alpha)$ on $(0,\infty)$. Moreover, the set of minimum points of the function $h(\alpha)$ on $(0,\infty)$ is convex.

6. <u>LEMMA 8.4</u>. Let x_0, $g \in E_n$. In order that

$$f(x_0+\alpha g) = f(x_0) + \alpha \frac{\partial f(x_0)}{\partial g} \qquad \forall \alpha \geq 0 , \qquad (8.28)$$

it is necessary (sufficient) that

$$\frac{\partial f(x_0)}{\partial g} = \frac{\partial_\varepsilon f(x_0)}{\partial g} \qquad (8.29)$$

for any $\varepsilon > 0$ (at least for one $\varepsilon > 0$).

Proof. Necessity. By virtue of Theorem 8.1, we have from (8.28)

$$\frac{\partial_\varepsilon f(x_0)}{\partial g} = \inf_{\alpha>0} \alpha^{-1}[f(x_0+\alpha g) - f(x_0) + \varepsilon]$$

$$= \frac{\partial f(x_0)}{\partial g} + \inf_{\alpha>0} \alpha^{-1}\varepsilon = \frac{\partial f(x_0)}{\partial g} .$$

Sufficiency. For $\alpha = 0$, (8.28) is clear. Hence, we must prove the validity of (8.28) for $\alpha > 0$ under condition (8.29). Let us show that if (8.29) is valid, then the set

$$G = \left\{\alpha > 0 \mid f(x_0+\alpha g) = f(x_0) + \alpha \frac{\partial_\varepsilon f(x_0)}{\partial g} - \varepsilon\right\}$$

is empty. We assume that this is not true. Let

$$f(x_0+\lambda g) = f(x_0) + \lambda \frac{\partial_\varepsilon f(x_0)}{\partial g} - \varepsilon \tag{8.30}$$

for some $\lambda > 0$. For any $v \in \partial f(x_0)$

$$f(x_0+\alpha g) - f(x_0) \geq (v,\alpha g) .$$

Hence

$$f(x_0+\alpha g) - f(x_0) \geq \alpha \max_{v\in\partial f(x_0)} (v,g) = \alpha \frac{\partial f(x_0)}{\partial g} . \tag{8.31}$$

From (8.29), (8.30) and (8.31) we obtain

$$\frac{\partial f(x_0)}{\partial g} = \frac{\partial_\varepsilon f(x_0)}{\partial g} \geq \lambda^{-1}[f(x_0+\lambda g) - f(x_0) + \varepsilon]$$

$$\geq \lambda^{-1}[f(x_0) + \lambda \frac{\partial f(x_0)}{\partial g} - f(x_0) + \varepsilon] = \frac{\partial f(x_0)}{\partial g} + \lambda^{-1}\varepsilon ,$$

which is impossible. Thus, we conclude that $G = \emptyset$ and hence

$$\frac{\partial_\varepsilon f(x_0)}{\partial g} = \lim_{\alpha\to+\infty} \lambda^{-1}[f(x_0+\alpha g) - f(x_0) + \varepsilon] . \tag{8.32}$$

Assume that (8.28) is not valid, i.e., there exists a $\lambda > 0$ such that

$$f(x_0+\gamma g) > f(x_0) + \gamma \frac{\partial f(x_0)}{\partial g} . \tag{8.33}$$

For any $v \in \partial(x_0+\gamma g)$ we have

$$f(x_0) - f(x_0+\gamma g) \geq -\gamma(v,g) .$$

Hence,

$$\gamma^{-1}[f(x_0+\gamma g) - f(x_0)] \leq \max_{v\in\partial f(x_0+\gamma g)} (v,g) = \frac{\partial f(x_0+\gamma g)}{\partial g} . \tag{8.34}$$

From (8.34), (8.33) and (8.29), we obtain

$$\frac{\partial f(x_0+ g)}{\partial g} > \frac{\partial f(x_0)}{\partial g} = \frac{\partial_\varepsilon f(x_0)}{\partial g} . \tag{8.35}$$

On the other hand, the same procedure used to derive (8.31) allows us to obtain for $\alpha > \gamma$

$$f(x_0+\alpha g) - f(x_0+\gamma g) \geq (\alpha-\gamma) \frac{\partial f(x_0+\gamma g)}{\partial g} \tag{8.36}$$

and from (8.32) and (8.36)

$$\begin{aligned}\frac{\partial_\varepsilon f(x_0)}{\partial g} &= \lim_{\alpha\to+\infty} \alpha^{-1}[f(x_0+\alpha g) - f(x_0) + \varepsilon] \\ &\geq \lim_{\alpha\to+\infty} \alpha^{-1}[f(x_0+\gamma g) + (\alpha-\gamma)\frac{\partial f(x_0+\gamma g)}{\partial g} - f(x_0) + \varepsilon] \\ &= \frac{\partial f(x_0+\gamma g)}{\partial g} ,\end{aligned}$$

which contradicts (8.35). This contradiction proves the sufficiency

The lemma is proved. ∎

COROLLARY. If for some $x_0, g \in E_n$ and $\varepsilon > 0$

$$\frac{\partial f(x_0)}{\partial g} = \frac{\partial_\varepsilon f(x_0)}{\partial g} < 0 ,$$

then

$$\inf_{x\in E_n} f(x) = 0 .$$

REMARK 2. If there exists at least one $\varepsilon > 0$ for which (8.29) is valid, then (8.29) is valid for all $\varepsilon > 0$.

LEMMA 8.5. For any $x_0, g \in E_n$ and $\varepsilon > 0$, there exists a $\beta > 0$, depending on x_0, g and ε, such that

$$f(x_0+\alpha g) \leq f(x_0) + \alpha \frac{\partial_\varepsilon f(x_0)}{\partial g} \qquad \forall \alpha \in [0,\beta]. \qquad (8.37)$$

Proof. If $\frac{\partial_\varepsilon f(x_0)}{\partial g} = \frac{\partial f(x_0)}{\partial g}$, then, by Lemma 8.4, $\beta = +\infty$.

Since

$$\frac{\partial_\varepsilon f(x_0)}{\partial g} \geq \frac{\partial f(x_0)}{\partial g} ,$$

it remains to consider the case

$$\frac{\partial_\varepsilon f(x_0)}{\partial g} = \frac{\partial f(x_0)}{\partial g} + \delta , \qquad (8.38)$$

where

$$\delta > 0 . \qquad (8.39)$$

Assume that the assertion of the lemma is not valid. Then let us show that

$$f(x_0+\alpha g) > f(x_0) + \alpha \frac{\partial_\varepsilon f(x_0)}{\partial g} \qquad \forall \alpha > 0 , \qquad (8.40)$$

otherwise we can find a $\beta_0 > 0$ such that

$$f(x_0+\beta_0 g) \leq f(x_0) + \beta_0 \frac{\partial_\varepsilon f(x_0)}{\partial g} . \qquad (8.41)$$

Since $f(x)$ is a convex function, we have for $\gamma \in [0,1]$

$$f(x_0+\gamma\beta_0 g) = f(\gamma(x_0+\beta_0 g) + (1-\gamma)x_0) \leq \gamma f(x_0+\beta_0 g) + (1-\gamma)f(x_0) .$$

From (8.41)

$$f(x_0+\gamma\beta_0 g) \leq f(x_0) + \gamma\beta_0 \frac{\partial_\varepsilon f(x_0)}{\partial g} \qquad \forall \gamma \in [0,1] ,$$

i.e., there exists a $\beta_0 > 0$ such that

$$f(x_0+\alpha g) \leq f(x_0) + \alpha \frac{\partial_\varepsilon f(x_0)}{\partial g} \qquad \forall \alpha \in [0,\beta_0] ,$$

which contradicts the assumption. Thus, if the assertion of the lemma is not valid, then (8.40) should be valid. From (8.40) and (8.38)

$$\alpha^{-1}[f(x_0+\alpha g) - f(x_0)] > \frac{\partial f(x_0)}{\partial g} + \delta \qquad \forall \alpha > 0 . \qquad (8.42)$$

Since

$$\frac{\partial f(x_0)}{\partial g} = \inf_{\alpha>0} \alpha^{-1}[f(x_0+\alpha g) - f(x_0)]$$

(see Corollary of Theorem 4.2), then (8.42) implies that $\delta \le 0$. This is a contradiction. The lemma is proved. ∎

Lemma 8.5 geometrically implies that the straight line

$$L = \{(\alpha, z) \in E_2 \mid z = f(x_0) + \alpha \frac{\partial_\varepsilon f(x_0)}{\partial g}\}$$

is either the secant of the graph of the function $h(\alpha) = f(x_0+\alpha g)$ or coincides with the graph of the function $h(\alpha)$ for $\alpha \ge 0$.

LEMMA 8.6. Let $x_0, g \in E_n$, $\varepsilon > 0$ and define

$$G = \{\alpha \ge 0 \mid f(x_0+\alpha g) = f(x_0) + \alpha \frac{\partial_\varepsilon f(x_0)}{\partial g} - \varepsilon\},$$

$$\bar{\lambda} = \begin{cases} \varepsilon \left[\frac{\partial_\varepsilon f(x_0)}{\partial g} - \frac{\partial f(x_0)}{\partial g}\right]^{-1} & \text{if } G \text{ is bounded}, \\ +\infty & \text{if } G \text{ is empty or unbounded}. \end{cases}$$

Then

$$f(x_0+\alpha g) \le f(x_0) + \alpha \frac{\partial_\varepsilon f(x_0)}{\partial g} \qquad \forall \alpha \in [0, \bar{\lambda}]. \quad (8.43)$$

Proof. If $G \ne \emptyset$ and $\gamma \in G$, then, by virtue of the convexity of the function $f(x)$, (8.43) is satisfied for $\alpha \in [0, \gamma]$. Hence, if G is unbounded, then (8.43) is valid for any $\bar{\lambda} > 0$.

In the case where the set G is bounded, we obtain for any $\gamma \in G$

$$f(x_0) + \gamma \frac{\partial_\varepsilon f(x_0)}{\partial g} - \varepsilon = f(x_0+\gamma g) \ge f(x_0) + \gamma \frac{\partial f(x_0)}{\partial g}.$$

From this it follows that

$$\gamma \ge \varepsilon \left[\frac{\partial_\varepsilon f(x_0)}{\partial g} - \frac{\partial f(x_0)}{\partial g}\right]^{-1} = \bar{\lambda}.$$

Now, let us consider the case $G = \emptyset$.

First of all, we note in this case that

$$\frac{\partial_\varepsilon f(x_0)}{\partial g} = \inf_{\alpha>0} \alpha^{-1}[f(x_0+\alpha g) - f(x_0) + \varepsilon]$$

$$= \lim_{\alpha\to\infty} \alpha^{-1}[f(x_0+\alpha g) - f(x_0) + \varepsilon] \quad . \qquad (8.44)$$

Here, suppose that the assertion of the lemma is not correct in the case considered, i.e., for some $\gamma > 0$

$$\frac{\partial_\varepsilon f(x_0)}{\partial g} < \gamma^{-1}[f(x_0+\gamma g) - f(x_0)] \quad . \qquad (8.45)$$

Then from (8.34)

$$\frac{\partial_\varepsilon f(x_0)}{\partial g} < \frac{\partial f(x_0+\gamma g)}{\partial g} \quad . \qquad (8.46)$$

On the other hand, we have from (8.36)

$$f(x_0+\alpha g) - f(x_0+\gamma g) \geq (\alpha-\gamma)\frac{\partial f(x_0+\gamma g)}{\partial g} \qquad \forall \alpha \geq \gamma \ .$$

Noting this, (8.44) and (8.45), we obtain

$$\frac{\partial_\varepsilon f(x_0)}{\partial g} = \lim_{\alpha\to+\infty} \alpha^{-1}[f(x_0+\alpha g) - f(x_0) + \varepsilon]$$

$$\geq \lim_{\alpha\to+\infty} \alpha^{-1}\left[f(x_0+\gamma g) + (\alpha-\gamma)\frac{\partial f(x_0+\gamma g)}{\partial g} - f(x_0) + \varepsilon\right]$$

$$\geq \lim_{\alpha\to+\infty} \alpha^{-1}\left[\alpha\frac{\partial f(x_0+\gamma g)}{\partial g} + \gamma\left(\frac{\partial_\varepsilon f(x_0)}{\partial g} - \frac{\partial f(x_0+\gamma g)}{\partial g}\right) + \varepsilon\right]$$

$$= \frac{\partial f(x_0+\gamma g)}{\partial g} \quad ,$$

which contradicts (8.46). The lemma is proved. ∎

9. SOME PROPERTIES AND INEQUALITIES FOR CONVEX FUNCTIONS

1. Let $f(x)$ be a convex function on E_n, i.e.,

$$f(\alpha x_1 + (1-\alpha)x_2) \leq \alpha f(x_1) + (1-\alpha)f(x_2) \qquad (9.1)$$

$$\forall x_1, x_2 \in E_n \ , \qquad \forall \alpha \in [0,1] \quad .$$

By Lemma 4.1 we have for any $x_1, x_2 \in E_n$ and $\alpha \notin [0,1]$

$$f(\alpha x_1 + (1-\alpha)x_2) \geq \alpha f(x_1) + (1-\alpha)f(x_2) \quad . \tag{9.2}$$

DEFINITION. Let $x_0 \in E_n$. The set

$$D(x_0) = \{x \in E_n \mid f(x) \leq f(x_0)\} \tag{9.3}$$

is said to be the *Lebesgue set* of the function $f(x)$ at the point x_0

It is clear that this set is nonempty, because $x_0 \in D(x_0)$, convex and closed.

Take an arbitrary $v \in \partial f(x_0)$. Then, by definition,

$$f(x) - f(x_0) \geq (v, x-x_0) \qquad \forall x \in E_n \quad .$$

For $x \in D(x_0)$, we have $(v,x-x_0) \leq 0$, i.e., the hyperplane $P(x_0) = \{x \in E_n \mid (x-x_0, v)=0\}$ is a supporting plane for the set $D(x_0)$ at the point x_0. Let

$$M^* = \{x \in E_n \mid f(x) \leq f(z) \quad \forall z \in E_n\}. \tag{9.4}$$

The set M^* is the set of minimum points of the function $f(x)$ on E_n. This set can be empty (see Example 1 below).

LEMMA 9.1. If the set M^* is nonempty and bounded, then, for any $x_0 \in E_n$, the Lebesgue set $D(x_0)$ (see (9.3)) is bounded.

Proof. It has already been observed that the set $D(x_0)$ is nonempty closed and convex. Let us show that for any $\varepsilon > 0$

$$\psi_\varepsilon \equiv \min_{x^* \in M^*} \psi_\varepsilon(x^*) = \min_{x^* \in M^*} \min_{\|g\|=1} \frac{\partial_\varepsilon f(x^*)}{\partial g} > 0 \tag{9.5}$$

(see Figure 11). Assume that (9.5) is not valid. Then, for some $\varepsilon > 0$, $x^* \in M^*$ and $g \in E_n$, $\|g\| = 1$, it follows that

$$\frac{\partial_\varepsilon f(x^*)}{\partial g} \leq 0 \quad . \tag{9.6}$$

Since $\partial f(x^*) \subset \partial_\varepsilon f(x^*)$ and $x^* \in M^*$, then

$$\frac{\partial_\varepsilon f(x^*)}{\partial g} \geq \frac{\partial f(x^*)}{\partial g} \geq 0 \quad . \tag{9.7}$$

Using (9.6) and (9.7) we can conclude that $\frac{\partial_\varepsilon f(x^*)}{\partial g} = \frac{\partial f(x^*)}{\partial g} = 0$. Hence, by virtue of Lemma 8.4, we obtain

$$f(x^*+\alpha g) = f(x^*) + \alpha\frac{\partial f(x^*)}{\partial g} = f(x^*) \qquad \forall\alpha \geq 0 \; ,$$

which contradicts the boundedness of the set M^*.

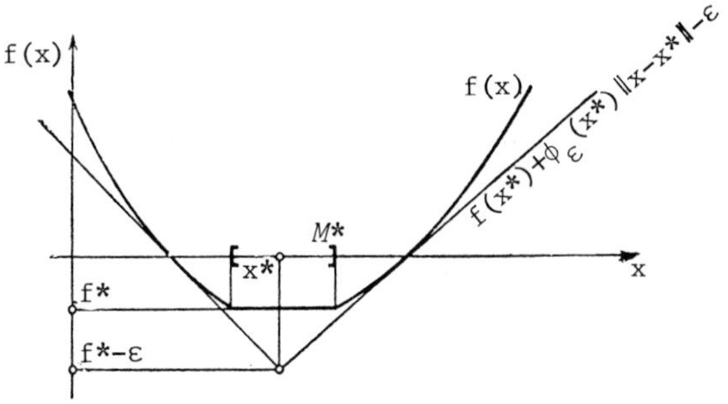

Figure 11

For any $x \in E_n$, $x^* \in M^*$ and $\varepsilon > 0$, we have

$$f(x) - f(x^*) \geq \max_{v \in \partial_\varepsilon f(x^*)} (v, x-x^*) - \varepsilon \geq \psi_\varepsilon(x^*)\|x-x^*\| - \varepsilon \; . \tag{9.8}$$

Let here $\varepsilon = f(x_0) - f^*$. Then it follows from (9.5) and (9.8) that

$$D(x_0) \subset S_{r_0}(M^*) \quad ,$$

where $r_0 = 2\varepsilon\psi_\varepsilon^{-1}$. ∎

COROLLARY 1. In order that the set of minimum points of a convex function be bounded, it is necessary that, for any $x_0 \in E_n$, the Lebesgue set $D(x_0)$ be bounded, and it is sufficient that this set be bounded for at least one $x_0 \in E_n$.

The necessity has just been proved and the sufficiency is obvious.

Let

$$D_\varepsilon = \{x \in E_n \mid f(x) \leq f^*+\varepsilon\} \; .$$

By virtue of Lemma 8.1, D_ε is the set of ε-stationary points of the function $f(x)$ on E_n.

COROLLARY 2. If the set M^* is nonempty and bounded, then the set D_ε is bounded for any fixed $\varepsilon > 0$.

Let $a_\varepsilon = \max_{x \in D_\varepsilon} \rho(x, M^*)$, where $\rho(x,M) = \min_{v \in M} \|x-v\|$.

COROLLARY 3. If the set M^* is nonempty and bounded, then $a_\varepsilon \xrightarrow[\varepsilon \to +0]{} 0$.

To prove this, assume the opposite. Then, there exist a number $b > 0$ and a sequence $\{x_k\}$ such that $x_k \in D_{\varepsilon_k}$, $\varepsilon_k \to +0$, $\rho(x_k, M^*) \ge b$.

Since the set D_ε is bounded for $\varepsilon \le \varepsilon_0$, we can assume without loss of generality that $x_k \to x^*$. Then, additionally, we have

$$\rho(x^*, M^*) \ge b . \tag{9.9}$$

On the other hand, we have $x_k \in D_{\varepsilon_k}$, i.e., $f(x_k) \le f^* + \varepsilon_k$.

Passing here to the limit, we get $f(x^*) = f^*$, which contradicts (9.9). ∎

COROLLARY 4. If the set M^* is nonempty and bounded, then for any $a > 0$ there exists an $\varepsilon = \varepsilon(a) > 0$ such that

$$D_\varepsilon \subset S_a(M^*) \equiv \{x \in E_n \mid \rho(x, M^*) \le a\} .$$

LEMMA 9.2. If the set M^* is nonempty and bounded, then for any $\varepsilon \ge 0$ and $r > 0$ there exists a $c_{\varepsilon r} > 0$ such that

$$\|v\| \ge c_{\varepsilon r} > 0 \quad \forall v \in \partial_\varepsilon f(x), \quad \forall x \notin S_r(D_\varepsilon) . \tag{9.10}$$

Proof. Let

$$G_{\varepsilon r} = \{x \in E_n \mid \rho(x, D_\varepsilon) = r\} .$$

Due to Corollary 2 of Lemma 9.1, the set D_ε is bounded. Then the set $G_{\varepsilon r}$ is also bounded. Since $D_\varepsilon \cap G_{\varepsilon r} = \emptyset$, we have

$$f_{\varepsilon r} = \min_{x \in G_{\varepsilon r}} f(x) > f^* + \varepsilon . \tag{9.11}$$

Let us fix $x^* \in M^*$ and take an arbitrary $x \in S_r(D_\varepsilon)$. Let $x_1 = x_1(x) \in G_\varepsilon \cap \mathrm{co}\{x^*, x\}$. Due to Lemma 4.1,

$$f(x) - f(x^*) \geq \|x_1 - x^*\|^{-1} \|x - x^*\| (f(x_1) - f(x^*))$$

$$\geq \frac{\|x - x^*\|}{\|x_1 - x^*\|} (f_{\varepsilon r} - f^*) \quad . \qquad (9.12)$$

Since $v \in \partial_\varepsilon f(x)$, then

$$f(x^*) - f(x) \geq (v, x^* - x) - \varepsilon \geq -\|v\| \, \|x^* - x\| - \varepsilon .$$

From this and (9.12) we get

$$\|v\| \geq \|x_1 - x^*\|^{-1}(f_{\varepsilon r} - f^*) - \|x - x^*\|^{-1}\varepsilon \quad .$$

Since $\|x - x^*\| \geq \|x_1 - x^*\|$, then $\|v\| \geq \|x_1 - x^*\|^{-1}(f_{\varepsilon r} - f^* - \varepsilon)$. However, $\|x_1 - x^*\| \leq d \equiv \mathrm{diam}\, S_r(D_\varepsilon)$. Hence, considering (9.11), we obtain

$$\|v\| \geq d^{-1}(f_{\varepsilon r} - f^* - \varepsilon) \equiv c_{\varepsilon r} > 0 \qquad (9.13)$$

$$\forall\, x \notin S_r(D_\varepsilon), \quad \forall\, v \in \partial_\varepsilon f(x) \; .$$

2. If $f(x)$ is a strictly (or strongly) convex function on a closed convex set $\Omega \subset E_n$, then we can conclude from the inequality

$$f\left(\frac{x_1 + x_2}{2}\right) < \frac{1}{2}[f(x_1) + f(x_2)] \qquad \forall x_1, x_2 \in \Omega, \quad x_1 \neq x_2 \; , \qquad (9.14)$$

that the function $f(x)$ attains its minimum value on Ω at one point at most.

<u>LEMMA 9.3</u>. If $f(x)$ is a strongly convex function on Ω, then it attains its minimum value on the set Ω (and, in addition, at a unique point).

<u>Proof</u>. The uniqueness has been mentioned above.

Take an arbitrary point $x_0 \in \Omega$. It follows from (8.14) that, for any $v \in \partial f(x_0)$,

$$\mathcal{D}(x_0) \equiv \{x \in \Omega \mid f(x) \le f(x_0)\} \subset \{x \in \Omega \mid 0 \ge -\|v\| \cdot \|x-x_0\| + m\|x-x_0\|^2\}$$

Hence, $\mathcal{D}(x_0)$ is compact. Since the function $f(x)$ is continuous, it attains the lower bound on $\mathcal{D}(x_0)$. ∎

REMARK 1. For strictly convex functions, the infimum cannot be necessarily attained. For example, the function $f(x) = e^{-x}$ given on E_1 is strictly convex and $\inf\limits_{x \in E_1} f(x) = 0$, however, this infimum is not attained.

3. We can obtain some estimates for strongly convex functions

Let $f(x)$ be a strongly convex function on E_n with a strong convexity constant $m > 0$, i.e.,

$$f(\alpha x_1 + (1-\alpha)x_2) \le \alpha f(x_1) + (1-\alpha)f(x_2) - \alpha(1-\alpha)m\|x_1-x_2\|^2 \tag{9.15}$$

$$\forall x_1, x_2 \in E_n, \quad \forall \alpha \in [0,1].$$

LEMMA 9.4. If $f(x)$ is a strongly convex function on E_n with a strong convexity constant $m > 0$, then, for arbitrary

$$x, x_0 \in E_n, \quad v(x_0) \in \partial_\varepsilon f(x_0), \quad \varepsilon \ge 0, \quad \alpha \in (0,1]$$

we have the inequality

$$f(x) - f(x_0) \ge (v(x_0), x-x_0) - \varepsilon\alpha^{-1} + (1-\alpha)m\|x_0-x\|^2. \tag{9.16}$$

Proof. Let $x_0 \in E_n$ and $\alpha \in (0,1]$ be fixed. Take an arbitrary $v(x_0) \in \partial_\varepsilon f(x_0)$. By the definition of an ε-subgradient, we have

$$f(z) \ge f(x_0) + (v(x_0), z-x_0) - \varepsilon \qquad \forall z \in E_n.$$

Let $z = \alpha x + (1-\alpha)x_0$. It follows from (9.15) and the above inequality that

$$\alpha f(x) + (1-\alpha)f(x_0) - \alpha(1-\alpha)m\|x_0-x\|^2$$

$$f(\alpha x + (1-\alpha)x_0) \equiv f(z) = f(x_0 + \alpha(x-x_0))$$

$$\ge f(x_0) + \alpha(v(x_0), x-x_0) - \varepsilon.$$

Inequality (9.16) is obvious.

COROLLARY 1. For any $x_0 \in E_n$ and $v(x_0) \in \partial f(x_0)$ we have the inequality

$$f(x) - f(x_0) \geq (v(x_0), x-x_0) + m\|x-x_0\|^2 \qquad \forall x \in E_n . \quad (9.17)$$

The validity of (9.17) follows from (9.16) if we put there $\varepsilon = 0$ and take the limit as $\alpha \to +0$.

COROLLARY 2. For any $x_1, x_2 \in E_n$, $v_1 \in \partial f(x_1)$, $v_2 \in \partial f(x_2)$, we have the inequality

$$(v_1-v_2, x_1-x_2) \geq 2m\|x_1-x_2\|^2 . \quad (9.18)$$

Proof. It follows from (9.17) that for any $x_1, x_2 \in E_n$, we have

$$f(x) - f(x_1) \geq (v_1, x-x_1) + m\|x-x_1\|^2 \qquad \forall x \in E_n ,$$

$$f(x) - f(x_2) \geq (v_2, x-x_2) + m\|x-x_2\|^2 \qquad \forall x \in E_n ,$$

where $v_1 \in \partial f(x_1)$, $v_2 \in \partial f(x_2)$.

Letting $x = x_2$ in the first inequality and $x = x_1$ in the second inequality and summing up the resulting inequalities, we obtain (9.18). ■

It is clear from (9.18) that

$$\|v_1-v_2\| \geq 2m\|x_1-x_2\| .$$

Hence, in particular, if $f(x)$ is a strongly convex function, then an arbitrary vector $v \in E_n$ can be its subgradient at one point at most.

PROBLEM 9.1. Prove that this proposition is valid also for strictly convex functions, i.e., if $f(x)$ is a strictly convex function, then an arbitrary vector can be its subgradient at one point at most.

COROLLARY 3. Let $x \neq x_0$. The function on the right-hand side of

(9.16) attains its maximum value with respect to α at $\alpha = \sqrt{\varepsilon m^{-1}}\,\|x-x_0\|^{-1}$. Moreover, if $\|x-x_0\| \geq \sqrt{\varepsilon m^{-1}}$, then from (9.16) we have the inequality

$$f(x) - f(x_0) \geq (v(x_0),\ x-x_0) + m\|x-x_0\|^2 - 2\sqrt{\varepsilon m}\,\|x-x_0\|^2 .$$

Recall that (9.17) is valid for any strongly convex function.

It turns out that the converse proposition is also valid.

LEMMA 9.5. Let a function $f(x)$ be given on E_n. If for any $x_0 \in E_n$ there exists a $v(x_0) \in E_n$ such that (9.17) is satisfied, then the function $f(x)$ is strongly convex on E_n with a strong convexity constant m.

Proof. Take arbitrary $x_1, x_2 \in E_n$ and $\alpha \in [0,1]$.

Let $x_0 = \alpha x_1 + (1-\alpha)x_2$. Due to the assumption, there exists a $v(x_0) \in E_n$, for which (9.17) is satisfied. For $x = x_1$, it follows from (9.17) that

$$f(x_1) \geq f(x_0) + (1-\alpha)(v(x_0),\ x_1-x_2) + m(1-\alpha)^2\|x_1-x_2\|^2 . \quad (9.19)$$

Now, letting $x = x_2$, we obtain analogously

$$f(x_2) \geq f(x_0) - \alpha(v(x_0),\ x_1-x_2) + m\alpha^2\|x_1-x_2\|^2 . \quad (9.20)$$

Multiplying (9.19) by α and (9.20) by $(1-\alpha)$ and summing yiel[d]

$$\begin{aligned} f(x_0) &= f(\alpha x_1 + (1-\alpha)x_2) \\ &\leq \alpha f(x_1) + (1-\alpha)f(x_2) - \alpha(1-\alpha)m\|x_1-x_2\|^2 . \end{aligned}$$

Here, since $x_1, x_2 \in E_n$ and $\alpha \in [0,1]$ are arbitrary, then, comparing with (9.15), we can conclude that $f(x)$ is a strongly convex function with a strong convexity function m. ∎

REMARK 2. If (9.17) is valid for some $x_0 \in E_n$ and $v(x_0) \in E_n$, then it is clear that $v(x_0) \in \partial f(x_0)$.

4. Let $f(x)$ be a strongly convex function with a strong convexity constant $m > 0$. Let $f^* = \inf_{x \in E_n} f(x)$. As established in Lemma 9.3, the value f^* is attained in addition, at a unique point, i.e., there exists a point $x^* \in E_n$ such that $f(x^*) = f^*$. Take an arbitrary $x_0 \in E_n$, $\varepsilon > 0$ and let

$$d_\varepsilon = d_\varepsilon(x_0) \equiv \min_{v \in \partial_\varepsilon f(x_0)} \|v\| = \|v_\varepsilon(x_0)\| .$$

LEMMA 9.6. We have the inequalities

$$0 \le f(x_0) - f^* \le \varepsilon + (d_\varepsilon^2 + 4 d_\varepsilon \sqrt{\varepsilon m})(4m)^{-1} , \tag{9.21}$$

$$\|x_0 - x^*\| \le (d_\varepsilon + 2\sqrt{\varepsilon m})(2m)^{-1} . \tag{9.22}$$

Proof. Let us establish first inequality (9.21). For $d_\varepsilon = 0$, it is obvious from Lemma 8.1. Let $d_\varepsilon > 0$ and take $v \in \partial_\varepsilon f(x_0)$. It follows from (9.16) that for all $x \in E_n$ and $\beta \in (0,1]$

$$\begin{aligned} f(x) &\ge f(x_0) + \max_{v \in \partial_\varepsilon f(x_0)} (v, x - x_0) - \varepsilon\beta^{-1} + (1-\beta) m \|x_0 - x\|^2 \\ &\ge \|x - x_0\| \min_{\|g\|=1} \max_{v \in \partial_\varepsilon f(x_0)} (v, g) - \varepsilon\beta^{-1} + (1-\beta) m \|x - x_0\|^2 . \end{aligned}$$

Due to Corollary 2 of Lemma 6.4,

$$d_\varepsilon = - \min_{\|g\|=1} \max_{v \in \partial_\varepsilon f(x_0)} (v, g) .$$

Hence

$$f(x) - f(x_0) \ge -d_\varepsilon \|x - x_0\| + (1-\beta) m \|x - x_0\|^2 - \varepsilon\beta^{-1}$$

$$\forall x \in E_n, \quad \forall \beta \in (0,1] .$$

Thus, we get

$$\begin{aligned} f^* \equiv \inf_{x \in E_n} f(x) &\ge f(x_0) + \inf_{x \in E_n} [(1-\beta) m \|x - x_0\|^2 - d_\varepsilon \|x - x_0\|] - \varepsilon\beta^{-1} \\ &= f(x_0) + \inf_{\alpha \ge 0} [(1-\beta) m \alpha^2 - d_\varepsilon \alpha] - \varepsilon\beta^{-1} \end{aligned} \tag{9.23}$$

$$\forall \beta \in (0,1] .$$

Here, the infimum is attained at $\alpha = 2^{-1}m^{-1}(1-\beta)^{-1}d_\varepsilon$ and is equal to $-d_\varepsilon^2[4m(1-\beta)]^{-1}$. From this and (9.23), it follows that

$$0 \le f(x_0) - f^* \le \varepsilon\beta^{-1} + d_\varepsilon^2[4m(1-\beta)]^{-1} . \qquad (9.24)$$

This inequality is valid for all $\beta \in (0,1]$. Let us find the minimum of the function

$$h(\beta) = \varepsilon\beta^{-1} + d_\varepsilon^2[4m(1-\beta)]^{-1} .$$

Since $h'(\beta) = [d_\varepsilon^2\beta^2 - 4m\varepsilon(1-\beta)^2][4m\beta^2(1-\beta)^2]^{-1}$, then we have $h'(\beta)=0$ for

$$\beta_1 = 2\sqrt{m\varepsilon}\,(d_\varepsilon + 2\sqrt{m\varepsilon})^{-1}, \qquad \beta_2 = 2\sqrt{m\varepsilon}\,(2\sqrt{m\varepsilon} - d_\varepsilon)^{-1} .$$

As $\beta_2 \notin (0,1)$, there exists a unique point β_1 in $(0,1)$, where $h(\beta)$ attains the extremum. It is not difficult to verify that $h''(\beta_1) > 0$, i.e., β_1 is the minimum point of the function $h(\beta)$.

Moreover, elementary calculations show that

$$h(\beta_1) = (4m)^{-1}(d_\varepsilon + 2\sqrt{\varepsilon m})^2 = \varepsilon + (4m)^{-1}(d_\varepsilon^2 + 4d_\varepsilon\sqrt{\varepsilon m}) .$$

From this and (9.24) we obtain (9.21).

It remains to prove (9.22). From (9.17) we have for the minimum point x^*

$$f(x_0) - f(x^*) \ge (v, x-x^*) + m||x_0-x^*||^2 \qquad \forall v \in \partial f(x^*).$$

Since x^* is a minimum point of $f(x)$ on E_n, then, by the necessary and sufficient condition for a minimum, $0 \in \partial f(x^*)$.

Hence,

$$f(x_0) - f(x^*) \ge m||x_0-x^*||^2 .$$

From this and (9.21) we obtain (9.22) straightforwardly. The lemma is proved. ∎

<u>COROLLARY</u>. For $\varepsilon = 0$ we have

$$0 \le f(x_0) - f^* \le (4m)^{-1}d^2 , \qquad ||x_0-x^*|| \le (2m)^{-1}d ,$$

where $d = d_0 = d_0(x_0) = \min_{v \in \partial f(x_0)} \|v\|$.

For $d_\varepsilon = 0$ (i.e., if x_0 is an ε-stationary point), we obtain from (9.21) and (9.22)

$$0 \le f(x_0) - f^* \le \varepsilon , \qquad \|x_0 - x^*\| \le \sqrt{\varepsilon m^{-1}} .$$

Thus, for any strongly convex function we can estimate how much the value of the function at the point x_0 differs from the minimum value (see inequality (9.21)), and also we can estimate the distance between the point x_0 and the minimum point (see inequality (9.22).

5. Let $\Omega \subset E_n$ be a convex compact set and $f(x)$ be a convex function on E_n.

LEMMA 9.7. If the function $f(x)$ is not constant on Ω, then it attains the maximum value on Ω only on the boundary set of Ω.

Proof. Assume the opposite. Then, for $x_0 \in \text{int}\,\Omega$, we have

$$f(x_0) \ge f(x) \qquad \forall x \in \Omega, \tag{9.25}$$

i.e., x_0 is a maximum point of $f(x)$ on Ω. Since $x_0 \in \text{int}\,\Omega$, there exists an $r > 0$ such that $S_r(x_0) \subset \Omega$. Take an arbitrary $\bar{x} \in \Omega$, $\bar{x} \ne x_0$.

Let

$$\tilde{x} = x_0 - r(\bar{x} - x_0)\|\bar{x} - x_0\|^{-1} . \tag{9.26}$$

Since $\tilde{x} \in S_r(x_0)$, then $\tilde{x} \in \Omega$ and, hence, from (9.25)

$$f(\tilde{x}) \le f(x_0) . \tag{9.27}$$

On the other hand, for any $v \in \partial f(x_0)$,

$$f(\tilde{x}) \ge f(x_0) + (v, \tilde{x} - x_0) . \tag{9.28}$$

Therefore, from (9.27) and (9.28), the inequality $(v, \tilde{x} - x_0) \le 0$ should be valid. From this inequality and (9.26) we obtain

$$(v, \bar{x} - x_0) \geq 0 . \tag{9.29}$$

Since, by the definition of a subgradient, we have $f(\bar{x}) \geq f(x_0) + (v, \bar{x} - x_0)$, then (9.29) implies $f(\bar{x}) \geq f(x_0)$. Hence, considering (9.25), we have the equality

$$f(\tilde{x}) = f(x_0) . \tag{9.30}$$

As $\bar{x}$ is an arbitrary point of Ω, we obtain from (9.30)

$$f(x) = f(x_0) \qquad \forall x \in \Omega,$$

which contradicts the assumption of the lemma. ■

REMARK 3. From the proof of this lemma, it is clear that if a set Ω is not bounded and a convex function $f(x)$ is not constant, then its maximum value on Ω cannot be attained except on the boundary of the set Ω (however, it cannot be attained at all if the function $f(x)$ is not bounded on Ω).

LEMMA 9.8. If a function $f(x)$ is strongly convex on E_n, then for any $v \in E_n$ there exists a unique point $x \in E_n$ such that $v \in \partial f(x)$.

Proof. Take an arbitrary $v_0 \in E_n$. Since the function $f(x)$ is strongly convex on E_n, then the function $h(x) = f(x) - (v_0, x)$ is also strongly convex because it is the sum of a strongly convex function and a linear function. By Lemma 9.3, there exists a unique point $x_0 \in E_n$ at which the function $h(x)$ attains its minimum value on E_n. The condition for a minimum of the function $h(x)$ on E_n implies the relation $0 \in \partial h(x_0)$. On the other hand, we have $\partial h(x_0) = \partial f(x_0) - \{v_0\}$ (since $h(x)$ is the sum of two convex functions, then, by Lemma 5.2, $\partial h(x_0) = \partial f(x_0) + \partial F(x_0)$, where $F(x_0) = -(v_0, x))$. Hence, $v_0 \in \partial f(x_0)$.

Thus, for any $v_0 \in E_n$, we can find a point $x \in E_n$ such that

$v_0 \in \partial f(x_0)$. The uniqueness of the point x_0 follows from Corollary 2 of Lemma 9.4.

REMARK 4. In Lemma 9.8, the strong convexity assumption on the function $f(x)$ is essential, that is evident from the following example.

EXAMPLE 1. Let $x \in E_1$, $f(x) = e^x$. The function $f(x)$ is strictly convex and differentiable. Hence

$$\partial f(x) = \{f'(x)\} \qquad \forall x \in E_1 .$$

Obviously, there does not exist a point $x_0 \in E_1$ such that $f'(x_0) = e^{x_0} = 0$, i.e., for $v_0 = 0$ we cannot find out the corresponding point x_0.

PROBLEM 9.2. Let $f(x)$ be a strictly convex function on E_n.

Prove that the following lemma is valid.

LEMMA 9.9. If $f(x)$ is strictly convex on E_n, then, for any $x_0 \in E_n$ and $v(x_0) \in \partial f(x_0)$, we have the inequality

$$f(x) - f(x_0) > (v(x_0), x - x_0) \qquad \forall x \in E_n, \quad x \neq x_0. \tag{9.31}$$

Conversely, if, for any $x_0 \in E_n$, there exists a $v(x_0) \in E_n$ such that (9.31) holds, then the function $f(x)$ is strictly convex.

10. CONDITIONAL $\mathcal{E}$-SUBDIFFERENTIALS

1. Definition and Properties of the Conditional Subdifferential and the Conditional ε-Subdifferential.

Let $\Omega \subset E_n$ be a convex set. Denote by $\Gamma(x)$ a cone of feasible directions of the set Ω at the point $x \in \Omega$, which is the closure of the set

$$\gamma(x) = \{v = \lambda(z - x) \mid \lambda > 0, \ z \in \Omega\}.$$

Let a finite convex function $f(x)$ be defined on the set Ω. For fixed $x_0 \in \Omega$ and $\varepsilon \geq 0$, define

$$\partial^{\Omega} f(x_0) = \{v \in E_n \mid f(z)-f(x_0) \geq (v, z-x_0) \quad \forall z \in \Omega\} ,$$

$$\partial_{\varepsilon}^{\Omega} f(x_0) = \{v \in E_n \mid f(z)-f(x_0) \geq (v, z-x_0) - \varepsilon \ \forall z \in \Omega\} .$$

The sets $\partial^{\Omega} f(x_0)$ and $\partial_{\varepsilon}^{\Omega} f(x_0)$ are called, respectively, the *conditional subdifferential* and the *conditional ε-subdifferential* of the function $f(x)$ at the point $x_0 \in \Omega$ with respect to the set Ω.

An element $v \in \partial^{\Omega} f(x_0)$ $(v \in \partial_{\varepsilon}^{\Omega} f(x_0))$ is said to be a *conditional subgradient* (*conditional ε-subgradient*) of the function $f(x)$ at the point $x_0 \in \Omega$ with respect to the set Ω.

If the function $f(x)$ is defined and convex on E_n, then the sets $\partial^{E_n} f(x_0) \equiv \partial f(x_0)$ and $\partial_{\varepsilon}^{E_n} f(x_0) \equiv \partial f(x_0)$ are called, as usual, the *subdifferential* and the *ε-subdifferential* of the function $f(x)$ at the point x_0 with respect to E_n.

It is obvious that if $\Omega_1 \supset \Omega_2$, then for $\varepsilon \geq 0$ we have

$$\partial_{\varepsilon}^{\Omega_1} f(x) \subset \partial_{\varepsilon}^{\Omega_2} f(x) \qquad \forall x \in \Omega_2 . \tag{10.1}$$

We mention one more property of the conditional ε-subdifferential: if $\bar{\Omega}$ is the closure of a convex set Ω and $f(x)$ is continuous on $\bar{\Omega}$, then

$$\partial_{\varepsilon}^{\Omega} f(x) = \partial_{\varepsilon}^{\bar{\Omega}} f(x) \qquad \forall x \in \Omega, \quad \varepsilon \geq 0 .$$

<u>LEMMA 10.1</u>. Let a function $f(x)$ be finite and convex on E_n.

If $x_0 \in \text{int}\,\Omega$, then for any fixed $\varepsilon \geq 0$ the set $\partial_{\varepsilon}^{\Omega} f(x_0)$ is nonempty, convex, closed and bounded.

<u>Proof</u>. The nonemptiness, convexity and closedness of the set $\partial_{\varepsilon}^{\Omega} f(x_0)$ are obvious. The proof of the boundedness is analogous to the proof of the boundedness of $\partial_{\varepsilon} f(x_0)$ in Theorem 7.1. ∎

<u>LEMMA 10.2</u>. If a function $f(x)$ is finite and convex on E_n, then

$$\partial^{\Omega} f(x_0) = \partial f(x_0) - \Gamma^{+}(x_0) , \tag{10.2}$$

where $\Gamma^+(x_0)$ is the cone which is conjugate to the cone of feasible directions of the set Ω at the point $x_0 \in \Omega$.

Proof. We denote by D the right-hand side of (10.2). Let $v = v_1 + v_2$, where $v_1 \in \partial f(x_0)$, $-v_2 \in \Gamma^+(x_0)$. Then, $v \in D$ and

$$f(x) - f(x_0) \geq (v_1, x-x_0) \qquad \forall x \in E_n ,$$

$$(-v_2, x-x_0) \geq 0 \qquad \forall x \in \Omega .$$

Summing these inequalities, we obtain

$$f(x) - f(x_0) \geq (v_1+v_2, x-x_0) \qquad \forall x \in \Omega .$$

Hence, $v \in \partial^{\Omega} f(x_0)$, i.e.,

$$D \subset \partial^{\Omega} f(x_0) . \tag{10.3}$$

Now, let $v \in \partial^{\Omega} f(x_0)$. Then

$$f(x) - f(x_0) \geq (v, x-x_0) \qquad \forall x \in \Omega . \tag{10.4}$$

Define $h(x) = f(x) - f(x_0) - (v,x-x_0)$. Clearly, the function $h(x)$ is convex on Ω and it follows from (10.4) that $h(x) \geq 0$ for all $x \in \Omega$ and, moreover, $h(x_0) = 0$. It means that x_0 is a minimum point of the function $h(x)$ on the set Ω. Due to the necessary and sufficient condition for the minimum of the function $h(x)$ on Ω, we have

$$\partial h(x_0) \cap \Gamma^+(x_0) \neq \emptyset$$

(see Theorem 6.1).

On the other hand, $\partial h(x_0) = \partial f(x_0) - \{v\}$. Hence, there exist a $v_0 \in \partial f(x_0)$ and a $w_0 \in \Gamma^+(x_0)$ such that $v_0 - v = w_0$, i.e., $v = v_0 - w_0$.

Consequently, we have

$$\partial^{\Omega} f(x_0) \subset D . \tag{10.5}$$

From (10.3) and (10.5) we obtain (10.2). ∎

COROLLARY 1. If $x_0 \in \text{int}\,\Omega$ and a function $f(x)$ is finite and convex on E_n, then

$$\partial^{\Omega} f(x_0) = \partial f(x_0) . \tag{10.6}$$

Indeed, for $x_0 \in \text{int}\,\Omega$, we have

$$\Gamma(x_0) = E_n , \qquad \Gamma^{+}(x_0) = \{0\} .$$

Thus, the conditional subdifferential of a convex function f(x at interior points of a convex set Ω coincides with the subdifferential of the function $f(x)$.

Let

$$\Omega_{\delta} = S_{\delta}(x_0) = \{x \mid \|x-x_0\| \le \delta\} , \qquad \delta > 0 .$$

Then, as was proved above,

$$\partial f(x_0) = \partial^{\Omega_{\delta}} f(x_0) . \qquad \blacksquare$$

It is clear from this that the subdifferential of a convex function $f(x)$ is defined only by local properties of the function $f(x)$.

REMARK 1. Generally speaking, for $\varepsilon > 0$ it is possible that

$$\partial^{\Omega}_{\varepsilon} f(x_0) \ne \partial_{\varepsilon} f(x_0) .$$

This is seen from the following example. Let $f(x) = (a,x) + b$, where $a, x \in E_n$, $b \in E_1$. In this case, as was shown in Example 2 in Section 7, we have

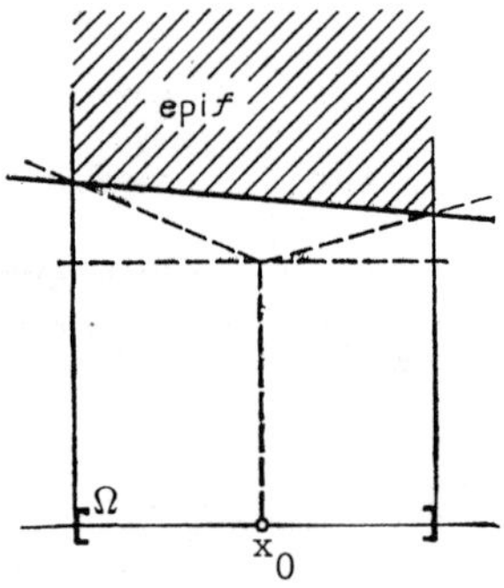

Figure 12

$$\partial_{\varepsilon} f(x) = \partial f(x) = \{a\} \quad \forall x \in E_n, \quad \forall \varepsilon \ge 0$$

(see Figure 12). Let $\Omega = \{z \in E_n \mid \|z-x_0\| \le 1\}$. We shall show that

$$\partial^{\Omega}_{\varepsilon} f(x_0) = \{v = a+r \mid r \in E_n, \ \|r\| \le \varepsilon\} . \tag{10.7}$$

By the definition of $\partial_\varepsilon^\Omega f(x_0)$, we have for $v \in \partial_\varepsilon^\Omega f(x_0)$

$$f(z) - f(x_0) = (a, z-x_0) \geq (v, z-x_0) - \varepsilon \qquad \forall z \in \Omega ,$$

i.e.,

$$(v-a,\ z-x_0) \leq \varepsilon \qquad \forall z \in \Omega . \qquad (10.8)$$

However,

$$\max_{z\in\Omega} (v-a,\ z-x_0) = (v-a,\ z_0-x_0) ,$$

where $z_0 - x_0 = \|v-a\|^{-1}(v-a)$. Hence, we have from (10.8)

$$\max_{z\in\Omega} (v-a,\ z-x_0) = \|v-a\| \leq \varepsilon \qquad \forall v \in \partial_\varepsilon^\Omega f(x_0) ,$$

implying in turn inequality (10.7).

Thus, for $\varepsilon > 0$ the ε-subdifferential and the conditional ε-subdifferential of linear functions do not coincide.

COROLLARY 2. If a function $f(x)$ is finite and convex on E_n, then

$$\partial^\Omega f(x_0) = \partial^{K_\Omega} f(x_0) ,$$

where

$$K_\Omega(x_0) = \{x = x_0+v \mid v \in \Gamma(x_0)\} , \qquad \Gamma(x_0) = \bar{\gamma}(x_0) ,$$

$$\gamma(x_0) = \{v = \lambda(x-x_0) \mid \lambda > 0,\ x \in \Omega\}$$

(i.e., $K_\Omega(x_0)$ is the cone of feasible directions of the set Ω at the point x_0, which is translated to the point x_0).

This result follows from (10.2).

COROLLARY 3. If $g \in \Gamma(x_0)$, then

$$\frac{\partial f(x_0)}{\partial g} = \max_{v\in\partial f(x_0)} (v,g) = \sup_{v\in\partial^\Omega f(x_0)} (v,g) = \max_{v\in\partial^\Omega f(x_0)} (v,g) .$$

However, if $g \notin \Gamma(x_0)$, then $\sup_{v\in\partial^\Omega f(x_0)} (v,g) = +\infty$.

Proof. Let $v \in \partial^\Omega f(x_0)$, $g \in \Gamma(x_0)$.

Then, by Lemma 10.2, $v = v_1 - w$, where $v_1 \in \partial f(x_0)$, $w \in \Gamma^+(x_0)$.

Since $(w,g) \geq 0$ for all $g \in \Gamma(x_0)$, then

$$(v,g) = (v_1,g) - (w,g) \leq (v_1,g) .$$

Hence,

$$\sup_{v\in\partial^{\Omega}f(x_0)} (v,g) \leq \max_{v\in\partial f(x_0)} (v,g) .$$

Since $\partial^{\Omega}f(x_0) \supset \partial f(x_0)$, the inequality

$$\max_{v\in\partial f(x_0)} (v,g) \leq \sup_{v\in\partial^{\Omega}f(v_0)} (v,g) .$$

Thus

$$\sup_{v\in\partial^{\Omega}f(x_0)} (v,g) = \max_{v\in\partial f(x_0)} (v,g) .$$

The compactness of the set $\partial f(x_0)$ implies that the maximum in the last equality is attained and, therefore, the supremum with respect to $\partial^{\Omega}f(x_0)$ is also attained.

Now, let $g \notin \Gamma(x_0)$. Then, there exists a $w_0 \in \Gamma^{+}(x_0)$ such that $(g,w_0) = -a < 0$. Take an arbitrary $v_0 \in \partial f(x_0)$. By Lemma 10.2 we have for any $\lambda > 0$

$$v_\lambda = [v_0 - \lambda w_0] \in \partial^{\Omega}f(x_0) .$$

However,

$$(v_\lambda,g) = (v_0,g) + \lambda a \xrightarrow[\lambda\to+\infty]{} +\infty .$$

Corollary 3 has completely been proved. ∎

<u>COROLLARY 4</u>. If functions $f_1(x)$ and $f_2(x)$ are finite and convex on E_n and a point x_0 belongs to the set Ω, then, for the function $f(x) = f_1(x) + f_2(x)$, we have the relation

$$\partial^{\Omega}f(x_0) = \partial^{\Omega}f_1(x_0) + \partial^{\Omega}f_2(x_0) . \tag{10.9}$$

In fact, since

$$\partial f(x_0) = \partial f_1(x_0) + \partial f_2(x_0)$$

(see Lemma 5.2), we obtain

$$\begin{aligned}\partial^{\Omega} f(x_0) &= \partial f(x_0) - \Gamma^+(x_0) = \partial f_1(x_0) + \partial f_2(x_0) - \Gamma^+(x_0)\\ &= [\partial f_1(x_0) - \Gamma^+(x_0)] + [\partial f_2(x_0) - \Gamma^+(x_0)]\\ &= \partial^{\Omega} f_1(x_0) + \partial^{\Omega} f_2(x_0) \quad .\end{aligned}$$

We have used the equality

$$\Gamma^+(x_0) = \Gamma^+(x_0) + \Gamma^+(x_0) \quad ,$$

which is valid because $\Gamma^+(x_0)$ is a convex cone. ■

REMARK 2. From the proof of formula (10.9) it is clear that the relations

$$\partial^{\Omega} f(x_0) = \partial^{\Omega} f_1(x_0) + \partial f_2(x_0) = \partial f_1(x_0) + \partial^{\Omega} f_2(x_0) \tag{10.10}$$

are valid, i.e., for one of the functions $f_1(x)$ or $f_2(x)$ we can take the subdifferential with respect to E_n.

2. Representation of the Conditional ε-Subdifferential.

Let $\varepsilon > 0$ be fixed. For $x_0 \in \Omega$, let

$$B_{\varepsilon}^{\Omega} f(x_0) = \{v \in E_n \mid \exists x \in \Omega;\ v \in \partial^{\Omega} f(x),\ f(x) - f(x_0) \geq (v, x - x_0) - \varepsilon\} \ ,$$

THEOREM 10.1. We have the relation

$$\partial_{\varepsilon}^{\Omega} f(x_0) = \overline{B_{\varepsilon}^{\Omega} f(x_0)} \quad . \tag{10.11}$$

Proof. Take an arbitrary $v \in \overline{B_{\varepsilon}^{\Omega} f(x_0)}$. Then

$$v = \lim_{k\to\infty} v_k \ , \qquad v_k \in B_{\varepsilon}^{\Omega} f(x_0) \qquad \forall\, k \quad .$$

By the definition of the set $\overline{B_{\varepsilon}^{\Omega} f(x_0)}$, there exist $x_k \in \Omega$ such that $v_k \in \partial^{\Omega} f(x_k)$,

$$f(x_k) - f(x_0) \geq (v_k,\ x_k - x_0) - \varepsilon \quad .$$

Since $v_k \in \partial^{\Omega} f(x_k)$, we have

$$f(z) - f(x_k) \geq (v_k,\ z - x_k) \qquad \forall\, z \in \Omega \quad .$$

Summing these two last inequalities yields

$$f(z) - f(x_0) \geq (v_k, z-x_0) - \varepsilon \qquad \forall z \in \Omega .$$

Taking here the limit as $k \to +\infty$ for any fixed $z \in \Omega$, we obtain

$$f(z) - f(x_0) \geq (v, z-x_0) - \varepsilon \qquad \forall z \in \Omega .$$

Thus, $v \in \partial_\varepsilon^\Omega f(x_0)$, therefore we conclude that

$$B_\varepsilon^\Omega f(x_0) \subset \partial_\varepsilon^\Omega f(x_0) . \tag{10.12}$$

Now let $v \in \partial_\varepsilon^\Omega f(x_0)$, i.e.,

$$f(z) - f(x_0) \geq (v, z-x_0) - \varepsilon \qquad \forall z \in \Omega . \tag{10.13}$$

Define $h(z) = f(z) - f(x_0) - (v, z-x_0)$. Then, it is clear from (10.13) that

$$h(z) \geq -\varepsilon \qquad \forall z \in \Omega . \tag{10.14}$$

Let us fix an arbitrary $\delta > 0$ and define the function

$$h_\delta(z) = h(z) + \delta \max_{i \in 1:n} |z^{(i)} - x_0^{(i)}| \equiv h(z) + \delta q(z) ,$$

where

$$z = (z^{(1)}, \ldots, z^{(n)}), \qquad x_0 = (x_0^{(1)}, \ldots, x_0^{(n)}) ,$$

$$q(z) = \max_{i \in 1:n} |z^{(i)} - x_0^{(i)}| .$$

The functions $h(z)$, $h_\delta(z)$, $g(z)$ are convex on Ω. From the definition of $h_\delta(z)$ and (10.14) it follows that $h_\delta(z) \to +\infty$ as $||z|| \to +\infty$. Therefore, for any fixed $\delta > 0$, the function $h(z)$ attains the minimum value on Ω at some point z_δ.

Due to the necessary and sufficient condition for a minimum of the function $h(z)$ on the set Ω, we have

$$\partial h_\delta(z_\delta) \cap \Gamma^+(z_\delta) \neq \emptyset \tag{10.15}$$

(see Theorem 6.1).

On the other hand, we have

$$\partial h_\delta(z) = \partial h(z) + \delta \partial q(z) , \qquad (10.16)$$

where (see Example 2 in Section 5)

$$\partial h(z) = \partial f(z) - \{v\} , \qquad \partial q(z) = \text{co } \{w_i \mid i \in R(z)\} ,$$

$$R(z) = \{i \in 1:n \mid |z^{(i)} - x_0^{(i)}| = q(z)\} ,$$

$$w_i = \begin{cases} \text{co } \{e_i, -e_i\} & \text{if } z^{(i)} - x_0^{(i)} = 0 , \\ e_i & \text{if } z^{(i)} - x_0^{(i)} > 0 , \\ -e_i & \text{if } z^{(i)} - x_0^{(i)} < 0 , \end{cases}$$

and e_i are unit vectors. From (10.15) and (10.16), there exists a $w_\delta \in \partial q(z_\delta)$ such that

$$v_\delta = (v - \delta w_\delta) \in [\partial f(z_\delta) - \Gamma^+(z_\delta)] = \partial^\Omega f(z_\delta) . \qquad (10.17)$$

Clearly,

$$v_\delta \to v \quad \text{as} \quad \delta \to 0 . \qquad (10.18)$$

From the definition of $h_\delta(z)$ and (10.14) it follows that $h_\delta(z_\delta) \geq h(z_\delta) \geq -\varepsilon$. Hence, we obtain

$$f(z_\delta) - f(x_0) + (v - \delta w_\delta, x_0 - z_0) \geq -\varepsilon , \qquad (10.19)$$

because

$$\max_{i \in 1:n} |z^{(i)} - x_0^{(i)}| = (w_\delta, z_\delta - x_0)$$

(see Example 2 in Section 5). From (10.17) and (10.19) it is clear that $v_\delta \in B_\delta^\Omega f(x_0)$ and from (10.18) it follows that $v \in B_\varepsilon^\Omega f(x_0)$. Thus

$$\partial_\varepsilon^\Omega f(x_0) \subset \overline{B_\varepsilon^\Omega f(x_0)} . \qquad (10.20)$$

From (10.12) and (10.20) we obtain (10.11). The theorem has been proved. ∎

In Lemma 7.2 it was proved that the relation

$$\partial_\varepsilon f(x_0) = B_\varepsilon f(x_0)$$

holds in the case of a strongly convex function $f(x)$. Analogously, it can be proved that if a function $f(x)$ is strongly convex on the set Ω, then

$$\partial_{\varepsilon}^{\Omega} f(x_0) = B_{\varepsilon} f(x_0) .$$

PROBLEM 10.1. Prove that if a function $f(x)$ is finite and convex on E_n and a set Ω is a hyperplane in E_n, then we have the equality

$$\partial_{\varepsilon}^{\Omega} f(x_0) = \partial_{\varepsilon}^{\Omega} f(x_0) - \Gamma^{+}(x_0) \qquad \forall \varepsilon \geq 0 . \tag{10.21}$$

3. Conditional ε-Subdifferential of a Sum of Functions.

Let functions $f_1(x)$ and $f_2(x)$ be convex on E_n. Let $f(x) = f_1(x) + f_2(x)$. Take $x_0 \in \Omega$.

THEOREM 10.2. The following relation holds:

$$\partial_{\varepsilon}^{\Omega} f(x_0) = \bigcup_{\substack{\varepsilon_1+\varepsilon_2=\varepsilon \\ \varepsilon_1 \geq 0, \varepsilon_2 \geq 0}} [\partial_{\varepsilon_1}^{\Omega} f_1(x_0) + \partial_{\varepsilon_2}^{\Omega} f_2(x_0)] . \tag{10.22}$$

Proof. Denote by D the right-hand side of (10.22). The inclusion

$$D \subset \partial_{\varepsilon}^{\Omega} f(x_0) \tag{10.23}$$

is obvious. Next let us show the converse inclusion.

Let $v \in \partial_{\varepsilon}^{\Omega} f(x_0)$. By Theorem 10.1, there exists a sequence $\{v_k\}$ such that

$$v_k \in B_{\varepsilon}^{\Omega} f(x_0) , \qquad v_k \xrightarrow[k\to\infty]{} v , \tag{10.24}$$

i.e., for each v_k there exists a $x_k \in \Omega$ such that

$$v_k \in \partial^{\Omega} f(x_k) , \qquad f(x_k) - f(x_0) + (v_k, x_0 - x_k) \geq -\varepsilon .$$

It follows from (10.10) that $v_k = v_k' + v_k''$, where

$$v_k' \in \partial f_1(x_k) , \qquad v_k'' \in \partial^{\Omega} f_2(x_k) . \tag{10.25}$$

Hence, we have

$$[(v_k', x_0 - x_k) + f_1(x_k) - f_1(x_0)] + [(v_k'', x_0 - x_k) + f_2(x_k) - f_2(x_0)] \geq -\varepsilon, \tag{10.26}$$

and, considering (10.25), we conclude that the terms in the square brackets of (10.26) are nonpositive. Therefore, inequality (10.26) can be satisfied iff there exist $\varepsilon_{k1} \geq 0$, $\varepsilon_{k2} \geq 0$ such that $\varepsilon_{k1} + \varepsilon_{k2} \leq \varepsilon$ and

$$(v_k', x_0 - x_k) + f_1(x_k) - f_1(x_0) \geq -\varepsilon_{k1},$$

$$(v_k'', x_0 - x_k) + f_2(x_k) - f_2(x_0) \geq -\varepsilon_{k2}.$$

From these inequalities and (10.25) we have

$$v_k' \in B_{\varepsilon_{k1}} f_1(x_0), \qquad v_k'' \in B_{\varepsilon_{k2}} f_2(x_0).$$

The sequence $\{v_k'\}$ is bounded because

$$B_{\varepsilon_{k1}} f_1(x_0) \subset \partial_{\varepsilon_{k1}} f_1(x_0) \subset \partial_\varepsilon f_1(x_0),$$

and the set $\partial_\varepsilon f_1(x_0)$ is compact in E_n. From (10.24), the sequence $\{v_k''\}$ is also bounded. Without loss of generality, we can assume that

$$v_k' \to v', \qquad v_k'' \to v'', \qquad \varepsilon_{k1} \to \varepsilon' \geq 0, \qquad \varepsilon_{k2} \to \varepsilon'' \geq 0,$$

where $\varepsilon' + \varepsilon'' \leq \varepsilon$. Since $v_k' \in \partial_{\varepsilon_{k_1}} f_1(x_0)$ and $v'' \in \partial^\Omega_{\varepsilon_{k_2}} f_2(x_0)$, then we have

$$f_1(x) - f_1(x_0) \geq (v_k', x - x_0) - \varepsilon_{k1} \qquad \forall x \in E_n,$$

$$f_2(z) - f_2(x_0) \geq (v_k'', z - x_0) - \varepsilon_{k2} \qquad \forall z \in \Omega.$$

By taking the limit in these inequalities as $k \to +\infty$ for fixed $x \in E_n$, $z \in \Omega$, we make sure that $v' \in \partial_{\varepsilon'} f_1(x_0)$, $v'' \in \partial^\Omega_{\varepsilon''} f_2(x_0)$.

Hence, from (10.24) and $v_k = v_k' + v_k''$ it follows that

$$v = (v' + v'') \in [\partial_{\varepsilon'} f_1(x_0) + \partial_{\varepsilon''} f_2(x_0)]$$
$$\subset [\partial^\Omega_{\varepsilon'} f_1(x_0) + \partial^\Omega_{\varepsilon''} f_2(x_0)],$$

i.e., $\partial^\Omega_\varepsilon f(x_0) \subset D$. From this and (10.23) we have (10.22). ∎

COROLLARY. From the proof of Theorem 10.2 it is clear that instead of (10.22) we can write

$$\partial_{\varepsilon}^{\Omega} f(x_0) = \bigcup_{\substack{\varepsilon_1+\varepsilon_2=\varepsilon \\ \varepsilon_1\ge 0,\varepsilon_2\ge 0}} [\partial_{\varepsilon_1}^{\Omega} f_1(x_0) + \partial_{\varepsilon_2} f_2(x_0)]$$

$$= \bigcup_{\substack{\varepsilon_1+\varepsilon_2=\varepsilon \\ \varepsilon_1\ge 0,\varepsilon_2\ge 0}} [\partial_{\varepsilon_2} f_1(x_0) + \partial_{\varepsilon_2}^{\Omega} f_2(x_0)] \quad , \tag{10.27}$$

i.e., for one of the functions $f_1(x)$ or $f_2(x)$ we can take the ε-subdifferential with respect to E_n instead of the conditional ε-subdifferential.

4. Extension of a Convex Function.

Let $\Omega \subset E_n$ be a convex set, $G \subset \text{int}\,\Omega$ be a convex compact set and $f(x)$ be a convex function on Ω. Define the function

$$F(x) = \sup_{z\in G} [f(z) + (v(z), x-z)] \quad , \tag{10.28}$$

where $v(z)$ is an arbitrary (but fixed for each z) vector of $\partial^{\Omega} f(z)$

LEMMA 10.3. The function $F(z)$ is convex and finite on E_n and coincides with the function $f(x)$ on G, i.e.,

$$F(x) = f(x) \qquad \forall x \in G \ . \tag{10.29}$$

Proof. The convexity of $F(x)$ is obvious. We shall prove (10.29). Let $x \in G$. Then, of course, $x \in \Omega$ and, by the definition of the set $\partial^{\Omega} f(x)$ we have

$$f(x) - f(z) \ge (v(z), x-z) \qquad \forall z \in G \ ,$$

i.e.,

$$f(x) \ge \sup_{z\in G} [f(z) + (v(z), x-z)] \equiv F(x) \quad .$$

Hence, the function $F(x)$ is finite. On the other hand, we have $F(x) \ge f(x)$. From these inequalities we get (10.29). ∎

Thus, any finite convex function defined on Ω can be extended (while preserving the convexity and the finiteness) on E_n, and, moreover, this function coincides with $f(x)$ on G. In addition, the supremum in (10.28) is achieved, i.e.,

$$F(x) = \max_{z\in\Omega} [f(z) + (v(z), x-z)] .$$

REMARK 3. In Lemma 10.3, we assumed that the function $f(x)$ is given on the set Ω and the set $G \subset \operatorname{int}\Omega$ is closed and bounded.

This condition cannot be dropped, as one can clearly see in the following example.

Let $x \in E_1$, $f(x) = 1-\sqrt{1-x^2}$, $G = [0,1]$. Then, the function $f(x)$ is differentiable on $[0,1)$. Since

$$f''(x) = (1-x^2)^{-3/2} > 0 \qquad \forall x \in [0,1) ,$$

the function $f(x)$ is convex. It is bounded on $[0,1]$. However, $f'(x)$ is not bounded on $[0,1)$, hence the function

$$F(x) = \sup_{z\in[0,1)} [f(z)+(f'(z),x-z)]$$

is no longer finite for $x > 1$ as we have $F(x) = +\infty$ for any fixed $x > 1$.

11. CONDITIONAL DIRECTIONAL DERIVATIVES. CONTINUITY OF THE CONDITIONAL ε-SUBDIFFERENTIAL MAPPING

1. Let a function $f(x)$ be convex and finite on E_n and $\Omega \subset E_n$ be a convex set. Let $x_0 \in \Omega$, $\varepsilon \geq 0$, $g \in E_n$, $\|g\| = 1$. Moreover, define

$$\frac{\partial^{\Omega} f(x_0)}{\partial g} = \sup_{v\in\partial^{\Omega} f(x_0)} (v,g) , \qquad \frac{\partial^{\Omega}_{\varepsilon} f(x_0)}{\partial g} = \sup_{v\in\partial^{\Omega}_{\varepsilon} f(x_0)} (v,g) .$$

Let us call the quantity $\frac{\partial^{\Omega} f(x_0)}{\partial g}$ the conditional derivative of the function $f(x)$ at the point x_0 in the direction g with respect to the set Ω and the quantity $\dfrac{\partial_\varepsilon^\Omega f(x_0)}{\partial g}$ the conditional ε-derivative of the function $f(x)$ at the point x_0 in the direction g with respect to Ω.

From (6.5) (with $C = \partial^\Omega f(x_0)$, $B = \partial f(x_0)$), we have

$$\frac{\partial^\Omega f(x_0)}{\partial g} = \begin{cases} \dfrac{\partial f(x_0)}{\partial g} & \text{if } g \in \Gamma(x_0), \\ +\infty & \text{if } g \notin \Gamma(x_0). \end{cases} \tag{11.1}$$

Since the inclusion $\partial^\Omega f(x_0) \subset \partial_\varepsilon^\Omega f(x_0)$ is valid, then

$$\frac{\partial_\varepsilon^\Omega f(x_0)}{\partial g} = +\infty \quad \text{for } g \notin \Gamma(x_0).$$

LEMMA 11.1. Let K be a convex cone, $g \in \operatorname{int} K$. If the points g and $-g$ both belong to K, then

$$K = E_n . \tag{11.2}$$

Proof. Since $g \in \operatorname{int} K$, there exists an $r > 0$ such that

$$x_q = g + q \in K \qquad \forall q \in S_r(0).$$

Since K is a convex cone and $g_1 = -g \in K$, we have

$$y_q = [0.5g_1 + 0.5x_q] \in K \quad \forall q \in S_r(0). \tag{11.3}$$

Since $y_q = 0.5q$ for all $q \in S_r(0)$, then (11.3) implies that $S_{0.5r}(0) \subset K$, i.e., the sphere centered at the origin with radius $0.5r$ is contained in the cone K. Therefore $K = E_n$. ■

COROLLARY. Let $\bar{x}$ be a boundary point of the set Ω and let $g \in E_n$, $||g|| = 1$. If $g \in \operatorname{int} \Gamma(\bar{x})$, then

$$g_1 = -g \notin \Gamma(\bar{x}) . \tag{11.4}$$

To show this, assume the opposite. Let $g_1 \in \Gamma(\bar{x})$. Then, it follows from Lemma 11.1 that

$$\Gamma(\bar{x}) = E_n ,$$

which is impossible because $\bar{x}$ is a boundary point. ■

THEOREM 11.1. If $x_0 \in \Omega$, $g \in \gamma(x_0)$ and $\|g\| = 1$, then

$$\frac{\partial_\varepsilon^\Omega f(x_0)}{\partial g} = \sup_{v \in \partial_\varepsilon^\Omega f(x_0)} (v,g) = \inf_{\substack{\alpha>0 \\ x_0+\alpha g \in \Omega}} \alpha^{-1}[f(x_0+\alpha g) - f(x_0) + \varepsilon] \ . \tag{11.5}$$

Proof. First of all, we note that if the infimum in (11.5) is attained at some α_1, then $\alpha_1 > 0$ because $h(\alpha) \xrightarrow[\alpha\to+0]{} +\infty$, where

$$h(\alpha) = \alpha^{-1}[f(x_0+\alpha g) - f(x_0) + \varepsilon] \ .$$

By the definition of $\partial_\varepsilon^\Omega f(x_0)$, we have

$$f(x_\alpha) - f(x_0) \geq \alpha(v,g) - \varepsilon \qquad \forall\, v \in \partial_\varepsilon^\Omega f(x_0), \quad \forall\, x_\alpha \in \Omega ,$$

where $x_\alpha = x_0 + \alpha g$. Hence,

$$h_\Omega^* \equiv \inf_{\alpha>0,\, x_\alpha \in \Omega} h(\alpha) \geq \sup_{v \in \partial_\varepsilon^\Omega f(x_0)} (v,g) \ . \tag{11.6}$$

We shall establish the converse inequality. Since

$$\partial_\varepsilon^\Omega f(x) = \partial_\varepsilon^{\bar\Omega} f(x) \qquad \forall\, x \in \Omega$$

$$h^* = h_{\bar\Omega}^* \equiv \inf_{\alpha>0,\, x_\alpha \in \bar\Omega} h(\alpha) \ ,$$

then, without loss of generality, we can assume that the set Ω is closed.

Recall (see the Corollary of Lemma 8.3) that if α_1 is a local minimum point of the function $h(\alpha)$ on $(0,\infty)$, then

$$h^* \equiv \inf_{\alpha>0} h(\alpha) = h(\alpha_1) \ . \tag{11.7}$$

The following two cases are possible:

(a) $h^* = h_\Omega^*$; (11.8)

(b) $h^* < h_\Omega^*$. (11.9)

Consider the case (a). It follows from (8.2) that

$$h^* = \max_{v \in \partial_\varepsilon f(x_0)} (v,g) \ . \tag{11.10}$$

From the fact that $\partial_\varepsilon f(x_0) \subset \partial_\varepsilon^\Omega f(x_0)$, and from (11.8) and (11.
we have

$$h_\Omega^* \leq \sup_{v\ \partial_\varepsilon^\Omega f(x_0)} (v,g) \quad ,$$

which, together with (11.6), gives us (11.5).

Now consider the case (b). Let

$$h_\Omega^* = h(\alpha_1) \tag{11.11}$$

and, moreover, let $x_\alpha \in \Omega$ for all $\alpha \in (0,\alpha_1)$.

Then $x_\alpha \notin \Omega$ for all $\alpha > \alpha_1$, because otherwise, by the convexity of the set Ω, the point α_1 should be a local minimum point of the function $h(\alpha)$ on $(0,\infty)$ and, hence, by virtue of (11.7) and (11.11), (11.8) is valid, which contradicts (11.9).

Thus, x_{α_1} must be a boundary point of the set Ω.

Since $h(\alpha_1) = \min_{\alpha\in(0,\alpha_1]} h(\alpha)$, then, due to Remark 1 of Section
the inequality

$$h'_-(\alpha_1) \equiv \lim_{\alpha\to-0} \alpha^{-1}[h(\alpha_1+\alpha) - h(\alpha_1)] \geq 0$$

should hold. However, it is easy to check that

$$h'_-(\alpha_1) = \alpha_1^{-1} \frac{\partial f(x_1)}{\partial g_1} + \alpha_1^{-2}[f(x_1) - f(x_0) + \varepsilon] \quad ,$$

where

$$g_1 = -g \ , \qquad x_1 = x_{\alpha_1} \ .$$

Hence, there exists a $v_0 \in \partial f(x_1)$ such that

$$\alpha_1^{-1}[(v_0,g_1) + \alpha_1^{-1}(f(x_1)-f(x_0)+\varepsilon)] \geq 0 \quad ,$$

i.e.,

$$\alpha_1^{-1}[f(x_1) - f(x_0) + \varepsilon] \geq (v_0,g) \ . \tag{11.12}$$

We assume first that $x_0 \in \operatorname{int}\Omega$. Then, $-g \in \operatorname{int}\gamma(x_1)$ and, due to Corollary of Lemma 11.1, $g \notin \Gamma(x_1)$. Therefore there exists a $w \in \Gamma^+(x_1)$ such that $(w,g) < 0$. Since $\Gamma^+(x_1)$ is a cone, there exists a $\lambda \geq 0$ such that $w_0 = \lambda w \in \Gamma^+(x_1)$ and we have

$$f(x_1) - f(x_0) = \alpha_1(v_0 - w_0, g) - \varepsilon \tag{11.13}$$

instead of (11.12).

Since

$$v_1 = (v_0 - w_0) \in \partial^{\Omega} f(x_1), \qquad g = \alpha_1^{-1}(x_1 - x_0),$$

then from (11.13) we obtain $f(x_1) - f(x_0) = (v_1, x_1 - x_0) - \varepsilon$. Hence, by the definition of the set $B_{\varepsilon}^{\Omega} f(x_0)$ and Theorem 10.1, we have $v_1 \in B_{\varepsilon}^{\Omega} f(x_0) \subset \partial_{\varepsilon}^{\Omega} f(x_0)$.

Thus, from (11.11) and (11.13) it follows that

$$h_{\Omega}^* = h(\alpha_1) = (v_1, g) \leq \sup_{v \in \partial_{\varepsilon}^{\Omega} f(x_0)} (v, g). \tag{11.14}$$

We assume now that x_0 is a boundary point of the set Ω. Define

$$\Omega_\delta = \operatorname{co}\{\Omega \cup S_\delta(x_0)\}, \qquad h_{\Omega_\delta}^* = \inf_{\alpha > 0,\, x_\alpha \in \Omega_\delta} h(\alpha), \qquad \delta > 0.$$

In this case (Figure 13), $\Omega_\delta \cap \{x = x_0 + \alpha g \mid \alpha \geq 0\} = [x_0, x_{1\delta}]$, where $x_{1\delta} = x_0 + \alpha_{1\delta} g$. Let us show that

$$x_{1\delta} \xrightarrow[\delta \to +0]{} x_1 = x_0 + \alpha_1 g.$$

Since $g \notin \gamma(x_1)$, then, for any fixed $\alpha > \alpha_1$, $x_\alpha = x_0 + \alpha g \notin \Omega$.

By the separation theorem (Theorem 1.2), there exist a $g_0 \in E_n$ and a number $a > 0$ such that

$$(x - x_\alpha, g_0) \leq -a \qquad \forall\, x \in \Omega.$$

For sufficiently small $\delta > 0$, we have

$$(x - x_\alpha, g_0) \leq -0.5a \qquad \forall\, x \in \Omega \cup S_\delta(x_\alpha);$$

and then the inequality

$$(x - x_\alpha, g_0) \leq -0.5a \qquad \forall\, x \in \Omega_\delta$$

is valid, i.e., $x_\alpha \in \Omega_\delta$ and $x_{1\delta} \in [x_1, x_\alpha]$ for sufficiently small $\delta > 0$.

Hence, $x_{1\delta} \xrightarrow[\delta\to+0]{} x_1$ and the continuity of $h(\alpha)$ implies that $h^*_{\Omega_\delta} \to h^*_\Omega$. From this and (11.9) it follows that $h^* < h^*_{\Omega_\delta}$ for sufficiently small δ. Thus, we are again under the same conditions as in the case (b), however, not for the set Ω but for the set Ω_δ. Accordingly, we have

$$h^*_{\Omega_\delta} \le \sup_{v\in\partial_\varepsilon^{\Omega_\delta} f(x_0)} (v,g)$$

instead of (11.14).

From this and the relation $\partial_\varepsilon^{\Omega_\delta} f(x_0) \subset \partial_\varepsilon^{\Omega} f(x_0)$, which follows from $\Omega \subset \Omega_\delta$, we conclude that

$$h^*_{\Omega_\delta} \le \sup_{v\in\partial_\varepsilon^{\Omega} f(x_0)} (v,g) \quad .$$

Passing here to the limit as $\delta \to +0$, we get again (11.14).

Thus, (11.14) is valid for the case (b) irrespective of whether the point x_0 is a boundary point or an interior point of the set Ω.

Inequalities (11.14) and (11.6) imply (11.5). ∎

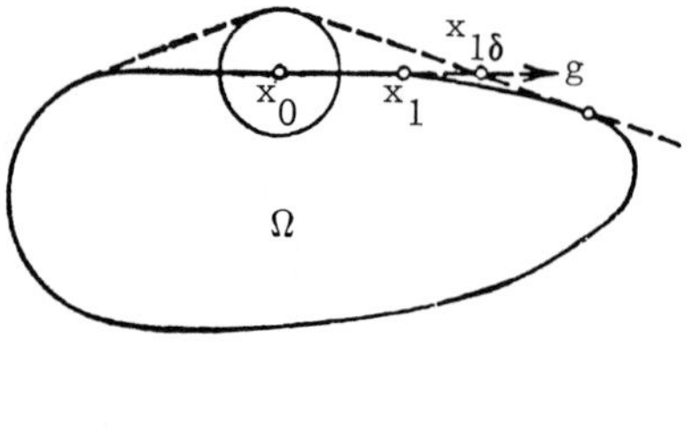

Figure 13

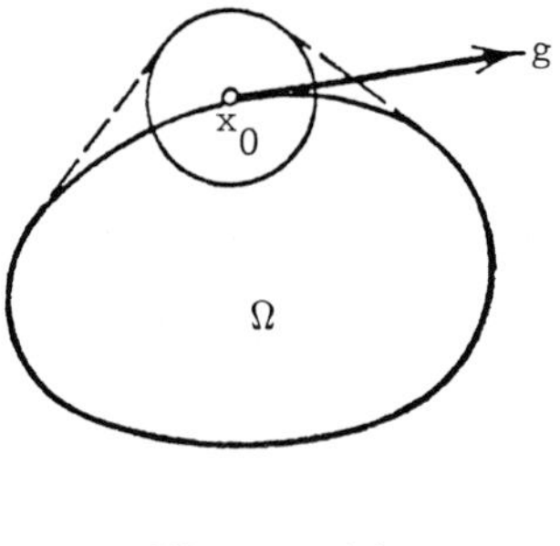

Figure 14

<u>COROLLARY 1</u>. For any $\varepsilon > 0$, the following relation holds:

$$\frac{\partial_\varepsilon^\Omega f(x_0)}{\partial g} = +\infty \qquad \forall g \notin \gamma(x_0) \quad .$$

To show this, take $x_0 \in \Omega$, $\varepsilon > 0$. It has been shown (see (11.1) that

$$\frac{\partial_\varepsilon^\Omega f(x_0)}{\partial g} = +\infty$$

for $g \notin \Gamma(x_0)$. Now let us consider the case $g \in \Gamma(x_0) \setminus \gamma(x_0)$ (Figure 14), i.e., $g \notin \gamma(x_0)$, however, $g = \lim_{k\to\infty} g_k$, $g_k \in \gamma(x_0)$. Consider the set $\Omega_\delta = \text{co}\,\{\Omega \cup S_\delta(x_0)\} \supset \Omega$, where $\delta > 0$. Then $g \in \gamma_\delta(x_0)$, where

$$\gamma_\delta(x_0) = \{v = \lambda(x-x_0) \mid \lambda > 0,\ x \in \Omega_\delta\} .$$

By Theorem 11.1 we have

$$h_\delta \equiv \inf_{\alpha>0,\, x_\alpha\in\Omega_\delta} h(\alpha) = \sup_{v\in\partial_\varepsilon^{\Omega_\delta} f(x_0)} (v,g) \le \sup_{v\in\partial_\varepsilon^{\Omega} f(x_0)} (v,g). \quad (11.15)$$

On the other hand, we have $h_\delta = \inf_{\alpha\in(0,\alpha_\delta)} h(\alpha)$. Since $g \notin \gamma(x_0)$, then $\alpha_\delta \xrightarrow[\delta\to+0]{} 0$, and, hence, $h_\delta \xrightarrow[\delta\to+0]{} +\infty$. From (11.15), we get

$$\frac{\partial_\varepsilon^\Omega f(x_0)}{\partial g} = +\infty .$$

<u>COROLLARY 2</u>. If, for some $g_0 \in E_n$, $\|g_0\| = 1$, the inequality

$$\frac{\partial_\varepsilon^\Omega f(x_0)}{\partial g_0} < 0 \quad (11.16)$$

is valid, then

$$\inf_{\alpha>0,\, x_\alpha\in\Omega} f(x_0+\alpha g_0) < f(x_0) - \varepsilon . \quad (11.17)$$

<u>Proof</u>. By virtue of (11.16), there exists an $\alpha_1 > 0$ such that

$$h(\alpha_1) = \alpha_1^{-1}[f(x_0+\alpha_1 g_0) - f(x_0) + \varepsilon] < 0 , \quad (11.18)$$

$x_0 + \alpha_1 g_0 \in \Omega$. Since $\alpha_1 > 0$, then we have from (11.8)

$$f(x_0+\alpha_1 g_0) - f(x_0) + \varepsilon < 0 ,$$

which implies (11.17).

<u>COROLLARY 3</u>. If $f(x)$ is strongly convex and

$$\frac{\partial_\varepsilon^\Omega f(x_0)}{\partial g} \le 0 ,$$

then

$$\inf_{\alpha>0,\, x_0+\alpha g_0\in\Omega} f(x_0+\alpha g_0) \le f(x_0) - \varepsilon .$$

Proof. The case $\frac{\partial_\varepsilon^\Omega f(x_0)}{\partial g_0} < 0$ was considered in Corollary 2. Now let

$$\frac{\partial_\varepsilon^\Omega f(x_0)}{\partial g_0} = 0 .$$

Then there exist sequences $\{\alpha_k\}$, $\{\beta_k\}$ such that

$$\alpha_k > 0 , \qquad x_0 + \alpha_k g_0 \in \Omega , \qquad h(\alpha_k) \le \beta_k , \qquad \beta_k \to 0 ,$$

where

$$h(\alpha) = \alpha^{-1}[f(x_0+\alpha g_0) - f(x_0) + \varepsilon] .$$

The strong convexity of the function $f(x)$ implies that there exists an $a_0 > 0$ such that $\alpha_k \le \alpha_0$ for all k. Hence,

$$f(x_0+\alpha_k g_0) - f(x_0) + \varepsilon \le \alpha_k\beta_k \le \alpha_0\beta_k .$$

Finally, we obtain

$$\inf_{\alpha>0,\, x_0+\alpha g_0\in\Omega} f(x_0+\alpha g_0) - f(x_0) + \varepsilon \le 0 ,$$

which proves the corollary. ∎

We note that the assumption of the strong convexity is essential, as can be seen from the following example.

Let $f(x) = \text{const.}\ \forall x$ and $\Omega \in E_n$. Then, for any $g_0 \in E_n$, $||g|| = 1$, we have

$$\inf_{\alpha>0} \alpha^{-1}[f(x_0+\alpha g_0) - f(x_0) + \varepsilon] = \inf_{\alpha>0} \alpha^{-1}\varepsilon = 0 ,$$

however, the function $f(x)$ cannot be decreased.

2. Let us consider the point-to-set mapping

$$\partial_\varepsilon^\Omega f : (0,\infty) \times \Omega \to \Pi(E_n) ,$$

where, by definition, for fixed $x \in \Omega$ and $\varepsilon > 0$

$$\partial_\varepsilon^\Omega f(x) = \{v \in E_n \mid f(z)-f(x) \ge (v,z-x) - \varepsilon \quad \forall z \in \Omega\} .$$

The mapping $\partial_{\varepsilon}^{\Omega} f(x)$, as well as the mapping $\partial^{\Omega} f(x)$, is not bounded.

Let a function $f(x)$ be finite and convex on E_n.

THEOREM 11.2. If $\Omega \subset E_n$ is a closed convex set, then the mapping $\partial_{\varepsilon}^{\Omega} f(x)$ is continuous in the Kakutani sense with respect to ε and x on the set $(0,\infty) \times \Omega$ (i.e., both upper and lower semicontinuous).

Proof. The upper semicontinuity of the mapping $\partial_{\varepsilon}^{\Omega} f(x)$ follows directly from the definition. We shall prove the lower semicontinuity of this mapping. Let $\varepsilon_0 > 0$, $x_0 \in \Omega$, $v_0 \in \partial_{\varepsilon_0}^{\Omega} f(x_0)$ and arbitrary sequences $\{\varepsilon_k\}$, $\{x_k\}$ such that $\varepsilon \to \varepsilon_0$, $x_k \in \Omega$, $x_k \to x_0$ be given.

We have to show that there exist $v_k \in \partial_{\varepsilon_k}^{\Omega} f(x_k)$ such that $v_k \to v_0$.

Otherwise, there exists an $a > 0$ such that

$$\inf_{v \in \partial_{\varepsilon_k}^{\Omega} f(x_k)} \|v - v_0\| = \|v_k - v_0\| \geq a > 0 \quad \forall k . \tag{11.19}$$

The infimum in (11.19) is attained because the sets $\partial_{\varepsilon_k}^{\Omega} f(x_k)$ are closed. Without loss of generality, we assume that

$$v_k \to v^* , \qquad g_k \equiv \|v_0 - v_k\|^{-1} (v_0 - v_k) \to g_0 , \qquad \|g_0\| = 1 .$$

Then (11.19) implies that

$$(v - v_0, g_k) \leq (v_k - v_0, g_k) \leq -a \qquad \forall v \in \partial_{\varepsilon_k}^{\Omega} f(x_k) .$$

Therefore,

$$\frac{\partial_{\varepsilon_k}^{\Omega} f(x_k)}{\partial g_k} \equiv \sup_{v \in \partial_{\varepsilon_k}^{\Omega} f(x_k)} (v, g_k) = (v_k, g_k) \tag{11.20}$$

and $(v_k, g_k) \leq (v_0, g_k) - a$. From this inequality and (11.20) it is seen that there exists a number K such that

$$\frac{\partial_{\varepsilon_k}^{\Omega} f(x_k)}{\partial g_k} = (v_k, g_k) \leq (v_0, g_0) - 0.5a \qquad \forall k > K . \tag{11.21}$$

Hence,
$$\frac{\partial^{\Omega}_{\varepsilon_0} f(x_0)}{\partial g_0} \geq \frac{\partial^{\Omega}_{\varepsilon_k} f(x_k)}{\partial g_k} + 0.5a \qquad \forall\, k > K \,. \tag{11.22}$$

By Theorem 11.1,

$$\frac{\partial^{\Omega}_{\varepsilon_k} f(x_k)}{\partial g_k} = \inf_{\alpha>0,\, x_k+\alpha g_k\in\Omega} \alpha^{-1}[f(x_k+\alpha g_k) - f(x_k) + \varepsilon_k]$$

$$\equiv \inf_{\alpha>0,\, x_k+\alpha g_k\in\Omega} h(x_k, g_k, \alpha, \varepsilon_k) \;.$$

From (11.21) it is seen that there exists an $\alpha_0 > 0$ such that

$$\frac{\partial^{\Omega}_{\varepsilon_k} f(x_k)}{\partial g_k} = \inf_{\alpha\geq\alpha_0,\, x_k+\alpha g_k\in\Omega} h(x_k, g_k, \alpha, \varepsilon_k) \;. \tag{11.23}$$

We assume first that the function $f(x)$ is strongly convex. Then, for each k, the infimum in (11.23) is attained at some point α_k, i.e.,

$$\frac{\partial^{\Omega}_{\varepsilon_k} f(x_k)}{\partial g_k} = h(x_k, g_k, \alpha_k, \varepsilon_k) \quad , \tag{11.24}$$

and, in addition, $\alpha_k \leq \beta_0$, where $0 < \beta_0 < +\infty$ and β_0 is the same for all k.

We can assume that $\alpha_k \to \alpha' \geq \alpha_0$. Then, from (11.22) and (11.24) we have

$$\frac{\partial^{\Omega}_{\varepsilon_k} f(x_k)}{\partial g_k} \to h(x_0, g_0, \alpha', \varepsilon_0) \leq \frac{\partial^{\Omega}_{\varepsilon_0} f(x_0)}{\partial g_0} - 0.5a \;. \tag{11.25}$$

Since $x_k + \alpha_k g_k \in \Omega$, then, surely, $x_0 + \alpha' g_0 \in \Omega$. Therefore,

$$\frac{\partial^{\Omega}_{\varepsilon_0} f(x_0)}{\partial g_0} = \inf_{\alpha\geq\alpha_0,\, x_0+\alpha g_0\in\Omega} h(x_0, g_0, \alpha, \varepsilon_0) \leq h(x_0, g_0, \alpha', \varepsilon_0) \;,$$

which contradicts (11.25). This contradiction proves the lower semicontinuity of the mapping $\partial^{\Omega}_{\varepsilon} f(x)$ at the point $[\varepsilon_0, x_0]$.

Now, let $f(x)$ be an arbitrary finite convex function. Define the function $F_\delta(x) = f(x) + \delta(x-x_0)^2$, where $\delta > 0$. The mapping $\partial_\varepsilon^\Omega F_\delta(x)$ is lower semicontinuous at the point $[\varepsilon_0, x_0]$ since the function $F_\delta(x)$ is strongly convex. From (10.27) we have

$$\partial_\varepsilon^\Omega F_\delta(x) = \bigcup_{\substack{\varepsilon_1+\varepsilon_2=\varepsilon \\ \varepsilon_1\ge 0,\, \varepsilon_2\ge 0}} [\partial_{\varepsilon_1}^\Omega f(x) + \partial_{\varepsilon_2}^\Omega f_\delta(x)] \quad ,$$

where $f_\delta(x) = \delta(x-x_0)^2$. As follows from Example 1 in Section 7 (see (7.7)),

$$\partial_\varepsilon f_\delta(x) = \{v = 2\delta(x-x_0)+r \mid r \in E_n,\ \|r\| \le 2\sqrt{\varepsilon\delta}\ \} .$$

Here, we take an arbitrary $v_0 \in \partial_\varepsilon^\Omega f(x_0)$ and the sequences $\{\varepsilon_k\}$ and $\{x_k\}$ such that $\varepsilon_k \to \varepsilon_0$, $x_k \to x_0$. Since $0 \in \partial_\varepsilon f_\delta(x_0)$, then $v_0 \in \partial_\varepsilon^\Omega F_\delta(x_0)$ for any $\delta > 0$. The lower semicontinuity of the mapping $\partial_\varepsilon^\Omega F_\delta(x)$ at the point $[\varepsilon_0, x_0]$ implies that there exists a sequence $\{v_{k\delta}\}$ such that

$$v_{k\delta} \in \partial_{\varepsilon_k}^\Omega F_\delta(x_k) \ , \qquad v_{k\delta} \to v_0 \ .$$

Here, according to the above representation of $\partial_\varepsilon^\Omega F_\delta(x)$, we have $v_{k\delta} = v'_{k\delta} + v''_{k\delta}$, where $v'_{k\delta} \in \partial_{\varepsilon_k}^\Omega f(x_k)$, $v''_{k\delta} \in \partial_{\varepsilon_k} f(x_k)$. Since $\|v''_k\| \xrightarrow[\delta\to 0]{} 0$, then we can choose subsequences $\{\delta_k\}$ and $\{v'_{k\delta_k}\}$ such that $\delta_k \xrightarrow[k\to\infty]{} +0$, $v'_{k\delta_k} \xrightarrow[k\to\infty]{} v_0$. It means that the mapping $\partial_\varepsilon^\Omega f(x)$ is lower semicontinuous at the point $[\varepsilon_0, x_0]$. By the arbitrariness of the point $[\varepsilon_0, x_0]$, the mapping $\partial_\varepsilon^\Omega f(x)$ is lower semicontinuous on the set $(0,\infty) \times \Omega$. The theorem has been proved. ∎

<u>COROLLARY</u>. If $x_0 \in \operatorname{int} \Omega$ and $\varepsilon_0 > 0$, then the mapping $\partial_\varepsilon^\Omega f(x)$ is continuous in the Hausdorff sense at the point $[\varepsilon_0, x_0]$.

<u>Proof</u>. Follows from Lemma 10.1.

3. Necessary and Sufficient Conditions for a Constrained Minimum.

Let $\Omega \subset E_n$ be a closed convex set and let a convex function $f(x)$ be given on the set Ω. Let us consider the problem of minimizing the function $f(x)$ on the set Ω.

The following proposition is obvious.

LEMMA 11.2. For the function $f(x)$ to attain the minimum value on the set Ω at a point $x^* \in \Omega$, it is necessary and sufficient that

$$0 \in \partial^{\Omega} f(x^*) . \tag{11.26}$$

REMARK. If the function $f(x)$ is finite and convex on E_n, then, by Lemma 10.2,

$$\partial^{\Omega} f(x^*) = \partial f(x^*) - \Gamma^+(x^*) ,$$

and condition (11.26) is equivalent to the relation

$$\partial f(x^*) \cap \Gamma^+(x^*) \neq \emptyset$$

(see (6.16)). If $\Omega = E_n$, then condition (11.26) takes the form

$$0 \in \partial f(x^*) .$$

Let $\varepsilon \geq 0$. The point $\bar{x} \in \Omega$ is said to be an ε-*stationary point* of the function $f(x)$ on the set Ω if

$$0 \in \partial_{\varepsilon}^{\Omega} f(\bar{x}) . \tag{11.27}$$

LEMMA 11.3. Relation (11.27) is equivalent to the inequality

$$0 \leq f(\bar{x}) - f^* \leq \varepsilon , \tag{11.28}$$

where $f^* = \inf_{x \in \Omega} f(x)$.

Proof. By definition, relation (11.27) is equivalent to

$$f(x) \geq f(\bar{x}) + (0, x-\bar{x}) - \varepsilon = f(\bar{x}) - \varepsilon \quad \forall x \in \Omega,$$

which proves the lemma. ∎

Let $x \in \Omega$. Define

$$\psi_\varepsilon(x) = \inf_{\|g\|=1} \frac{\partial_\varepsilon^\Omega f(x)}{\partial g} .$$

According to Corollary 1 of Theorem 11.1, we have

$$\psi_\varepsilon(x) = \min_{\|g\|=1, g\in\gamma(x)} \frac{\partial_\varepsilon^\Omega f(x)}{\partial g} .$$

COROLLARY 1. Condition (11.27) (or, equivalently, condition (11.28)) can be represented as

$$\psi_\varepsilon(\bar{x}) \geq 0 .$$

This follows from Corollary 2 of Lemma 6.4.

COROLLARY 2. If $0 \notin \partial_\varepsilon^\Omega f(\bar{x})$, then there exists a direction $g_0 \in \gamma(\bar{x})$, $\|g_0\| = 1$, such that

$$\frac{\partial_\varepsilon^\Omega f(\bar{x})}{\partial g_0} < 0 .$$

Moreover, the inequality

$$\inf_{\alpha>0, \bar{x}+\alpha g_0 \in\Omega} f(\bar{x} + \alpha g_0) < f(\bar{x}) - \varepsilon$$

is valid (see Corollary 2 of Theorem 11.1).

Let $x_0 \in \Omega$, $\varepsilon \geq 0$ and $\psi_\varepsilon(x_0) < 0$.

DEFINITION. A vector $g_0 \in E_n$, $\|g_0\| = 1$, such that

$$\frac{\partial_\varepsilon^\Omega f(x_0)}{\partial g_0} = \inf_{\|g\|=1} \frac{\partial_\varepsilon^\Omega f(x_0)}{\partial g} = \psi_\varepsilon(x_0)$$

is called the *conditional* ε-*steepest* (*steepest*, if $\varepsilon = 0$) *descent direction* of the function $f(x)$ at the point $x_0 \in \Omega$.

According to the Corollary of Lemma 6.3, this vector exists and is unique.

Moreover, if we have $\|v(x_0)\| = \min_{v\in\partial_\varepsilon^\Omega f(x_0)} \|v\|$ for $v(x_0) \in \partial_\varepsilon^\Omega f(x_0)$, then the vector $g_0 = -\|v(x_0)\|^{-1} v(x_0)$ is the conditional ε-steepest descent direction of the function $f(x)$ at the point $x_0 \in \Omega$.

It follows from Corollary 1 of Theorem 11.1 and formula (11.1) that

$$\psi_\varepsilon(x_0) = \min_{\|g\|=1, g\in\gamma(x_0)} \frac{\partial_\varepsilon^\Omega f(x_0)}{\partial g}, \qquad g_0 \in \gamma(x_0),$$

and, if $\varepsilon = 0$,

$$\psi(x_0) \equiv \psi_0(x_0) = \min_{\|g\|=1, g\in\Gamma(x_0)} \frac{\partial_\varepsilon^\Omega f(x_0)}{\partial g}, \quad g_0 \in \Gamma(x_0).$$

Finally, from Corollary 2 of Theorem 11.1 we conclude this section with the fact that if

$$\frac{\partial_\varepsilon^\Omega f(x_0)}{\partial g_0} < 0,$$

then

$$\inf_{\alpha>0, x_0+\alpha g_0\in\Omega} f(x_0 + \alpha g_0) < f(x_0) - \varepsilon. \tag{11.29}$$

12. REPRESENTATION OF A CONVEX SET BY MEANS OF INEQUALITIES

1. Let $\Omega \subset E_n$ be a closed convex set. For any $x_0 \in \Omega$, consider the *normal cone*

$$N(x_0) = \{v \in E_n \mid (v, x-x_0) \le 0 \ \forall x \in \Omega\}. \tag{12.1}$$

If $x_0 \in \operatorname{int} \Omega$, then it is clear that $N(x_0) = \{0\}$. If x_0 is a boundary point of Ω, then $N(x_0)\setminus\{0\} \ne \emptyset$ (see Corollary 1 of Theorem 1.2).

Let

$$P = \{x \in E_n \mid (v(x_0), x-x_0) \le 0 \quad \forall v(x_0) \in N(x_0), \quad \forall x_0 \in \Omega\}. \tag{12.2}$$

REMARK 1. It is obvious that if $\Omega \ne E_n$, then

$$P = \{x \in E_n \mid (v(x_0), x-x_0) \le 0 \quad \forall v(x_0) \in N(x_0), \ \|v(x_0)\| = 1, \quad \forall x_0 \in \Omega_{fr}\},$$

where Ω_{fr} is the set of the boundary points of Ω.

LEMMA 12.1. The following relation holds:

$$\Omega = P . \tag{12.3}$$

Proof. Let $x \in \Omega$. Then it is clear from (12.1) that $x \in P$, i.e.,

$$\Omega \subset P . \tag{12.4}$$

Now let $z \in P$. We shall show that $z \in \Omega$. Assume that this is not the case, i.e., $z \notin \Omega$. Then

$$\min_{z \in \Omega} \|x-z\| = \|x_0 - z\| > 0 . \tag{12.5}$$

Clearly, x_0 is a boundary point of Ω. Let $g = \|z-x_0\|^{-1}(z-x_0)$. As seen from the proof of Theorem 1.2, we have

$$(x-x_0, g) \leq 0 \qquad x \in \Omega ,$$

i.e., $g \in N(x_0)$. From (12.5) it follows that $(g,z-x_0) = \|z-x_0\| > 0$, and this contradicts the assumption that $z \in P$. Hence, $z \in \Omega$, i.e., $P \subset \Omega$.

From this and (12.4), we get (12.3). ∎

In the definition of the set P, all vectors of the set $N(x_0)$ enter (12.2). Let us establish the conditions under which it is sufficient to take only one nonzero vector from $N(x_0)$ for each x_0. Let

$$P_1 = \{x \in E_n \mid (A(x_0), x-x_0) \leq 0 \quad x_0 \in \Omega_{fr}\} , \tag{12.6}$$

where Ω_{fr} is the set of boundary points of Ω and $A(x_0) \in N(x_0)$, $\|A(x_0)\| = 1$ (i.e., it is sufficient to arbitrarily choose one and only one unit vector from $N(x_0)$).

LEMMA 12.2. If $\text{int}\,\Omega \neq 0$, then

$$\Omega = P_1 . \tag{12.7}$$

Proof. The inclusion

$$\Omega \subset P_1 \tag{12.8}$$

is obvious. Now let $z \in P_1$. It is necessary to prove that $z \in \Omega$. Assume that $z \notin \Omega$. Take $\bar{x} \in \operatorname{int} \Omega$. Then there exists an $r > 0$ such that $S_r(\bar{x}) \subset \Omega$. On the segment $[\bar{x}, z]$, we find a boundary point $x_1 \in \Omega_{fr}$ (this point exists and is unique). Then $(A(x_1), x - x_1) \le 0$ $\forall x \in \Omega$. In particular, $(A(x_1), x - x_1) \le 0$ $\forall x \in S_r(\bar{x})$.

Then,

$$(A(x_1), \bar{x} + v - x_1) \le 0 \qquad \forall v \in S_r(0) .$$

Hence,

$$(A(x_1), \bar{x} - x_1) \le (-v, A(x_1)) \qquad \forall v \in S_r(0) . \tag{12.9}$$

Let us take a $v = rA(x_1)$. From (12.9) we have $(A(x_1), \bar{x} - x_1) \le -r < 0$. On the other hand, $z - x_1 = -\alpha(\bar{x} - x_1)$, where $\alpha > 0$. Therefore

$$(A(x_1), z - x_1) = -\alpha(A(x_1), \bar{x} - x_1) \ge \alpha r > 0 . \tag{12.10}$$

Inequality (12.10) contradicts the assumption that $z \in P_1$. It follows from this contradiction that $z \in \Omega$, i.e., $P_1 \subset \Omega$. This, together with (12.8), completes the proof of the lemma. ∎

<u>REMARK 2</u>. The assumption of the existence of interior points in Lemma 12.2 (this is Slater's condition) is essential, that is evident from the following example.

Let $x = (x^{(1)}, x^{(2)}, x^{(3)}) \in E_3$. We take a polygon Ω on the plane $Q = \{x \in E_3 \mid x^{(3)} = 0\}$. The set Ω is convex but it has no interior points in the space E_3. In this case, $e_3 = (0,0,1) \in N(x_0)$ for all $x_0 \in \Omega$.

Hence, we can take $A(x_0) = e_3$ for all $x_0 \in \Omega_{fr}$ and then we obtain

$$P_1 = \{x \in E_3 \mid (e_3, x - x_0) \le 0\} = \{x \in E_3 \mid x^{(3)} \le 0\} .$$

However, it is clear that $P_1 \neq \Omega$.

From Lemma 12.1 and 12.2 we have:

COROLLARY. The following representation holds:

$$\Omega = \{x \in E_n \mid h(x) \le 0\} \quad ,$$

where

$$h(x) = \begin{cases} h_1(x) & \text{if} \quad \operatorname{int} \Omega = \emptyset \ , \\ h_2(x) & \text{if} \quad \operatorname{int} \Omega = \emptyset \ , \end{cases}$$

$$h_1(x) = \sup_{x_0 \in \Omega_{fr}} \ \sup_{\substack{A(x_0) \in N(x_0) \\ \|A(x_0)\| = 1}} (A(x_0), x - x_0) \ ,$$

$$h_2(x) = \sup_{x_0 \in \Omega_{fr}} (A(x_0), x - x_0) \ .$$

In the definition of $h_2(x)$, it is enough to arbitrarily choose one and only one vector $A(x_0)$ from $N(x_0)$, where $\|A(x_0)\| = 1$.

The function $h(x)$ is defined on E_n. Clearly, $h(x)$ is a finite convex function on Ω, and if Ω is a bounded set, then $h(x)$ is a finite function on E_n.

2. Let

$$\Omega = \{x \in E_n \mid h(x) \le 0\} \quad , \tag{12.11}$$

where $h(x)$ is some convex function. Clearly, the set Ω given by (12.11) is always convex (if it is nonempty), and if $h(x)$ is a finite function, then the set Ω is also closed.

Let us consider the set

$$\Omega = \{x \in E_n \mid h_1(x,y) \le 0 \quad \forall y \in \omega\} \ , \tag{12.12}$$

where $\omega \subset E_p$ is closed bounded and the function $h_1(x,y)$ is convex in x on E_n for any $y \in \omega$.

The set Ω defined by (12.12) is also convex and, moreover, it can be represented in the form of (12.11) if we let $h(x) = \max_{y \in \omega} h_1(x,y)$. The function $h(x)$ is convex.

Thus, let Ω be defined by (12.11), where $h(x)$ is a finite convex function on E_n. Let us assume that *Slater's condition* is satisfied: i.e., there exists an $\bar{x} \in E_n$ such that

$$h(\bar{x}) < 0 . \tag{12.13}$$

Take an $x_0 \in \Omega$ and let

$$\gamma(x_0) = \{v \in E_n \mid v = \lambda(x-x_0),\ \lambda > 0,\ x \in \Omega\} , \tag{12.14}$$

$$\Gamma(x_0) = \bar{\gamma}(x_0) .$$

The set $\gamma(x_0)$ can also be represented in the form

$$\gamma(x_0) = \{v \in E_n \mid \exists \alpha_0 = \alpha_0(v) > 0 :\ h(x_0+\alpha v) \le 0 \quad \forall \alpha \in [0,\alpha_0]\} , \tag{12.15}$$

which is equivalent to (12.14).

The sets $\Gamma(x_0)$ and $\Gamma^+(x_0)$ were examined in Section 3.

Let

$$B(x_0) = \begin{cases} E_n & \text{if } h(x_0) < 0 , \\ \{v \in E_n \mid (w,v) \le 0 \quad \forall w \in \partial h(x_0)\} & \text{if } h(x_0) = 0 , \end{cases}$$

where $\partial h(x_0)$ is the subdifferential of the function $h(x)$ at the point x_0, i.e.,

$$\partial h(x_0) = \{w \in E_n \mid h(x)-h(x_0) \ge (v,x-x_0) \quad \forall x \in E_n\} .$$

It is clear that if $h(x_0) < 0$, then $x_0 \in \operatorname{int} \Omega$. It follows from (12.13) that if $x_0 \in \Omega$ but $x_0 \notin \operatorname{int} \Omega$ (i.e., $h(x_0) = 0$), then

$$0 \notin h(x_0) , \tag{12.16}$$

because x_0 is not a minimum point of $h(x)$ on E_n.

<u>LEMMA 12.3</u>. If Slater's condition (12.13) is satisfied, then

$$\Gamma(x_0) = B(x_0) . \tag{12.17}$$

<u>Proof</u>. Relation (12.17) is obvious if $h(x_0) < 0$. Now let

$$h(x_0) = 0 . \tag{12.18}$$

We shall first show that $\Gamma(x_0) \subset B(x_0)$. Let $v \in \Gamma(x_0)$. Then there exists a sequence $\{v_k\}$ such that

$$v = \lim_{k\to\infty} v_k , \qquad v_k \in \gamma(x_0) .$$

Here we fix an arbitrary k. Then (see (12.15)) there exists an $\alpha_0(v_k) > 0$ such that $h(x_0+\alpha v_k) \le 0$ $\forall v \in (0,\alpha_0(v_k))$. From this and the definition of $\partial h(x_0)$, the inequality

$$(w,v_k) \le \alpha^{-1}[h(x_0+\alpha v_k) - h(x_0)] \le 0$$

follows for all $w \in \partial h(x_0)$ because $h(x_0) = 0$ by virtue of (12.18).

Fixing an arbitrary $w \in \partial h(x_0)$ and taking the limit as $k \to \infty$ yield

$$(w,v) \le 0 \qquad \forall w \in \partial h(x_0) ,$$

i.e., $\Gamma(x_0) \subset B(x_0)$.

Next we prove the converse inclusion. Let $v \in B(x_0)$. We assume first that

$$\max_{w\in\partial h(x_0)} (w,v) = -a < 0 . \tag{12.19}$$

Then

$$h(x_0+\alpha v) = h(x_0) + \alpha \frac{\partial h(x_0)}{\partial v} + o(\alpha)$$

$$= \alpha \max_{w\in\partial h(x_0)} (w,v) + o(\alpha) = -\alpha a + o(\alpha) .$$

Clearly, $h(x_0+\alpha v) < 0$ for sufficiently small $\alpha > 0$, i.e., $x_0 + \alpha v \in \Omega$. Thus, $v \in \gamma(x_0) \subset \Gamma(x_0)$.

Now we assume that $\max_{w\in\partial h(x_0)} (w,v) = 0$. Then, take the vector $v_\varepsilon = v + \varepsilon(\bar{x}-x_0)$, where $\varepsilon > 0$ and $\bar{x}$ is a point satisfying Slater's condition (12.13). From the definition of $\partial h(x_0)$ we have

$$h(\bar{x}) - h(x_0) \ge (w, \bar{x}-x_0) \qquad \forall w \in \partial h(x_0) ,$$

i.e., $(w,\bar{x}-x_0) \le h(\bar{x}) < 0$. Hence, since

$$(w, v_\varepsilon) = (w,v) + \varepsilon(w, \bar{x}-x_0) ,$$

we obtain

$$\max_{w\in\partial h(x_0)} (w,v_\varepsilon) \le \max_{w\in\partial h(x_0)} (w,v) + \varepsilon \max_{w\in\partial h(x_0)} (w,\bar{x}-x_0)$$

$$\le \varepsilon h(\bar{x}) < 0 .$$

Thus, for v_ε, (12.19) is satisfied and in this case, as shown above, $v_\varepsilon \in \gamma(x_0)$. Clearly, we have $v_\varepsilon \to v$, $v_\varepsilon \in \gamma(x_0)$ as $\varepsilon \to 0$. It means that $v \in \Gamma(x_0)$. The lemma has been proved. ■

Let $x_0 \in \Omega$ again. Define

$$K(x_0) = \begin{cases} \{0\} & \text{if } h(x_0) < 0 , \\ \{v = \lambda w \mid \lambda \le 0,\ w \in \partial h(x_0)\} & \text{if } h(x_0) = 0 . \end{cases}$$

LEMMA 12.4. If Slater's condition (12.13) is satisfied, then

$$\Gamma^+(x_0) = K(x_0) . \tag{12.20}$$

Proof. This proposition is obvious if $h(x_0) < 0$. Let us consider the case:

$$h(x_0) = 0 . \tag{12.21}$$

Then, by Lemma 12.3,

$$\Gamma(x_0) = \{v \in E_n \mid (w,v) \le 0 \ \ \forall w \in \partial h(x_0)\} .$$

According to (12.15), for any $v \in \gamma(x_0)$ there exists an $\alpha_0(v) > 0$ such that

$$h(x_0+\alpha v) \le 0 \quad \forall \alpha \in [0,\alpha_0(v)] . \tag{12.22}$$

For any $w \in \partial h(x_0)$, by the definition of the subdifferential, we have

$$h(x_0+\alpha v) - h(x_0) \ge \alpha(w,v) .$$

Considering (12.22) and (12.21) yields $(w,v) < 0$. Hence,

$$(\lambda w,v) \ge 0 \quad \forall \lambda \le 0, \quad \forall v \in \gamma(x_0) .$$

Since $\Gamma(x_0) = \overline{\gamma(x_0)}$, we have

$$(\lambda w, v) \geq 0 \qquad \forall v \in \Gamma(x_0) ,$$

and this implies that $\lambda w \in \Gamma^+(x_0)$ (see the definition of the conjugate cone in Section 3). Thus

$$K(x_0) \subset \Gamma^+(x_0) . \tag{12.23}$$

Now we show the converse inclusion. Let $v \in \Gamma^+(x_0)$, i.e.,

$$(v,g) \geq 0 \qquad \forall\, g \in \Gamma(x_0) . \tag{12.24}$$

Consider the ray $L = \{w = -\alpha v \mid \alpha \geq 0\}$. If $L \cap \partial h(x_0) \neq \emptyset$, there exists an α_0 such that $w = -\alpha_0 v \in \partial h(x_0)$. Clearly, $\alpha_0 > 0$ because if $\alpha_0 = 0$, we get $0 \in \partial h(x_0)$, and this contradicts (12.16).

Therefore $v = -\alpha_0^{-1} w$ and this implies that $v \in K(x_0)$ since $\lambda = -\alpha_0^{-1} < 0$. Let us show that the relation

$$L \cap \partial h(x_0) \neq \emptyset \tag{12.25}$$

is always valid. Assume that $L \cap \partial h(x_0) = \emptyset$. As in the proof of Lemma 6.2, there exists a z, $||z|| = 1$, such that

$$(z,v) = 0 , \qquad (z,g) = -a < 0 \qquad \forall\, g \in \partial h(x_0) . \tag{12.26}$$

By (12.26) we can find z' such that

$$(z',g) \leq -0.5a \qquad \forall\, g \in \partial h(x_0) , \tag{12.27}$$

$$(z',v) < 0 \tag{12.28}$$

(see Figure 15).

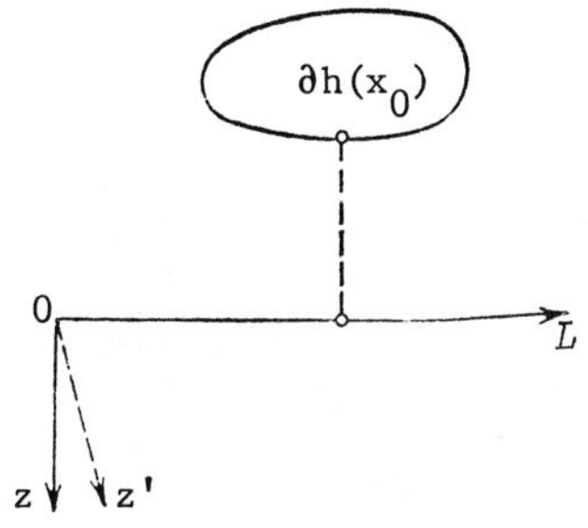

Figure 15

From (12.27) and Lemma 12.3 we conclude that $z' \in \Gamma$. Then, however, (12.28) contradicts (12.24). Thus, (12.25) is always valid and, therefore, as we have already established, $v \in K(x_0)$, i.e., $\Gamma^+(x_0) \subset K(x_0)$. From this and

(12.23) we obtain (12.20). The lemma has been proved. ∎

3. Let us consider the case where the set Ω is

$$\Omega = \{x \in E_n \mid h(x) \le 0\} , \tag{12.29}$$

where

$$h(x) = \max_{i \in I} h_i(x) \qquad I = 1:N , \tag{12.30}$$

and the functions $h_i(x)$ are convex and continuously differentiable on E_n.

Assume that Slater's condition (12.13) is satisfied; i.e., there exists an $\bar{x} \in E_n$ such that

$$h_i(\bar{x}) < 0 \quad \forall i \in I . \tag{12.31}$$

It is obvious that $\bar{x} \in \operatorname{int} \Omega$ and, hence, $\operatorname{int} \Omega \neq \emptyset$.

PROBLEM 12.1. Prove that Slater's condition is equivalent to the condition: there exist points $x_i \in \Omega$, $i \in I$, such that

$$h_i(x_i) < 0 \qquad \forall i \in I .$$

Let $x \in \Omega$ and define $Q(x) = \{i \in I \mid h_i(x) = 0\}$. It is clear that $h_i(x) < 0$ $\forall i \in Q(x)$. We note that if $Q(x) = \emptyset$, then $x \in \operatorname{int} \Omega$.

However, the converse does not hold since $Q(x) = \emptyset$ no longer follows from the fact that $x \in \operatorname{int} \Omega$.

From Lemma 5.4 and 12.3 we have the following:

LEMMA 12.5. If $x_0 \in \Omega$, the set Ω is given by relations (12.29), (12.30) and Slater's condition (12.31) is satisfied, then

$$\Gamma(x_0) = B(x_0) , \tag{12.32}$$

where

$$B(x_0) = \begin{cases} E_n & \text{if } Q(x_0) = \emptyset , \\ \{v \in E_n \mid (v, h_i'(x_0)) \le 0 \ \forall i \in Q(x_0)\} & \text{if } Q(x_0) \neq \emptyset . \end{cases} \tag{12.33}$$

Lemmas 5.4 and 12.4 imply the following:

LEMMA 12.6. Under the conditions of Lemma 12.5, we have the relation

$$\Gamma^{+}(x_0) = A(x_0) , \tag{12.34}$$

where

$$A(x_0) = \begin{cases} \{0\} & \text{if } Q(x_0) = \emptyset \\ \{v = -\sum_{i \in Q(x_0)} \lambda_i h_i'(x_0) \mid \lambda_i \geq 0 \ \forall i \in Q(x_0)\} & \text{if } Q(x_0) \neq \emptyset . \end{cases} \tag{12.35}$$

13. NORMAL CONES. CONICAL MAPPINGS

1. Let $\Omega \subset E_n$ be a closed convex set. In Section 12, we introduced the normal cone for $x_0 \in \Omega$:

$$N(x_0) = \{v \in E_n \mid (v, x - x_0) \leq 0 \quad \forall x \in \Omega\} . \tag{13.1}$$

Let x_0 be an arbitrary point of E_n (x_0 does not necessarily belong to Ω).

LEMMA 13.1. For any $x_0 \in E_n$, the set $N(x_0)$ is a nonempty, convex and closed cone. If $x_0 \notin \text{int}\,\Omega$, then

$$N(x_0) \setminus \{0\} \neq \emptyset . \tag{13.2}$$

If $x_0 \in \text{int}\,\Omega$, then $N(x_0) = \{0\}$.

Proof. All properties except (13.2) are obvious. Relation (13.2) follows from Theorem 1.2 (if $x_0 \notin \Omega$) or from its Corollary 1 (if x_0 is a boundary point of Ω). ∎

THEOREM 13.1. If x_0 is not a boundary point of the set Ω, then the point-to-set mapping $N(x)$ is continuous in the Kakutani sense at the point x_0.

Proof. If $x \in \text{int}\,\Omega$, then, as noted in Subsection 12.1, we have $N(x) = \{0\}$ and, hence, the continuity at the point $x_0 \in \text{int}\,\Omega$ is obvious. Now let $x_0 \notin \text{int}\,\Omega$. It is necessary to establish the upper and lower semicontinuity of $N(x)$ at the point x_0.

For any sequences $\{x_k\}$ and $\{v_k\}$ such that $x_k \to x_0$, $v_k \to v_0$, $v_k \in N(x_k)$, it follows from (13.1) that $v_0 \in N(x_0)$, i.e., the mapping $N(x)$ is upper semicontinuous at the point x_0.

Next let us take an arbitrary $v_0 \in N(x_0)$ and any sequence $\{x_k\}$ such that $x_k \to x_0$. Find $\min_{z \in \Omega} \|z-x_k\| = \|z_k - x_k\|$. We assume without loss of generality that $x_k \notin \Omega$ and

$$\|z_k - x_k\| \geq a > 0 , \tag{13.3}$$

where $2a = \min_{z \in \Omega} \|z-x_0\|$. As follows from Remark 3 in Section 1,

$$(x-x_k, z_k-x_k) \geq \|z_k - x_k\|^2 \quad \forall x \in \Omega . \tag{13.4}$$

Let

$$\delta_k = 2\|z_k - x_k\|^{-1}|(v_0,x_0-x_k)| ,$$

$$v_k = v_0 + \delta_k\|x_k - z_k\|^{-1}(x_k-z_k) . \tag{13.5}$$

Clearly, $\delta_k \to 0$. Hence, we have for $x \in \Omega$

$$(x-x_k,v_k) = (x-x_k, v_0 + \delta_k\|x_k-z_k\|^{-1}(x_k - z_k)) \tag{13.6}$$

$$= (x-x_0,v_0) + (x_0-x_k,v_0) + \delta_k(x-x_k, \|x_k-z_k\|^{-1}(x_k-z_k)) .$$

Since $v_0 \in N(x_0)$, then

$$(x-x_0, v_0) \leq 0 \quad \forall x \in \Omega . \tag{13.7}$$

It follows from (13.4) - (13.7) that

$$(x-x_k, v_k) \leq -|(v_0,x_0-x_k)| \leq 0 \quad \forall x \in \Omega , \tag{13.8}$$

i.e., $v_k \in N(x_k)$. Since $\delta_k \to 0$, then $v_k \to v_0$, and this implies that the mapping $N(x)$ is lower semicontinuous as well as upper semicontinuous at the point $x_0 \notin \Omega$. The theorem has been proved. ∎

2. Let $f(x)$ be a convex function on E_n. In the space E_{n+1} we consider the set

$$G = \text{epi } f \equiv \{\bar{x} = [\beta,x] \mid \beta \in E_1, \beta \geq f(x), x \in E_n\}$$

(see Subsection 4.1). This is a closed convex set. Take a point $\bar{x}_0 = [f(x_0)-\varepsilon_0, x_0]$, where $\varepsilon_0 > 0$. Clearly, $\bar{x}_0 \notin G$. We proved in Theorem 13.1 that the mapping $N(\bar{x}_0) = \{\bar{v} \in E_{n+1} \mid (\bar{v}, \bar{x}-\bar{x}_0) \le 0 \ \forall \bar{x} \in G\}$ is continuous in the Kakutani sense (or K-continuous).

Let

$$Q(\bar{x}_0) = \{v \in E_n \mid [-1,v] \in N(\bar{x}_0)\} .$$

<u>LEMMA 13.2</u>. We have the relation

$$Q(x_0) = \partial_{\varepsilon_0} f(x_0) . \qquad (13.9)$$

<u>Proof</u>. Take an arbitrary $v_0 \in \partial_{\varepsilon_0} f(x_0)$. Then

$$f(x) - f(x_0) \ge (v_0, x-x_0) - \varepsilon_0 \qquad \forall x \in E_n .$$

From this inequality we obtain for $\beta \ge f(x)$

$$(\beta - [f(x_0)-\varepsilon_0])(-1) + (v_0, x-x_0) \le 0 \qquad \forall x \in E_n .$$

Moreover, letting $\bar{v}_0 = [-1, v_0]$, $\bar{x} = [\beta, x]$ and $\bar{x}_0 = [f(x_0)-\varepsilon_0, x_0]$, we get

$$(\bar{v}_0, \bar{x}-\bar{x}_0) \le 0 \qquad \forall \bar{x} \in G , \quad (13.10)$$

because $\bar{x} = [\beta, x] \in G$ for $\beta \ge f(x)$. It follows from (13.10) that $\bar{v}_0 \in N(\bar{x}_0)$ and, hence, $v_0 \in Q(x_0)$. Therefore

$$\partial_{\varepsilon_0} f(x_0) \subset Q(\bar{x}_0) . \qquad (13.11)$$

Now, let $v_0 \in Q(\bar{x}_0)$, then $\bar{v}_0 = [-1, v_0] \in N(\bar{x}_0)$, i.e.,

$$(\bar{v}_0, \bar{x}-\bar{x}_0) \le 0 \qquad \forall \bar{x} \in G . \quad (13.12)$$

Since

$$\bar{x} = [\beta, x] , \qquad \beta \ge f(x) , \qquad \bar{x}_0 = [f(x_0)-\varepsilon_0, x_0] ,$$

then from (13.12)

$$[(-1)(\beta - f(x_0)+\varepsilon_0) + (v_0, x-x_0)] \le 0 \qquad \forall x \in E_n .$$

For $\beta = f(x)$ we get

$$f(x) - f(x_0) \ge (v_0, x-x_0) - \varepsilon_0 \qquad \forall x \in E_n ,$$

and this implies that $v_0 \in \partial_{\varepsilon_0} f(x_0)$. Thus, $Q(x_0) \subset \partial_{\varepsilon_0} f(x_0)$.

From this and (13.11) we obtain (13.9). ■

Lemma 13.2 and Theorem 13.1 yield the following:

COROLLARY. The mapping $\partial_\varepsilon f(x)$ is K-continuous with respect to ε and x on $(0,\infty) \times E_n$.

Proof. The upper semicontinuity of the mapping $\partial_\varepsilon f(x)$ is obvious. Hence we show the lower semicontinuity. Let $x_k \to x_0$, $v_0 \in \partial_\varepsilon f(x_0) = Q(\bar{x}_0)$, $\varepsilon_k \to \varepsilon$ and $\bar{x}_k = [f(x_k)-\varepsilon_k, x_k]$. Since $\bar{x}_k \to \bar{x}_0 \notin G$ and $\bar{v}_0 = [-1, v_0] \in N(\bar{x}_0)$, then, by virtue of the lower semicontinuity of $N(\bar{x})$, there exists a sequence $\{\bar{v}_k\}$ such that $\bar{v}_k \in N(\bar{x}_k)$, $\bar{v}_k \to \bar{v}_0$.

Since $\bar{v}_k \equiv [\beta_k, v_k] \in N(\bar{x}_k)$, then we also have

$$\bar{v}_k' = |\beta_k|^{-1} \bar{v}_k = [-1, |\beta_k|^{-1} v_k] \in N(\bar{x}_k),$$

moreover, $\bar{v}_k' \to \bar{v}_0$. Therefore, $v_k' \equiv |\beta_k|^{-1} v_k \to v_0$ and $v_k' \in Q(x_k)$ $(= \partial_\varepsilon f(x_k))$, which proves the theorem. ■

The mapping $\partial_\varepsilon f(x)$ is bounded and closed. Hence it is H-continuous, due to Theorem 2.1, i.e., continuous in the Hausdorff sense. The H-continuity of the mapping $\partial_\varepsilon f(x)$ was already established in Section 8 via a different method.

REMARK. Let a function $f(x)$ be defined only on a closed convex set Ω. Then, letting

$$G = \{\bar{x} = [\beta, x] \in E_{n+1} \mid x \in \Omega, \ \beta \geq f(x)\}$$

and using the same arguments as in proving Lemma 13.2, we conclude that the relation

$$Q(\bar{x}_0) = \partial^{\Omega}_{\varepsilon_0} f(x_0) \tag{13.13}$$

is valid for $\bar{x}_0 = [f(x_0)-\varepsilon_0, x_0]$, where $\varepsilon_0 > 0$ and $x_0 \in \Omega$. The K-continuity of the conditional ε-subdifferential follows from

(13.13) and Theorem 13.1. This fact was also established in Section 11 via a different method.

3. Let $K: G \to \Pi(E_n)$ be a point-to-set mapping given on $G \subset E_n$ and, moreover, let the set $K(x)$ be a closed convex cone with the vertex at the origin for each $x \in \Omega$. Such a mapping is called *conical.*

Denote by $K^+(x)$ the *conjugate cone:*

$$K^+(x) = \{w \in E_n \mid (v,w) \geq 0 \ \forall v \in K(x)\} .$$

The properties of conjugate cones were studied in Section 3. In particular, it was established that $K^+(x)$ is a closed convex cone and, in addition, that

$$(K^+(x))^+ \equiv K^{++}(x) = K(x) . \qquad (13.14)$$

Clearly, the mapping $K^+(x)$ is also conical.

LEMMA 13.3. If a mapping $K(x)$ is upper semicontinuous at a point $x_0 \in G$, then the mapping $K^+(x)$ is lower semicontinuous at that same point.

Proof. Let $w_0 \in K^+(x_0)$ and let a sequence $\{x_k\}$ be such that $x_k \in G$, $x_k \to x_0$. We have to prove that there exist $z_k \in K^+(x_k)$ such that $z_k \to w_0$. Assume that this is not the case. Then, without loss of generality, we can assume that there exists an $a > 0$ such that

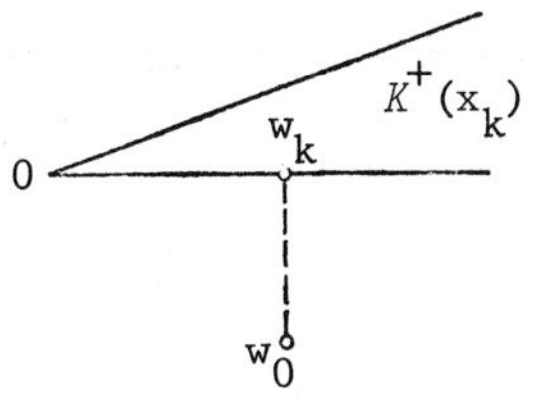

Figure 16

$$\min_{w \in K^+(x_k)} \|w - w_0\| \equiv \|w_k - w_0\| \equiv a_k \geq a > 0 \qquad \forall k .$$

We note that $\|w_k\|$ are bounded because $\|w_k\| \leq \|w_0\|$ (see Figure 16). Due to the necessary and sufficient condition for a minimum,

$$(w_k-w_0,\ w-w_0) \ \geq\ a_k^2 \ \geq\ a^2 \ >\ 0 \qquad \forall w \in K^+(x_k) \quad (13.15)$$

(see Remark 3 of Section 1). Define $v_k = w_k - w_0$. As in the proof of Theorem 3.1 (see (3.7)), we have $(v_k, w_k) = 0$. Hence, from (13.15)

$$(v_k,w) = (v_k,\ w-w_0) + (v_k,\ w_0-w_k) + (v_k,w_k) \ \geq\ a^2 - a^2 = 0 \qquad \forall w \in K^+(x_k) ,$$

i.e., by virtue of (13.14)

$$v_k \in (K^+(x_k))^+ \equiv K^{++}(x_k) = K(x_k) .$$

Without loss of generality, we can assume that $v_k \to v_0$. Since $K(x)$ is upper semicontinuous at the point x_0, then $v_0 \in K(x_0)$. On the other hand, for $w = 0$, we obtain from (13.15) $(v_k,w_0) \leq -a_k^2 \leq -a^2 < 0$. Passing to the limit we get

$$(v_0,w_0) \ \leq\ -a^2 . \qquad (13.16)$$

Consequently, (13.16) and $v_0 \in K(x_0)$ contradict the assumption that $w_0 \in K^+(x_0)$.

This contradiction proves the lemma. ■

LEMMA 13.4. If a point-to-set mapping $K(x)$ is lower semicontinuous at the point $x_0 \in G$, then the point-to-set mapping $K^+(x)$ is upper semicontinuous at the point x_0.

Proof. Assume that this is not the case. Then, there exist sequences $\{x_k\}$ and $\{w_k\}$ such that

$$x_k \in G , \qquad x_k \to x_0 , \qquad w_k \in K^+(x_k) , \qquad w_k \to w_0 ,$$
$$\text{but} \qquad w_0 \notin K^+(x_0) .$$

Find

$$\min_{w \in K^+(x_0)} ||w-w_0|| = ||\bar{w}-w_0|| \equiv ||\bar{v}|| .$$

Then

$$||\bar{v}|| = a > 0 , \qquad (\bar{v},\ w-w_0) \geq a^2 > 0 \qquad \forall w \in K^+(x_0) ,$$

i.e., $\bar{v} \in (K^+(x_0))^+ = K(x_0)$. In addition, we have

$$(\bar{v}, w_0) = -\|\bar{v}\|^2 = -a^2 < 0 . \tag{13.17}$$

Since the point-to-set mapping $K(x)$ is lower semicontinuous at the point x_0, then, for the point $\bar{v}$ and the sequence $\{x_k\}$, there exist $v_k \in K(x_k)$ such that $v_k \to \bar{v}$. Then

$$(v_k, w) \geq 0 \qquad \forall w \in K^+(x_k) .$$

In particular, we have for $w_k \in K^+(x_k)$

$$(v_k, w_k) \geq 0 .$$

Here, taking the limit yields $(\bar{v}, w_0) \geq 0$, which contradicts (13.17). The lemma has been proved. ■

Lemmas 13.3 and 13.4 imply the following:

COROLLARY. If $K(x)$ is a conical mapping continuous in the Kakutani sense at the point x_0, then the point-to-set mapping $K^+(x)$ is also continuous in the Kakutani sense at the point x_0.

14. DIRECTIONAL DIFFERENTIABILITY OF A SUPREMUM FUNCTION

In Section 5, we already considered an example of a maximum function constructed from a finite number of convex functions and established the formula for the directional derivatives.

Below, in Subsection 1, we shall study in more detail the directional differentiability of a supremum function constructed from differentiable but not necessarily convex functions, and in Subsection 2 consider a supremum function constructed from convex functions.

1. Let

$$f(x) = \sup_{y \in G} \phi(x, y) , \tag{14.1}$$

where $x \in E_n$, $y \in E_p$ and $G \subset E_p$ is an arbitrary set. We assume that the one-parameter family of the functions $\phi(x,y)$ and $\phi'_x(x,y)$, where $y \in G$ is a parameter, is equicontinuous in x at the point x_0, i.e., for $\varepsilon > 0$ there exists $\delta > 0$ such that

$$|\phi(x,y) - \phi(x_0,y)| \leq \varepsilon , \qquad \|\phi'_x(x,y) - \phi'_x(x_0,y)\| \leq \varepsilon$$
$$\forall x: \|x-x_0\| \leq \delta , \quad \forall y \in G .$$

Moreover, we assume that

$$|\phi(x_0,y)| \leq K_0 < \infty , \qquad \|\phi'_x(x_0,y)\| \leq K < \infty \quad \forall y \in G . \tag{14.2}$$

It is easy to show that $f(x)$ is a continuous function at the point x_0. Take $\varepsilon > 0$ and introduce the sets

$$R_\varepsilon(x) = \{y \in G \mid f(x)-\phi(x,y) \leq \varepsilon\} ,$$

$$H(x) = \{v \in E_n \mid \exists \varepsilon_i \to +0,\ y_i \in R_{\varepsilon_i}(x):\ \phi'_x(x,y_i) \to v\} .$$

We note that the set $R_\varepsilon(x)$ is not necessarily closed. The set $H(x_0)$ is nonempty, closed and bounded. The point-to-set mapping $H(x_0)$ is upper semicontinuous at the point x_0, i.e., from $v_i \in H(x_i)$, $x_i \to x_0$, $v_i \to v$, it follows that $v \in H(x_0)$.

<u>THEOREM 14.1</u>. The function $f(x)$ given by relation (14.1) is differentiable in any direction $g \in E_n$ at the point x_0 and, moreover,

$$\frac{\partial f(x_0)}{\partial g} = \max_{v \in H(x_0)} (v,g) . \tag{14.3}$$

<u>Proof</u>. Define

$$h(\alpha) = \alpha^{-1}[f(x_0+\alpha g) - f(x_0)] .$$

Let a sequence $\{\alpha_i\}$ be such that $\alpha_i \to +0$ and

$$h(\alpha_i) \to \varliminf_{\alpha \to +0} h(\alpha) . \tag{14.4}$$

For any $v \in H(x_0)$, there exist sequences $\{\varepsilon_i\}$ and $\{y_i\}$ such that

$$\varepsilon_i \to +0 , \qquad y_i \in R_{\varepsilon_i}(x_0) , \qquad \phi'_x(x_0,y_i) \to v . \tag{14.5}$$

We can assume without loss of generality that

$$\alpha_i^{-1}\varepsilon_i \xrightarrow[i\to\infty]{} 0 . \tag{14.6}$$

Then

$$f(x_0+\alpha_i g) \ge \phi(x_0+\alpha_i g,\ y_i) = \phi(x_0,y_i) + \alpha_i(\phi'_x(x_0+\theta_i g,\ y_i),\ g) ,$$

where $\theta_i \in (0,\alpha_i)$. Since the family of functions $\phi'_x(x,y)$ is equicontinuous at the point x_0, then we have

$$f(x_0+\alpha_i g) \ge \phi(x_0,y_i) + \alpha_i(\phi'_x(x_0,y_i),\ g) + o_i(\alpha_i) ,$$

where

$$\frac{o_i(\alpha_i)}{\alpha_i} \xrightarrow[i\to\infty]{} 0 . \tag{14.7}$$

Since $y_i \in R_{\varepsilon_i}(x_0)$, then

$$f(x_0+\alpha_i g) \ge f(x_0) - \varepsilon_i + \alpha_i(\phi'_x(x_0,y_i),\ g) + o_i(\alpha_i) .$$

Hence,

$$h(\alpha_i) \ge (\phi'_x(x_0,y_i),\ g) + \frac{o_i(\alpha_i) - \varepsilon_i}{\alpha_i} .$$

Considering (14.4) - (14.7), we obtain $\varliminf_{\alpha\to+0} h(\alpha) \ge (v,g)$, which holds for all $v \in H(x_0)$. Therefore,

$$\varliminf_{\alpha\to+0} h(\alpha) \ge \max_{v\in H(x_0)} (v,g) . \tag{14.8}$$

For any $\varepsilon > 0$ there exists an $\alpha_0 > 0$ such that

$$f(x_0+\alpha g) = \sup_{y\in R_\varepsilon(x_0)} \phi(x_0+\alpha g,\ y)$$

for all $\alpha \in [0,\alpha_0]$. Since

$$\phi(x_0+\alpha g,\ y) = \phi(x_0,y) + \alpha(\phi'_x(x_0,y),\ g) + o(\alpha,y) ,$$

where $\frac{o(\alpha,y)}{\alpha} \xrightarrow[\alpha\to+0]{} 0$ uniformly with respect to $y \in G$, then

$$f(x_0+\alpha g) \le \sup_{y\in R_\varepsilon(x_0)} \phi(x_0,y) + \alpha \sup_{y\in R_\varepsilon(x_0)} (\phi'_x(x_0,y),\ g) + o(\alpha) .$$

Recalling that $\sup_{y\in R_\varepsilon(x_0)} \phi(x_0,y) = f(x_0)$, we obtain

$$h(\alpha) \le \sup_{y\in R_\varepsilon(x_0)} (\phi'_x(x_0,y),\ g) + \frac{o(\alpha)}{\alpha} . \tag{14.9}$$

Hence, taking the upper limit as $\alpha \to +0$ yields

$$\overline{\lim_{\alpha\to+0}}\, h(\alpha) \le \sup_{y\in R_\varepsilon(x_0)} (\phi'_x(x_0,y),\ g) . \tag{14.10}$$

Inequality (14.10) is valid for all $\varepsilon > 0$.

Let $\varepsilon_k \to +0$ and let $y_k \in R_\varepsilon(x_0)$ be such as

$$\sup_{y\in R_{\varepsilon_k}(x_0)} (\phi'_x(x_0,y),\ g) \le (\phi'_x(x_0,y_k),\ g) + \delta_k ,$$

where $\delta_k \xrightarrow[k\to\infty]{} +0$. Since the set $\{\phi'_x(x_0,y) \mid y \in G\}$ is bounded due to (14.2), then, without loss of generality, we can assume that $\phi'_x(x_0,y_k) \to \bar{v}$. Clearly, $\bar{v} \in H(x_0)$. We conclude from (14.10) that

$$\overline{\lim_{\alpha\to+0}}\, h(\alpha) \le (\bar{v},g) \le \max_{v\in H(x_0)} (v,g) . \tag{14.11}$$

Combining (14.8) with (14.11), we get

$$\max_{v\in H(x_0)} (v,g) \le \underline{\lim_{\alpha\to+0}}\, h(\alpha) \le \overline{\lim_{\alpha\to+0}}\, h(\alpha) \le \max_{v\in H(x)} (v,g) .$$

From this, the existence $\frac{\partial f(x_0)}{\partial g} \equiv \lim_{\alpha\to+0} h(\alpha)$ as well as formula (14.3) follow. The theorem has been proved. ■

<u>REMARK 1</u>. As is seen from the proof, it is sufficient that the functions ϕ and ϕ'_x be equicontinuous in x at the point x_0 for $y \in R_{\varepsilon_1}(x_0)$, where $\varepsilon_1 > 0$ is an arbitrary fixed number.

<u>REMARK 2</u>. Obviously, we can write (14.3) as

$$\frac{\partial f(x_0)}{\partial g} = \max_{v\in \operatorname{co} H(x_0)} (v,g) .$$

2. Let

$$f(x) = \sup_{y \in G} \phi(x,y) ,$$

where $x \in E_n$, $y \in E_p$, $G \subset E_p$ is an arbitrary set and the function $\psi_y(x) \equiv \phi(x,y)$ is convex in x for any fixed $y \in G$. Take $g \in E_n$, $x_0 \in E_n$.

Then

$$\phi(x_0+\alpha g,\ y) = \phi(x_0,y) + \alpha \frac{\partial \phi(x_0,y)}{\partial g} + o(\alpha,g,y) , \tag{14.12}$$

where

$$\frac{o(\alpha,g,y)}{\alpha} \xrightarrow[\alpha\to+0]{} 0 ,$$

$$\frac{\partial \phi(x_0,y)}{\partial g} = \max_{v \in \partial \phi(x_0,y)} (v,g) , \tag{14.13}$$

and here the set

$$\partial \phi(x_0,y) = \{v \in E_n \mid \phi(z,y) \geq \phi(x_0,y)+(v,z-x_0) \quad \forall z \in E_n\}$$

is the subdifferential of the function $\psi_y(x) \equiv \phi(x,y)$ at the point $x_0 \in E_n$ for a fixed $y \in G$.

Let us assume that the sets $\partial\phi(x_0,y)$ are uniformly bounded, i.e.,

$$||v|| \leq K < \infty \qquad \forall v \in \partial\phi(x_0,y) , \qquad \forall y \in G . \tag{14.14}$$

Moreover, assume that the following condition is satisfied: in (14.12), the convergence $\frac{o(\alpha,g,y)}{\alpha} \xrightarrow[\alpha\to+0]{} 0$ is uniform with respect to $y \in R_{\varepsilon_0}(x_0)$, where $\varepsilon_0 > 0$ is fixed.

It is easy to see that the function $f(x)$ is convex and hence differentiable in all directions (see Section 4): Take $\varepsilon > 0$, $\varepsilon \leq \varepsilon_0$ and introduce the sets

$$R_\varepsilon(x_0) = \{y \in G \mid f(x_0)-\phi(x_0,y) \leq \varepsilon\} ,$$

$$H(x_0) = \{v \in E_n \mid \exists \varepsilon_i \to +0,\ y_i \in R_{\varepsilon_i}(x_0),\ v_i \in \partial\phi(x_0,y_i) :\ v_i \to v\} .$$

Under assumptions mentioned above, the set $H(x_0)$ is nonempty, closed and bounded.

THEOREM 14.2. The function $f(x)$ is differentiable at the point x_0 in any direction g and, moreover,

$$\frac{\partial f(x_0)}{\partial g} = \max_{v \in H(x_0)} (v,g) . \tag{14.15}$$

Proof. As noted above, the directional differentiability of $f(x)$ follows from the convexity of the function $f(x)$. It remains only to find a formula for the directional derivative. Let

$$h(\alpha) = \alpha^{-1}[f(x_0+\alpha g) - f(x_0)] .$$

Then, for any $v \in H(x_0)$, there exist sequences $\{\varepsilon_i\}$, $\{y_i\}$ and $\{v_i\}$ such that $\varepsilon \to +0$, $y_i \in R_{\varepsilon_i}(x_0)$, $y_i \in \partial\phi(x_0,y_i)$, $v_i \to v$.

Take a sequence $\{\alpha_i\}$ such that $\alpha_i \to +0$. As in Subsection 1, we can assume that

$$\alpha_i^{-1}\varepsilon_i \xrightarrow[i\to\infty]{} 0 . \tag{14.16}$$

Then, by the definition of $\partial\phi(x_0,y_i)$,

$$f(x_0+\alpha_i g) \geq \phi(x_0+\alpha_i g, y_i) \geq \phi(x_0,y_i) + \alpha_i(v_i,g) . \tag{14.17}$$

Since $y_i \in R_{\varepsilon_i}(x_0)$, then $f(x_0+\alpha_i g) \geq f(x_0) - \varepsilon_i + \alpha_i(v_i,g)$.

Hence, $h(\alpha_i) \geq (v_i,g) - \alpha_i^{-1}\varepsilon_i$. Moreover, considering (14.16), we obtain the inequality $\frac{\partial f(x_0)}{\partial g} \geq (v,g)$. Since this inequality is valid for all $v \in H(x_0)$, we have

$$\frac{\partial f(x_0)}{\partial g} \geq \max_{v \in H(x_0)} (v,g) . \tag{14.18}$$

Next we prove the converse inequality. For any $\varepsilon > 0$ and sufficiently small $\alpha > 0$, the equality

$$f(x_0+\alpha g) = \sup_{y \in R_\varepsilon(x_0)} \phi(x_0+\alpha g, y)$$

is valid. Hence, from (14.12) we have

$$f(x_0+\alpha g) \le \sup_{y\in R_\varepsilon(x_0)} \phi(x_0,y) + \alpha \sup_{y\in R_\varepsilon(x_0)} \frac{\partial\phi(x_0,y)}{\partial g} + o(\alpha) .$$

Here, since $\sup_{y\in R_\varepsilon(x_0)} \phi(x_0,y) = f(x_0)$, then

$$h(\alpha) \le \sup_{y\in R_\varepsilon(x_0)} \frac{\partial\phi(x_0,y)}{\partial g} + \frac{o(\alpha)}{\alpha} .$$

Taking the limit as $\alpha \to +0$ yields

$$\frac{\partial f(x_0)}{\partial g} \le \sup_{y\in R_\varepsilon(x_0)} \frac{\partial\phi(x_0,y)}{\partial g} . \tag{14.19}$$

Let $\varepsilon_k \to +0$ and let $y_k \in R_{\varepsilon_k}(x_0)$ be such as

$$\sup_{y\in R_{\varepsilon_k}(x_0)} \frac{\partial\phi(x_0,y)}{\partial g} \le \frac{\partial\phi(x_0,y_k)}{\partial g} + \delta_k ,$$

where $\delta_k \xrightarrow[k\to\infty]{} +0$. Since

$$\frac{\partial\phi(x_0,y_k)}{\partial g} = \max_{v\in\partial\phi(x_0,y_k)} (v,g) = (v_k,g) ,$$

then, without loss of generality, we can assume by virtue of (14.14) that $v_k \to \bar{v}$. Clearly, $\bar{v} \in H(x_0)$. Consequently, we conclude that

$$\frac{\partial f(x_0)}{\partial g} \le (\bar{v},g) \le \max_{v\in H(x_0)} (v,g) . \tag{14.20}$$

Relations (14.18) and (14.20) imply (14.15). The theorem has been proved. ∎

Here, let us introduce the set $L(x_0) = \text{co}\, H(x_0)$.

<u>COROLLARY 1</u>. We have the equality

$$\partial f(x_0) = L(x_0).$$

<u>Proof</u>. It follows from (14.15) that

$$\frac{\partial f(x_0)}{\partial g} = \max_{v\in L(x_0)} (v,g) . \tag{14.21}$$

On the other hand, we have

$$\frac{\partial f(x_0)}{\partial g} = \max_{v \in \partial f(x_0)} (v,g) \quad , \qquad (14.22)$$

where $\partial f(x_0) = \{v \in E_n \mid f(x)-f(x_0) \geq (v,x-x_0) \ \forall x \in E_n\}$ is the subdifferential of the function $f(x)$ at the point x_0. Since (14.21) and (14.22) are valid for any $g \in E_n$ and the sets $L(x_0)$ and $\partial f(x_0)$ are convex and compact, then $\partial f(x_0) = L(x_0)$ (see Corollary of Lemma 2. ∎

<u>COROLLARY 2</u>. If $f(x) = \max_{y \in G} \phi(x,y)$, where G is a compact set, and the function $\phi(x,y)$ is convex in x for any fixed $y \in G$ and continuous in y, then

$$\partial f(x) = \text{co}\,\{\phi'_x(x,y) \mid y \in R(x)\} \quad ,$$

$$R(x) = \{y \in G \mid f(x) = \phi(x,y)\} \quad .$$

<u>EXAMPLE 1</u>. Let $x \in E_1$ and $f(x) = \max_{i \in 1:2} f_i(x)$, where $f_1(x) = (x-1)^2$, $f_2(x) = (x+1)^2$. Take the point $x_0 = 0$. Since $f(x)$ is a maximum function, then

$$\frac{\partial f(x_0)}{\partial g} = \max_{i \in R(x_0)} (f'_i(x_0),\ g)$$

(see Example 2 in Section 5), where

$R(x_0) = \{i \in 1:2 \mid f_i(x_0) = f(x_0)\} = 1:2.$

Hence,

$$\frac{\partial f(x_0)}{\partial g} = \max\,\{-2g, 2g\} \quad .$$

Here, $||g|| = 1$. In the space E_1, there exist only two such directions

$$g_1 = +1 \ , \qquad g_2 = -1 \ .$$

Therefore we get

$$\frac{\partial f(x_0)}{\partial g} = +2 \qquad (14.23)$$

for both $g = g_1$ and $g = g_2$.

On the other hand, since $f(x)$ is a convex function, it can be represented as

$$f(x) = \max_{y \in E_1} [f(y) + (v(y), x-y)] \tag{14.24}$$

(see (5.11)), where $v(y) \in \partial f(y)$ (for each y, one $v(y)$ is taken from $\partial f(y)$).

It follows from the definition of $f(x)$ that

$$f(x) = \begin{cases} f_1(x) & \text{if } x \le 0 , \\ f_2(x) & \text{if } x \ge 0 . \end{cases}$$

Hence,

$$\partial f(x) = \begin{cases} \{2(x-1)\} & \text{if } x < 0 , \\ \{2(x+1)\} & \text{if } x > 0 , \end{cases}$$

and $\partial f(0) = [-2,2]$ for $x = 0$.

In (14.24), let us take $v(0) = 0$. Since (14.24) implies that $f(x) = \max_{y \in E1} \{1-y^2\}$, therefore $R(0) = \{0\}$, hence $\max_{y \in R(0)} (v(y), g) = 0$. From this, it is clear that the formula

$$\frac{\partial f(x_0)}{\partial g} = \max_{y \in R(x_0)} (v(y), g)$$

is incorrect because it contradicts (14.23).

In (14.15) we obtain $H(x_0) = \{-2,2\}$.

15. DIFFERENTIABILITY OF A CONVEX FUNCTION

1. In this section we study the differentiability of a convex function given on E_n. It turns out that the set of points where the function has the differential is sufficiently massive.

We consider first the case where $n = 1$.

Let a function $f(x)$ be given on the line E_1 and let $x_0 \in E_1$. The limits

$$f'_+(x_0) = \lim_{\alpha \to +0} \alpha^{-1}[f(x_0+\alpha) - f(x_0)]$$

and

$$f'_-(x_0) = \lim_{\alpha \to -0} \alpha^{-1}[f(x_0+\alpha) - f(x_0)]$$

are called the *left* and *right derivatives* of the function $f(x)$ at the point x_0, respectively. On the line E_1, there exist only two unit directions.

One of them is given by the vector $g_1 = +1$ and the other is given by the vector $g_2 = -1$. It is obvious that

$$f'_+(x_0) = \lim_{\alpha \to +0} \alpha^{-1}[f(x_0+\alpha g_1) - f(x_0)] = \frac{\partial f(x_0)}{\partial g_1},$$

$$f'_-(x_0) = \lim_{\alpha \to +0} \alpha^{-1}[f(x_0) - f(x_0+\alpha g_2)] = \frac{\partial f(x_0)}{\partial g_2}.$$

If the function $f(x)$ is convex, then from Theorem 4.2, the directional derivatives $\frac{\partial f(x)}{\partial g_1}$ and $\frac{\partial f(x)}{\partial g_2}$ exist.

More precisely, the following lemma holds.

LEMMA 15.1. If $f(x)$ is a convex function given on E_1, then the left and right derivatives exist at each point $x_0 \in E_1$ and, moreover

$$f'_+(x_0) = \frac{\partial f(x_0)}{\partial g_1}, \qquad f'_-(x_0) = \frac{\partial f(x_0)}{\partial g_2},$$

where $g_1 = +1$, $g_2 = -1$.

Next, let us derive inequalities for the left as well as right derivatives of a convex function.

LEMMA 15.2. Let $f(x)$ be a convex function on E_1 and let $x_0, x \in E_1$, where $x_1 > x_0$. Then

$$f'_+(x_0) \le (x_1 - x_0)^{-1}(f(x_1) - f(x_0)) \le f'_-(x_1) .$$

Proof. Due to the Corollary of Theorem 4.2,

$$\frac{\partial f(x_0)}{\partial g_1} = \inf_{\alpha>0} \alpha^{-1}[f(x_0+\alpha) - f(x_0)] ,$$

$$\frac{\partial f(x_0)}{\partial g_2} = \inf_{\alpha>0} \alpha^{-1}[f(x_1-\alpha) - f(x_1)] ,$$

where $g_1 = +1$, $g_2 = -1$. Letting $\alpha = x_1 - x_0$, we obtain

$$f'_+(x_0) = \frac{\partial f(x_0)}{\partial g_1} \le \alpha^{-1}[f(x_1) - f(x_0)] \le -\frac{\partial f(x_1)}{\partial g_2} = f'_-(x_1) ,$$

which proves the lemma. ■

<u>LEMMA 15.3</u>. If $f(x)$ is a convex function on E_1 and $x_0 \in E_1$, then

$$f'_-(x_0) \le f'_+(x_0) .$$

<u>Proof</u>. For $g_1 = +1$, $g_2 = -1$, we have

$$f'_+(x_0) - f'_-(x_0) = \frac{\partial f(x_0)}{\partial g_1} + \frac{\partial f(x_0)}{\partial g_2} = \max_{v\in\partial f(x_0)} v + \max_{v\in\partial f(x_0)} (-v)$$

$$\ge \max_{v\in\partial f(x_0)} (v + (-v)) = 0 .$$ ■

Now, let us establish some fundamental properties of one-sided derivatives.

<u>THEOREM 15.1</u>. If $f(x)$ is a convex function on E_1, then its one-sided derivatives do not decrease. Moreover, for all $x \in E_1$, we have the equalities

$$\lim_{y\to x+0} f'_+(y) = f'_+(x) , \qquad \lim_{y\to x+0} f'_-(y) = f'_+(x) , \qquad (15.1)$$

$$\lim_{z\to x-0} f'_+(z) = f'_-(x) , \qquad \lim_{z\to x-0} f'_-(z) = f'_-(x) . \qquad (15.2)$$

<u>Proof</u>. Let $y > x$. Lemmas 15.2 and 15.3 imply

$$f'_-(x) \le f'_+(x) \le (y-x)^{-1}(f(y) - f(x)) \le f'_-(y) \le f'_+(y) . \qquad (15.3)$$

Hence, the functions $f'_+(x)$ and $f'_-(x)$ do not decrease. The continuity of the function $f(x)$ implies that

$$(y-x)^{-1}(f(y)-f(x)) = \lim_{t\to x-0}(y-t)^{-1}(f(y)-f(t)) .$$

Since, due to Lemma 15.2, the inequality

$$(y-x)^{-1}(f(y)-f(t)) \geq f'_+(t)$$

is valid for $y > t$, then

$$(y-x)^{-1}(f(y)-f(x)) \geq \lim_{t\to x+0} f'_+(t)$$

(the existence of the limit is guaranteed by the monotonicity of the function $f'_+(x)$). Hence,

$$f'_+(x) = \lim_{y\to x+0}(y-x)^{-1}(f(y)-f(x)) \geq \lim_{t\to x+0} f'_+(t) .$$

Replacing t by y in the latter limit, we get finally

$$f'_+(x) \geq \lim_{y\to x+0} f'_+(y) . \tag{15.4}$$

On the other hand, as follows from (15.3),

$$f'_+(x) \leq f'_-(y) \leq f'_+(y) ,$$

and hence, taking the limit, which exists due to the monotonicity of f'_+ and f'_-, we have

$$f'_+(x) \leq \lim_{y\to x+0} f'_-(y) \leq \lim_{y\to x+0} f'_+(y) . \tag{15.5}$$

Inequalities (15.4) and (15.5) imply equalities (15.1). Equalities (15.2) can be proved in an analogous way. The theorem has been proved. ■

It is well known that the existence of the derivative at some point is equivalent (in the case of a function of one variable) to the fact that the left as well as right derivatives at this point coincide (the existence of one-sided derivatives follows from Lemma 15.1). Let

$$\psi(x) = f'_+(x) - f'_-(x) .$$

By virtue of Lemma 15.3, $\psi(x) \geq 0$ for all x. It then follows that the function $f(x)$ is not differentiable at the point x_0 iff $\psi(x_0) > 0$.

We say that $f(x)$ *is not differentiable at a point* x_0 *to within an accuracy of* ε (where $\varepsilon > 0$ is fixed) if $\psi(x_0) \geq \varepsilon$. Geometrically, this implies that the difference of the slope of the tangent line from the right and the slope of the tangent line from the left at the point x_0 is not less than ε.

LEMMA 15.4. The set of points, where $f(x)$ is not differentiable up to ε, is either empty or finite on any bounded interval (a,b), for each $\varepsilon > 0$.

Proof. We assume that this set is infinite. Then, we can choose a monotone sequence $\{x_k\}$ of its elements which converges to some point $x_0 \in [a,b]$. To be more precise, let us suppose that the sequence $\{x_k\}$ decreases. Then, by virtue of Theorem 15.1,

$$\lim_{k\to\infty} (f'_+(x_k) - f'_-(x_k)) = \lim_{y\to x_0+0} f'_+(y) - \lim_{y\to x_0+0} f'_-(y) = 0 ,$$

which contradicts the inequality

$$f'_+(x_k) - f'_-(x_k) > \varepsilon \quad \forall k .$$

The lemma has been proved. ■

THEOREM 15.2. A convex function $f(x)$ given on E_1 is differentiable at all points of E_1, except, perhaps, for a countable set at most. If a function $f(x)$ is differentiable at some point x_0, then both left and right derivatives are continuous at that point.

In particular, if $f(x)$ is differentiable on some interval, then its derivative is continuous there.

Proof. Let

$$A_{mk} = \{x \in [-m,m] \mid \psi(x) > k^{-1}\} .$$

Then the set $A = \{x \in E_1 \mid \psi(x) > 0\}$ consisting of all points at which $f(x)$ is not differentiable is represented in the form

$$A = \bigcup_{m=1}^{\infty} \bigcup_{k=1}^{\infty} A_{mk} .$$

By virtue of Lemma 15.4, each set A_{mk} is finite or empty. Hence, for each m, the set $A_m = \bigcup_{i=1}^{\infty} A_{mk}$, which is the countable union of the sets A_{mk}, is at most countable. Consequently, the set $A = \bigcup_{k=1}^{\infty} A_m$ is at most countable.

Now, assume that $f(x)$ is differentiable at a point x_0. Then $f'_+(x_0) = f'_-(x_0)$ and hence, all four limits in Theorem 15.1 coincide. In particular, this implies the continuity of the left as well as right derivatives at x_0. If $f(x)$ is differentiable on an interval, then its derivative coincides with the right (or left) derivative on the entire interval and, hence, the derivative of $f(x)$ is also continuous. The theorem has been proved. ■

2. Now, let us examine the general case. Let a function $f(x)$ be defined on E_n (n is an arbitrary natural number). A function $f(x)$ defined on E_n is called *differentiable at a point* $x_0 \in E_n$ if there exists a vector $A \in E_n$ for which

$$\lim_{x \to x_0} \|x - x_0\|^{-1} [f(x) - f(x_0) - (A, x - x_0)] = 0$$

or, equivalently,

$$f(x) = f(x_0) + (A, x - x_0) + o(\|x - x_0\|) .$$

The vector A is called the *gradient* of the function $f(x)$ at the point x_0 and is denoted by $f'(x_0)$. It is well known that

$$f'(x_0) = \left(\frac{\partial f(x_0)}{\partial x^{(1)}}, \ldots, \frac{\partial f(x_0)}{\partial x^{(n)}} \right) ,$$

where

$$x = (x^{(1)}, \ldots, x^{(n)})$$

and $\frac{\partial f(x_0)}{\partial x^{(i)}}$ is the partial derivative of the function $f(x)$ with respect to the ith component of x. Thus, the differentiability implies the existence of partial derivatives. The converse does not hold, i.e., the existence of partial derivatives does not imply the differentiability of a function. It is also known that the existence of continuous partial derivatives guarantees the differentiability. Let us show that, in the case of a convex function, the differentiability is equivalent to the existence of partial derivatives. But first we mention the following trivial equalities:

$$\frac{\partial f(x)}{\partial x^{(i)}} = \frac{\partial f(x)}{\partial e_i}, \qquad -\frac{\partial f(x)}{\partial x^{(i)}} = \frac{\partial f(x)}{\partial(-e_i)}, \tag{15.6}$$

where e_i is the ith unit vector of the space E_n.

<u>THEOREM 15.3</u>. Let $f(x)$ be a convex function on E_n. The function $f(x)$ is differentiable at a point $x_0 \in E_n$ iff the partial derivatives $\frac{\partial f(x_0)}{\partial x^{(1)}}, \ldots, \frac{\partial f(x_0)}{\partial x^{(n)}}$ exist.

If $f(x)$ is differentiable at a point x_0, then

$$\partial f(x_0) = \{f'(x_0)\} .$$

<u>Proof</u>. If the function $f(x)$ is differentiable, then it has the partial derivatives. Moreover, for $g \in E_n$, we have

$$f(x_0+\alpha g) = f(x_0) + \alpha(f'(x_0), g) + o(\alpha) .$$

From Lemma 5.3 we obtain

$$\partial f(x_0) = \{f'(x_0)\} .$$

Next, let us suppose that the partial derivatives $\frac{\partial f(x_0)}{\partial x^{(1)}}, \ldots, \frac{\partial f(x_0)}{\partial x^{(n)}}$ exist. We shall show that the vector

$$f'(x_0) = \left\{\frac{\partial f(x_0)}{\partial x^{(i)}}, \ldots, \frac{\partial f(x_0)}{\partial x^{(n)}}\right\}$$ is a subgradient of the function $f(x)$ at the point x_0.

For $v = (v^{(1)},\ldots,v^{(n)}) \in \partial f(x_0)$ we have

$$\frac{\partial f(x_0)}{\partial x^{(i)}} = \frac{\partial f(x_0)}{\partial e_i} = \max_{w \in \partial f(x_0)} (w,e_i) \geq (v,e_i) = v^{(i)} \quad \forall i \in 1:n ,$$

$$-\frac{\partial f(x_0)}{\partial x^{(i)}} = \frac{\partial f(x_0)}{\partial(-e_i)} = \max_{w \in \partial f(x_0)} (w,-e_i) \geq (-v,e_i) = -v^{(i)} \quad \forall i \in 1:n .$$

Thus, $v^{(i)} = \dfrac{\partial f(x_0)}{\partial x^{(i)}}$ and, therefore, $f'(x_0)$ is a subgradient.

In addition, we have $\partial f(x_0) = \{f'(x_0)\}$. Therefore, for all $u \in E_n$, the inequality

$$f(x_0 + u) \geq f(x_0) + \sum_{i=1}^{n} \frac{\partial f(x_0)}{\partial x^{(i)}} u^{(i)} \tag{15.7}$$

is satisfied. Here $u = (u^{(1)},\ldots,u^{(n)})$.

Let us verify that the function $f(x)$ is differentiable at the point x_0. Consider the vectors $g_1,\ldots,g_n,g_{n+1},\ldots,g_{2n}$, where $g_{n+1} = -e_i$ $\forall i \in 1:n$, $g_i = e_i$ $\forall i \in 1:n$. Define

$$z_i = \begin{cases} \max\,(u^{(i)},0) , & i \in 1:n , \\ \max\,(-u^{(i)},0) , & i \in (n+1):2n ; \end{cases}$$

$$I_+ = \{i \in 1:n \mid u^{(i)} \geq 0\} ,$$

$$I_- = \{i \in 1:n \mid u^{(i)} < 0\} .$$

Then

$$u = \sum_{i=1}^{n} u^{(i)} e_i = \sum_{i \in I_+} u^{(i)} e_i + \sum_{i \in I_-} u^{(i)} e_i$$

$$= \sum_{i=1}^{n} z^{(i)} g_i + \sum_{i=n+1}^{2n} z^{(i)} g_i = \sum_{i=1}^{2n} z^{(i)} g_i .$$

Denote by $\lambda(u)$ the sum $\sum_{i=1}^{2n} z^{(i)} = \sum_{i=1}^{n} u^{(i)}$. From the

definition, it follows directly that $\sum_{i=1}^{2n} \frac{z^{(i)}}{\lambda(u)} = 1 \quad (u \neq 0)$.
Moreover, $[\lambda(u)]^{-1} z^{(i)} \geq 0$.

Hence, for any vectors $v_1, \ldots, v_{2n}$, we have

$$f\left(\sum_{i=1}^{2n} [\lambda(u)]^{-1} z^{(i)} v_i\right) \leq \sum_{i=1}^{2n} [\lambda(u)]^{-1} z^{(i)} f(v_i) \quad . \tag{15.8}$$

Consider the quantities $o_i(\alpha)$ defined at the point x_0 by the following equalities:

$$f(x_0+\alpha g_i) = f(x_0) + \alpha \frac{\partial f(x_0)}{\partial g_i} + o_i(\alpha) \qquad \forall i \in 1:2n \ . \tag{15.9}$$

Since $f(x)$ is a convex function, then $o_i(\alpha) \geq 0$. Now, we estimate the value of $f(x_0+u)$, assuming that $u \neq 0$.

Using relations (15.8) with $v_i = x_0 + \lambda(u) g_i$ and (15.9) with

$$\begin{aligned} f(x_0+u) &= f\left[x_0 + \sum_{i=1}^{2n} z^{(i)} g_i\right] = f\left[x_0 + \lambda(u) \sum_{i=1}^{2n} [\lambda(u)]^{-1} z^{(i)} g_i\right] \\ &= f\left[\sum_{i=1}^{2n} [\lambda(u)]^{-1} z^{(i)} (x_0 + \lambda(u) g_i)\right] \\ &\leq \sum_{i=1}^{2n} [\lambda(u)]^{-1} z^{(i)} f(x_0 + \lambda(u) g_i) \\ &= \sum_{i=1}^{2n} [\lambda(u)]^{-1} z^{(i)} \left[f(x_0) + \lambda(u) \frac{\partial f(x_0)}{\partial g_i} + o_i(\lambda(u))\right] \\ &= f(x_0) + \sum_{i=1}^{2n} z^{(i)} \frac{\partial f(x_0)}{\partial g_i} + \sum_{i=1}^{2n} o_i(\lambda(u)) [\lambda(u)]^{-1} z^{(i)} \ . \end{aligned} \tag{15.10}$$

We show that the last term of (15.10) converges to zero faster than $||u||$ as $||u|| \to 0$. Indeed, we have

$$\begin{aligned} ||u||^{-1} \sum_{i=1}^{2n} [\lambda(u)]^{-1} z^{(i)} o_i(\lambda(u)) &= \sum_{i=1}^{2n} [\lambda(u)]^{-1} o_i(\lambda(u)) ||u||^{-1} z^{(i)} \\ &\leq \sum_{i=1}^{2n} [\lambda(u)]^{-1} o_i(\lambda(u)) \xrightarrow[||u|| \to 0]{} 0 \ , \end{aligned}$$

where we have used the obvious relation $z^{(i)} \le |u^{(i)}| \le ||u||$ and the nonnegativity of the quantity $o_i(\lambda(u))$. The second term on the right-hand side of (15.10) has the form

$$\sum_{i=1}^{2n} z^{(i)} \frac{\partial f(x_0)}{\partial g_i} = \sum_{i=1}^{n} u^{(i)} \frac{\partial f(x_0)}{\partial x^{(i)}} ,$$

which follows from equality (15.6) and the definition of $z^{(i)}$.

Hence, we obtain from (15.10)

$$f(x_0+u) \le f(x_0) + \sum_{i=1}^{n} u^{(i)} \frac{\partial f(x_0)}{\partial x^{(i)}} + o(u) . \qquad (15.11)$$

Consequently, the differentiability of the function $f(x)$ follows from (15.7) and (15.11). The theorem has been proved. ∎

THEOREM 15.4. A convex function $f(x)$ is differentiable at a point $x_0 \in E_n$ iff its subdifferential $\partial f(x_0)$ consists of a unique point.

Proof. If the function $f(x)$ is differentiable, then

$$\partial f(x_0) = \{f'(x_0)\} .$$

Hence, the subdifferential $\partial f(x_0)$ consists of a unique point. Now we prove the "if" part. Let $\partial f(x_0) = \{v\}$, $v = (v^{(1)},\dots,v^{(n)})$. Then

$$\lim_{\alpha\to+0} \alpha^{-1}[f(x_0+\alpha e_i) - f(x_0)]$$

$$= \frac{\partial f(x_0)}{\partial e_i} = \max_{w\in\partial f(x_0)} (w,e_i) = (v,e_i) = v^{(i)} ,$$

$$\lim_{\alpha\to-0} \alpha^{-1}[f(x_0+\alpha e_i) - f(x_0)] = \lim_{\beta\to+0} \beta^{-1}[f(x_0) - f(x_0+\beta(-e_i))]$$

$$= -\frac{\partial f(x_0)}{\partial(-e_i)} = -\max_{w\in\partial f(x_0)} (w,-e_i) = -(v,-e_i) = v^{(i)} .$$

Thus, the partial derivatives

$$\frac{\partial f(x_0)}{\partial x^{(i)}} = \lim_{\alpha\to 0} \alpha^{-1}[f(x_0+\alpha e_i) - f(x_0)]$$

exist and, as follows from Theorem 15.3, the function $f(x)$ is differentiable at the point x_0. The theorem has been proved. ■

Let us now describe the structure of the set where the convex function $f(x)$ is not differentiable. Denote this set by A and consider the intersection A_m of the set A and the open cube

$$K_m = \{x \in E_n \mid |x^{(i)}| < m \quad i \in 1:n\} ,$$

where m is a natural number. Denote by A_{mk} $(k = 1:n)$ the set of points $x \in K_m$ where the partial derivative $\frac{\partial f(x)}{\partial x^{(k)}}$ does not exist.

From Theorem 15.3 it follows that

$$A_m = A_{1m} \cup A_{2m} \cup \cdots \cup A_{nm} .$$

The relation

$$A = \bigcup_{m=1}^{\infty} A_m = \bigcup_{m=1}^{\infty} \bigcup_{k=1}^{n} A_{km}$$

reduces the study of the set A to that of the set A_{km}. Moreover, we can, in turn, decompose these sets into simply constructed subsets. We show this in the sequel. Consider both the right and left partial derivatives $f'_{+k}(x_0)$ and $f'_{-k}(x_0)$ of the function $f(x)$ at the point x_0 with respect to the kth variable. By definition,

$$f'_{+k}(x_0) = \lim_{\alpha \to +0} \alpha^{-1}[f(x_0+\alpha e_k) - f(x_0)] ,$$

$$f'_{-k}(x_0) = \lim_{\alpha \to -0} \alpha^{-1}[f(x_0+\alpha e_k) - f(x_0)] .$$

If we define the function $h(\alpha)$ of one real variable by the relation $h(\alpha) = f(x_0+\alpha e_k)$, then

$$f'_{+k}(x_0) = h'_+(0) , \qquad f'_{-k}(x_0) = h'_-(0) .$$

The convexity of $f(x)$ implies the convexity of $h(\alpha)$ and, hence, due to Lemma 15.3,

$$f'_{-k}(x_0) \le f'_{+k}(x_0) .$$

Let

$$\psi_k(x_0) = f'_{+k}(x_0) - f'_{-k}(x_0) .$$

Then, $\psi_k(x_0) \geq 0$ and, moreover, the existence of partial derivatives is equivalent to the equality $\psi_k(x_0) = 0$. In other words, we have

$$A_{km} = \{x \in K_m \mid \psi_k(x) > 0\} .$$

By letting

$$A_{kmi} = \{x \in K_m \mid \psi_k(x) \geq i^{-1}\} , \qquad i \in 1:\infty ,$$

the set A_{km} is finally represented as follows:

$$A_{km} = \bigcup_{i=1}^{\infty} A_{kmi} .$$

It is quite simple to describe the structure of the set A_{kmi} by clarifying the property of the intersection of this set with the corresponding straight line in the direction e_k. To do this, consider the hyperplane $H_k = \{x \in E_n \mid (x, e_k) = 0\}$ and the intersection H_{km} of this hyperplane and the cube K_m. Let

$$x = (x^{(1)}, \ldots, x^{(k-1)}, 0, x^{(k+1)}, \ldots, x^{(k)}) \in H_{km}$$

and $\Pi_x = \{x + \lambda e_k \mid -\infty < \lambda \leq \infty\}$, where Π_x is the straight line passin through x in the direction e_k. Define $\phi(\lambda) = f(x_\lambda)$, where

$$x_\lambda = x + \lambda e_k = (x^{(1)}, \ldots, x^{(k-1)}, \lambda, x^{(k+1)}, \ldots, x^{(n)}) .$$

Then

$$\phi'_+(\lambda) = f'_{+k}(x_\lambda) , \qquad \phi'_-(\lambda) = f'_{-k}(x_\lambda) ,$$

and, hence, as easily verified, the intersection of the set A_{kmi} and the straight line Π_x coincided with the set of points x_λ for which

$$\phi'_+(\lambda) - \phi'_-(\lambda) \geq i^{-1} .$$

In addition, the intersection of each cube K_m and the straight line Π_x is the interval $\{x_\lambda \mid |\lambda| < m\}$. From Lemma 15.4, the set of numbers λ on the interval $(-m,m)$, where the function $\phi(x)$ is not differentiable up to $\varepsilon = i^{-1}$, is finite or empty.

Hence, it follows straightforwardly from this fact that the set of points x on the interval $\{x_\lambda \mid |\lambda| < m\}$, where

$$\phi'_+(\lambda) - \phi'_-(\lambda) = f'_{+k}(x_\lambda) - f'_{-k}(x_\lambda) \geq i^{-1},$$

is also finite or empty. In other words, the intersection of the set A_{kmi} and the straight line Π_x for any $x \in H_{km}$ is finite or empty.

Consequently, the set A_{km} is represented as the countable union of the sets A_{kmi} and the intersection of each A_{kmi} and any straight lines parallel to the vector e_k is finite or empty. This property of the sets A_{kmi} shows that they are in some sense «small» or «sparse».

The family of the sets $\{A_{kmi} \mid k \in 1:n,\ m \in 1:\infty,\ i \in 1:\infty\}$ is countable. Hence, the set $A = \bigcup_{k=1}^{n} \bigcup_{m=1}^{\infty} \bigcup_{i=1}^{\infty} A_{kmi}$ of points, where the function $f(x)$ is not differentiable, is a countable union of sets which are «small» in the above-mentioned sense. The countable union of «small» sets itself is «not too large». In this sense, the set of points, where the function $f(x)$ is differentiable, is sufficiently «massive».

Finally, we note that if the function $f(x)$ is differentiable on some open set, then its gradient is continuous on this set.

3. Let us explain the above argument in more detail for the reader who would be familiar with the measure theory (see, for example, [85]).

Consider the sets

$$B_{ki} = \{x \in E_n \mid f'_{+k}(x) - f'_{-k}(x) \geq i^{-1}\}, \qquad k \in 1:n, \qquad i \in 1:\infty .$$

Since

$$f'_{+k}(x) - f'_{-k}(x) = \frac{\partial f(x)}{\partial e_k} + \frac{\partial f(x)}{\partial(-e_k)}$$

$$= \max_{v \in \partial f(x)} (v, e_k) + \max_{v \in \partial f(x)} (v, -e_k),$$

then we have

$$B_{ki} = \left\{x \in E_n \mid \max_{v \in \partial f(x)} (v, e_k) + \max_{v \in \partial f(x)} (v, -e_k) \geq i^{-1}\right\} .$$

Let us show that this equality and the upper semicontinuity of the mapping $\partial f(x)$ imply the closedness of B_{ki}. Indeed, let $x_j \in B_{ki}$, $x_j \to x_0$. Find the vectors $v'_j \in \partial f(x_j)$ and $v''_j \in \partial f(x_j)$ such that

$$\max_{v \in \partial f(x_j)} (v, e_k) = (v'_j, e_k)$$

$$\max_{v \in \partial f(x_j)} (v, -e_k) = (v''_j, -e_k) .$$

Due to the Corollary of Theorem 5.1, the mapping $\partial f(x)$ is bounded in the neighborhood of the point x_0. Hence, without loss of generality, we can assume that the limits

$$v' = \lim_{j \to \infty} v'_j , \qquad v'' = \lim_{j \to \infty} v''_j$$

exist. In addition, $v' \in \partial f(x_0)$, $v'' \in \partial f(x_0)$. Since

$$(v'_j, e_k) + (v''_j, -e_k) \geq i^{-1} ,$$

we have

$$\max_{v \in \partial f(x_0)} (v, e_k) + \max_{v \in \partial f(x_0)} (v, -e_k) \geq (v', e_k) + (v'', -e_k) \geq i^{-1} ,$$

i.e., $x_0 \in B_{ki}$. Thus, B_{ki} is a closed and, therefore, measurable set. The relation $A_{kmi} = B_{ki} \cap K_m$ shows that the set A_{kmi} is also measurable. Let us find its measure. Let

$$A_{kmi}(x) = A_{kmi} \cap \Pi_x \qquad \forall x \in H_{km} .$$

As shown above, the set $A_{kmi}(x)$ is finite or empty and, hence, its measure $\mu A_{kmi}(x)$ is equal to zero. The equality

$$\mu A_{kmi} = \int_{H_{km}} \mu A_{kmi}(x)\, dx ,$$

which constructs the measures of A_{kmi} by the measure of its cross-section $A_{kmi}(x)$, shows that the set A_{kmi} also has zero measure.

Since the set A of points where f is nondifferentiable can be represented as the countable union of the sets A_{kmi}, then we have also $\mu A = 0$.

Thus, the following theorem has been proved.

THEOREM 15.5. Let $f(x)$ be a convex function on E_n.

Then $f(x)$ is differentiable almost everywhere on E_n.

4. We can obtain now the following interesting result. Let a function $f(x)$ be defined on E_n. Due to Theorem 15.5, the function $f(x)$ is differentiable almost everywhere on E_n. As above, denote by A the set of points where the function $f(x)$ is not differentiable.

THEOREM 15.6. We have the representation

$$\partial f(x_0) = \text{co} \left\{ z \in E_n \mid z = \lim_{k\to\infty} f'(x_k),\ x_k \to x_0,\ x_k \notin A \right\}. \qquad (15.12)$$

Proof. We denote by B the set on the right-hand side of (15.12). The inclusion

$$B \subset \partial f(x_0) \qquad (15.13)$$

is obvious from the fact that, due to Theorem 15.3, $\partial f(x_k) = \{f'(x_k)\}$ for $x_k \notin A$ and the upper semicontinuity of the mapping $\partial f(x)$.

Now we show the converse inclusion. Let $v \in \partial f(x_0)$. It is necessary to prove that $v \in B$. Assume that this is not the case, i.e., $v \notin B$. Then, there exist a number $a > 0$ and a vector $g_0 \in E_n$, $\|g_0\| = 1$ such that

$$(v-z,\ g_0) \ \geq\ a\ >\ 0 \qquad \forall\, z \in B\ . \tag{15.14}$$

Due to the Corollary of Theorem 5.1, the mapping $\partial f(x)$ is bounded on any bounded set. Since $\partial f(x) = \{f'(x)\}$ for $x \notin A$ and the measure of the set A is equal to zero, there exist a sequence $\{x_k\}$ and a point $z_0 \in B$ such that

$$\begin{aligned} x_k &= x_0 + \alpha_k g_k\ , \qquad \alpha_k \to +0\ , \qquad g_k \to g_0\ , \\ x_k &\notin A\ , \qquad z_k = f'(x_k) \to z_0 \in B\ . \end{aligned} \tag{15.15}$$

From $v \in \partial f(x_0)$ we have

$$f(x) - f(x_0) \ \geq\ (v,\ x-x_0) \qquad \forall\, x \in E_n\ .$$

Hence, considering (15.14) and the boundedness of B (this is obvious from (15.13)), we obtain for sufficiently large k

$$h(\alpha_k) \ \equiv\ \alpha_k^{-1}(f(x_0+\alpha_k g_k) - f(x_0)) \ \geq\ (v, g_k) \ \geq\ (z, g_k) + 0.5a \qquad \forall\, z \in B\ .$$

In particular, $h(\alpha_k) \geq (z_0, g_k) + 0.5a$. From (15.15), we have for sufficiently large k

$$h(\alpha_k) \ \geq\ (z_k, g_k) + 0.25a\ . \tag{15.16}$$

Since $\{z_k\} = \partial f(x_k)$,

$$f(x) - f(x_k) \ \geq\ (z_k,\ x-x_k) \qquad \forall\, x \in E_n\ .$$

Hence,

$$\alpha_k^{-1}(f(x_0) - f(x_k)) \ \geq\ (-z_k, g_k)\ ,$$

i.e.,

$$h(\alpha_k) \ \equiv\ \alpha_k^{-1}(f(x_0+\alpha_k g_k) - f(x_0)) \ =\ \alpha_k^{-1}(f(x_k) - f(x_0)) \ \leq\ (z_k, g_k)\ ,$$

which contradicts (15.16). Thus, we get $v \in B$, i.e., $\partial f(x_0) \subset B$.

From this and (15.13), the assertion of the theorem follows. ∎

16. CONJUGATE FUNCTIONS

Conjugate functions provide a useful tool for investigating properties of convex functions and convex sets. A great number of results presented above can be written in terms of these functions. In what follows, let us recall elementary results from the theory of conjugate functions.

A convex function f defined on a convex set $\Omega \subset E_n$, is said to be *closed* if its epigraph

$$\text{epi } f = \{[\lambda, x] \in E_1 \times \Omega \mid \lambda \geq f(x)\}$$

is a closed set. It is not difficult to show that every continuous convex function defined on a closed convex set is closed. The converse, however, does not hold. In particular, a closed function can be defined also on a nonclosed set. As an example, we introduce the function $f(x) = \tan x$ which is defined on the interval $[0, \frac{\pi}{2})$. Here we note that

$$\lim_{x \to \frac{\pi}{2} - 0} \text{tg} x = +\infty \quad .$$

As the next lemma shows, all closed functions defined on a nonclosed set have a similar property.

<u>LEMMA 16.1</u>. Let f be a closed convex function defined on a set Ω and x be a limit point of Ω. Additionally, let $x \notin \Omega$.

Then

$$\lim_{y \to x} f(y) = +\infty \quad .$$

<u>Proof</u>. We assume that the lemma does not hold. Then, there exists a sequence $y_k \to x$, $y_k \in \Omega$ such that the inequality

$$f(y_k) \leq \lambda \qquad \forall k$$

is satisfied for some number λ. This implies that $[\lambda, y_k] \in \text{epi } f$.

Since $[\lambda, y_k] \to [\lambda, x]$ and the set epi f is closed, we have $[\lambda, x] \in \text{epi } f$. Hence by the definition of an epigraph, we obtain $x \in \Omega$ which contradicts the condition of the lemma.

PROBLEM 16.1. Show that a convex function defined on a convex set is closed iff it has the following properties: (a) f is lower semicontinuous at each point $x \in \Omega$, i.e., $f(x) \le \underline{\lim}\, f(x_k)$ if $x_k \in \Omega$, $x_k \to x$; (b) $\lim_{y \to x} f(y) = +\infty$ if x is a limit point of Ω which does not belong to Ω. Recall that if the interior of Ω is nonempty, then f is continuous at interior points of Ω. However, a closed function can be discontinuous at boundary points of Ω.

PROBLEM 16.2. Construct an example of a convex closed function defined on the segment $[0,1] \subset E_1$.

Let f be a convex function defined on a convex set Ω. The set

$$U_f = \{[\mu, \ell] \in E_1 \times E_n \mid (\ell, x) - \mu \le f(x) \quad \forall x \in \Omega\}$$

is called a *support set* of the function f. We can say that each element of a support set defines a linear function $h(x) = (\ell, x) - \mu$ which does not exceed the function f. We note that, for each subgradient v of the function f at some point $x_0 \in \Omega$, there exists a number μ such that $[\mu, v] \in U_f$. Indeed, by the definition of a subgradient we have

$$f(x) - f(x_0) \ge (v, x) - (v, x_0) ,$$

i.e., the inequality $(v, x) - \mu \le f(x)$ is satisfied for $\mu \ge (v, x_0) - f(x_0)$. This implies the inclusion $[\mu, v] \in U_f$.

Let us establish a relationship between a convex closed function f and its support set U_f.

LEMMA 16.2. If f is a convex closed function defined on a convex set $\Omega \subset E_n$, then

$$f(x) = \sup_{[\mu, \ell] \in U_f} \{(\ell, x) - \mu\} \qquad \forall x \in \Omega . \quad (16.1)$$

Proof. For any $\varepsilon > 0$, let us consider the point $(f(x)-\varepsilon,\ x)$ in the space $E_1 \times E_n$. Since this point is not in the closed set epi f, then, applying the separation theorem, we can find a vector $[c,h] \in E_1 \times E_n$ for which the inequality

$$c(f(x)-\varepsilon) + (h,x) < c\lambda + (h,z) \quad \forall (\lambda,z) \in \text{epi } f \qquad (16.2)$$

is satisfied. Here, letting $z = x$ and $\lambda = f(x)$, we get $-c\varepsilon > 0$, i.e., $c < 0$. Hence, letting $\lambda = f(z)$ in (16.2) yields

$$f(z) > (-c^{-1}h,\ z) + f(x) - \varepsilon + c^{-1}(h,x) \qquad \forall z \in \Omega .$$

Here, let $-c^{-1}h = \ell_0$, $f(x) - \varepsilon + c^{-1}(h,x) = -\mu_0$. Then,

$$f(z)(h,z) > (\ell_0,z) - \mu_0 \qquad \forall z \in \Omega ,$$

i.e., the vector $[\mu_0,\ell_0]$ belongs to U_f. In addition, we have $(\ell_0,x) - \mu_0 = f(x) - \varepsilon$. Consequently,

$$\sup_{[\mu,\ell]\in U_f} \{(\ell,x) - \mu\} \geq (\ell_0,x) - \mu_0 = f(x) - \varepsilon .$$

By virtue of the arbitrariness of ε, we obtain

$\sup_{[\mu,\ell]\in U_f} (\ell,x) - \mu \geq f(x)$.

The converse inequality follows directly from the definition of a support set. The lemma has been proved. ■

REMARK 1. If x_0 is an interior point of the set Ω, then there exists an element of a support set $[\mu_0,\ell_0]$ such that $f(x_0) = (\ell_0,x_0) - \mu_0$, i.e., the supremum in formula (16.1) is attained.

This follows directly from formula (5.13). If x_0 is a boundary point of Ω, then such an element does not necessarily exist.

PROBLEM 16.3. Let $\Omega = \{x = (x^{(1)},x^{(2)}) \in E_2 \mid x^{(1)} \geq 0,\ x^{(2)} \geq 0\}$, and $f(x) = -\sqrt{x^{(1)}x^{(2)}}$. Show that for the point $x_0 = (1,0)$, there exists no element $[\mu_0,\ell_0] \in U_f$ such that the equality $f(x_0) = (\ell_0,x_0) - \mu_0$ is satisfied.

To continue, it is first necessary to calculate the supremum with respect to a support set for any element $x \in E_n$ which does not necessarily belong to Ω.

LEMMA 16.3. Let f be a convex closed function defined on a convex set Ω. Then, if $x \notin \Omega$, we have

$$\sup_{[\mu,\ell] \in U_f} \{(\ell,x) - \mu\} = +\infty .$$

Proof. The proof is divided into two steps.

1. Let x be a limit point of Ω. Take some number λ. Since, as established in Lemma 16.1, we have $\lim_{y \in x} f(x) = +\infty$, then there exists a convex neighborhood V of the point x such that the inequality $f(y) > \lambda$ is satisfied for its element y $(y \in \Omega)$. Take a point $y \in V \cap \Omega$ and consider the segment L with endpoints (λ,x) and (λ,y). Let us show that this segment and the epigraph epi f of the function f do not intersect. If $(\nu,v) \in L$, then $\nu = \lambda$, $v = \alpha x + (1-\alpha)y$ for some $\alpha \in [0,1]$. Since $x \in V$, $y \in V$, then $v \in V$ and, hence, $f(v) > \lambda$. This implies that $[\lambda,v] \in$ epi f. Since (i) the sets L and epi f do not intersect, (ii) they are closed and convex and, (iii) moreover, the segment L is bounded, then applying the separation theorem (see Theorem 1.3), we can find a vector $[c,v] \in E_1 \times E_n$ which has the following properties:

$$c\nu + (h,z) > c\lambda + (h,x) \qquad \forall [\nu,z] \in \text{epi } f , \tag{16.3}$$

$$c\nu + (h,z) > c\lambda + (h,y) \qquad \forall [\nu,z] \in \text{epi } f . \tag{16.4}$$

By letting $z = y$ and $\nu = f(y)$ in (16.4), we obtain $cf(y) > c\lambda$ and, in addition, the inequality $c > 0$ (since $f(y) > \lambda$). Letting $\nu = f(z)$ in (16.3) yields the inequality

$$f(z) > (-c^{-1}h,z) + \lambda + (c^{-1}h,\ x) \qquad \forall z \in \Omega .$$

Here, let $\ell_0 = -c^{-1}h$, $-\mu_0 = \lambda + c^{-1}(h,x)$. Then, $f(z) > (\ell_0, z) - \mu_0$, i.e., $[\mu_0, \ell_0] \in U_f$. At the same time, we get $(\ell_0, x) - \mu_0 = \lambda$.

Thus

$$\sup_{[\mu,\ell]\in U_f} \{(\ell,x) - \mu\} \geq (\ell_0,x) - \mu_0 = \lambda .$$

Since λ is arbitrary, this inequality implies

$$\sup_{[\mu,\ell]\in U_f} \{(\ell,x) - \mu\} = +\infty .$$

2. Let $x \notin \bar{\Omega}$, where $\bar{\Omega}$ is the closure of Ω. Fix some point $y \in \Omega$ and consider the segment $\{\alpha x+(1-\alpha)y \mid \alpha\in[0,1]\}$ with the end-points at x and y. Since $x \notin \bar{\Omega}$, then there exists a number $\alpha_0 \in [0,1]$ such that, for $\alpha > \alpha_0$, the points $\alpha x + (1-\alpha)y$ do not belong to Ω.

Now, take some λ and show that there exists a δ such that the segment L in the space $E_1 \times E_n$ with endpoints $[\lambda,x]$ and $[\delta,y]$ and the epigraph epi f of the function f do not intersect. Indeed, we can take as δ an arbitrary number smaller than

$$\delta_0 = \inf_{0\leq\alpha\leq\alpha_0} (1-\alpha)^{-1}[f(\alpha x + (1-\alpha)y) - \alpha\lambda]$$

(note that $\delta_0 > -\infty$). We show this. If $L \cap \text{epi } f \neq \emptyset$, then there exists an $\alpha \in [0,\alpha_0]$ such that

$$\alpha\lambda + (1-\alpha)\delta \geq f(\alpha x + (1-\alpha)y) .$$

This implies that $\delta \geq \delta_0$. Thus, there exists a number δ such that the sets L and epi f do not intersect. Applying the separation theorem, we can find a vector $[c,h] \in E_1\times E_n$ which has the following properties:

$$c\nu + (h,z) > c\lambda + (h,x) \quad \forall(\nu,z) \in \text{epi } f , \qquad (16.5)$$

$$c\nu + (h,z) > c\delta + (h,y) \quad \forall(\nu,z) \in \text{epi } f . \qquad (16.6)$$

Letting $z = y$ in (16.6), we get $c(\nu-\delta) > 0$. Since $\nu - \delta > 0$ for sufficiently large ν, then $c > 0$. Let $\ell_0 = -c^{-1}h$, $-\mu_0 = \lambda + (c^{-1}h, x)$.

Next, letting $\nu = f(z)$ in (16.5), we obtain $f(z) > (\ell_0, z) - \mu_0$, i.e., $[\mu_0, \ell_0] \in U_f$. At the same time $(\ell_0, z) - \mu_0 = \lambda$. Thus, $\sup_{[\mu,\ell]\in U_f} \{(\ell,x) - \mu\} \geq \lambda$, which, by virtue of the arbitrariness of λ, proves the lemma. ∎

Lemmas 16.2 and 16.3 show that it is convenient to consider convex closed functions which are defined on the entire space and take, perhaps, the value $+\infty$. In fact, for such functions, the equality $f(x) = \sup_{[\mu,\ell]\in U_f} \{(\ell,x) - \mu\}$ is valid for all x.

Let us introduce the corresponding definitions. A function f defined on the entire space E_n and taking the value on the half-extended real line $(-\infty, +\infty]$ is said to be *convex* if

$$f(\alpha x + (1-\alpha)y) \leq \alpha f(x) + (1-\alpha)f(y)$$

for all $x, y \in E_n$, $\alpha \in [0,1]$. Here, we assume that $0 \cdot (+\infty) = +\infty$.

The set

$$\text{dom } f = \{x \in E_n \mid f(x) < +\infty\}$$

is called the *effective set* of the function f. If $\text{dom } f = \emptyset$, then f is said to be a *proper convex* function. Similarly, for functions assuming finite values, the sets

$$\text{epi } f = \{[\lambda, x] \in E_1 \times E_n \mid f(x) \geq \lambda\}$$

and

$$U_f = \{[\mu, \ell] \in E_1 \times E_n \mid (\ell, x) - \mu \leq f(x) \ \forall x \in E_n\}$$

are called, respectively, the *epigraph* and the *support set* of the function f. If the epigraph $\text{epi } f$ is closed, then the function f is said to be *closed*. Let f be a convex function defined on the set Ω and taking only finite values. Define

$$\tilde{f}(x) = \begin{cases} f(x) , & x \in \Omega , \\ +\infty , & x \notin \Omega . \end{cases}$$

Then, $\operatorname{dom} \tilde{f} = \Omega$, $\operatorname{epi} \tilde{f} = \operatorname{epi} f$, $U_{\tilde{f}} = U_f$. It is clear that the study of the function f is reduced to that of the function $\tilde{f}$.

Since $\operatorname{epi} \tilde{f} = \operatorname{epi} f$, the closedness of the function f is equivalent to that of the function $\tilde{f}$. Throughout the remainder of this section, we shall assume, unless mentioned otherwise, that f is defined on the entire space and may take the value $+\infty$; however, not identically equal to $+\infty$, i.e., f is a proper convex function.

Lemmas 16.2 and 16.3 define an essential property of the functions under consideration. Let us formulate this as a theorem.

THEOREM 16.1. Let f be a convex closed function.

Then, for all $x \in E_n$, we have the equality

$$f(x) = \sup_{[\mu,\ell] \in U_f} \{(\ell,x) - \mu\} .$$

The converse proposition holds as well. We introduce it in the following form.

THEOREM 16.2. Suppose that for a function f the equality

$$f(x) = \sup_{[\mu,\ell] \in U} \{(\ell,x) - \mu\}$$

is satisfied for some set $U \subset E_1 \times E_n$ and all $x \in E_n$.

Then the function f is convex and closed.

Proof. 1. The function f is convex.

Indeed, if $x,y \in E_n$, $\alpha \in [0,1]$, then

$$\begin{aligned} f(\alpha x+(1-\alpha)y) &= \sup_{[\mu,\ell]\in U} \{(\ell,\ \alpha x+(1-\alpha)y) - \mu\} \\ &= \sup_{[\mu,\ell]\in U} \{\alpha[(\ell,x)-\mu] + (1-\alpha)[(\ell,y)-\mu]\} \\ &\le \alpha[\sup_{[\mu,\ell]\in U} \{(\ell,x)-\mu\}] + (1-\alpha)[\sup_{[\mu,\ell]\in U} \{(\ell,y)-\mu\}] \\ &= \alpha f(x) + (1-\alpha)f(y) . \end{aligned}$$

2. The function f is closed.

For each element $[\mu,\ell]$ of the set U, let us consider the subset

$$H_{\mu\ell} = \{[\lambda,x]\in E_1\times E_n \mid (\ell,x)-\lambda \le \mu\}$$

in the space $E_1\times E_n$. It is clear that $H_{\mu\ell}$ is a closed set (in fact, $H_{\mu\ell}$ is a half-space). Let us show that

$$\operatorname{epi} f = \bigcap_{(\mu,\ell)\in U} H_{\mu\ell} . \tag{16.7}$$

This equality also proves the theorem, because the intersection of closed sets is again closed.

We proceed to prove (16.7). Let $[\lambda,x]\in \operatorname{epi} f$ and let $[\mu,\ell]\in U$. Then

$$\lambda \ge f(x) \ge (\ell,x) - \mu \quad \forall\, x\in E_n ,$$

i.e., the inclusion $[\lambda,x]\in H_{\mu\ell}$ holds. Since $[\lambda,x]$ is an arbitrary element of epi f, then $\operatorname{epi} f \subset H_{\mu\ell}$. Moreover, since $[\mu,\ell]$ is an arbitrary element of U, then

$$\operatorname{epi} f \subset \bigcap_{[\mu,\ell]\in U} H_{\mu\ell} .$$

Now, let $[\lambda_0,x_0]\in H_{\mu\ell}$ for all $(\mu,\ell)\in U$. Then

$$\lambda_0 \ge \sup_{[\mu,\ell]\in U} \{(\ell,x_0) - \mu\} = f(x_0)$$

and, consequently, $[\lambda_0,x_0]\in \operatorname{epi} f$. The theorem has been proved. ∎

<u>PROBLEM 16.4</u>. Let the set $U\subset E_1\times E_n$ be such that

$f(x) = \sup_{[\mu,\ell]\in U} \{(\ell,x) - \mu\}$. Show that in this case the convex closed hull of U coincides with the support set U_f of the function f.

Next, let us define a conjugate function.

But first we prove the following lemma.

LEMMA 16.4. A set U in the space $E_1 \times E_n$ is an epigraph of some closed function f iff U is nonempty, convex, closed and includes a <<vertical>> ray $\{[\lambda',x] \mid \lambda' > \lambda\}$ together with each point $[\lambda,x]$ and, moreover, does not include a vertical line $\{[\lambda,x] \mid -\infty < \lambda < +\infty\}$. In addition, the equality

$$f(x) = \inf \{\lambda \mid [\lambda,x] \in U\} \qquad (16.8)$$

is valid for all $x \in E_n$. (As usual, we agree that inf of the empty set is equal to $+\infty$).

Proof. It is clear that if U is the epigraph $\text{epi}\, f$ of a closed convex function f, then it has the properties stated in this lemma. Now, let the set U have these properties and let us define the function f by formula (16.8). Since U does not include vertical lines, then $f(x) > -\infty$ for all x. It is verified directly that the epigraph $\text{epi}\, f$ coincides with U. Moreover, the effective set $\text{dom}\, f$ coincides with the set of points $x \in E_n$ such that the real line $\{\lambda x \mid \lambda \in (-\infty,+\infty)\}$ intersects the set U. Hence, we have $\text{dom}\, f \neq \emptyset$ because the set U is nonempty. The convexity and closedness of the function f follow from the convexity and closedness of the set $U = \text{epi}\, f$.

The lemma has been proved. ∎

Every closed convex function f is associated with two sets: the epigraph $\text{epi}\, f$ and the support set

$$U_f = \{[\mu,\ell] \in E_1 \times E_n \mid f(x) \geq (\ell,x) - \mu \ \forall x \in E_n\} .$$

By definition, the set U_f is convex and closed. By virtue of

Theorem 16.1, this set is nonempty.

If $[\mu,\ell] \in U_f$ and $\mu' \geq \mu$, then $[\mu',\ell] \in U_f$. In fact, the set U_f includes vertical rays. Moreover, U_f does not include vertical real lines because if $x_0 \in \operatorname{dom} f$ and $[\mu,\ell] \in U_f$, then $\mu \geq (\ell,x_0) - f(x_0) > -\infty$. Hence, as follows from Lemma 16.4, the support set U_f is the epigraph of some closed convex function. This function is said to be *conjugate* of f and denoted by f^*. Thus, the conjugate function f^* is defined by the equality

$$\operatorname{epi} f^* = U_f . \tag{16.9}$$

Now, let us derive an explicit expression of this function. Using formula (16.8), we have

$$\begin{aligned} f^*(\ell) &= \inf \{\mu \mid [\mu,\ell] \in U_f\} \\ &= \inf \{\mu \mid (\ell,x)-\mu \leq f(x) \ \forall x \in E_n\} \\ &= \inf \{\mu \mid \mu \geq (\ell,x)-f(x) \ \forall x \in E_n\} . \end{aligned}$$

If the set of real numbers $\{(\ell,x)-f(x) \mid x \in E_n\}$ is bounded from above, then $f^*(\ell) = \sup\limits_{x \in E_n} [(\ell,x) - f(x)]$.

Otherwise, the set $\{\mu \mid \mu \geq (\ell,x)-f(x) \ \forall x \in E_n\}$ is empty and hence

$$\begin{aligned} &\inf \{\mu \mid \mu \geq (\ell,x)-f(x) \ \forall x \in E_n\} \\ &\quad = +\infty = \sup \{(\ell,x) - f(x) \mid x \in E_n\} . \end{aligned}$$

Thus, the equality

$$f^*(\ell) = \sup_{x \in E_n} [(\ell,x) - f(x)] \tag{16.10}$$

is valid in all cases.

The following proposition follows immediately from (16.10) (*Young's inequality**/): for any $x,\ell \in E_n$, we have the relation

---

*/ Sometimes this inequality is called *Fenchel's inequality*.

$$(\ell,x) \le f(x) + f^*(\ell) .$$

Since the function f^* is convex and closed, then it makes sense to consider the conjugate of f^*. The conjugate is denoted by f^{**} and is called the *second conjugate of* f.

One of the most important results in the theory of conjugate functions is the following theorem, which is usually called *Fenchel-Moreau's theorem*.

THEOREM 16.3. If f is a closed convex function, then $f^{**} = f$.

Proof. Let $[\mu',\ell'] \in U_f$. Then, $(\ell',x) - f(x) \le \mu'$ for all $x \in E_n$ and hence $f^*(\ell') = \sup_{x\in E_n} [(\ell',x) - f(x)] \le \mu'$.

Therefore

$$f^{**}(x) = \sup_{\ell \in E_n} [(\ell,x) - f^*(\ell)] \ge (\ell',x) - f^*(\ell') \ge (\ell',x) - \mu' .$$

Applying Theorem 16.1, we have

$$f^{**}(x) \ge \sup_{[\mu',\ell']\in U_f} [(\ell',x) - \mu'] = f(x) .$$

On the other hand, as follows directly from Young's inequality, we have

$$f^{**}(x) = \sup_{\ell\in E_n} [(\ell,x) - f^*(\ell)] \le f(x) .$$

Hence, the theorem has been proved. ■

REMARK 2. It follows from Theorem 16.3 and from (16.9) that

$$U_{f^*} = \text{epi } f^{**} = \text{epi } f .$$

Thus, if U is a set in $E_1 \times E_n$ satisfying the hypotheses of Lemma 16.4, then it can be treated either as the epigraph of some closed convex function f or as the support set of the closed convex function f^*. In addition, for the set U the conjugate set V is defined, which is the support set of f and, at the same time, the "epigraph" of f^*. We can say that, in general, the theory of closed convex functions coincides with the theory of special convex sets, viz. the sets described in Lemma 16.4.

REMARK 3. Let f be an arbitrary, generally speaking, nonclosed and nonconvex function defined on E_n and taking its values on $(-\infty,+\infty)$. Formula (16.10) defines the function f^* which is also called the conjugate of f. Let

$$U = \{[\mu,x] \in E_1 \times E_n \mid x \in \operatorname{dom} f,\ \mu = f(x)\} .$$

Since $f^*(\ell) = \sup_{[u,x]\in U} [(\ell,x) - \mu]$, then, by virtue of Theorem 16. f^* is also a closed convex function. It is easy to verify that the function f^{**} coincides with a "closed convexification" of the function f, that is, with the largest convex function not exceeding f. In other words, the epigraph $\operatorname{epi} f^{**}$ of the function f^{**} is the closed convex hull of the epigraph $\operatorname{epi} f = \{[\lambda,x] \mid \lambda \geq f(x)\}$ of the function f.

We shall consider some examples.

1. Let f be a constant function, i.e., $f(x) = c$ for all $x \in E_n$ Then

$$f^*(\ell) = \sup_{x\in E_n} [(\ell,x) - f(x)] = \sup_{x\in E_n} [(\ell,x) - c] = \begin{cases} -c, & \ell = 0 \\ +\infty, & \ell \neq 0. \end{cases}$$

In particular, if $f \equiv 0$, then

$$f^*(\ell) = \begin{cases} 0, & \ell = 0, \\ +\infty, & \ell \neq 0. \end{cases}$$

2. Let f be a linear function, i.e., $f(x) = (h,x)$ for all $x \in E_n$. Then

$$f^*(\ell) = \sup_{x\in E_n} [(\ell,x) - (h,x)] = \sup_{x\in E_n} [(\ell-h,x)] = \begin{cases} 0, & \ell = h, \\ +\infty, & \ell \neq h. \end{cases}$$

3. Let U be a closed convex subset in the space E_n. Introduce the following two functions related to U:

$$\delta_U(x) = \begin{cases} 0 \quad , & x \in U \ , \\ +\infty \quad , & x \notin U \ , \end{cases}$$

$$P_U(x) = \sup_{\ell \in U} (\ell, x) \quad .$$

Usually, δ_U is called the *indicator function* of the set U and P_U is called the *support function* of this set. It is obvious that δ_U and P_U are closed convex functions. Clearly,

$$\delta_U^*(\ell) = \sup_{x \in E_n} [(\ell, x) - \delta_U(x)] = \sup_{x \in U} (\ell, x) = P_U(\ell) \ .$$

Thus, $\delta_U^* = P_U$. It follows from Theorem 16.3 that $P_U^* = \delta_U^{**} = \delta_U$.

PROBLEM 16.5. Consider the following closed convex functions defined on E_1: $f_1(x) = e^x$, $f_2(x) = p^{-1}|x|^p$ $(p > 1)$,

$$f_3(x) = \begin{cases} -0.5 - \ln x, & x > 0 \ , \\ +\infty, & x \le 0 \ . \end{cases}$$

Show that

$$f_1^*(\ell) = \begin{cases} \ell \ln \ell - \ell, & \ell > 0 \ , \\ 0 \ , & \ell = 0 \ , \\ +\infty \ , & \ell < 0 \ , \end{cases}$$

$$f_2^*(\ell) = q^{-1}|\ell|^q \ , \qquad \text{where} \quad p^{-1} + q^{-1} = 1 \ ,$$

$$f_3^*(\ell) = f_3(-\ell) = \begin{cases} -0.5 - \ln(-\ell) \ , & \ell < 0 \ , \\ +\infty \ , & \ell \ge 0 \ . \end{cases}$$

Conjugate functions provide a useful tool of investigating convex functions. In particular, it is convenient to study subdifferentials in terms of conjugate functions. In that case, the following proposition plays a crucial role.

THEOREM 16.4. Let f be a closed convex function and let $x_0, \ell_0 \in E_n$. Then the following conditions are equivalent:

(a) $\ell_0 \in \partial f(x_0)$;

(b) $x_0 \in \partial f^*(\ell_0)$;

(c) $(\ell_0, x_0) = f(x_0) + f^*(\ell_0)$

(relation (c) shows that equality in Young's inequality is obtained for x_0 and ℓ_0).

Proof. Let $\ell_0 \in \partial f(x_0)$. Then, by the definition of a subgradient, we have

$$(\ell_0, x_0) - f(x_0) \geq (\ell_0, x) - f(x) .$$

Hence

$$f^*(\ell_0) = \sup_{x \in E_n} [(\ell_0, x) - f(x)] = (\ell_0, x_0) - f(x_0) .$$

Conversely, the equality $f^*(\ell_0) = (\ell_0, x_0) - f(x_0)$ implies the relation $(\ell_0, x_0) - f(x_0) \geq (\ell_0, x) - f(x)$. This shows the equivalence of (a) and (c). Arguing analogously in the case of the function f^*, we obtain that the inclusion $x_0 \in \partial f^*(\ell_0)$ is equivalent to the equality

$$f^{**}(x_0) = (\ell_0, x_0) - f^*(\ell_0) .$$

Since, due to Theorem 16.3, we have $f^{**} = f$, then (b) and (c) are equivalent as well. The theorem has been proved. ∎

It is convenient to formulate the extremum properties of a convex function f in terms of conjugate functions. First of all, we observe that

$$f^*(0) = \sup_{x \in E_n} (-f(x)) = -\inf_{x \in E_n} f(x) .$$

Thus,

$$\inf_{x \in E_n} f(x) = -f^*(0) .$$

This implies that f is bounded from below iff $0 \in \operatorname{dom} f^*$. Let

$$D = \{x \in E_n \mid f(x) = -f^*(0)\} = \{x \in E_n \mid f(x) = \inf_{y \in E_n} f(y)\} .$$

This is the set of minimum points of the function f. The relation $x \in D$ is equivalent to the fact that equality in Young's inequality is obtained for the pair of elements $[x, 0]$, i.e.,

$$0 = (x,0) = f(x) + f^*(0) .$$

Hence, as follows from Theorem 16.4, the relation $x \in D$ is equivalent to the inclusion $0 \in \partial f(x)$ and this is the well-known condition for a minimum (see Section 6).

On the other hand, the same theorem implies that the relations $0 \in \partial f(x)$ and $x \in \partial f^*(0)$ are equivalent. Therefore, the set of minimum points D can be expressed in terms of the conjugate function f^*:

$$D = \partial f^*(0) .$$

In particular, this implies that the function f attains the minimum iff the conjugate function f^* has nonempty subdifferential at zero (this can be guaranteed, for example, in the case where 0 is an interior point of the effective set $\operatorname{dom} f^*$).

A further application of conjugate functions is based on theorems which enable us to calculate conjugate functions of sums, multiplications by a number, the maximum of closed convex functions, etc. In that case, we have to consider not necessarily only closed functions, but also functions which take, perhaps, the value $-\infty$. We can investigate almost all problems in convex analysis using the above-mentioned theorem stated in terms of conjugate functions.

Let us illustrate this with an example. Let f be a closed convex function and let Ω be a closed convex set included in $\operatorname{dom} f$. Consider the indicator function δ_Ω of the set Ω

$$\delta_\Omega(x) = \begin{cases} 0 , & x \in \Omega , \\ +\infty , & x \notin \Omega , \end{cases}$$

and let $f_1 = f + \delta_\Omega$. It is clear that

$$f_1(x) = \begin{cases} f(x) , & x \in \Omega , \\ +\infty , & x \notin \Omega , \end{cases}$$

i.e., dom $f_1 = \Omega$. Hence, the computation of the conditional subdiffe- ential of the function f on the set Ω is reduced to the computati of the ordinary subdifferential of the sum $f_1 = f + \delta_\Omega$, and it can be done using the conjugate function method. The detailed statement of many results in convex analysis in terms of conjugate function theory can be found in R. Rockafellar [110] and also in [57,76].

17. COMPUTATION OF ε-SUBGRADIENTS OF SOME CLASSES OF CONVEX FUNCTIONS

The computation of the whole set $\partial_\varepsilon f(x)$, $\varepsilon > 0$ is, undoubtedly, a very difficult problem because, by the definition of $\partial_\varepsilon f(x)$, we need information about the global behavior of the function $f(x)$ on the entire space E_n. The problem of computing individual ε-sub-gradients of $f(x)$ is, in general, simpler than that of the subgra-dient of f, because we have $\partial f(x) \subset \partial_\varepsilon f(x)$ and the set $\partial_\varepsilon f(x)$ can be richer than $\partial f(x)$. However, this is still difficult to do because we again need to know the behavior of f on the entire space E_n.

We shall show how to calculate ε-subgradients of smooth strong-ly convex functions and of maximum functions of smooth strongly con-vex functions.

<u>LEMMA 17.1</u>. Let a function $f(x)$ be strongly convex on E_n with a strong convexity constant m and let $\varepsilon \geq 0$.

Then, we have the inclusion

$$\partial f(x) + S_{2\sqrt{\varepsilon m}}(0) \subset \partial_\varepsilon f(x)\ , \tag{17.1}$$

where

$$S_\delta(0) = \{v \in E_n \mid \|v\| \leq \delta\}\ , \qquad \delta \geq 0\ .$$

<u>Proof</u>. Let us show that the relation

$$(v(x_0) + r) \in \partial_\varepsilon f(x_0) \tag{17.2}$$

holds for any $v(x_0) \in \partial f(x_0)$ and some $r \in E_n$. By the definition of $\partial_\varepsilon f(x_0)$, in order that (17.2) be satisfied, it is necessary and sufficient that

$$f(z) - f(x_0) \geq (v(x_0), z-x_0) + (r, z-x_0) - \varepsilon$$

$$\equiv A(z) \qquad \forall z \in E_n. \quad (17.3)$$

Since the function $f(x)$ is strongly convex on E_n, we have

$$f(z) - f(x_0) \geq (v(x_0), z-x_0) + m\|z-x_0\|^2 \equiv B(z) \qquad \forall z \in E_n$$

(see (9.17)). This implies that (17.3) is satisfied if

$$A(z) \leq B(z) \qquad \forall z \in E_n, \quad (17.4)$$

i.e., if

$$m\|z-x_0\|^2 \geq (r, z-x_0) - \varepsilon \qquad \forall z \in E_n .$$

We rewrite the last inequality:

$$h(r,z) \equiv (r, z-x_0) - m\|z-x_0\|^2 \leq \varepsilon \quad \forall z \in E_n .$$

We can see that (17.4) is satisfied if

$$\max_{z \in E_n} h(r,z) \leq \varepsilon \quad . \qquad (17.5)$$

Now, let us find $\max_{z \in E_n} h(r,z)$. To do this, we equate to zero the derivative of the function $h(r,z)$ with respect to z. In fact, we have $r - 2m(z-x_0) = 0$. Therefore, $(z-x_0) = r2^{-1}m^{-1}$.

Hence, $\max_{z \in E_n} h(r,z) = \|r\|^2 4^{-1}m^{-1}$. From this, it is clear that the inequality

$$\|r\| \leq 2\sqrt{\varepsilon m}$$

implies, in turn, (17.5), (17.4), (17.3) and (17.2).

Thus, inclusion (17.1) has been proved. ∎

It follows from Lemma 17.1 that, for a strongly convex function $f(x)$, any subgradient $v(x) \in \partial f(x)$ belongs to the ε-subdifferential $\partial_\varepsilon f(x)$ together with its own neighborhood of radius $2\sqrt{\varepsilon m}$.

COROLLARY 1. Let $x, g \in E_n$, $\|g\| = 1$.

Then,

$$\frac{\partial f(x)}{\partial g} + 2\sqrt{\varepsilon m} \le \frac{\partial_\varepsilon f(x)}{\partial g} .$$

Indeed, this follows from

$$\frac{\partial_\varepsilon f(x)}{\partial g} = \max_{v \in \partial_\varepsilon f(x)} (v,g) \ge \max_{v \in \partial f(x) + S_{2\sqrt{\varepsilon m}}(0)} (v,g)$$

$$= \max_{v \in \partial f(x)} (v,g) + \max_{v \in S_{2\sqrt{\varepsilon m}}(0)} (v,g) = \frac{\partial f(x)}{\partial g} + 2\sqrt{\varepsilon m} .$$

COROLLARY 2. Let $f(x) + (Ax,x) + (a,x)$, where $a \in E_n$, A is an $n \times n$ symmetric positive definite matrix and m is its minimum eigenvalue. Then

$$(2Ax + a + r) \in \partial_\varepsilon f(x) \qquad r \in S_{2\sqrt{\varepsilon m}}(0) .$$

LEMMA 17.2. Let a function $f(x)$ be differentiable on E_n and let its derivative $f'(x)$ satisfy a Lipschitz condition on E_n with a constant $L < \infty$, i.e.,

$$\|f'(y) - f'(x)\| \le L\|y-x\| \qquad \forall x,y \in E_n .$$

Then,

$$|f(y) - f(x) + (f'(x), y-x)| \le 0.5L\|y-x\|^2 \qquad \forall x,y \in E_n .$$

Proof. For a smooth function $f(x)$,

$$f(y) = f(x) + \int_0^1 (f'(x + t(y-x)), y-x)\, dt .$$

Hence,

$$|f(y) - f(x) - (f'(x), y-x)| = \left| \int_0^1 (f'(x + t(y-x)) - f'(x), y-x)\, dt \right|$$

$$\le \int_0^1 \|f'(x + t(y-x)) - f'(x)\| \cdot \|y-x\|\, dt$$

$$\le L\|y-x\|^2 \int_0^1 t\, dt = 0.5L\|y-x\|^2 .$$

∎

Let a function $f(x)$ be strongly convex and differentiable on E_n, its derivative $f'(x)$ satisfy a Lipschitz condition on E_n, m be a strong convexity constant of $f(x)$ and L be a Lipschitz constant of the derivative of $f(x)$. Let

$$v_0^{(i)} = (f_\nu(x_0+he_i) - f_\nu(x_0)) \cdot h^{-1}, \qquad i \in 1:n, \qquad (17.6)$$

where e_i is the ith coordinate vector, $h > 0$ and $\nu \geq 0$ is the accuracy of computation of the function $f(x)$, i.e.,

$$f(x) - \nu \leq f_\nu(x) \leq f(x) + \nu .$$

<u>LEMMA 17.3</u>. If

$$0.5Lh + h^{-1}2\nu \leq 2\sqrt{\varepsilon m n^{-1}}, \qquad (17.7)$$

then the vector $v_0 = (v_0^{(1)}, \ldots, v_0^{(n)})$ with coordinates (17.6) is an ε-subgradient of $f(x)$ at the point x_0, i.e., $v_0 \in \partial_\varepsilon f(x_0)$.

<u>Proof</u>. It follows from Lemma 17.2 that

$$|f(x+\Delta x) - f(x) - (f'(x),\Delta x)| \leq 0.5L||\Delta x||^2 \qquad \forall\, x,\Delta x \in E_n. \qquad (17.8)$$

Let $f'(x_0) = \left(\frac{\partial f(x_0)}{\partial x^{(1)}}, \ldots, \frac{\partial f(x_0)}{\partial x^{(n)}}\right)$ be the gradient of the function $f(x)$ at the point x_0. Let us estimate $\left|v_0^{(i)} - \frac{\partial f(x_0)}{\partial x^{(i)}}\right|$. Since

$$\frac{\partial f(x_0)}{\partial x^{(i)}} = (f'(x_0),e_i) \qquad \forall\, i \in 1:n,$$

then

$$|(f_\nu(x_0+he_i) - f_\nu(x_0))h^{-1} - (f'(x_0), e_i)|$$
$$\leq |f(x_0+he_i) - f(x_0) - h(f'(x_0),e_i)|h^{-1} + 2\nu h^{-1}$$

(see (17.6)). From this, according to (17.8), we obtain

$$\left|v_0^{(i)} - \frac{\partial f(x_0)}{\partial x^{(i)}}\right| \leq 0.5Lh + 2\nu h^{-1} .$$

Hence, by virtue of (17.7),

$$\|v_0 - f'(x_0)\| \le \sqrt{n(0.5Lh + 2\nu h^{-1})^2} = (0.5Lh + 2\nu h^{-1})\sqrt{n} \le 2\sqrt{\varepsilon m} .$$

Finally, it follows from Lemma 17.1 that $v_0 \in \partial_\varepsilon f(x_0)$.

The lemma has been proved. ∎

Thus, in Lemma 17.3 we gave a method for constructing an ε-subgradient of the function $f(x)$ at the point x_0 by using approximate values of the function $f(x)$ at x_0 and $x_0 + he_i$, $i \in 1:n$.

Note that the approach described in Lemma 17.3 is frequently used in many practical computations, where it is difficult to find an explicit formula for the gradient of the function $f(x)$.

Now, let us consider the problem of computing an ε-subgradient of a maximum function of convex functions. Let

$$f(x) = \max_{y \in G} \phi(x,y) , \tag{17.9}$$

where $G \subset E_p$ is a closed bounded set, the function $\phi(x,y)$ is continuous together with $\frac{\partial\phi(x,y)}{\partial x}$ on $E_n \times G$ and is convex with respec[t] to x on E_n for each $y \in G$.

Let us introduce the notation:

$$R_\varepsilon(x) = \{y \in G \mid f(x) - \phi(x,y) \le \varepsilon\} , \qquad \varepsilon \ge 0, \qquad R(x) = R_0(x) ,$$

$$y(x) \in R(x) , \qquad y_\varepsilon(x) \in R_\varepsilon(x) ,$$

$$L_{\varepsilon\mu}(x) = \text{co} \{\partial_\mu \phi(x,y) \mid y \in R_\varepsilon(x)\} ,$$

$$\varepsilon \ge 0, \qquad \mu \ge 0, \qquad L(x) = L_{00}(x) .$$

It is known that $\partial f(x) = L(x)$ (see Section 5). Hence, if the problem of finding $\max_{y \in Q} \phi(x,y)$ is complicated, then to compute the function $f(x)$ we need an infinite process of successive approximations. The same of course is true for the computation of the derivative $\frac{\partial\phi(x,y)}{\partial x}$.

LEMMA 17.4. We have the inclusion

$$L_{\varepsilon\mu}(x) \subset \partial_{\varepsilon+\mu} f(x) \qquad \forall x \in E_n, \quad \forall \varepsilon \ge 0, \quad \mu \ge 0 . \tag{17.10}$$

Proof. Fix $\varepsilon \ge 0$, $\mu \ge 0$ and $x_0 \in E_n$. Let $y_\varepsilon(x_0) \in R_\varepsilon(x_0)$, i.e., $\phi(x_0, y_\varepsilon(x_0)) \ge f(x_0) - \varepsilon$ and let $v_\mu(x_0, y_\varepsilon(x_0)) \in \partial_\mu \phi(x_0, y_\varepsilon(x_0))$. We need to prove that

$$f(z) - f(x_0) \ge (v_\mu(x_0, y_\varepsilon(x_0)), z - x_0) - (\varepsilon + \mu) \qquad \forall z \in E_n .$$

Using the definitions of $f(x)$ and $\partial_\mu \phi(x,y)$, we obtain

$$\begin{aligned} f(z) - f(x_0) &\ge \phi(z, y(z)) - \phi(x_0, y_\varepsilon(x_0)) - \varepsilon \\ &\ge \phi(z, y_\varepsilon(x_0)) - \phi(x_0, y_\varepsilon(x_0)) - \varepsilon \\ &\ge (v_\mu(x_0, y_\varepsilon(x_0)), z - x_0) - \mu - \varepsilon \qquad \forall z \in E_n . \end{aligned}$$

The lemma has been proved. ∎

Lemmas 17.3 and 17.4 provide the following method for computing $(\varepsilon+\mu)$-subgradients of the function $f(x)$ defined by relation (17.9). Fix $\varepsilon > 0$, $\mu > 0$ and x_0 and find $y_\varepsilon(x_0) \in R_\varepsilon(x_0)$.

To do this, it is sufficient to solve the problem of maximizing the function $\phi(x_0, y)$ with respect to y on G up to ε, and this is, in general, achieved in a finite number of steps of an infinite procedure for the computation of $f(x_0)$.

Let us construct the vector $v_0 = (v_0^{(1)}, \ldots, v_0^{(n)})$ with the components

$$v_0^{(i)} = h^{-1}[\phi_\nu(x_0 + he_i, y_\varepsilon(x_0)) - \phi_\nu(x_0, y_\varepsilon(x_0))] , \qquad i \in 1:n . \tag{17.11}$$

THEOREM 17.1. Assume that a function $\phi(x,y)$ is strongly convex with respect to x on E_n for each $y \in G$ with a strong convexity constant m and that

$$\left\| \frac{\partial \phi(x,y)}{\partial x} - \frac{\partial \phi(x',y)}{\partial x} \right\| \le L\|x - x'\| \qquad \forall x \in E_n, \quad \forall x' \in S_h(x), \quad \forall y \in G.$$

If

$$0.5Lh + 2\nu h^{-1} \le 2\sqrt{\mu m n^{-1}} ,$$

then the relation

$$v_0 \in \partial_{\varepsilon+\mu} f(x_0) \tag{17.12}$$

is valid for the vector v_0 with the components given by (17.11).

Proof. Indeed, by virtue of Lemma 17.3, $v_0 \in \partial_\mu \phi(x_0, y_\varepsilon(x_0))$, hence, Lemma 17.4 implies (17.12). ∎

Chapter 2

QUASIDIFFERENTIABLE FUNCTIONS

1. DEFINITION AND EXAMPLES OF QUASIDIFFERENTIABLE FUNCTIONS

1. In Chapter 1, we studied in detail nonsmooth convex functions: in particular, we established that if $f(x)$ is a finite convex function on an open convex set $S \subset E_n$, then it is directionally differentiable at any point $x_0 \in S$ and, furthermore,

$$\frac{\partial f(x_0)}{\partial g} = \max_{v \in \partial f(x)} (v,g) , \tag{1.1}$$

where $\partial f(x_0)$ is the subdifferential of the function $f(x)$ at the point x_0. The set $\partial f(x_0) \subset E_n$ is a compact convex set.

In Section 1.14 we established under some assumptions the directional differentiability of supremum functions and derived the formula

$$\frac{\partial f(x_0)}{\partial g} = \max_{v \in \partial f(x_0)} (v,g) , \tag{1.2}$$

where $\partial f(x_0) = \text{co}\, H(x_0) \subset E_n$ is a convex compact set (see Remark 2 of Section 1.14). This set $\partial f(x_0)$ will also be called the *subdifferential* of the function $f(x)$ at the point x_0.

Thus, if f is a convex function or a supremum function, then every point x_0 is associated with the set $\partial f(x_0)$, i.e., the subdifferential of f.

Moreover, we have the following properties (which were established in Chapter 1 but hold as well for supremum functions):

1. The directional derivative is expressed by formulas (1.1) and (1.2) in terms of $\partial f(x_0)$.

2. The necessary condition for a minimum on E_n is given by

$$0 \in \partial f(x^*) , \tag{1.3}$$

where x^* is a minimum point of $f(x)$ on E_n.

3. If $0 \notin \partial f(x_0)$, then the direction given by

$$g(x_0) = -v(x_0) \|v(x_0)\|^{-1}$$

is the steepest descent direction of the function $f(x)$ at the point x_0, where $\|v(x_0)\| = \min_{v \in \partial f(x_0)} \|v\|$.

In the following chapters, we shall indicate how we can use the steepest descent direction to construct a numerical method for the minimization of $f(x)$.

As a result, one can see that the concept of a subdifferential is useful for the investigation of functions, in particular, for solving optimization problems.

These considerations led to attempts to find a wider class of functions for which we can construct something analogous to the subdifferential. Among these attempts, we mention F.H. Clarke [133], B.N. Pshenichnyj [104], and N.Z. Shor [126].

As early as 1969, Pshenichnyj [103] considered functions (which he called quasidifferentiable) for which at each point x_0 the set $\partial f(x_0)$ exists and the directional derivative is calculated by formula (1.1) in terms of the set $\partial f(x_0)$. The class of such functions is not a linear space (the function $f_1(x) = -f(x)$ is no longer quasidifferentiable in the Pshenichnyj sense.

Clarke [133] introduced the concept of an upper subdifferential. Let $f(x)$ be a function continuous in a neighborhood of x_0.

Define

$$\frac{\partial_{Cl} f(x_0)}{\partial g} = \overline{\lim_{\substack{x' \to x_0 \\ \alpha \to +0}}} \frac{f(x'+\alpha g) - f(x')}{\alpha} .$$

This quantity, if it exists and is finite, is called the *upper subdifferential* (or *Clarke derivative*) of the function $f(x)$ at a point x_0 in the direction g, and we shall say that the function $f(x)$ is *differentiable in the Clarke sense* at the point x_0.

It was established that if a function $f(x)$ satisfies a Lipschitz condition in the neighborhood of a point x_0, then it is differentiable in the Clarke sense at the point x_0 and, also, there exists a convex compact set $\partial_{Cl} f(x_0) \subset E_n$ such that

$$\frac{\partial_{Cl} f(x_0)}{\partial g} = \max_{v \in \partial_{Cl} f(x_0)} (v,g).$$

The set $\partial_{Cl} f(x_0)$ is called the *Clarke subdifferential* of $f(x)$ at x_0. We denote by $\Lambda(f) \subset S$ the set of points at which the function $f(x)$ is differentiable.

Clarke [133] showed that

$$\partial_{Cl} f(x_0) = \text{co } \{v \mid \exists\{x_k\}: x_k \to x_0,\ x_k \in \Lambda(f),\ f'(x_k) \to v\}. \quad (1.3)$$

In addition, if $S = E_n$, then for $x^* \in E_n$ to be a minimum point of the function $f(x)$ on E_n it is necessary that

$$0 \in \partial_{Cl} f(x^*) . \quad (1.4)$$

The Clarke subdifferential has many important and interesting properties (see [165], [9] and also [65, 136, 140]).

<u>EXAMPLE 1</u>. Let $x = (x^{(1)}, x^{(2)}) \in E_2$, $x_2 = 0 = (0,0)$.

Consider the functions

$$f_1(x) = |x^{(1)}| + |x^{(2)}| , \qquad f_2(x) = |x^{(1)}| - |x^{(2)}| ,$$

$$f_3(x) = -|x^{(1)}| - |x^{(2)}| , \qquad f_4(x) = -|x^{(1)}| + |x^{(2)}| ,$$

$$f_5(x) = \max \{f_2(x), f_4(x)\} .$$

It is easy to show that, for all functions $f_i(x)$, $i \in 1:5$, the following property is valid: the plane E_2 is divided into four parts and in each of them the function $f_i(x)$ coincides with one of the following functions:

$$x^{(1)} + x^{(2)}, \qquad x^{(1)} - x^{(2)}, \qquad -x^{(1)} + x^{(2)}, \qquad -x^{(1)} - x^{(2)} .$$

For example, for the function $f_1(x)$ we have

$$f_1(x) = \begin{cases} x^{(1)} + x^{(2)} & \text{if } x^{(1)} \geq 0, \ x^{(2)} \geq 0, \\ x^{(1)} - x^{(2)} & \text{if } x^{(1)} \geq 0, \ x^{(2)} \leq 0, \\ -x^{(1)} + x^{(2)} & \text{if } x^{(1)} \leq 0, \ x^{(2)} \geq 0, \\ -x^{(1)} - x^{(2)} & \text{if } x^{(1)} \leq 0, \ x^{(2)} \leq 0. \end{cases}$$

Hence, we conclude from (1.3) that

$$\partial_{Cl} f_i(x_0) = \text{co} \{(1,1), (1,-1), (-1,1) (-1,-1)\} \quad \forall\, i \in 1:5 .$$

For all functions $f_i(x)$, $i \in 1:5$, the necessary condition (1.4 is satisfied at the point $x_0 = 0$, though x_0 is a minimum point onl for the functions $f_1(x)$ and $f_5(x)$.

In addition, we note that the Clarke subdifferential is suitable neither for the calculation of the directional derivative (because the Clarke derivative is not always equal to the directional derivative) nor for finding the steepest descent direction.

Thus even for a concave function on E_n, $\partial_{Cl} f(x_0) = \bar{\partial} f(x_0)$ (se Subection 1.5.4); however,

$$\frac{\partial f(x_0)}{\partial g} = \min_{v \in \bar{\partial} f(x_0)} (v,g) = \min_{v \in \partial_{Cl} f(x_0)} (v,g) ,$$

i.e., formula (1.1) is no longer valid.

In this chapter, we do not investigate the class of functions that are sufficiently rich and have <<good>> properties instead of a wide class of functions satisfying a Lipschitz condition.

2. Let a finite function $f(x)$ be given on some open set $S \subset E_n$.

We say that the function $f(x)$ is *quasidifferentiable* at a point $x_0 \in S$ if it is differentiable at x_0 in any direction $g \in E_n$ and there exist convex compact sets $\underline{\partial} f(x_0) \subset E_n$ and $\overline{\partial} f(x_0) \subset E_n$ such that

$$\frac{\partial f(x_0)}{\partial g} \equiv \lim_{\alpha \to +0} \alpha^{-1}[f(x_0+\alpha g) - f(x_0)]$$

$$= \max_{v \in \underline{\partial} f(x_0)} (v,g) + \max_{w \in \overline{\partial} f(x_0)} (w,g) \quad . \qquad (1.5)$$

The pair of sets

$$Df(x_0) = [\underline{\partial} f(x_0), \overline{\partial} f(x_0)]$$

is called a *quasidifferential* (QD) of the function $f(x)$ at the point x_0 and the sets $\underline{\partial} f(x_0)$ and $\overline{\partial} f(x_0)$ are called, respectively, a *subdifferential* and a *superdifferential* of the function $f(x)$ at the point x_0.

From formula (1.5) it is seen that QD is not uniquely defined: the pair of sets $A(x_0) = [\underline{\partial} f(x_0) + B, \overline{\partial} f(x_0) - B]$, is also a quasidifferential, where $B \subset E_n$ is an arbitrary convex compact set.

If among quasidifferentials of the function $f(x)$ at a point x_0 there is an element of the form

$$Df(x_0) = [\underline{\partial} f(x_0), \{0\}] \equiv [\underline{\partial} f(x_0), 0] \quad ,$$

then the function $f(x)$ is called *subdifferentiable* at the point x_0.

Analogously, if among quasidifferentials of the function $f(x)$

at a point x_0 there is an element of the form

$$\mathcal{D}f(x_0) = [\{0\}, \overline{\partial}f(x_0)] \equiv [0, \overline{\partial}f(x_0)] ,$$

then the function $f(x)$ is called *superdifferentiable* at x_0.

Let $\Omega \subset S$. If a function $f(x)$ is quasidifferentiable at each point $x_0 \in \Omega$, then it is called *quasidifferentiable* on the set Ω.

We introduce very simple examples of quasidifferentiable functions.

EXAMPLE 2. Let a function $f(x)$ be continuously differentiable on an open set $S \subset E_n$ and let $x_0 \in S$. Since

$$\frac{\partial f(x_0)}{\partial g} = (f'(x_0), g) ,$$

then it is clear that $f(x)$ is a quasidifferentiable function on S and, moreover, we can take the pair of sets

$$\mathcal{D}f(x_0) = [f'(x_0), 0] ,$$

i.e.,

$$\underline{\partial}f(x_0) = \{f'(x_0)\} , \qquad \overline{\partial}f(x_0) = \{0\} ,$$

as the QD of the function $f(x)$ at x_0.

Obviously, the pair of sets

$$[0, f'(x_0)]$$

is also the QD of the function $f(x)$ at x_0.

Thus, a function $f(x)$ differentiable at a point x_0 is both subdifferentiable and superdifferentiable at that point.

EXAMPLE 3. Let $f(x)$ be a finite convex function on an open convex set $S \subset E_n$ and let $x_0 \in S$. Then, $f(x)$ is directionally differentiable. By Theorem 5.2 of Chapter 1, we have

$$\frac{\partial f(x_0)}{\partial g} = \max_{v \in \partial f(x_0)} (v,g) ,$$

where $\partial f(x_0) = \{v \in E_n \mid f(x)-f(x_0) \geq (v,x-x_0) \ \forall x \in S\}$.

It is then seen that the function $f(x)$ is quasidifferentiable at the point x_0 and, moreover, we can take

$$\mathcal{D}f(x_0) = [\partial f(x_0), 0]$$

as its QD, i.e., $\underline{\partial} f(x_0) = \partial f(x_0)$, $\overline{\partial} f(x_0) = \{0\}$.

Since $\overline{\partial} f(x_0) = \{0\}$, then a convex function is subdifferentiable at a point x_0 (also on S).

<u>EXAMPLE 4</u>. Let $f(x)$ be a finite concave function on an open convex set $S \subset E_n$ and let $x_0 \in S$. The function $f(x)$ is directionally differentiable at a point x_0. We showed in Subsection 1.5.4 that

$$\frac{\partial f(x_0)}{\partial g} = \min_{w \in \overline{\partial} f(x_0)} (w,g) ,$$

where

$$\overline{\partial} f(x_0) = \{w \in E_n \mid f(x)-f(x_0) \le (w,x-x_0) \ \forall x \in S\} .$$

It is then obvious that the function $f(x)$ is quasidifferentiable at the point x_0 and, moreover, we can take

$$\mathcal{D}f(x_0) = [0, \overline{\partial} f(x_0)]$$

as its QD.

Since we have here $\underline{\partial} f(x_0) = \{0\}$, then a concave function is superdifferentiable at the point x_0 (also on S).

<u>EXAMPLE 5</u>. Let

$$f(x) = \sup_{y \in G} \phi(x,y) ,$$

where $x \in E_n$, $y \in E_p$ and $G \subset E_p$ is an arbitrary set. Let the families of functions $\phi(x,y)$ and $\phi'_x(x,y)$, where $y \in G$, be equicontinuous in x at a point x_0. In addition, we assume that

$$|\phi(x_0,y)| \le K < \infty , \qquad |\phi'_x(x_0,y)| \le K < \infty \qquad \forall y \in G .$$

As shown in Theorem 14.1 of Chapter 1, the function $f(x)$ is subdifferentiable and, of course, quasidifferentiable. We can take a pair of sets

$$Df(x_0) = [\text{co } H(x_0), \{0\}]$$

as its QD, where $H(x_0)$ is the same as in Subsection 1.14.1.

2. BASIC PROPERTIES OF QUASIDIFFERENTIABLE FUNCTIONS. BASIC FORMULAS OF QUASIDIFFERENTIAL CALCULUS

1. At first we shall establish some elementary properties of quasidifferentiable functions. As before, let all functions considered below be given on an open set $S \subset E_n$.

1. If a function $f_1(x)$ is quasidifferentiable at a point $x_0 \in S$, then the function $f(x) = cf_1(x)$, where $c \geq 0$, is also quasidifferentiable at x_0.

Indeed, since

$$\frac{\partial f(x_0)}{\partial g} = c\frac{\partial f_1(x_0)}{\partial g} = c\left[\max_{v \in \underline{\partial} f_1(x_0)} (v,g) + \min_{w \in \overline{\partial} f_1(x_0)} (w,g)\right]$$

$$= \max_{v \in [c\underline{\partial} f_1(x_0)]} (v,g) + \min_{w \in [c\overline{\partial} f_1(x_0)]} (w,g) ,$$

then it is clear that the function $f(x)$ is quasidifferentiable at the point x_0 and, in this case, we have

$$Df(x_0) = [\underline{\partial} f(x_0), \overline{\partial} f(x_0)] ,$$

where $\underline{\partial} f(x_0) = c\underline{\partial} f_1(x_0)$, $\overline{\partial} f(x_0) = c\overline{\partial} f_1(x_0)$. ∎

2. If a function $f_1(x)$ is quasidifferentiable at a point $x_0 \in S$, then the function $f(x) = -f_1(x)$ is also quasidifferentiable at x_0.

We show this. First we note that

$$\frac{\partial f(x_0)}{\partial g} = -\frac{\partial f_1(x_0)}{\partial g} = -\left[\max_{v \in \underline{\partial} f_1(x_0)} (v,g) + \min_{w \in \overline{\partial} f_1(x_0)} (w,g)\right] . \quad (2.1)$$

Since

$$-\max_{v\in\underline{\partial}f_1(x_0)}(v,g) = \min_{v\in\underline{\partial}f_1(x_0)}(-v,g) = \min_{w\in[-\underline{\partial}f_1(x_0)]}(w,g),$$

$$-\min_{w\in\overline{\partial}f_1(x_0)}(w,g) = \max_{v\in\overline{\partial}f_1(x_0)}(-w,g) = \max_{v\in[-\overline{\partial}f_1(x_0)]}(v,g),$$

then from (2.1) we have

$$\frac{\partial f(x_0)}{\partial g} = \max_{v\in[-\overline{\partial}f_1(x_0)]}(v,g) + \min_{w\in[-\underline{\partial}f_1(x_0)]}(w,g) .$$

Thus, the function $f(x)$ is quasidifferentiable at the point x_0 and, moreover,

$$\mathcal{D}f(x_0) = [\underline{\partial}f(x_0), \overline{\partial}f(x_0)] ,$$

where

$$\underline{\partial}f(x_0) = -\overline{\partial}f_1(x_0) , \qquad \overline{\partial}f(x_0) = -\underline{\partial}f_1(x_0) .$$

3. If functions $f_1(x)$ and $f_2(x)$ are quasidifferentiable at a point $x_0\in S$, then the function $f(x) = f_1(x)+f_2(x)$ is also quasidifferentiable at the point x_0.

Indeed, since

$$\frac{\partial f(x_0)}{\partial g} = \frac{\partial f_1(x_0)}{\partial g} + \frac{\partial f_2(x_0)}{\partial g} = \max_{v\in\underline{\partial}f_1(x_0)}(v,g) + \min_{w\in\overline{\partial}f_1(x_0)}(w,g)$$

$$+ \max_{v\in\underline{\partial}f_2(x_0)}(v,g) + \min_{w\in\overline{\partial}f_2(x_0)}(w,g)$$

$$= \max_{v\in[\underline{\partial}f_1(x_0)+\underline{\partial}f_2(x_0)]}(v,g) + \min_{w\in[\overline{\partial}f_1(x_0)+\overline{\partial}f_2(x_0)]}(w,g) ,$$

then it is clear that the function $f(x)$ is quasidifferentiable at x_0 and, in this case, we have

$$\mathcal{D}f(x_0) = [\underline{\partial}f(x_0), \overline{\partial}f(x_0)],$$

where

$$\underline{\partial}f(x_0) = [\underline{\partial}f_1(x_0)+\underline{\partial}f_2(x_0)] , \qquad \overline{\partial}f(x_0) = [\overline{\partial}f_1(x_0)+\overline{\partial}f_2(x_0)].$$

■

4. If functions $f_1(x), \ldots, f_k(x)$ are quasidifferentiable at a point $x_0\in S$, then the function $f(x) = \sum_{i=1}^{k} c_i f_i(x)$, where $c_i\in E_1$,

is also quasidifferentiable at x_0.

This is obvious from properties (1.) - (3.).

It is not difficult to write down formulas for a subdifferentia[l] and a superdifferential of the function $f(x)$.

Now, from properties (1.) - (4.) we have

LEMMA 2.1. The set of quasidifferentiable functions is a linear space.

2. Let $S \subset E_n$ be an open set and let $f_i(x)$, $i \in I \equiv 1:N$, be quasidifferentiable functions at a point $x_0 \in S$. Define the function

$$f(x) = \max_{i \in I} f_i(x) . \tag{2.2}$$

First of all, we note that if the functions $f_i(x)$ are differentiable in a direction $g \in E_n$ at a point x_0, then the function $f(x)$ is also differentiable in the direction g at the point x_0 and, moreover,

$$\frac{\partial f(x_0)}{\partial g} = \max_{i \in R(x_0)} \frac{\partial f_i(x_0)}{\partial g} , \tag{2.3}$$

where $R(x) = \{i \in I \mid f_i(x)=f(x)\}$.

We show this: since

$$f_i(x_0+\alpha g) = f_i(x_0) + \alpha \frac{\partial_i(x_0)}{\partial g} + o_i(\alpha) , \tag{2.4}$$

where $\frac{o_i(\alpha)}{\alpha} \xrightarrow[\alpha \to +0]{} 0$, then, for sufficiently small $\alpha > 0$, we have

$$f(x_0+\alpha g) \equiv \max_{i \in I} f_i(x_0+\alpha g) = \max_{i \in R(x_0)} f_i(x_0+\alpha g) .$$

Hence, from (2.4) we get

$$f(x_0+\alpha g) - f(x_0) = \max_{i \in R(x_0)} \left[\alpha \frac{\partial f_i(x_0)}{\partial g} + o_i(\alpha)\right] .$$

Now, we have

$$\underline{o}(\alpha) \leq f(x_0+\alpha g) - f(x_0) - \alpha \max_{i \in R(x_0)} \frac{\partial f_i(x_0)}{\partial g} \leq \overline{o}(\alpha) , \tag{2.5}$$

where

$$\underline{o}(\alpha) = \min_{i \in I} o_i(\alpha) , \qquad \overline{o}(\alpha) = \max_{i \in I} o_i(\alpha) ,$$

$$\frac{\underline{o}(\alpha)}{\alpha} \xrightarrow[\alpha \to +0]{} 0 , \qquad \frac{\overline{o}(\alpha)}{\alpha} \xrightarrow[\alpha \to +0]{} 0 .$$

Let $h(\alpha) = \alpha^{-1}[f(x_0+\alpha g) - f(x_0)]$. From (2.5) it follows that

$$\left| h(\alpha) - \max_{i \in R(x_0)} \frac{\partial f_i(x_0)}{\partial g} \right| \xrightarrow[\alpha \to +0]{} 0 ,$$

implying in turn (2.3). ∎

LEMMA 2.2. If the functions $f_i(x)$, $i \in I \equiv 1:N$, are quasidifferentiable at the point $x_0 \in S$, then the function

$$f(x) = \max_{i \in I} f_i(x)$$

is quasidifferentiable at x_0 and, moreover,

$$\mathcal{D}f(x_0) = [\underline{\partial} f(x_0), \overline{\partial} f(x_0)] ,$$

where

$$\underline{\partial} f(x_0) = \text{co}\left\{\underline{\partial} f_k(x_0) - \sum_{\substack{i \in R(x_0) \\ i \neq k}} \overline{\partial} f_i(x_0) \;\middle|\; k \in R(x_0)\right\} , \tag{2.6}$$

$$\overline{\partial} f(x_0) = \sum_{k \in R(x_0)} \overline{\partial} f_k(x_0) . \tag{2.7}$$

Proof. The functions $f_i(x)$ are quasidifferentaible at the point x_0, i.e.,

$$\frac{\partial f_i(x_0)}{\partial g} = \max_{v \in \underline{\partial} f_i(x_0)} (v,g) + \min_{w \in \overline{\partial} f_i(x_0)} (w,g) . \tag{2.8}$$

Let

$$A_i = \underline{\partial} f_i(x_0) , \qquad B_i = \overline{\partial} f_i(x_0) . \tag{2.9}$$

Without loss of generality, we can assume that $R(x_0) \equiv 1:m$, where $m \leq N$. From (2.3) and (2.8) we have

$$\frac{\partial f(x_0)}{\partial g} = \max_{i \in 1:m} \left[\max_{v \in A_i} (v,g) + \min_{w \in B_i} (w,g)\right] . \tag{2.10}$$

We shall consider first the case $m = 2$. Let a, b, c, d, r be real numbers. Since

$$\max\{a+b, c+d\} = \max\{a+b-r, c+d-r\} + r ,$$

then, letting $r = b + d$, we have

$$\max\{a+b, c+d\} = \max\{a-d, c-b\} + (b+d) .$$

Hence, from (2.10)

$$H_2 \equiv \max_{i\in 1:2} \left[\max_{v\in A_i}(v,g) + \min_{w\in B_i}(w,g)\right]$$

$$= \max\left\{\max_{v\in A_1}(v,g) - \min_{w\in B_2}(w,g),\ \max_{v\in A_2}(v,g) - \min_{w\in B_1}(w,g)\right\}$$

$$+ \left[\min_{w\in B_1}(w,g) + \min_{w\in B_2}(w,g)\right] . \qquad (2.11)$$

Since

$$-\min_{w\in B_i}(w,g) = \max_{v\in B_i}(-w,g) = \max_{v\in[-B_i]}(v,g) , \qquad i \in 1:2 ,$$

then from (2.11)

$$H_2 = \max\left\{\max_{v\in A_1}(v,g) + \max_{v\in[-B_2]}(v,g),\ \max_{v\in A_2}(v,g) + \max_{v\in[-B_1]}(v,g)\right\}$$

$$+ \min_{w\in B_1}(w,g) + \min_{w\in B_1}(w,g) . \qquad (2.12)$$

On the other hand, we have

$$\max_{v\in A}(v,g) + \max_{v\in B}(v,g) = \max_{v\in[A+B]}(v,g) ,$$

$$\min_{w\in A}(w,g) + \min_{w\in B}(w,g) = \min_{w\in[A+B]}(w,g) .$$

Therefore, from (2.12)

$$H_2 = \max\left\{\max_{v\in[A_1-B_2]}(v,g);\ \max_{v\in[A_2-B_1]}(v,g)\right\} + \min_{w\in[B_1+B_2]}(w,g) .$$

Since

$$\max\left\{\max_{v\in A}(v,g),\ \max_{v\in B}(v,g)\right\} = \max_{v\in co\{A\cup B\}}(v,g) ,$$

then

$$H_2 = \max_{v\in G_{21}}(v,g) + \min_{W\in G_{22}}(w,g) , \qquad (2.13)$$

where $G_{21} = co\{A_1-B_2,\ A_2-B_1\}$, $G_{22} = B_1+B_2$.

To complete the proof, we use induction. Assume that the relation

$$H_\ell \equiv \max_{i\in 1:\ell}\left|\max_{v\in A_i}(v,g) + \min_{w\in B_i}(w,g)\right| = \max_{v\in G_{\ell 1}}(v,g) + \min_{w\in G_{\ell 2}}(w,g) \qquad (2.14)$$

has already been established, where

$$G_{\ell 1} = co\left\{A_k - \sum_{\substack{i\in 1:\ell\\ i\neq k}} B_i \mid k\in 1:\ell\right\} , \qquad (2.15)$$

$$G_{\ell 2} = \sum_{i=1}^{\ell} B_i . \qquad (2.16)$$

Then we get

$$\begin{aligned} H_{\ell+1} &\equiv \max_{i\in 1:(\ell+1)}\left|\max_{v\in A_i}(v,g) + \min_{w\in B_i}(w,g)\right| \\ &= \max\left\{H_\ell,\ \max_{v\in A_{\ell+1}}(v,g) + \min_{w\in B_{\ell+1}}(w,g)\right\} \\ &= \max\left\{\max_{v\in G_{\ell 1}}(v,g) + \min_{w\in G_{\ell 2}}(w,g),\ \max_{v\in A_{\ell+1}}(v,g) + \min_{w\in B_{\ell+1}}(w,g)\right\} . \end{aligned}$$

Using the same arguments as in the derivation of (2.13), we obtain

$$H_{\ell+1} = \max_{v\in A}(v,g) + \min_{w\in B}(w,g) ,$$

where $A = co\{G_{\ell 1}-B_{\ell+1},\ A_{\ell+1}-G_{\ell 2}\}$, $B = G_{\ell 2}+B_{\ell+1}$.

Finally, we observe

$$A = co\left\{A_k - \sum_{\substack{i\in 1:(\ell+1)\\ i\neq k}} B_i \mid k\in 1:(\ell+1)\right\} \equiv G_{\ell+1,1} ,$$

$$B = \sum_{i=1}^{\ell+1} B_i \equiv G_{\ell+1,2} .$$

Thus, it is proved that

$$H_m = \max_{v \in G_{m1}} (v,g) + \min_{w \in G_{m2}} (w,g) .$$

Now, the validity of the lemma follows from (2.15), (2.16), (2.9) and (2.10).

LEMMA 2.3. If the functions $f_i(x)$, $i \in I \equiv 1:N$, are quasidifferentiable at a point $x_0 \in S$, then the function

$$f(x) = \min_{i \in I} f_i(x)$$

is quasidifferentiable at x_0 and, moreover,

$$Df(x_0) = [\underline{\partial} f(x_0), \overline{\partial} f(x_0)] ,$$

where

$$\underline{\partial} f(x_0) = \sum_{k \in Q(x_0)} \underline{\partial} f_k(x_0) ,$$

$$\overline{\partial} f(x_0) = \text{co} \left\{ \overline{\partial} f_k(x_0) - \sum_{\substack{i \in Q(x_0) \\ i \neq k}} \underline{\partial} f_i(x_0) \mid k \in Q(x_0) \right\} ,$$

$$Q(x) = \{i \in I \mid f_i(x) = f(x)\} , \tag{2.17}$$

$$\text{co} \{A_k \mid k \in Q\} \equiv \text{co} \{\cup \{A_k \mid k \in Q\}\} .$$

Proof. Since $f(x) = -\max_{i \in I} h_i(x)$, where $h_i(x) = -f_i(x)$, then using Lemma 2.2 and property (2.) in Subsection 1, we obtain the assertion of the lemma directly.

From Lemmas 2.2 and 2.3 we have

COROLLARY. If functions

$$f_{ij}(x) \qquad (i \in I \equiv 1:N, \quad j \in J_i \equiv 1:N_i)$$

are quasidifferentiable at a point $x_0 \in S$, then the function

$$f(x) = \max_{i \in I} \min_{j \in J_i} f_{ij}(x) \tag{2.18}$$

is also quasidifferentiable at x_0.

Analogously, we can establish that a sequential maxmin function, i.e.,

$$f(x) = \max_i \min_j \max_k \dots \min_\ell f_{ijk\dots\ell}(x)$$

is quasidifferentiable if $f_{ijk\dots\ell}(x)$ are quasidifferentiable functions.

PROBLEM 2.1. Find the quasidifferential of the function (2.18).

EXAMPLE 1. Let $f(x)$ be a quasidifferentiable function on an open set $S \in E_n$. Then, the function

$$F(x) = \text{sat } f(x) \equiv \begin{cases} +1 & \text{if} \quad f(x) > 1 , \\ f(x) & \text{if} \quad -1 \le f(x) \le 1 , \\ -1 & \text{if} \quad f(x) < -1 , \end{cases} \tag{2.19}$$

is also quasidifferentiable on S.

To show this, let

$$F(x) = \max \{f_1(x), -1\} ,$$

where $f_1(x) = \min \{f(x), +1\}$.

The function $f_1(x)$ is quasidifferentiable due to Lemma 2.3 and, hence, the function $F(x)$ is quasidifferentiable due to Lemma 2.2.

PROBLEM 2.2. Find a quasidifferential of the function (2.19).

3. Let $S \subset E_n$ be an open set and let a function $f(x)$ be quasidifferentiable at a point $x_0 \in S$ and continuous in a neighborhood of x_0.

LEMMA 2.4. If $f_1(x_0) \neq 0$, then the function $f(x) = f_1^{-1}(x)$ is quasidifferentiable at the point x_0 and, moreover,

$$\mathcal{D}f(x_0) = [\underline{\partial} f(x_0), \overline{\partial} f(x_0)] ,$$

where $\underline{\partial} f(x_0) = -f_1^{-2}(x_0)\overline{\partial} f_1(x_0)$, $\overline{\partial} f(x_0) = -f_1^{-2}(x_0)\underline{\partial} f_1(x_0)$.

Here and below, $f_1^{-1}(x) = (f_1(x))^{-1}$ and $f_1^{-2}(x) = (f_1(x))^{-2}$.

Proof. Let us note that

$$\frac{\partial f(x_0)}{\partial g} \equiv \lim_{\alpha\to+0} \alpha^{-1}[f(x_0+\alpha g) - f(x_0)]$$

$$= \lim_{\alpha\to+0} \alpha^{-1}[f_1^{-1}(x_0+\alpha g) - f_1^{-1}(x_0)]$$

$$= \lim_{\alpha\to+0} [f_1^{-1}(x_0)\, f_1^{-1}(x_0+\alpha g)\, \alpha^{-1}(f_1(x_0)-f_1(x_0+\alpha g))]$$

$$= -f_1^{-2}(x_0)\, \frac{\partial f_1(x_0)}{\partial g} .$$

Now, using properties (1.) and (2.) in Subsection 1, we complete the proof. ∎

LEMMA 2.5. If functions $f_1(x)$ and $f_2(x)$ are quasidifferentiable at a point $x_0 \in S$, then the function $f(x) = f_1(x)f_2(x)$ is also quasidifferentiable at that point and, moreover,

$$\mathcal{D}f(x_0) = [\underline{\partial} f(x_0), \overline{\partial} f(x_0)] ,$$

where

$$\underline{\partial} f(x_0) = \begin{cases} f_1\underline{\partial} f_2+f_2\underline{\partial} f_1 & \text{if } f_1 \ge 0, \ f_2 \ge 0 , \\ f_1\overline{\partial} f_2+f_2\underline{\partial} f_1 & \text{if } f_1 \le 0, \ f_2 \ge 0 , \\ f_1\overline{\partial} f_2+f_2\overline{\partial} f_1 & \text{if } f_1 \le 0, \ f_2 \le 0 , \\ f_1\underline{\partial} f_2+f_2\overline{\partial} f_1 & \text{if } f_1 \ge 0, \ f_2 \le 0 , \end{cases}$$

$$\overline{\partial} f(x_0) = \begin{cases} f_1\overline{\partial} f_2+f_2\overline{\partial} f_1 & \text{if } f_1 \ge 0, \ f_2 \ge 0 , \\ f_1\underline{\partial} f_2+f_2\overline{\partial} f_1 & \text{if } f_1 \le 0, \ f_2 \ge 0 , \\ f_1\underline{\partial} f_2+f_2\underline{\partial} f_1 & \text{if } f_1 \le 0, \ f_2 \le 0 , \\ f_1\overline{\partial} f_2+f_2\underline{\partial} f_1 & \text{if } f_1 \ge 0, \ f_2 \le 0 . \end{cases}$$

Here, all values of the functions $f_1(x)$, $f_2(x)$ and their sub- and superdifferentials are calculated at the point x_0.

Proof. It is obvious that

$$\frac{\partial f(x_0)}{\partial g} = f_1(x_0)\, \frac{\partial f_2(x_0)}{\partial g} + f_2(x_0)\, \frac{\partial f_1(x_0)}{\partial g} . \tag{2.20}$$

Since

$$\frac{\partial f_i(x_0)}{\partial g} = \max_{v \in \underline{\partial} f_i(x_0)} (v,g) + \min_{w \in \overline{\partial} f_i(x_0)} (w,g), \qquad i \in 1:2,$$

then from (2.20)

$$\frac{\partial f(x_0)}{\partial g} = f_1(x_0) \max_{v \in \underline{\partial} f_2(x_0)} (v,g) + f_1(x_0) \min_{w \in \overline{\partial} f_2(x_0)} (w,g)$$

$$+ f_2(x_0) \max_{v \in \underline{\partial} f_1(x_0)} (v,g) + f_2(x_0) \min_{w \in \overline{\partial} f_1(x_0)} (w,g) .$$

Using properties (1.) and (2.) in Subsection 1, we get the proof. ■

4. Let $F(y_1, y_2, \ldots, y_s)$ be a continuously differentiable function on $\Omega \subset E_s$ and let $f_i(x)$, $i \in 1:s$, be quasidifferentiable functions at a point $x_0 \in S$. Assume that

$$y_0 = (y_0^{(1)}, \ldots, y_0^{(s)}) \equiv (f_1(x_0), \ldots, f_s(x_0)) \in \Omega .$$

<u>LEMMA 2.6</u>. The function $f(x) = F(x_1(x), \ldots, f_s(x))$ is quasidifferentiable at the point x_0.

<u>Proof</u>. Is obvious if we take into account that

$$\frac{\partial f(x_0)}{\partial g} = \frac{\partial F(y_0)}{\partial y^{(1)}} \frac{\partial f_1(x_0)}{\partial g} + \frac{\partial F(y_0)}{\partial y^{(2)}} \frac{\partial f_2(y_0)}{\partial g}$$

$$+ \cdots + \frac{\partial F(y_0)}{\partial y^{(s)}} \frac{\partial f_s(y_0)}{\partial g}$$

And now, following the lines of the proof of Lemma 2.5, we complete the proof of Lemma 2.6.

Thus, the set of quasidifferentiable functions is a linear space, which is closed with respect to all <<algebraic>> operations: the operations of taking the pointwise maximum, minimum and sequential maximum.

<u>REMARK</u>. Let $D = [A,B]$ be a pair of sets in E_n. Define the operations of multiplication by a real number λ and of addition as follows:

$$D\lambda = \begin{cases} [\lambda A, \lambda B] & \text{if } \lambda \geq 0 , \\ [\lambda B, \lambda A] & \text{if } \lambda < 0 . \end{cases}$$

If $D_1 = [A_1, B_1]$, $D_2 = [A_2, B_2]$, then $D_1 + D_2 = [A_1 + A_2, B_1 + B_2]$.

Then, the formulas for a quasidifferential of a linear combination, a product and a quotient of quasidifferentiable functions can be written in more compact forms:

$$D\left[\sum_{i=1}^{N} \lambda_i f_i(x_0)\right] = \sum_{i=1}^{N} \lambda_i Df_i(x_0) ,$$

$$D[f_1(x_0) f_2(x_0)] = f_1(x_0) Df_2(x_0) + f_2(x_0) Df_1(x_0) ,$$

$$D\left[\frac{f_1(x_0)}{f_2(x_0)}\right] = \frac{1}{f_2^2(x_0)} [f_2(x_0) Df_1(x_0) - f_1(x_0) Df_2(x_0)]$$

(in the last formula, it is assumed that $f_2(x_0) \neq 0$).

These representations are a generalization of ordinary formulas in differential calculus.

3. CALCULATING QUASIDIFFERENTIALS: EXAMPLES

Simple examples of the calculation of quasidifferentials will be given only for the sake of illustration and the reader familiar with the subject matter can view them as an exercise. Let

$$x = (x^{(1)}, x^{(2)}) \in E_2 , \qquad x_0 = (0,0) = 0 ,$$

$$A_1 = (1,0) , \qquad A_2 = (-1,0) ,$$

$$B_1 = (0,1) , \qquad B_2 = (0,-1) ,$$

$$C_1 = (1,1) , \qquad C_2 = (-1,-1) .$$

<u>EXAMPLE 1</u>. Let $f(x) = f(x^{(1)}, x^{(2)}) = |x^{(1)}|$. We have

$$|x^{(1)}| = \max \{x^{(1)}, -x^{(1)}\} = \max \{f_1(x), f_2(x)\} ,$$

where $f_1(x) = x^{(1)}$, $f_2(x) = -x^{(1)}$. The functions $f_1(x)$ and $f_2(x)$ are continuously differentiable and hence (see Example 2 of Section 1)

$$Df_1(x_0) = [\underline{\partial} f_1(x_0), \overline{\partial} f_1(x_0)] ,$$

where

$$\underline{\partial} f_1(x_0) = \{A_1\} , \qquad \overline{\partial} f_1(x_0) = \{0\} ,$$

and

$$Df_2(x_0) = [\underline{\partial} f_2(x_0), \overline{\partial} f_2(x_0)] ,$$

where

$$\underline{\partial} f_2(x_0) = \{A_2\} , \qquad \overline{\partial} f_2(x_0) = \{0\} .$$

Due to Lemma 2.2, we have

$$Df(x_0) = [\underline{\partial} f(x_0), \overline{\partial} f(x_0)] ,$$

where

$$\underline{\partial} f(x_0) = \text{co}\ \{A_2, A_2\} , \qquad \overline{\partial} f(x_0) = \{0\}$$

(Figure 17).

<u>EXAMPLE 2</u>. Let $f(x) = -|x_1| = -f_1(x)$. From the above example and property (2.) in Subsection 2.2,

$$Df(x_0) = [\underline{\partial} f(x_0), \overline{\partial} f(x_0)] ,$$

where

$$\underline{\partial} f(x_0) = \{0\} , \qquad \overline{\partial} f(x_0) = -\text{co}\ \{A_1, A_2\} .$$

In this case, we have

$$-\text{co}\ \{A_1, A_2\} = \text{co}\ \{A_1, A_2\}$$

(Figure 17).

<u>EXAMPLE 3</u>. Let $f(x) = |x^{(2)}|$. As in Example 1, we obtain

$$Df(x_0) = [\underline{\partial} f(x_0), \overline{\partial} f(x_0)] ,$$

where

$$\underline{\partial} f(x_0) = \text{co}\ \{B_1, B_2\} , \qquad \overline{\partial} f(x_0) = \{0\} .$$

<u>EXAMPLE 4</u>. Let $f(x) = -|x^{(2)}|$. As in Example 2, we have

$$Df(x_0) = [\underline{\partial} f(x_0), \overline{\partial} f(x_0)] ,$$

where

$$\underline{\partial}f(x_0) = \{0\} , \qquad \overline{\partial}f(x_0) = \text{co } \{B_1,B_2\} .$$

EXAMPLE 5. Let $f(x) = |x^{(1)}| - |x^{(2)}|$. Due to property (3.) in Subsection 2.1, we have from Examples 1 and 4:

$$\mathcal{D}f(x_0) = [\underline{\partial}f(x_0), \overline{\partial}f(x_0)] ,$$

where

$$\underline{\partial}f(x_0) = \text{co } \{A_1,A_2\} , \qquad \overline{\partial}f(x_0) = \text{co } \{B_1,B_2\}$$

(Figure 18).

EXAMPLE 6. Let $f(x) = |x^{(1)}| + |x^{(2)}|$. Due to property (3.), we have from Subsection 2.1:

$$\mathcal{D}f(x_0) = [\underline{\partial}f(x_0), \overline{\partial}f(x_0)] ,$$

where

$$\underline{\partial}f(x_0) = \text{co } \{A_1,A_2\} + \text{co } \{B_1,B_2\} , \qquad \overline{\partial}f(x_0) = \{0\}$$

(Figure 19).

EXAMPLE 7. Let $f_1(x) = |x^{(1)}| - |x^{(2)}|$, $f_2(x) = |x^{(2)}| - |x^{(1)}|$.

Define

$$f(x) = \max \{f_1(x),f_2(x)\} = ||x^{(1)}| - |x^{(2)}|| .$$

From Example 5 we have

$$\mathcal{D}f_1(x_0) = [\underline{\partial}f_1(x_0), \overline{\partial}f_1(x_0)] ,$$

where

$$\underline{\partial}f_1(x_0) = \text{co } \{A_1,A_2\} , \qquad \overline{\partial}f_1(x_0) = \text{co } \{B_1,B_2\} .$$

From property (2.) in Subsection 2.1, we have

$$\mathcal{D}f_2(x_0) = [\underline{\partial}f_2(x_0), \overline{\partial}f_2(x_0)] ,$$

where

$$\underline{\partial}f_2(x_0) = -\overline{\partial}f_1(x_0) , \qquad \overline{\partial}f_2(x_0) = -\underline{\partial}f_1(x_0) .$$

Hence, in our case we have (Figure 18)

$$\underline{\partial}f_2(x_0) = -\text{co } \{B_1,B_2\} = \text{co } \{B_1,B_2\} ,$$

$$\overline{\partial}f_2(x_0) = -\text{co } \{A_1,A_2\} = \text{co } \{A_1,A_2\} .$$

Applying Lemma 2.2, we obtain

$$\mathcal{D}f(x_0) = [\underline{\partial}f(x_0),\ \overline{\partial}f(x_0)] \quad ,$$

where

$$\begin{aligned}\underline{\partial}f(x_0) &= \text{co}\ \{\underline{\partial}f_1 - \overline{\partial}f_2,\ \underline{\partial}f_2 - \overline{\partial}f_1\} \\ &= \text{co}\ \{\text{co}\{A_1,A_2\} + \text{co}\{A_1,A_2\},\ \text{co}\{B_1,B_2\} + \text{co}\{B_1,B_2\}\} \ ,\end{aligned}$$

$$\overline{\partial}f(x_0) = \overline{\partial}f_1 + \overline{\partial}f_2 = \text{co}\{B_1,B_2\} + \text{co}\{A_1,A_2\} \ .$$

The sets $\underline{\partial}f(x_0)$ and $\overline{\partial}f(x_0)$ are shown in Figure 20 where the bold solid line is the boundary of $\underline{\partial}f(x_0)$ and the dashed line is the boundary of $\overline{\partial}f(x_0)$.

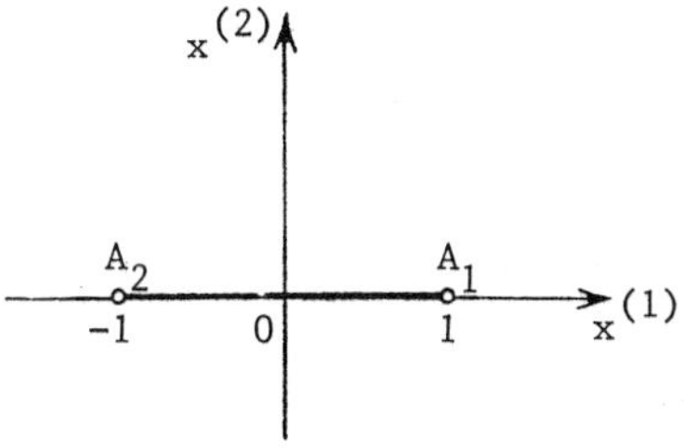

Figure 17

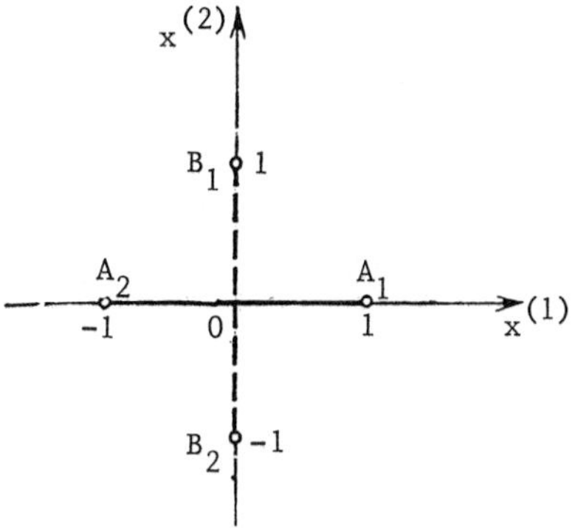

Figure 18

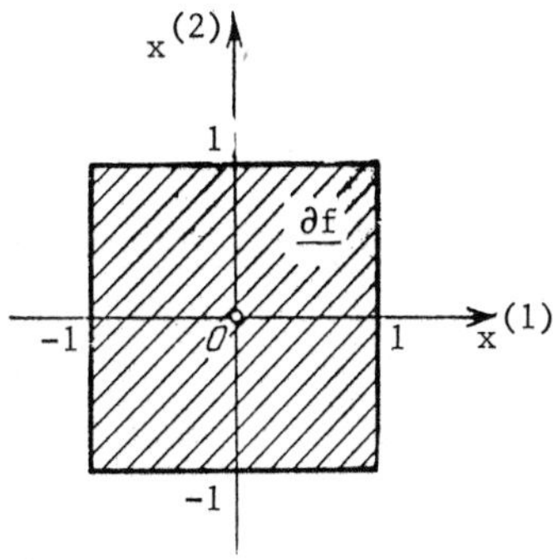

Figure 19

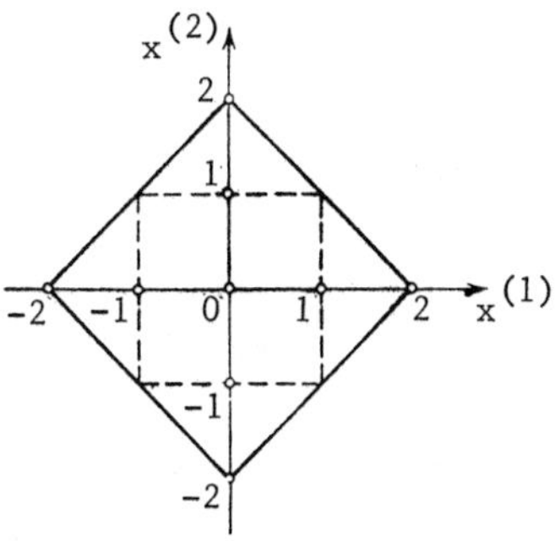

Figure 20

EXAMPLE 8. Let $f_1(x) = |x^{(1)}| + |x^{(2)}|$, $f_2(x) = |x^{(1)}| - |x^{(2)}|$.

Define

$$f(x) = \max \{f_1(x), f_2(x)\} .$$

Using Lemma 2.2, we have from Examples 5 and 6:

$$\mathcal{D}f(x_0) = [\underline{\partial}f(x_0), \overline{\partial}f(x_0)] ,$$

where

$$\underline{\partial}f(x_0) = \text{co } \{\underline{\partial}f_1 - \overline{\partial}f_2, \underline{\partial}f_2 - \overline{\partial}f_1\}$$

$$= \text{co } \{\text{co}\{A_1,A_2\}, \text{co}\{A_1,A_2\} + \text{co}\{B_1,B_2\} - \text{co}\{B_1,B_2\}\} ,$$

$$\overline{\partial}f(x_0) = \text{co } \{B_1,B_2\}$$

(Figure 21).

EXAMPLE 9. Let $f_1(x) = |x^{(1)}| + |x^{(2)}|$, $f_2(x) = |x^{(1)}| - |x^{(2)}|$.

Define

$$f(x) = \min \{f_1(x), f_2(x)\} .$$

From Examples 5 and 6 and Lemma 2.3 we have

$$\mathcal{D}f(x_0) = [\underline{\partial}f(x_0), \overline{\partial}f(x_0)] ,$$

where

$$\underline{\partial}f(x_0) = \underline{\partial}f_1 + \underline{\partial}f_2 = \text{co}\{A_1,A_2\} + \text{co}\{A_1,A_2\} + \text{co}\{B_1,B_2\} ,$$

$$\overline{\partial}f(x_0) = \text{co } \{\overline{\partial}f_1 - \underline{\partial}f_2, \overline{\partial}f_2 - \underline{\partial}f_1\}$$

$$= \text{co } \{\text{co}\{B_1,B_2\} - \text{co}\{A_1,A_2\} - \text{co}\{B_1,B_2\}, \text{co}\{A_1,A_2\}\}$$

$$= \text{co } \{\text{co}\{B_1,B_2\} + \text{co}\{A_1,A_2\} + \text{co}\{B_1,B_2\}, \text{co}\{A_1,A_2\}\} ;$$

the last equality holds by virtue of the fact that

$$\text{co } \{A_1,A_2\} = -\text{co } \{A_1,A_2\} ,$$

$$\text{co } \{B_1,B_2\} = -\text{co } \{B_1,B_2\} .$$

The sets $\underline{\partial}f(x_0)$ and $\overline{\partial}f(x_0)$ are shown in Figure 22.

EXAMPLE 10. Let $f(x) = ||x^{(1)}| + x^{(2)}|$. We have

$$f(x) = \max \{f_1(x), f_2(x)\} ,$$

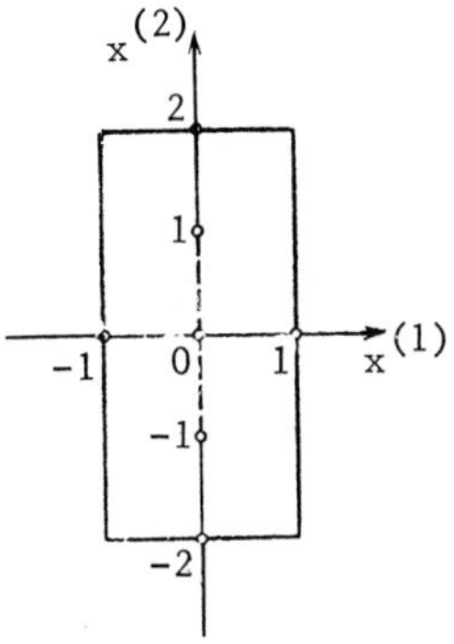

Figure 21

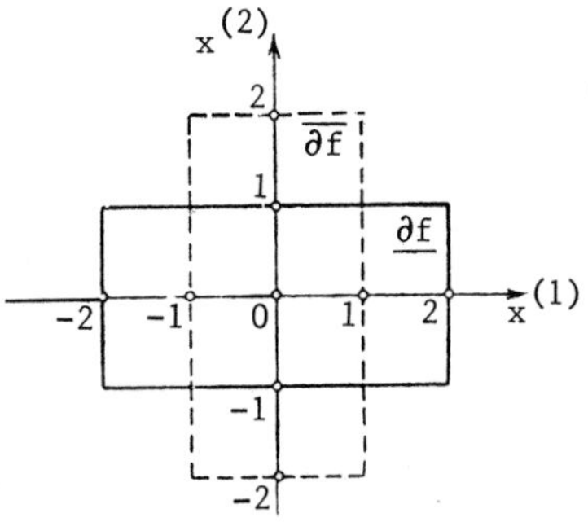

Figure 22

where

$$f_1(x) = |x^{(1)}| + x^{(2)} , \qquad f_2(x) = -|x^{(1)}| - x^{(2)} .$$

Since for the function $f_3(x) = x^{(2)}$ we get

$$Df_3(x_0) = [\underline{\partial} f_3(x_0), \overline{\partial} f_3(x_0)] ,$$

where

$$\underline{\partial} f_3(x_0) = \{B_1\} , \qquad \overline{\partial} f_3(x_0) = \{0\} ,$$

then it follows from Examples 1 and 2 that

$$Df_1(x_0) = [\underline{\partial} f_1(x_0), \overline{\partial} f_1(x_0)] ,$$

where

$$\underline{\partial} f_1(x_0) = \text{co}\ \{A_1, A_2\} + B_1 , \qquad \overline{\partial} f_1(x_0) = \{0\},$$

and

$$Df_2(x_0) = [\underline{\partial} f_2(x_0), \overline{\partial} f_2(x_0)] ,$$

where

$$\underline{\partial} f_2(x_0) = \{0\} , \qquad \overline{\partial} f_2(x_0) = -\text{co}\{A_1, A_2\} - B_1 .$$

Due to Lemma 2.2,

$$Df(x_0) = [\underline{\partial} f(x_0), \overline{\partial} f(x_0)] ,$$

where

$$\begin{aligned} \underline{\partial} f(x_0) &= \text{co}\ \{\underline{\partial} f_1 - \overline{\partial} f_2, \underline{\partial} f_2 - \overline{\partial} f_1\} \\ &= \text{co}\ \{\text{co}\{A_1, A_2\} + B_1 + \text{co}\{A_1, A_2\} + B_1, 0\} , \end{aligned}$$

$$\partial f(x_0) \;=\; \overline{\partial} f_1 + \overline{\partial} f_2 \;=\; -\text{co}\{A_1, A_2\} - B_1 \;=\; \text{co}\{A_1, A_2\} - B_1$$

(Figure 23).

<u>EXAMPLE 11</u>. Let $f(x) = |x^{(1)} + x^{(2)}|$.

Since

$$f(x) \;=\; \max\, \{f_1(x), f_2(x)\}\ ,$$

where

$$f_1(x) \;=\; x^{(1)} + x^{(2)}\ , \qquad f_2(x) \;=\; -x^{(1)} - x^{(2)}\ ,$$

then

$$\mathcal{D}f(x_0) \;=\; [\underline{\partial} f(x_0),\ \overline{\partial} f(x_0)]\ ,$$

$$\underline{\partial} f(x_0) \;=\; \text{co}\ \{C_1, C_2\}\ , \qquad \overline{\partial} f(x_0) \;=\; \{0\}$$

(Figure 24).

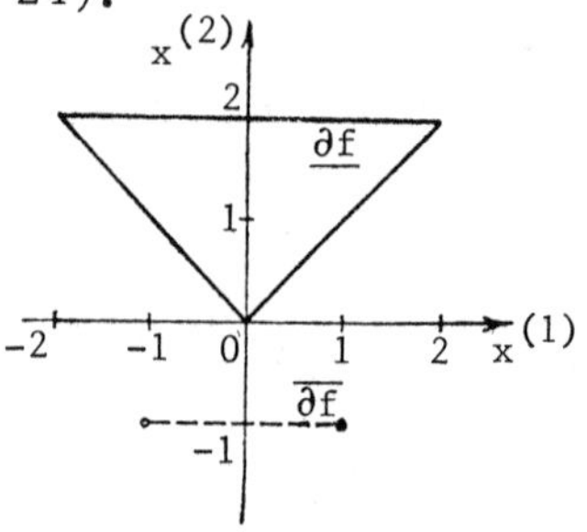

Figure 23

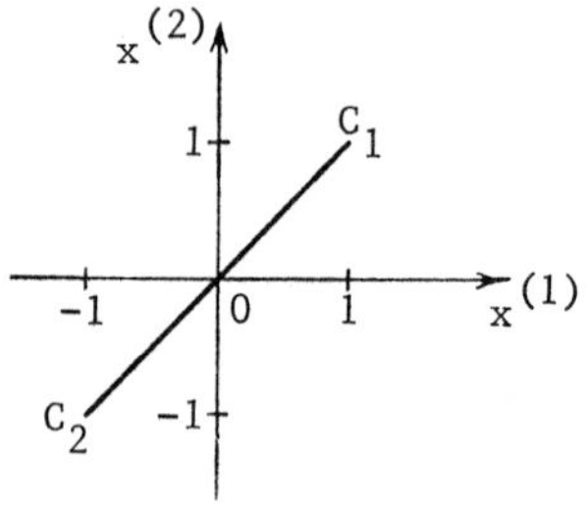

Figure 24

<u>EXAMPLE 12</u>. Let $f_1(x) = 1 + |x^{(1)}|$, $f_2(x) = 1 + |x^{(1)} + x^{(2)}|$.

Define

$$f(x) \;=\; f_1(x) \cdot f_2(x)\ .$$

If c is a constant, then

$$\mathcal{D}[f(x) + c] \;=\; \mathcal{D}f(x)\ .$$

Hence

$$\mathcal{D}f_1(x_0) \;=\; [\underline{\partial} f_1(x_0),\ \overline{\partial} f_1(x_0)]\ ,$$

$$\underline{\partial} f_1(x_0) \;=\; \text{co}\ \{A_1, A_2\}\ , \qquad \overline{\partial} f_1(x_0) \;=\; \{0\}\ ,$$

$$\mathcal{D}f_2(x_0) \;=\; [\underline{\partial} f_2(x_0),\ \overline{\partial} f_2(x_0)]\ ,$$

$$\underline{\partial} f_2(x_0) \;=\; \text{co}\ \{C_1, C_2\}\ , \qquad \overline{\partial} f_2(x_0) \;=\; \{0\}\ .$$

Consequently, due to Lemma 2.5, we have

$$\mathcal{D}f(x_0) = [\underline{\partial}f(x_0), \overline{\partial}f(x_0)] ,$$

where

$$\underline{\partial}f(x_0) = f_1(x_0)\underline{\partial}f_2(x_0) + f_2(x_0)\underline{\partial}f_1(x_0) = \text{co}\{A_1,A_2\} + \text{co}\{C_1,C_2\},$$

$$\overline{\partial}f(x_0) = \{0\}$$

(Figure 25).

<u>EXAMPLE 13</u>. Let $f_1(x) = |x^{(1)}| + x^{(2)} + 1$, $f_2(x) = [f_3(x)]^{-1}$, $f_3(x) = |x^{(1)}| - |x^{(2)}| + 1$. Define

$$f(x) = f_1(x) \cdot f_2(x) .$$

Since $\mathcal{D}f_1(x) = \mathcal{D}f_4(x)$, where $f_4(x) = |x^{(1)}| + x^{(2)}$, then

$$\mathcal{D}f_1(x_0) = [\underline{\partial}f_1(x_0), \overline{\partial}f_1(x_0)] ,$$

$$\underline{\partial}f_1(x_0) = \text{co } \{A_1,A_2\} + B_1 , \qquad \overline{\partial}f_1(x_0) = \{0\}$$

(see Example 10). Since $\mathcal{D}f_3(x) = \mathcal{D}f_5(x)$, where $f_5(x) = |x^{(1)}| - |x^{(2)}|$, we get

$$\mathcal{D}f_3(x_0) = [\underline{\partial}f_3(x_0), \overline{\partial}f_3(x_0)] ,$$

$$\underline{\partial}f_3(x_0) = \text{co } \{A_1,A_2\} , \qquad \overline{\partial}f_3(x_0) = \text{co } \{B_1,B_2\}$$

(see Example 5).

Due to Lemma 2.4,

$$\mathcal{D}f_2(x_0) = [\underline{\partial}f_2(x_0), \overline{\partial}f_2(x_0)] ,$$

where

$$\underline{\partial}f_2(x_0) = -\overline{\partial}f_3(x_0) = -\text{co } \{B_1,B_2\} = \text{co } \{B_1,B_2\} ,$$

$$\overline{\partial}f_2(x_0) = -\underline{\partial}f_3(x_0) = -\text{co } \{A_1,A_2\} = \text{co } \{A_1,A_2\} .$$

Due to Lemma 2.5, we finally obtain (Figure 26)

$$\mathcal{D}f(x_0) = [\underline{\partial}f(x_0), \overline{\partial}f(x_0)] ,$$

where

$$\underline{\partial}f(x_0) = f_1(x_0)\underline{\partial}f_2(x_0) + f_2(x_0)\underline{\partial}f_1(x_0)$$

$$= \text{co } \{B_1,B_2\} + \text{co } \{A_1,A_2\} + B_1 ,$$

$$\overline{\partial} f(x_0) \;=\; f_1(x_0)\overline{\partial} f_2(x_0) + f_2(x_0)\overline{\partial} f_1(x_0)$$

$$= \; \text{co}\,\{A_1, A_2\} + \{0\} \;=\; \text{co}\,\{A_1, A_2\} \;.$$

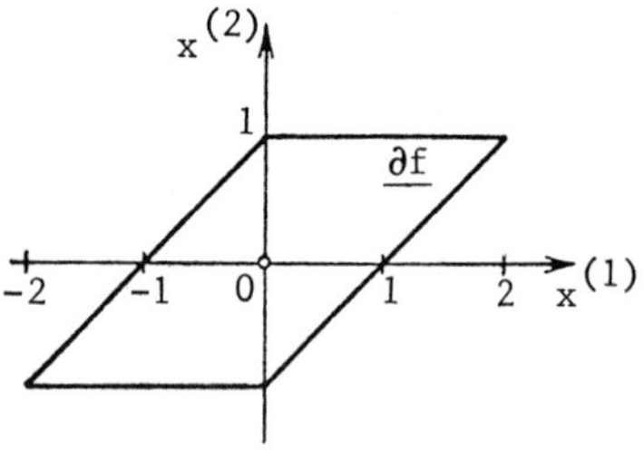

Figure 25

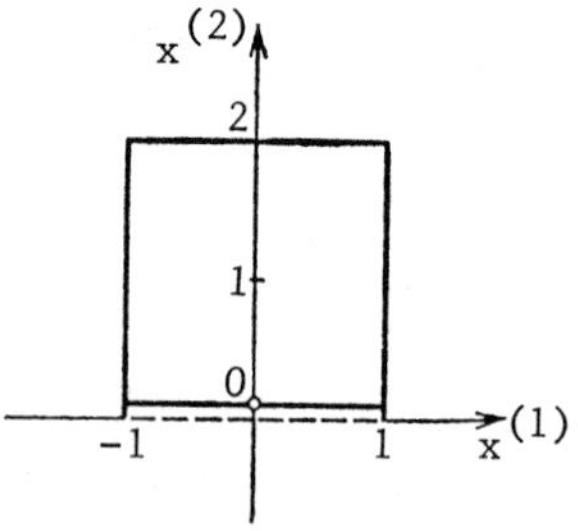

Figure 26

EXAMPLE 14. Let $f_1(x) = 1 + |x^{(1)}| + x^{(2)}$. Define the function $f(x) = [f_1(x)]^2$. In the above example, we already noticed that

$$Df_1(x_0) \;=\; [\underline{\partial} f_1(x_0),\; \overline{\partial} f_1(x_0)] \;,$$

where

$$\underline{\partial} f_1(x_0) \;=\; \text{co}\,\{A_1, A_2\} + B_1 \;, \qquad \overline{\partial} f_1(x_0) \;=\; \{0\} \;.$$

Due to Lemma 2.5, we get

$$Df(x_0) \;=\; [\underline{\partial} f(x_0,\; \overline{\partial} f(x_0)] \;,$$

where

$$\underline{\partial} f(x_0) \;=\; f_1(x_0)\underline{\partial} f_1(x_0) + f_1(x_0)\underline{\partial} f_1(x_0) \;=\; \underline{\partial} f_1(x_0) + \underline{\partial} f_1(x_0)$$

$$= \; \text{co}\,\{A_1, A_2\} + B_1 + \text{co}\,\{A_1, A_2\} + B_1 \;,$$

$$\overline{\partial} f(x_0) \;=\; f_1(x_0)\overline{\partial} f_1(x_0) + f_1(x_0)\overline{\partial} f_1(x_0) \;=\; \{0\}$$

(Figure 27).

EXAMPLE 15. Let $f_1(x) = 2 + |x^{(1)}| + x^{(2)}$.

Define $f(x) = [f_1(x)]^2$. Since

$$Df_1(x_0) \;=\; [\underline{\partial} f_1(x_0),\; \overline{\partial} f_1(x_0)]$$

where

$$\underline{\partial} f_1(x_0) \;=\; \text{co}\,\{A_1, A_2\} + B_1 \;, \qquad \overline{\partial} f_1(x_0) \;=\; \{0\}$$

(see Example 13), then, due to Lemma 2.5,

$$\mathcal{D}f(x_0) = [\underline{\partial}f(x_0),\ \overline{\partial}f(x_0)] \quad .$$

Here,

$$\underline{\partial}f(x_0) = f_1(x_0)\underline{\partial}f_1(x_0) + f_1(x_0)\underline{\partial}f_1(x_0) = 2\underline{\partial}f_1(x_0) + 2\underline{\partial}f_1(x_0)$$

$$= 2\ \mathrm{co}\,\{A_1, A_2\} + 2B_1 + 2\ \mathrm{co}\,\{A_1, A_2\} + 2B_1 \quad ,$$

$$\overline{\partial}f(x_0) = f_1(x_0)\overline{\partial}f_1(x_0) + f_1(x_0)\overline{\partial}f_1(x_0) = \{0\}$$

(see Figure 28).

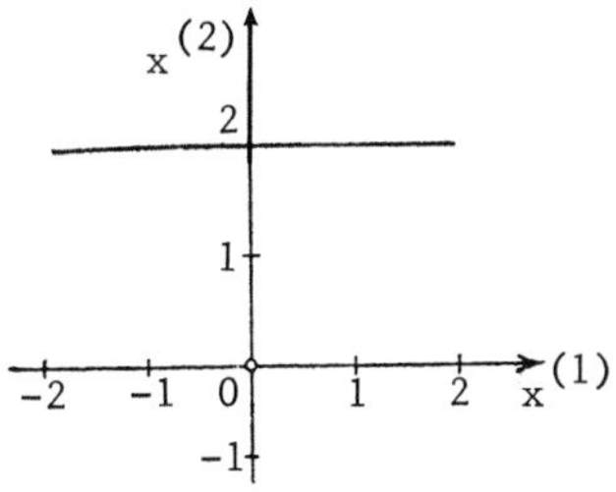

Figure 27

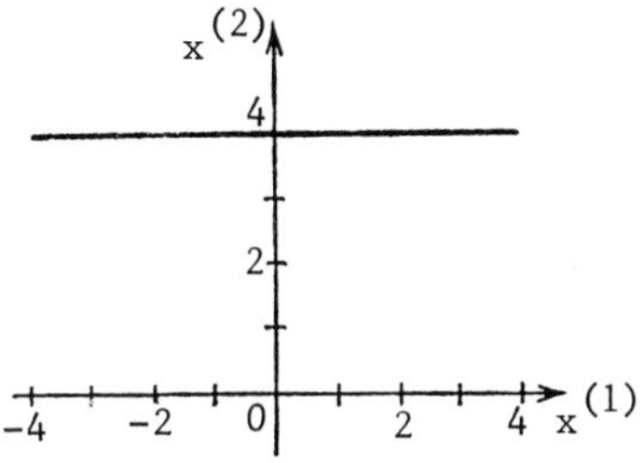

Figure 28

EXAMPLE 16. Let $f(x) = ||x||$, where $||x|| = \sqrt{(x^{(1)})^2 + (x^{(2)})^2}$.

Clearly, $\dfrac{\partial f(x_0)}{\partial g} = \lim\limits_{\alpha\to+0} \alpha^{-1}[f(x_0+\alpha g) - f(x_0)] = \alpha\cdot\alpha^{-1}||g|| = ||g||$.

Therefore, $\dfrac{\partial f(x_0)}{\partial g} = \max\limits_{v\in S_1(0)} (v,g)$, where $S_1(0) = \{v\in E_2 \mid ||v|| \le 1\}$.

Hence, the function $f(x)$ is quasidifferentiable (and, moreover, subdifferentiable) and, in addition, $\mathcal{D}f(x_0) = [\underline{\partial}f(x_0), \overline{\partial}f(x_0)]$, where $\underline{\partial}f(x_0) = S_1(0)$, $\overline{\partial}f(x_0) = \{0\}$.

REMARK 1. If $x = (x^{(1)}, \ldots, x^{(n)}) \in E_n$, $f(x) = ||x||$, where

$$||x|| = \sqrt{\sum (x^{(ij)})^2} \quad ,$$

then, analogously, we can show that

$$\mathcal{D}f(x_0) = [\underline{\partial}f(x_0),\ \overline{\partial}f(x_0)] \quad ,$$

where

$$\underline{\partial}f(x_0) = S_1(0) \equiv \{v \in E_n \mid ||v|| \le 1\} \quad , \qquad \overline{\partial}f(x_0) = \{0\} \ .$$

REMARK 2. Everywhere above, we constructed one of possible quasi-differentials for each function. The function in Example 8 coincide with the function in Example 6 and the function in Example 9 with the function in Example 5; however, quasidifferentials of the coincident functions are different.

PROBLEM 3.1.* Let $A \subset E_n$ be a convex compact set. Define the mapping $H_A\colon E_n \to \Pi(E_n)$:

$$H_A(g) = \{v \in A \mid (v,g) = \max_{z \in A}(z,g)\} .$$

It is clear that $H_A(0) = A$ and that it is enough to consider H_A on the unit sphere $C_1 = \{g \in E_n \mid \|g\| = 1\}$, because

$$H_A(\lambda g) = H_A(g) \qquad \forall\, \lambda > 0 .$$

It is possible to show that for almost all $g \in C_1$ the set $H_A(g)$ consists of a unique point.

Let $f(x)$ be a quasidifferentiable function at a point x_0 and let $Df(x_0) = [\underline{\partial} f(x_0), \overline{\partial} f(x_0)]$ be its quasidifferential at x_0.

Introduce the mapping $Q\colon C_1 \to \Pi(E_n)$ by the following relation:

$$Q(g) = H_{\underline{\partial} f(x_0)}(g) - H_{-\overline{\partial} f(x_0)}(g) .$$

The set $Q(g)$ also consists of a unique point for almost all $g \in C_1$.

Denote by $\bar{C}_1$ the set of $g \in E_1$ for which $Q(g)$ consists of a unique point. Prove or disprove the following proposition:

If a function $f(x)$ satisfies a Lipschitz condition and is quasidifferentiable in a neighborhood of the point x_0, then the representation

$$\partial_{Cl} f(x_0) = \text{co}\,\{Q(g) \mid g \in \bar{C}_1\} \tag{3.1}$$

holds for the Clarke subdifferential (see (1.3)) of this function.

Verify that relation (3.1) is satisfied in all examples considered in this Section.

4. QUASIDIFFERENTIABILITY OF CONVEXO-CONCAVE FUNCTIONS

1. Let a function $f(x) \equiv f(x,y)$ be given and finite on $S = S_1 \times S_2 \subset E_n$, where $x = [y,z] \subset E_n$, $n = n_1 + n_2$ and $S_1 \subset E_{n_1}$, $S_2 \subset E_{n_2}$ are open convex sets in the corresponding spaces. A function $f(x)$ is said to be convexo-concave on S if the function $f(y,z)$ is convex with respect to y on S_1 for each fixed $z \in S_2$ and concave with respect to z on S_2 for any fixed $y \in S_1$. Let

$$\partial f_y(y_0,z_0) = \{v \in E_{n_1} \mid f(y,z_0) - f(y_0,z_0) \geq (v,y-y_0) \ \forall y \in S_1\}, \quad (4.1)$$

$$\partial f_z(y_0,z_0) = \{w \in E_{n_2} \mid f(y_0,z) - f(y_0,z_0) \leq (w,z-z_0) \ \forall z \in S_2\}. \quad (4.2)$$

The set $\partial f_y(y_0,z_0)$ is the subdifferential of the convex function $h_1(y) = f(y,z_0)$ at a point y_0 and the set $\partial f_z(y_0,z_0)$ is the superdifferential of the concave function $h_2(z) = f(y_0,z)$ at a point z_0. These sets are nonempty, convex, closed and bounded.

Let us establish the continuity of a convexo-concave function $f(x)$ jointly in both variables (the continuity of $f(x)$ in each of variables y and z follows from the properties of convex and concave functions).

First of all, let us prove a lemma which is of independent interest.

<u>LEMMA 4.1</u>. If a function $f(y,z)$ is convex with respect to y on S_1 for any fixed $z \in S_2$ and continuous in z on S_2 for each fixed $y \in S_1$, then $f(y,z)$ is continuous jointly in both variables.

<u>Proof</u>. Let $x_0 = [y_0,z_0] \subset S$ and $\Omega = \Omega_1 \times \Omega_2$, where $\Omega_1 \subset S_1$, $\Omega_2 \subset S_2$ are convex compact sets such that $y_0 \in \operatorname{int} \Omega_1$, $z_0 \in \operatorname{int} \Omega_2$.

We establish first that the function $f(x)$ is bounded on Ω.

To do this, we consider the function $\phi(y) = \sup_{z \in \Omega_2} f(y,z)$.

For each fixed $y \in S_1$, the function $f(x,z)$ is continuous in z on Ω_2 and, since Ω_2 is compact, then the function is, of course, bounded there. Hence, $\phi(y)$ is a finite function and, furthermore, we can write $\phi(y) = \max_{z \in \Omega_2} f(y,z)$. Now, it is easy to see that $\phi(y)$ is a convex function on S_1. In fact, since $\phi(y,z)$ is a convex function with respect to y, we have

$$\begin{aligned}\phi(\alpha y_1 + (1-\alpha)y_2) &= \max_{z \in \Omega_2} f(\alpha y_1 + (1-\alpha)y_2, z) \\ &\le \max_{z \in \Omega_2} [\alpha f(y_1,z) + (1-\alpha)f(y_2,z)] \\ &\le \alpha \max_{z \in \Omega_2} f(y_1,z) + (1-\alpha) \max_{z \in \Omega_2} f(y_2,z) \\ &= \alpha\phi(y_1) + (1-\alpha)\phi(y_2) \qquad \forall \alpha \in [0,1] ,\end{aligned}$$

and this implies the convexity of $\phi(y)$ on S_1. On the other hand, a finite convex function is continuous on S_1 (see Section 1.4), then it is also bounded on any compact set $\Omega' \subset S_1$. Hence, we can conclude that $f(y,z)$ is a bounded function on $\Omega' \times \Omega_2$. Now, let us fix an arbitrary $\Omega' \subset S_1$ such that

$$\Omega_1 \subset \operatorname{int} \Omega' . \tag{4.3}$$

We show here that the set of subdifferentials $f_y(y,z)$ is bounded on $\Omega = \Omega_1 \times \Omega_2$. We assume the opposite, i.e., that there exists a sequence $\{v_k\}$ such that

$$v_k \in \partial f_y(y_k,z_k) , \qquad ||v_k|| \to +\infty , \qquad x_k = [y_k,z_k] \in \Omega . \tag{4.4}$$

By the definition (see (4.1)),

$$f(y,z_k) - f(y_k,z_k) \ge (v_k, y-y_k) \qquad \forall y \in S_1 . \tag{4.5}$$

By virtue of (4.3), there exists a $\beta > 0$ such that

$$y'_k = y_k + \beta v_k \|v_k\|^{-1} \in \Omega' \subset S_1 \qquad \forall k .$$

From (4.5) we have

$$f(y'_k, z_k) \geq f(y_k, z_k) + \beta \|v_k\| . \tag{4.6}$$

Since the $f(y_k, z_k)$ are bounded on $\Omega' \times \Omega_2$, then it follows from (4.4) and (4.6) that $f(y'_k, z_k) \to +\infty$, which contradicts the boundedness of $f(y,z)$ on $\Omega' \times \Omega_2$. Thus, there exists an $L < \infty$ such that

$$\|v\| \leq L \qquad \forall v \in \partial f_y(y,z), \quad \forall [y,z] \in \Omega . \tag{4.7}$$

Now, let

$$x' \to x_0 , \qquad x' \in \Omega , \qquad x' = (y', z') , \qquad x_0 = (y_0, z_0) .$$

Then we have

$$\begin{aligned} |f(x') - f(x_0)| &= |f(y', z') - f(y_0, z_0)| \\ &= |f(y', z') - f(y_0, z') + f(y_0, z') - f(y_0, z_0)| \\ &\leq |f(y' z') - f(y_0, z')| + |f(y_0, z') - f(y_0, z_0)| . \end{aligned} \tag{4.8}$$

From (4.5) and (4.7),

$$f(y', z') - f(y_0, z') \geq (v_0, y' - y_0) \geq -L\|y' - y_0\| , \tag{4.9}$$

where $v_0 \in \partial f_y(y_0, z')$. Similarly,

$$f(y_0, z') - f(y', z') \geq (v', y_0 - y') \geq -L\|y' - y_0\| , \tag{4.10}$$

where $v' \in \partial f_y(y_0, z')$. It follows from (4.9) and (4.10) that

$$|f(h', z') - f(y_0, z')| \leq L\|y' - y_0\| . \tag{4.11}$$

The function $f(y_0, z)$ is continuous in z. Hence

$$f(y_0, z') \xrightarrow[z' \to z_0]{} f(y_0, z_0) .$$

Therefore, from (4.11) and (4.8) we have $f(x') \to f(x_0)$, i.e., the function $f(x)$ is continuous jointly in both variables.

THEOREM 4.1. If $f(y,z)$ is a convexo-concave function on S, then it is continuous on S jointly in both variables.

Proof. The function $f(y,z)$ is concave with respect to z on S_2 for any fixed $y \in S_1$ and, therefore, it is continuous in z for any fixed y. The assertion of the theorem now follows from Lemma 4.1.

COROLLARY. A convexo-concave function $f(y,z)$ is Lipschitzian on any set $\Omega = \Omega_1 \times \Omega_2$, where $\Omega_1 \subset S_1$, $\Omega_2 \subset S_2$ and Ω_1, Ω_2 are convex compact sets.

Proof. As in the proof of Lemma 4.1, we can show that the set of superdifferentials $\partial f_z(y,z)$ is bounded on Ω, i.e., there exists an $L_1 < \infty$ such that

$$||w|| \leq L_1 \qquad \forall w \in \partial f_z(y,z), \quad \forall [y,z] \in \Omega . \tag{4.12}$$

Hence, from (4.8), (4.11) and (4.12), we conclude that the function $f(y,z)$ is Lipschitzian jointly in both variables on the set $\Omega = \Omega_1 \times \Omega_2$.

LEMMA 4.2. Let a function $f(y,z)$ be convex with respect to y on S_1 for any fixed $z \in S_2$ and be continuous in z on S_2 for each fixed $y \in S_1$ and let $x_0 \in [y_0,z_0] \in S = S_1 \times S_2$. Then, for any $\varepsilon > 0$, there exists a $\delta > 0$ (depending on x_0) such that

$$\partial f_y(y,z) \subset \partial f_y(y_0,z_0) + S_{1\varepsilon} \qquad \forall [y,z] \subset S_\delta(y_0,z_0), \tag{4.13}$$

where

$$S_{1\varepsilon} = \{v \in E_{n_1} \mid ||v|| \leq \varepsilon\} ,$$

$$S_\delta(y_0,z_0) = \{[y,z] \in S \mid ||y-y_0|| + ||z-z_0|| \leq \delta\} .$$

Proof. Assume the opposite. Then, there exist an $\varepsilon_0 > 0$ and a sequence $\{v_k\}$ such that

$$v_k \in \partial f_y(y_k,z_k) , \qquad [y_k,z_k] \to [y_0,z_0] = x_0 ,$$

$$\rho(v_k, \partial f_y(y_0,z_0)) = \min_{v \in \partial f_y(y_0,z_0)} ||v_k - v|| \geq \varepsilon_0 . \tag{4.14}$$

Since the sets $\partial f_y(y,z)$ are uniformly bounded (this fact was established in the proof of Lemma 4.1), then, without loss of generality, we can assume that

$$v_k \to v_0 . \tag{4.15}$$

From (4.14) it is obvious that

$$v_0 \notin \partial f_y(y_0,z_0) . \tag{4.16}$$

Since $v_k \in \partial f_y(y_k,z_k)$, we have

$$f(y,z_k) - f(y_k,z_k) \geq (v_k, y-y_k) \quad \forall\, y \in S_1 .$$

Hence,

$$f(y,z_0) - f(y_0,z_0) \geq (v_k, y-y_0) + (v_k, y_0-y_k) + f(y,z_0) - f(y,z_k)$$
$$+ f(y_k,z_k) - f(y_0,z_0) \quad \forall\, y \in S_1 . \tag{4.17}$$

Let us fix an arbitrary $y \in S_1$. Due to Lemma 4.1, the function $f(y,z)$ is continuous jointly in both variables.

Therefore, from (4.17) and (4.15) we have

$$f(y,z_0) - f(y_0,z_0) \geq (v_0, y-y_0)$$

as $k \to \infty$. Since the resulting inequality holds for each $y \in S_1$, we obtain from (4.1) that $v_0 \in \partial f_y(y_0,z_0)$, which contradicts (4.16).

The lemma has been proved. ∎

<u>COROLLARY 1</u>. Let $f(y,z)$ be a convexo-concave function on $S = S_1 \times S_2$ and let $x_0 = [y_0,z_0] \in S$. Then, for any $\varepsilon > 0$, there exists a $\delta > 0$ such that

$$\begin{aligned} \partial f_y(y,z) &\subset \partial f_y(y_0,z_0) + S_{1\varepsilon} , \\ \partial f_z(y,z) &\subset \partial f_z(y_0,z_0) + S_{2\varepsilon} , \end{aligned} \tag{4.18}$$

for all $[y,z] \in S_\delta(y_0,z_0)$, where

$$S_{1\varepsilon} = \{v \in E_{n_1} \mid \|v\| \leq \varepsilon\}, \qquad S_{2\varepsilon} = \{w \in E_{n_2} \mid \|w\| \leq \varepsilon\} ,$$

$$S_\delta(y_0,z_0) = \{[y,z] \in S \mid \|y-y_0\| + \|z-z_0\| \leq \delta\} .$$

Proof. Is analogous to that of Lemma 4.2.

COROLLARY 2. The mappings $\partial f_y(y,z)$ and $\partial f_z(y,z)$ are upper semi-continuous on S.

2. Now, let us find the directional derivative of a convexo-concave function. Let

$$x_0 = [y_0,z_0] \in S = S_1 \times S_2 , \qquad g = [g_1,g_2] \in E_{n_1} \times E_{n_2} .$$

THEOREM 4.2. A function $f(x) = f(y,z)$ is differentiable in any direction $g = [g_1,g_2] \in E_{n_1} \times E_{n_2}$ at a point $x_0 \in S$ and, moreover,

$$\frac{\partial f(x_0)}{\partial g} = \max_{v \in \partial f_y(y_0,z_0)} (v,g_1) + \min_{w \in \partial f_z(y_0,z_0)} (w,g_2) . \tag{4.19}$$

Proof. Denote by B the right-hand side of (4.19) and define

$$h(\alpha) = \alpha^{-1}[f(y_0+\alpha g_1,\ z_0+\alpha g_2) - f(y_0,z_0)] . \tag{4.20}$$

It is necessary to show that the limit exists:

$$\lim_{\alpha \to +0} h(\alpha) \equiv \frac{\partial f(x_0)}{\partial g} . \tag{4.21}$$

We have

$$\begin{aligned} f(y_0+\alpha g_1,\ z_0+\alpha g_2) - f(y_0,z_0) &= [f(y_0+\alpha g_1,z_0+\alpha g_2) - f(y_0+\alpha g_1,z_0)] \\ &\quad + [f(y_0+\alpha g_1,z_0) - f(y_0,z_0)] \\ &\equiv A_1 + A_2 . \end{aligned} \tag{4.22}$$

Since the function $f(y,z_0)$ is convex with respect to y, then

$$A_2 \equiv f(y_0+\alpha g_1,z_0) - f(y_0,z_0) = \alpha \frac{\partial f(y_0,z_0)}{\partial g_1} + o_1(\alpha) , \tag{4.23}$$

where $o_1(\alpha) \ge 0$ and

$$\frac{\partial f(y_0,x_0)}{\partial g_1} = \lim_{\alpha \to +0} \alpha^{-1}[f(y_0+\alpha g_1,z_0) - f(y_0,z_0)] = \max_{v \in \partial f_y(y_0,z_0)} (v,g_1) .$$

Since the function $f(y,z)$ is concave with respect to z, then from relation (5.45) in Chapter 1 we have

$$f(y_0+\alpha g_1, z_0+\alpha g_2) = f(y_0+\alpha g_1, z_0) + \int_0^\alpha \frac{\partial f(y_0+\alpha g_1, z_0+\tau g_2)}{\partial g_2}\, d\tau , \qquad (4.24)$$

where

$$\frac{\partial f(y,z)}{\partial g_2} \equiv \lim_{\alpha\to+0} \alpha^{-1}[f(y,z+\alpha g_2) - f(y,z)] = \min_{w\in\partial f_z(y,z)} (w,g_2). \qquad (4.25)$$

Let us fix $\varepsilon > 0$. It follows from (4.18) that there exists an $\alpha(\varepsilon) > 0$ such that

$$\partial f_z(y_0+\alpha g_1, z_0+\tau g_2) \subset \partial f_z(y_0,z_0) + S_{2\varepsilon}$$
$$\forall\ \tau \in [0,\alpha(\varepsilon)], \quad \forall\ \alpha \in [0,\alpha(\varepsilon)] . \qquad (4.26)$$

If $G \subset G_1+G_2$, then

$$\min_{w\in G_1} (w,g_2) \geq \min_{w\in G_1+G_2} (w,g_2) = \min_{w\in G_1} (w,g_2) + \min_{w\in G_2} (w,g_2) .$$

Hence, from (4.24) and (4.26) we get

$$A_1 \equiv f(y_0+\alpha g_1, z_0+\alpha g_2) - f(y_0+\alpha g_1, z_0)$$
$$\geq \alpha \min_{w\in\partial f_z(y_0,z_0)} (w,g_2) - \alpha\varepsilon \|g_2\| .$$

Since $\min_{w\in S_{2\varepsilon}} (w,g_2) = -\varepsilon\|g_2\|$, then from (4.22) and (4.23)

$$h(\alpha) \geq \max_{v\in\partial f_y(y_0,z_0)} (v,g_1) + \min_{w\in\partial f_z(y_0,z_0)} (w,g_2) - \varepsilon\|g_2\|$$
$$= B - \varepsilon\|g_2\| .$$

It means that

$$\underline{\lim}_{\alpha\to+0}\, h(\alpha) \geq B - \varepsilon\|g_2\|.$$

Therefore, by virtue of the arbitrariness of ε, we obtain

$$\underline{\lim}_{\alpha\to+0}\, h(\alpha) \geq B . \qquad (4.27)$$

Now, we note that, on the other hand,

$$f(y_0+\alpha g_1, z_0+\alpha g_2) - f(y_0,z_0) = [f(y_0+\alpha g_1, z_0+\alpha g_2) - f(y_0, z_0+\alpha g_2)] + [f(y_0,z_0+\alpha g_2) - f(y_0,z_0)]$$

$$\equiv A_3 + A_4 \quad . \tag{4.28}$$

Since $f(y,z)$ is a concave function, we have

$$A_4 \equiv f(y_0,z_0+\alpha g_2) - f(y_0,z_0) = \alpha \frac{\partial f(y_0,z_0)}{\partial g_2} + o_2(\alpha) \ , \tag{4.29}$$

where $o_2(\varepsilon) \leq 0$ and

$$\frac{\partial f(y_0,z_0)}{\partial g_2} = \lim_{\alpha\to+0} \alpha^{-1}[f(y_0,z_0+\alpha g_2) - f(y_0,z_0)] = \min_{w\in\partial f_z(y_0,z_0)} (w,g_2) \ .$$

The function $f(y,z_0+\alpha g_2)$ is convex with respect to y and, hence, from formula (5.26) in Chapter 1

$$A_3 \equiv f[(y_0+\alpha g_1, z_0+\alpha g_2) - f(y_0,z_0+\alpha g_2)] = \int_0^\alpha \frac{\partial f(y_0+\tau g_1, z_0+\tau g_2)}{\partial g_1} d\tau \ .$$

Here,

$$\frac{\partial f(y_0+\tau g_1, z_0+\alpha g_2)}{\partial g_1} = \max_{v\in\partial f_y(y_0+\tau g_1, z_0+\alpha g_2)} (v,g_1) \ . \tag{4.30}$$

Let us take $\varepsilon > 0$. Due to Lemma 4.2, there exists an $\alpha(\varepsilon) > 0$ such that

$$\partial f_y(y_0+\tau g_1, z_0+\alpha g_2) \subset \partial f_y(y_0,z_0) + S_{1\varepsilon}$$

$$\forall \tau \in [0,\alpha(\varepsilon)], \quad \forall \alpha \in [0,\alpha(\varepsilon)] \ .$$

Since, for $G \subset G_1+G_2$, we have

$$\max_{v\in G} (v,g_1) \leq \max_{v\in G_1+G_2} (v,g_1) = \max_{v\in G_1} (v,g_1) + \max_{v\in G_2} (v,g_1) \ ,$$

then, using (4.30), we obtain

$$A_3 \leq \alpha \max_{v\in\partial f_y(y_0,z_0)} (v,g_1) + \alpha\varepsilon\|g_1\| \ ,$$

because

$$\max_{v\in S_{1\varepsilon}} (v,g_1) = \varepsilon\|g_1\| \ .$$

From this and (4.29) we have

$$h(\alpha) \le \max_{v\in\partial f_y(y_0,z_0)} (v,g_1) + \min_{w\in\partial f_z(y_0,z_0)} (w,g_2) + \varepsilon\|g_2\|$$

$$\equiv B + \varepsilon\|g_2\| .$$

Hence

$$\overline{\lim_{\alpha\to+0}}\, h(\alpha) \le B + \varepsilon\|g_2\| .$$

By virtue of the arbitrariness of ε, we get

$$\overline{\lim_{\alpha\to+0}}\, h(\alpha) \le B . \tag{4.31}$$

Comparing (4.27) and (4.31), we conclude that the limit $\lim_{\alpha\to+\infty} h(\alpha)$ exists and

$$\lim_{\alpha\to+0} h(\alpha) = \frac{\partial f(x_0)}{\partial g} \equiv B ,$$

which proves the theorem. ∎

<u>COROLLARY</u>. If a function $f(x)$ is convexo-concave on S then it is quasidifferentiable and, moreover,

$$\mathcal{D}f(x_0) = [\underline{\partial} f(x_0), \overline{\partial} f(x_0)] , \qquad \forall x_0 \in S ,$$

where $\underline{\partial} f(x_0) = [\partial f_y(y_0,z_0), 0_{n_2}]$, $\overline{\partial} f(x_0) = [0_{n_1}, \partial f_z(y_0,z_0)]$.

This follows from the formula for the directional derivative (formula (4.19)).

5. NECESSARY CONDITIONS FOR AN EXTREMUM OF A QUASIDIFFERENTIABLE FUNCTION ON E_n

1. Let $f(x)$ be a quasidifferentiable function on E_n. First we shall derive necessary conditions for a minimum and a maximum.

<u>THEOREM 5.1</u>. In order that quasidifferentiable function $f(x)$ on E_n attain its minimum value on E_n at a point $x^* \in E_n$, it is necessary that

$$-\overline{\partial} f(x^*) \subset \underline{\partial} f(x^*) . \tag{5.1}$$

Proof. If the function $f(x)$ is directionally differentiable at a point x^*, where x^* is a minimum point of the function $f(x)$ on E_n, then

$$\frac{\partial f(x^*)}{\partial g} \geq 0 \qquad \forall g \in E_n \tag{5.2}$$

(see Lemma 6.1 of Chapter 1). Since, in our case,

$$\frac{\partial f(x^*)}{\partial g} = \max_{v \in \underline{\partial} f(x^*)} (v,g) + \min_{w \in \overline{\partial} f(x^*)} (w,g) , \tag{5.3}$$

then we have from (5.2) and (5.3)

$$\inf_{g \in S_1} \left[\max_{v \in \underline{\partial} f(x^*)} (v,g) + \min_{w \in \overline{\partial} f(x^*)} (w,g) \right] = 0 , \tag{5.4}$$

where (here and in the remainder of this section)

$$S_1 = \{g \in E_n \mid ||g|| \leq 1\} .$$

From (5.4),

$$\min_{g \in S_1} \min_{w \in \overline{\partial} f(x^*)} \max_{v \in \underline{\partial} f(x^*)} (v+w, g)$$

$$= \min_{g \in S_1} \min_{w \in \overline{\partial} f(x^*)} \max_{v \in [\underline{\partial} f(x^*)+w]} (v,g) = 0 . \tag{5.5}$$

Since the sets S_1 and $\overline{\partial} f(x^*)$ are convex compact, then from (5.5)

$$\min_{w \in \overline{\partial} f(x^*)} \min_{g \in S_1} \max_{v \in [\underline{\partial} f(x^*)+w]} (v,g) = 0 ,$$

i.e.,

$$\min_{g \in S_1} \max_{v \in [\underline{\partial} f(x^*)+w]} (v,g) = 0 \quad \forall w \in \overline{\partial} f(x^*) . \tag{5.6}$$

Due to Corollary of Lemma 6.1 as well as of Theorem 6.1 of Chapter 1, we obtain that condition (5.6) is equivalent to the condition

$$0 \in \underline{\partial} f(x^*) + w \qquad \forall w \in \overline{\partial} f(x^*) .$$

This implies in turn that

$$-w \in \underline{\partial} f(x^*) \qquad \forall w \in \overline{\partial} f(x^*) ,$$

i.e., $-\overline{\partial}f(x^*) \subset \underline{\partial}f(x^*)$, which proves the theorem. ∎

Analogously, we can prove the following:

THEOREM 5.2. In order that a quasidifferentiable function $f(x)$ on E_n attain its maximum value on E_n at a point $x^{**} \in E_n$, it is necessary that

$$-\underline{\partial}f(x^{**}) \subset \overline{\partial}f(x^{**}) . \qquad (5.7)$$

DEFINITION. A point $x^* \in E_n$ satisfying (5.1) is called an inf-*stationary* point of the function $f(x)$ on E_n and a point $x^{**} \in E_n$ satisfying (5.7) is called a sup-*stationary* point of the function $f(x)$ on E_n.

It is clear that if for some point $x_0 \in E_n$, the relation $-\underline{\partial}f(x_0) = \overline{\partial}f(x_0)$ is valid, then the point x_0 is both inf-stationary and sup-stationary.

2. Let x^* be an inf-stationary point of a function $f(x)$ on E_n, i.e., (5.1) is valid. By definition,

$$f(x^* + \alpha g) = f(x^*) + \alpha \frac{\partial f(x^*)}{\partial g} + o(\alpha, g) , \qquad (5.8)$$

where

$$\frac{o(\alpha, g)}{\alpha} \xrightarrow[\alpha \to +0]{} 0 . \qquad (5.9)$$

THEOREM 5.3. If x^* is an inf-stationary point of a function $f(x)$ on E_n and

$$-\overline{\partial}f(x^*) \subset \text{int}\ \underline{\partial}f(x^*) , \qquad (5.10)$$

and, moreover, relation (5.9) is satisfied uniformly in $g \in S_1$, then x^* is a strict local minimum of the function $f(x)$.

Proof. It follows from (5.10) that there exists an $r > 0$ such that

$$S_r \subset [\underline{\partial}f(x^*) + w] \qquad \forall w \in \overline{\partial}f(x^*) ,$$

where $S_r = \{z \in E_n \mid \|z\| \le r\}$.

Then

$$\max_{v \in [\underline{\partial} f(x^*)+w]} (v,g) \geq r \qquad \forall w \in \overline{\partial} f(x^*) , \qquad \forall g : \|g\| = 1 .$$

From (5.5),

$$\min_{\substack{g \in E_n \\ \|g\|=1}} \min_{w \in \overline{\partial} f(x^*)} \max_{v \in [\underline{\partial} f(x^*)+w]} (v,g) \geq r ,$$

i.e.,

$$\frac{\partial f(x^*)}{\partial g} \geq r \qquad \forall g \in E_n : \|g\| = 1 . \tag{5.11}$$

Since (5.9) is satisfied uniformly with respect to $g \in S_1$, then it follows from (5.8) and (5.11) that there exists a $\alpha_0 > 0$ such that

$$f(x^*+\alpha g) \geq f(x^*) + \alpha r 2^{-1} \qquad \forall \alpha \in [0,\alpha_0], \qquad \forall g \in E_n : \|g\|= 1,$$

and this implies that x^* is a strict local minimum of the function $f(x)$. ■

Analogously, we can prove the following:

<u>THEOREM 5.4</u>. If x^{**} is a sup-stationary point of a function $f(x)$ on E_n and

$$-\underline{\partial} f(x^{**}) \subset \text{int } \overline{\partial} f(x^{**}) \tag{5.12}$$

and, moreover, relation (5.9) is satisfied uniformly with respect to $g \in S_1$, then x^{**} is a strict local maximum of the function $f(x$

As examples, let us consider some functions discussed in Section 3.

The point $x_0 = (0,0)$ is inf-stationary (and, indeed, a minimum point) for the functions introduced in Examples 1, 3, 6, 7, 8, 10, 11, 12, 13 in Section 3. The point x_0 is sup-stationary for the functions introduced in Examples 2, 4. The point x_0 is a strict local minimum for the functions introduced in Examples 6 and 8.

3. Let a point x_0 be not inf-stationary, i.e., condition (5.1) is not satisfied. Then, condition (5.2) is not satisfied either, i.e.,

$$\inf_{g\in S_1} \frac{\partial f(x_0)}{\partial g} = -a < 0 . \tag{5.13}$$

From this and (5.3),

$$\min_{g\in S_1} \min_{w\in\overline{\partial} f(x_0)} \max_{v\in\underline{\partial} f(x_0)} (v+w,g) = \min_{w\in\overline{\partial} f(x_0)} \min_{g\in S_1} \max_{v\in\underline{\partial} f(x_0)} (v+w,g)$$

$$= -a . \tag{5.14}$$

As in the proof of Theorem 6.2 of Chapter 1, it can be shown that

$$\min_{g\in S_1} \max_{v\in\underline{\partial} f(x_0)} (v+w, g) = - \min_{v\in\underline{\partial} f(x_0)} \|v + w\| .$$

Hence, from (5.13) and (5.14), we get

$$a = - \max_{w\in\overline{\partial} f(x_0)} \min_{v\in\underline{\partial} f(x_0)} \|v + w\| . \tag{5.15}$$

Let us choose $w_0 \in \overline{\partial} f(x_0)$ and $v_0 \in \underline{\partial} f(x_0)$ such that

$$\max_{w\in\overline{\partial} f(x_0)} \min_{v\in\underline{\partial} f(x_0)} \|v+w\| = \min_{v\in\underline{\partial} f(x_0)} \|v+w_0\| = \|v_0+w_0\| .$$

Then, the direction $g_0 = -(v_0+w_0)\|v_0+w_0\|^{-1}$ is the steepest descent direction of the function $f(x)$ at the point x_0.

Possibly, it is nonunique. Analogously, it can be shown that if a point x_0 is not sup-stationary, i.e., $-\underline{\partial} f(x_0) \not\subset \overline{\partial} f(x_0)$, then

$$a \equiv \sup_{g\in S_1} \frac{\partial f(x_0)}{\partial g} = \max_{v\in\underline{\partial} f(x_0)} \min_{w\in\overline{\partial} f(x_0)} \|v+w\| . \tag{5.16}$$

If $a > 0$, then the direction $g_1 = (v_1+w_1)\| v_1+w_1\|^{-1}$, where

$$\min_{w\in\overline{\partial} f(x_0)} \|v_1+w\| = \max_{v\in\underline{\partial} f(x_0)} \min_{w\in\overline{\partial} f(x_0)} \|v+w\| ,$$

$$||v_1+w_1|| = \min_{w\in\overline{\partial}f(x_0)} ||v_1+w|| \quad ,$$

is the steepest ascent direction of the function $f(x)$ at the point x_0. Possibly, it is nonunique.

Now, let us find the distance between the sets $\underline{\partial}f(x_0)$ and $-\overline{\partial}f(x_0)$ in the Hausdorff metric (see (2.1) in Chapter 1):

$\rho(\underline{\partial}f(x_0), -\overline{\partial}f(x_0))$

$$\equiv \max_{w\in\overline{\partial}f(x_0)} \min_{v\in\underline{\partial}f(x_0)} ||v+w|| + \max_{v\in\underline{\partial}f(x_0)} \min_{w\in\overline{\partial}f(x_0)} ||v+w|| .$$

Here, the first term is the deviation of the set $-\overline{\partial}f(x_0)$ from the set $\underline{\partial}f(x_0)$ (in the Hausdorff sense) and the second term is the deviation of the set $\underline{\partial}f(x_0)$ from the set $-\overline{\partial}f(x_0)$.

Hence, from (5.15) and (5.16) we obtain that the sum of absolute values of the rates of steepest descent and steepest ascent of the function $f(x)$ at the point x_0 is equal to the distance between the sets $\underline{\partial}f(x_0)$ and $-\overline{\partial}f(x_0)$ in the Hausdorff metric.

EXAMPLE 1. Let $x = (x^{(1)},x^{(2)}) \in E_2$, $x_0 = (0,0)$.

Take the function

$$f(x) = \min\{|x^{(1)}|+|x^{(2)}|,\ |x^{(1)}|-|x^{(2)}|\}.$$

This function was considered in Example 9 of Section 3. It was shown there that $\underline{\partial}f(x_0)$ is a rectangle with vertices at (2,1), (-2,1), (-2,-1), (2,-1) and the set $\overline{\partial}f(x_0)$ is a rectangle with vertices at (1,2), (-1,2), (-1,-2), (1,-2). It is clear that, due to the symmetricity, $\overline{\partial}f(x_0) = -\overline{\partial}f(x_0)$. Applying formulas (5.15) and (5.16), we find that there exist two steepest descent directions, i.e., $g_0 = (0,1)$ and $g_0' = (0,1)$, and two steepest ascent directions, i.e., $g_1 = (1,0)$ and $g_1' = (-1,0)$. It is also obvious that

$$\frac{\partial f(x_0)}{\partial g_0} = \frac{\partial f(x_0)}{\partial g_0'} = -1 \ , \qquad \frac{\partial f(x_0)}{\partial g_1} = \frac{\partial f(x_0)}{\partial g_1'} = 1 \ .$$

EXAMPLE 2. Now let

$$x = (x^{(1)}, x^{(2)}) \in E_2 , \qquad x_0 = (0,0) , \qquad f(x) = \big|\,|x^{(1)}| + x^{(2)}\big| .$$

We considered this function in Example 10 in Section 3. We show that (Figure 30)

$$\underline{\partial} f(x_0) = \text{co}\ \{(0,0),\ (2,2),\ (-2,2)\} ,$$

$$\overline{\partial} f(x_0) = \text{co}\ \{(1,-1),\ (1-,-1)\} .$$

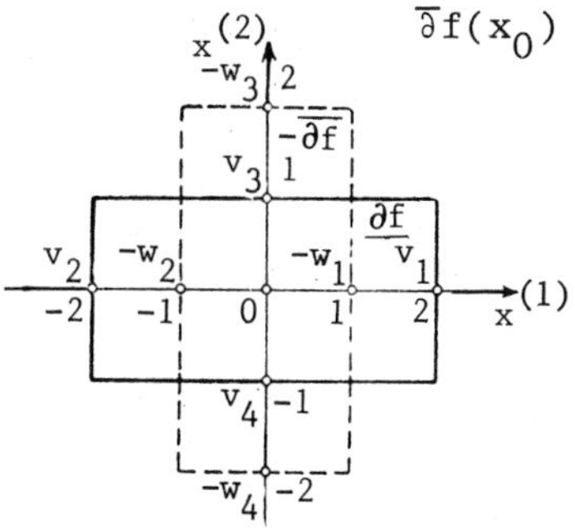

Figure 29

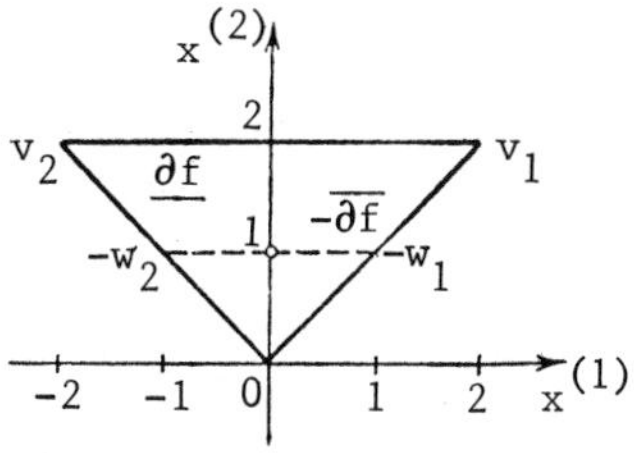

Figure 30

Then $-\overline{\partial} f(x_0) = \text{co}\ \{(1,1),\ (-1,1)\}$.

The point x_0 is inf-stationary, i.e., $\inf\limits_{g \in E_n} \dfrac{\partial f(x_0)}{\partial g} = 0.$

It is clear that there exist two steepest ascent directions

$$g_1 = \left(\frac{\sqrt{2}}{2},\ \frac{\sqrt{2}}{2}\right) , \qquad g_1' = \left(-\frac{\sqrt{2}}{2},\ \frac{\sqrt{2}}{2}\right)$$

and we have

$$\sup_{g \in S_1} \frac{\partial f(x_0)}{\partial g} = \frac{\partial f(x_0)}{\partial g_1} = \frac{\partial f(x_0)}{\partial g_1'} = \sqrt{2} .$$

Thus, the verification of the necessary conditions for a minimum is reduced to the problem of finding the deviation of the set $-\overline{\partial} f(x_0)$ from the set $\underline{\partial} f(x_0)$ (in the Hausdorff sense) and the verification of the necessary conditions for a maximum is reduced to to the problem of finding the deviation of the set $\underline{\partial} f(x_0)$ from the set $-\overline{\partial} f(x_0)$.

In the course of solving these problems, the directions of steepest descent or steepest ascent are identified simultaneously (depending on which conditions -- a minimum or a maximum -- is being verified).

6. QUASIDIFFERENTIABLE SETS

1. Let $\Omega \subset E_n$ be a closed set and let $x_0 \in \Omega$. A vector $v \in E_n$ is called a *feasible (in a broad sense) direction* of the set Ω and a point x_0 if there exist a number $\lambda \geq 0$ and a sequence $\{x_i\}$ such that

$$\begin{aligned} &x_i \in \Omega, \qquad x_i \neq x_0, \qquad x_i \to x_0, \\ &v_i = (x_i - x_0)\|x_i - x_0\|^{-1} \to v_0, \qquad v = \lambda v_0 . \end{aligned} \tag{6.1}$$

Clearly, the set of feasible (in a broad sense) directions is a closed cone with the vertex at the origin. Let us denote it by $\Gamma(x_0)$ and call it the *cone of feasible (in a broad sense) directions* of the set Ω at a point x_0.

Consider the case where Ω has the following form:

$$\Omega = \{x \in E_n \mid h(x) \leq 0\} ,$$

where $h(x)$ is a continuous and quasidifferentiable function on Ω.

Assume that $\Omega \neq \emptyset$. We shall say that the set Ω has a *quasidifferentiable boundary* or, simply, that it is *quasidifferentiable*.

Clearly, Ω is a closed set. Let $x_0 \in \Omega$ and define

$$\gamma(x_0) = \left\{g \in E_n \;\middle|\; \frac{\partial h(x_0)}{\partial g} < 0\right\} , \qquad \gamma_1(x_0) = \left\{g \in E_n \;\middle|\; \frac{\partial h(x_0)}{\partial g} \leq 0\right\}.$$

It is obvious that if $h(x_0) = 0$, then

$$\gamma(x_0) \subset \Gamma(x_0) . \tag{6.2}$$

Let $h(x_0) = 0$. We shall say that the *nondegeneracy condition* is satisfied at a point x_0 if

$$\bar{\gamma}(x_0) = \gamma_1(x_0) . \tag{6.3}$$

Here, $\bar{\gamma}$ is the closure of the set γ. It is easy to construct an example showing that condition (6.3) is not necessarily satisfied.

LEMMA 6.1. If $h(x_0) < 0$, then $\Gamma(x_0) = E_n$.

If $h(x_0) = 0$ and the nondegeneracy condition (6.3) is satisfied at a point x_0 and, moreover, the function $h(x)$ is Lipschitzian in a neighborhood of x_0, then

$$\Gamma(x_0) = \gamma_1(x_0) . \tag{6.4}$$

Proof. Clearly, if $h(x_0) < 0$, then $\Gamma(x_0) = E_n$ and this proves the first assertion of the lemma. Now let $h(x_0) = 0$. It follows from (6.2) and (6.3) that

$$\gamma_1(x_0) = \bar{\gamma}(x_0) \subset \Gamma(x_0) . \tag{6.5}$$

Below, we shall establish the converse inclusion. Let us take $g \in \Gamma(x_0)$. Then there exist a number $\lambda \geq 0$ and a sequence $\{x_i\}$ such that

$$\begin{gathered} x_i \in \Omega, \qquad x_i \neq x_0, \qquad x_i \to x_0 , \\ g_i = (x_i - x_0)\| x_i - x_0\|^{-1} \to g_0, \qquad g = \lambda g_0 . \end{gathered} \tag{6.6}$$

If $\lambda = 0$, then $g = 0$ and, obviously, $g = 0 \in \gamma_1(x_0)$. Now let $\lambda > 0$.

Now that we have

$$\frac{\partial h(x_0)}{\partial g} = \lim_{\alpha\to+0} \alpha^{-1}[h(x_0+\alpha g) - h(x_0)] \equiv \lim_{\alpha\to+0} H(\alpha) ,$$

where $H(\alpha) = \alpha^{-1}[h(x_0+\alpha g) + h(x_0)]$. Here we take

$$\alpha_i = \lambda^{-1}\|x_i - x_0\| . \tag{6.7}$$

Clearly, $\alpha_i \xrightarrow[i\to\infty]{} +0$. Moreover, we have

$$\begin{aligned} H(\alpha_i) &= \alpha_i^{-1}[h(x_0+\alpha_i g) - h(x_0)] \quad (6.8)\\ &= \alpha_i^{-1}[h(x_0+\alpha_i\lambda g_0) - h(x_0+\alpha_i\lambda g_i) + h(x_0+\alpha_i\lambda g_i) - h(x_0)]\\ &= \alpha_i^{-1}[h(x_0+\alpha_i\lambda g_0) - h(x_0+\alpha_i\lambda g_i)] + \alpha_i^{-1}[h(x_0+\alpha_i\lambda g_i) - h(x_0)]. \end{aligned}$$

Since the function $h(x)$ is Lipschitzian in a neighborhood of the point x_0, then there exist numbers $\delta > 0$ and $L < \infty$ such that

$$|h(x) - h(x')| \le L\|x-x'\| \quad \forall x, x' \in S_\delta(x_0) .$$

Indeed, for sufficiently large i,

$$|h(x_0+\alpha_i\lambda g_0) - h(x_0+\alpha_i\lambda g_i)| \le L\alpha_i\lambda\|g_i-g_0\| . \quad (6.9)$$

Since from (6.7) we have $x_0 + \alpha_i\lambda g_i = x_i \in \Omega$, then $h(x_0+\alpha_i\lambda g_i) \le 0 \quad (= h(x_0))$. Hence, from (6.8) and (6.9) we obtain

$$H(\alpha_i) \le \alpha_i^{-1}[h(x_0+\alpha_i\lambda g_0) - h(x_0+\alpha_i\lambda g_i)] \to 0 ,$$

i.e., $\dfrac{\partial h(x_0)}{\partial g} = \lim_{i\to\infty} H(\alpha_i) \le 0$, and this implies that $g \in \gamma_1(x_0)$. Thus,

$$\Gamma(x_0) \subset \gamma_1(x_0) . \quad (6.10)$$

The proof of the remaining part of the lemma follows from (6.5) and (6.10). ∎

<u>LEMMA 6.2</u>. The following relation holds:

$$\gamma_1(x_0) = \bigcup_{w\in\overline{\partial}h(x_0)} [-K^+(\underline{\partial}h(x_0) + w)] . \quad (6.11)$$

<u>Proof</u>. By the definition,

$$\gamma_1(x_0) = \left\{g \in E_n \;\middle|\; \frac{\partial h(x_0)}{\partial g} \le 0\right\} .$$

Since

$$\begin{aligned} \frac{\partial h(x_0)}{\partial g} &= \max_{v\in\underline{\partial}h(x_0)} (v,g) + \min_{w\in\overline{\partial}h(x_0)} (w,g)\\ &= \min_{w\in\overline{\partial}h(x_0)} \left[\max_{v\in[\underline{\partial}h(x_0)+w]} (v,g)\right] , \end{aligned}$$

then

$$\gamma_1(x_0) = \bigcup_{w \in \overline{\partial} h(x_0)} \Gamma_w , \qquad (6.12)$$

where

$$\Gamma_w = \{g \in E_n \mid (v,g) \le 0 \ \forall v \in [\underline{\partial} h(x_0)+w]\} .$$

Due to Lemma 3.9 of Chapter 1, we have

$$\Gamma_w^+ \equiv \{q \in E_n \mid (q,g) \ge 0 \ \forall g \in \Gamma_w\} = -\overline{K}(\text{co } (\underline{\partial} h(x_0)+w)) = -\overline{K}(\underline{\partial} h(x_0)+w).$$

Since Γ_w is a closed convex cone, then

$$\Gamma_w^{++} = \Gamma_w , \qquad \text{i.e.,} \quad \Gamma_w = -K^+(\underline{\partial} h(x_0) + w)$$

(see Lemma 3.3 of Chapter 1). Hence, (6.11) follows from (6.12). ∎

<u>LEMMA 6.3</u>. The following relation holds:

$$\gamma(x_0) = E_n \setminus \bigcup_{v \in \underline{\partial} h(x_0)} K^+(\overline{\partial} h(x_0) + v) . \qquad (6.13)$$

<u>Proof</u>. Let us denote by D the set on the right-hand side of (6.13). Take an arbitrary $g \in \gamma(x_0)$; then $\frac{\partial h(x_0)}{\partial g} < 0$, i.e.,

$$\max_{v \in \underline{\partial} h(x_0)} \ \min_{w \in [\overline{\partial} h(x_0)+v]} (w,g) < 0 .$$

Therefore,

$$\min_{w \in [\overline{\partial} h(x_0)+v]} (w,g) < 0 \qquad \forall w \in \underline{\partial} h(x_0) .$$

Hence

$$g \notin K^+(\overline{\partial} h(x_0) + v) \qquad \forall v \in \underline{\partial} h(x_0) ,$$

yielding

$$g \notin \bigcup_{v \in \underline{\partial} h(x_0)} K^+(\overline{\partial} h(x_0) + v) ,$$

i.e.,

$$g \in E_n \setminus \bigcup_{v \in \underline{\partial} h(x_0)} K^+(\overline{\partial} h(x_0) + v) \equiv D .$$

Since g is an arbitrary vector of $\gamma(x_0)$, then

$$\gamma(x_0) \subset D . \qquad (6.14)$$

Now, we take $g \in D$, i.e.,

$$g \notin \bigcup_{v \in \underline{\partial} h(x_0)} K^+(\overline{\partial} h(x_0) + v .$$

Then, we have

$$g \notin K^+(\overline{\partial} h(x_0) + v) \qquad \forall v \in \underline{\partial} h(x_0) .$$

Hence,

$$\min_{w \in [\overline{\partial} h(x_0)+v]} (w,g) < 0 \qquad \forall v \in \underline{\partial} h(x_0) . \qquad (6.15)$$

The set $\underline{\partial} h(x_0)$ is compact; therefore, from (6.15) we get

$$\max_{v \in \underline{\partial} h(x_0)} \min_{w \in \overline{\partial} h(x_0)+v} (w,g) < 0 ,$$

i.e., $\frac{\partial h(x_0)}{\partial g} < 0$. This implies that $g \in \gamma(x_0)$. Thus, $D \subset \gamma(x_0)$. From this and (6.14) we obtain (6.13). ∎

1. Let us illustrate the definitions of the sets $\Gamma(x_0)$, $\gamma(x_0)$ and $\gamma_1(x_0)$ introduced in Section 1.

<u>EXAMPLE 1</u>. Let $x = (x^{(1)}, x^{(2)}) \in E_2$. Take the points

$$x_1 = (1,1) , \quad x_2 = (-1,1) , \quad x_3 = (1-,-1) ,$$
$$x_4 = (1,-1) , \quad x_5 = (0,2) , \quad x_4 = (0,-2)$$

and construct the functions

$$h_1(x) = (x-x_1)^2 - 2 , \qquad h_2(x) = (x-x_2)^2 - 2 ,$$
$$h_3(x) = (x-x_3)^2 - 2 , \qquad h_4(x) = (x-x_4)^2 - 2 ,$$
$$h_5(x) = (x-x_5)^2 - 4 , \qquad h_6(x) = (x-x_6)^2 - 4 ,$$
$$h_7(x) = \max \{h_1(x), h_2(x)\} , \qquad h_8(x) = \max \{h_3(x), h_4(x)\} ,$$
$$h_9(x) = -h_7(x) , \qquad h_{10}(x) = -h_8(x) ,$$
$$h_{11}(x) = \max \{h_5(x), h_9(x)\} , \qquad h_{12}(x) = \max \{h_6(x), h_{10}(x)\},$$
$$h(x) = \min \{h_{11}(x), h_{12}(x)\} .$$

Define the set

$$\Omega = (x \in E_2 \mid h(x) \le 0\} .$$

This set is shown in Figure 31. Take the point $x_0 = (0,0)$ and find $\mathcal{D}h(x_0)$ and the cones $\gamma(x_0)$, $\gamma_1(x_0)$, $\Gamma(x_0)$.

First of all, we note that

$$\mathcal{D}h_1(x_0) = [-2x_1, 0], \qquad \mathcal{D}h_2(x_0) = [-2x_2, 0],$$

$$\mathcal{D}h_3(x_0) = [-2x_3, 0], \qquad \mathcal{D}h_4(x_0) = [-2x_4, 0],$$

$$\mathcal{D}h_5(x_0) = [-2x_5, 0], \qquad \mathcal{D}h_6(x_0) = [-2x_6, 0]$$

(more precisely, we have to write $\mathcal{D}h_1(x_0) = [\{-2x_1\}, \{0\}]$ and so on). According to Lemma 2.2,

$$\mathcal{D}h_7(x_0) = [\text{co }\{-2x_1, -2x_2\},\ 0] = [-\text{co }\{2x_1, 2x_2\},\ 0],$$

$$\mathcal{D}h_8(x_0) = [\text{co }\{-2x_3, -2x_4\},\ 0] = [-\text{co }\{2x_3, 2x_4\},\ 0].$$

Since $h_9(x) = -h_7(x)$, $h_{10}(x) = -h_8(x)$, then

$$\mathcal{D}h_9(x_0) = [0, A], \qquad \mathcal{D}h_{10}(x_0) = [0, B],$$

where

$$A = \text{co }\{2x_1, 2x_2\}, \qquad B = \text{co }\{2x_3, 2x_4\}.$$

Again, due to Lemma 2.2,

$$\mathcal{D}h_{11}(x_0) = [C, D], \qquad \mathcal{D}h_{12}(x_0) = [F, G],$$

where

$$C = \text{co }\{-2x_5 - A, 0\}, \qquad D = A,$$

$$F = \text{co }\{-2x_6 - B, 0\}, \qquad G = B.$$

The sets A, B, C, F are shown in Figure 32 (the sets A, B are segments and the sets C, F are triangles). Now, due to Lemma 2.3, we have

$$\mathcal{D}h(x_0) = [\underline{\partial} h(x_0), \overline{\partial} h(x_0)],$$

where

$$\underline{\partial} h(x_0) = \underline{\partial} h_{11}(x_0) + \underline{\partial} h_{12}(x_0) = C + F,$$

$$\overline{\partial} h(x_0) = \text{co }\{\overline{\partial} h_{11} - \underline{\partial} h_{12},\ \overline{\partial} h_{12} - \underline{\partial} h_{11}\} = \text{co }\{D-F\ \ G-C\}$$

$$= \text{co }\{A-F,\ B-C\}.$$

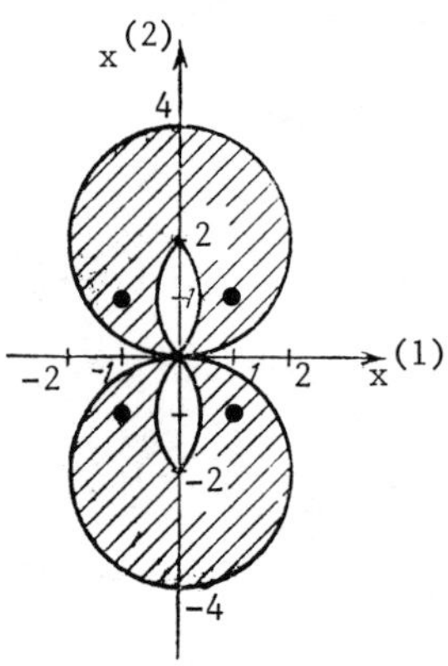

Figure 31

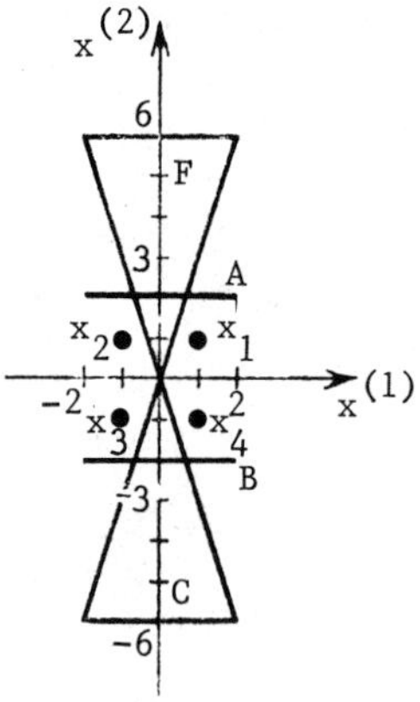

Figure 32

The sets $\underline{\partial}h(x_0)$ and $\overline{\partial}h(x_0)$ are shown in Figure 33 (the set $\underline{\partial}h(x_0)$ is restricted by the bold solid line and the set $\overline{\partial}h(x_0)$ by the dashed line). Since

$$\frac{\partial h(x_0)}{\partial g} = \max_{v\in\underline{\partial}h(x_0)} (v,g) + \min_{w\in\overline{\partial}h(x_0)} (w,g) ,$$

then

$$\gamma(x_0) = \{v=\lambda g \mid \lambda \geq 0,\ g=(1,\alpha),\ \alpha\in\{(0,1)\}\cup\{(0,-1)\}\}$$
$$\cup\ \{v=\lambda g \mid \lambda \geq 0,\ g=(-1,\alpha),\ \alpha\in\{(0,1)\}\cup\{(0,-1)\}\} ,$$

$$\gamma_1(x_0) = \{v=\lambda g \mid \lambda \geq 0,\ g=(1,\alpha),\ \alpha\in[-1,1]\}$$
$$\cup\ \{v=\lambda g \mid \lambda \geq 0,\ g=(-1,\alpha),\ \alpha\in[-1,1]\} .$$

It is obvious that $\gamma_1(x_0) = \gamma(x_0)$, i.e., the nondegeneracy condition is satisfied at the point x_0 and then, due to Lemma 6.1, $\Gamma(x_0) = \gamma_1(x_0)$.

EXAMPLE 2. Let $x = (x^{(1)},x^{(2)}) \in E_2$, $x_0 = (0,0)$, and

$$\Omega = \{x\in E_2 \mid h(x) \leq 0\} ,$$

where $h(x) = (x^{(1)}+1)^2 + (x^{(2)}+1)^2 - |x^{(1)}| - |x^{(2)}| - 2$.

It is clear that $x_0 \in \Omega$. Since $h(x) = h_1(x) + h_2(x)$, where

$$h_1(x) = (x^{(1)}+1)^2 + (x^{(2)}+1)^2 \quad ,$$

$$h_2(x) = -|x^{(1)}| - |x^{(2)}| \ ,$$

and the functions $h_1(x)$, $h_2(x)$ are, respectively, smooth and concave, then it is possible to take

$$\underline{\partial}h(x_0) = \{(2,2)\} \ ,$$

$$\overline{\partial}h(x_0) = \text{co}\ \{(1,1),\ (-1,1),\ (-1,-1),\ (1,-1)\}$$

(see Figure 34). To find $\gamma_1(x_0)$, it is better to use formula (6.13)

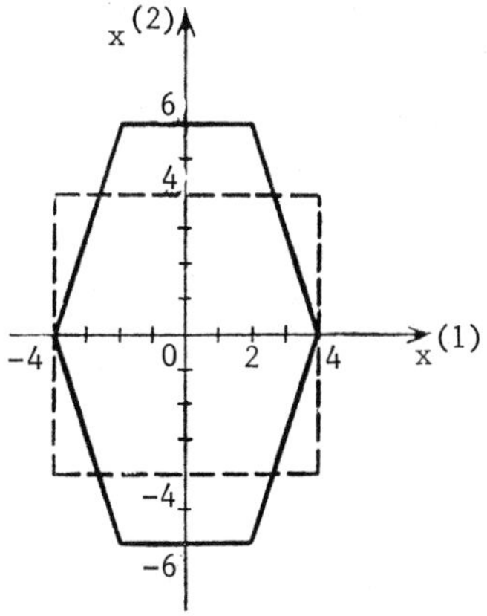

Figure 33

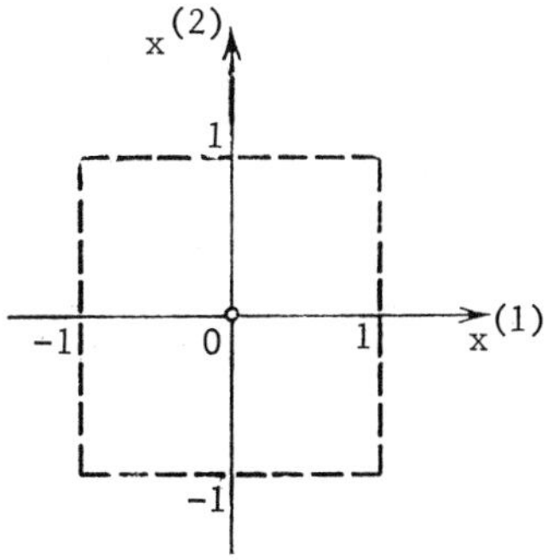

Figure 34

because, in our case, the subdifferential $\underline{\partial}h(x_0)$ consists of a unique point. In Figure 35, the cone $K(\overline{\partial}h(x_0)+v)$ (here, $v = (2,2)$) is restricted by the dashed line, the cone $K^+(\overline{\partial}h(x_0)+v)$ by the bold solid arc around the origin and the cone $\gamma_1(x_0)$:

$$\gamma_1(x_0) = \gamma(x_0) \cup \{\ell_1\} \cup \{\ell_2\}$$

by the double solid arcs around the origin, where

$\ell_1 = \{x = \lambda(3,-1) \mid \lambda \geq 0\}$, $\ell_2 = \{x = \lambda(-1,3) \mid \lambda \geq 0\}$.

<u>EXAMPLE 3</u>. Let

$$x = (x^{(1)},x^{(2)}) \in E_2 \ , \qquad x_0 = (0,0) \ , \qquad \Omega = \{x \in E_2 \mid h(x) \leq 0\} \ ,$$

where $h(x) = (x^{(1)}+1)^2 + (x^{(2)}+1)^2 + |x^{(1)}| + |x^{(2)}| - 2.$

The function $h(x)$ is convex and, in this case, as easily verified, it is possible to take

$$\underline{\partial}h(x_0) = \text{co}\ \{(1,1), (3,1), (3,3), (1,3)\}\ , \qquad \overline{\partial}h(x_0) = \{0\}\ .$$

To find $\gamma_1(x_0)$, it is better to use formula (6.11) because the superdifferential $\overline{\partial}h(x_0)$ consists of a unique point. The cones $K(\underline{\partial}h(x_0)+w)$, $K^+(\underline{\partial}h(x_0)+w)$ and $\gamma_1(x_0)$ are shown in Figure 36.

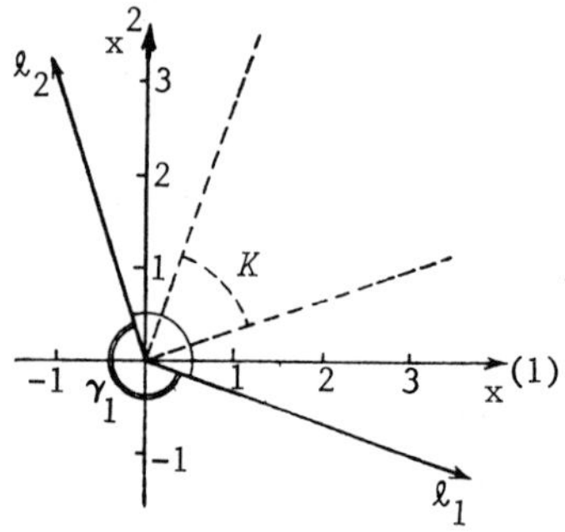

Figure 35

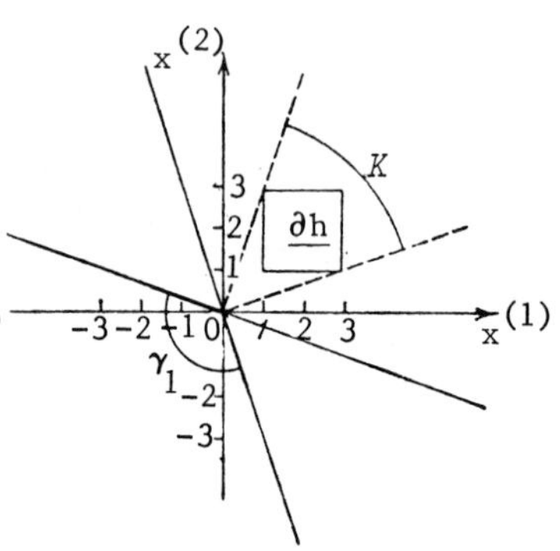

Figure 36

<u>PROBLEM 6.1.</u> Let

$$x = (x^{(1)}, x^{(2)}) \in E_2\ , \qquad x_0 = (0,0)\ ,$$
$$x_1 = (1,1)\ , \qquad x_2 = (-1,1)\ ,$$
$$x_3 = (-1,-1)\ , \qquad x_4 = (1,-1)\ ,$$
$$x_5 = (2,2)\ , \qquad x_6 = (-2,2)\ ,$$
$$x_7 = (-2,-2)\ , \qquad x_8 = (2,-2)\ .$$

Define the sets

$$S_1 = \{x \in E_2 \mid (x-x_1)^2 \le 2\}\ , \qquad S_2 = \{x \mid (x-x_2)^2 \le 2\}\ ,$$
$$S_3 = \{x \mid (x-x_3)^2 \le 2\}\ , \qquad S_4 = \{x \mid (x-x_4)^2 \le 2\}\ ,$$

$S_5 = \{x \mid (x-x_5)^2 \leq 8\}$, $S_6 = \{x \mid (x-x_6)^2 \leq 8\}$,

$S_7 = \{x \mid (x-x_7)^2 \leq 8\}$, $S_8 = \{x \mid (x-x_8)^2 \leq 8\}$.

Now we construct the set $\Omega = \Omega_1 \cup \Omega_2$, where $\Omega_1 = \overline{\Omega_3 \setminus \Omega_4}$, $\Omega_2 = \overline{\Omega_5 \setminus \Omega_6}$, $\Omega_3 = S_5 \cap S_6$, $\Omega_4 = S_1 \cap S_2$, $\Omega_5 = S_7 \cap S_8$, $\Omega_6 = S_3 \cap S_4$. The set Ω is shown in Figure 37.

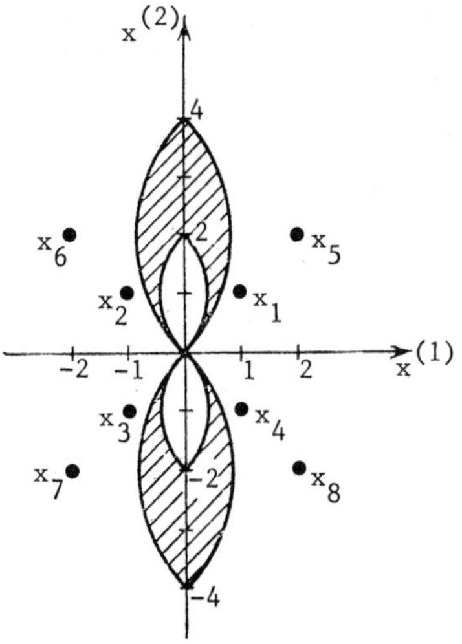

Figure 37

Find a function $h(x)$ such that

$$\Omega = \{x \in E_2 \mid h(x) \leq 0\} ,$$

and construct $\mathcal{D}h(x_0)$, $\gamma(x_0)$, $\gamma_1(x_0)$, $\Gamma(x_0)$. Verify whether the non-degeneracy condition is satisfied at the point x_0.

7. NECESSARY CONDITIONS FOR AN EXTREMUM OF A QUASIDIFFERENTIABLE FUNCTION ON A QUASIDIFFERENTIABLE SET

1. Let a finite continuous function $f(x)$ be given on some open set $S \subset E_n$ and let $\Omega \subset S$ be a closed set.

<u>LEMMA 7.1</u>. Let a function $f(x)$ be Lipschitzian in a neighborhood of a point $x^* \in \Omega$ and be directionally differentiable at x^*. For

the function $f(x)$ to attain its minimum on Ω at the point x^*, it is necessary that

$$\frac{\partial f(x^*)}{\partial g} \geq 0 \qquad \forall g \in \Gamma(x^*) , \tag{7.1}$$

where $\Gamma(x^*)$ is the cone of feasible (in the broad sense) directions of the set Ω at the point x^* (see Section 6).

<u>Proof</u>. Assume the opposite. Then there exists a $g_0 \in \Gamma(x^*)$ such that

$$\frac{\partial f(x^*)}{\partial g_0} = -a < 0 . \tag{7.2}$$

It follows from (6.1) that there exist a number $\lambda > 0$ and a sequence $\{x_i\}$ such that

$$x_i \in \Omega , \qquad x_i \neq x^* , \qquad x_i \to x^* ,$$

$$v_i = (x_i - x^*)\|x_i - x^*\|^{-1} \to v_0 , \qquad g_0 = \lambda v_0 .$$

Since

$$f(x_i) = f(x^* + x_i - x^*) = f(x^* + \|x_i - x^*\| v_i) = f(x^* + \alpha_i v_i) ,$$

where $\alpha_i = \|x_i - x^*\|$, then, due to Lemma 5.6 of Chapter 1,

$$\lim_{i\to\infty} \alpha_i^{-1}[f(x_i) - f(x^*)] = \lim_{i\to\infty} \alpha_i^{-1}[f(x^* + \alpha_i v_i) - f(x^*)] = \frac{\partial f(x^*)}{\partial v_0} .$$

Hence,

$$f(x_i) = f(x_0) + \alpha_i \frac{\partial f(x^*)}{\partial v_0} + o(\alpha_i) . \tag{7.3}$$

On the other hand, we have

$$\frac{\partial f(x^*)}{\partial v_0} = \lim_{\alpha\to+0} \alpha^{-1}[f(x^* + \alpha v_0) - f(x^*)]$$

$$= \lambda^{-1} \lim_{\alpha\to+0} \lambda\alpha^{-1}[f(x^* + \alpha\lambda^{-1}(\lambda v_0)) - f(x^*)] = \lambda^{-1}\frac{\partial f(x^*)}{\partial g_0} .$$

Now, from (7.2) and (7.3) we obtain that $f(x_i) < f(x^*)$ for sufficiently large i, which contradicts the assumption that x^* is a minimum point of the function $f(x)$ on the set Ω, since $x_i \in \Omega$. ■

Let $\Omega \subset E_n$ be a quasidifferentiable set, i.e.,

$$\Omega = \{x \in E_n \mid h(x) \le 0\} ,$$

where $h(x)$ is a continuous and quasidifferentiable function on E_n.

As before, we assume that $\Omega \subset S$. In the remainder of this section, we assume that the function $f(x)$ is quasidifferentiable on E_n.

<u>THEOREM 7.1</u>. Let a function $f(x)$ be Lipschitzian in a neighborhood of a point $x^* \in \Omega$. Moreover, if $h(x^*) = 0$, assume that the function $h(x)$ is Lipschitzian in a neighborhood of the point x^* and that the nondegeneracy condition (6.3) is satisfied.

Then, for the function $f(x)$ to attain its minimum on Ω at the point x^*, it is necessary that

$$-\overline{\partial} f(x^*) \subset \underline{\partial} f(x^*) \qquad \text{if} \quad h(x^*) < 0 , \tag{7.4}$$

and for all $w \in \overline{\partial} f(x^*)$, $w' \in \overline{\partial} h(x^*)$

$$(\underline{\partial} f(x^*) + w) \cap [-\overline{K}(\underline{\partial} h(x^*) + w')] \neq \emptyset \qquad \text{if} \quad h(x^*) = 0 . \tag{7.5}$$

Here, as usual, $K(A)$ is the conical hull of a set A.

<u>Proof</u>. For the case $h(x^*) < 0$, condition (7.4) follows from Theorem 5.1. Now, let $h(x^*) = 0$. Due to Lemma 7.1,

$$\frac{\partial f(x^*)}{\partial g} = \max_{v \in \underline{\partial} f(x^*)} (v,g) + \min_{w \in \overline{\partial} f(x^*)} (w,g) \ge 0 \qquad \forall g \in \Gamma(x^*) . \tag{7.6}$$

The cone $\Gamma(x^*)$ is, due to Lemma 6.1,

$$\Gamma(x^*) = \gamma_1(x^*) = \left\{ g \in E_n \;\middle|\; \frac{\partial h(x^*)}{\partial g} = \max_{v' \in \underline{\partial} h(x^*)} (v',g) + \min_{w' \in \overline{\partial} h(x^*)} (w',g) \le 0 \right\} . \tag{7.7}$$

From (7.6) and (7.7),

$$\min_{w \in \overline{\partial} f(x^*)} \; \max_{v \in [\underline{\partial} f(x^*) + w]} (v,g) \ge 0 \qquad \forall g \in \Gamma(x^*) , \tag{7.8}$$

$$\Gamma(x^*) = \left\{ g \in E_n \;\middle|\; \max_{w' \in \overline{\partial} h(x^*)} \; \min_{v' \in [\underline{\partial} h(x^*)+w']} (v',g) \le 0 \right\} . \quad (7.9)$$

From (7.9),

$$\Gamma(x^*) = \bigcup_{w' \in \overline{\partial} h(x^*)} \Gamma_{w'} , \quad (7.10)$$

where

$$\Gamma_{w'} = \left\{ g \in E_n \;\middle|\; \max_{v' \in [\underline{\partial} h(x^*)+w']} (v',g) \le 0 \right\} .$$

From (7.8) and (7.10) we obtain for arbitrary $w \in \overline{\partial} f(x^*)$ and $w' \in \overline{\partial} h(x^*)$

$$\max_{v \in [\underline{\partial} f(x^*)+w]} (v,g) \ge 0 \qquad \forall g \in \Gamma_{w'} . \quad (7.11)$$

Let

$$F_w(g) = \max_{v \in [\underline{\partial} f(x^*)+w]} (v,g) ,$$

$$H_{w'}(g) = \max_{v' \in [\underline{\partial} h(x^*)+w']} (v',g) .$$

Then $\Gamma_{w'} = \{g \in E_n \mid H_{w'}(g) \le 0\}$. The functions $F_w(g)$ and $H_{w'}(g)$ are convex. Find

$$\psi_{w,w'} = \min_{g \in \Gamma_{w'}} F_w(g) . \quad (7.12)$$

It follows from (7.11) that $\psi_{w,w'} = 0$, i.e., $g^* = 0$ is a solution of problem (7.12). According to Theorem 6.1 of Chapter 1, the relation

$$\partial F_w(g^*) \cap T^+_{w'}(g^*) \ne \emptyset \quad (7.13)$$

should hold, where $T^+_{w'}(g^*)$ is the cone conjugate to the cone of feasible directions of the set $\Gamma_{w'}$. The function $F_w(g)$ is a maximum function; hence, due to Corollary 2 of Theorem 14.2 of Chapter 1,

$$F_w(0) = \underline{\partial} f(x^*) + w . \quad (7.14)$$

Since the cone of feasible directions of the set $\Gamma_{w'}$ at the point $g^* = 0$ has the form

$$T_{w'} = \{q \in E_n \mid (q,v') \le 0 \quad \forall v' \in [\underline{\partial} h(x^*)+w']\} ,$$

then, due to Lemma 3.9 of Chapter 1,

$$T^{+}_{w'}(0) = -\overline{K}(\mathrm{co}\ (\underline{\partial}h(x^*)+w')) = -\overline{K}(\underline{\partial}h(x^*)+w') . \qquad (7.15)$$

From (7.13), (7.14) and (7.15), we obtain

$$(\underline{\partial}f(x^*)+w) \cap [-\overline{K}(\underline{\partial}h(x^*)+w')] \neq \emptyset \qquad \forall w \in \overline{\partial}f(x^*), \quad \forall w' \in \overline{\partial}h(x^*),$$

and this is the same as (7.5). The theorem has been proved. ■

<u>THEOREM 7.2</u>. Condition (7.5) is equivalent to the condition

$$-\overline{\partial}f(x^*) \subset \bigcap_{w \in \overline{\partial}h(x^*)} [\underline{\partial}f(x^*) + \overline{K}(\underline{\partial}h(x^*)+w)] . \qquad (7.16)$$

<u>Proof</u>. Let (7.5) hold. Take an arbitrary $w \in \overline{\partial}f(x^*)$ and $w' \in \overline{\partial}h(x^*)$. From (7.5), there exists a vector $v \in \underline{\partial}f(x^*)$ such that

$$v + w \in [-\overline{K}(\underline{\partial}h(x^*)+w')] .$$

Hence,

$$-w \in [v + \overline{K}(\underline{\partial}h(x^*)+w')] \subset \underline{\partial}f(X^*) + \overline{K}(\underline{\partial}h(x^*)+w') .$$

By virtue of the arbitrariness of $w' \in \overline{\partial}h(x^*)$, we have

$$-w \in \bigcap_{w' \in \overline{\partial}h(x^*)} [\underline{\partial}f(x^*) + \overline{K}(\underline{\partial}h(x^*)+w')] .$$

Since the vector $w \in \overline{\partial}f(x^*)$ is also arbitrary, then we obtain inclusion (7.16). Now, let (7.16) hold. Then, for any $w \in \overline{\partial}f(x^*)$ and $w' \in \overline{\partial}h(x^*)$,

$$-w \in [\underline{\partial}f(x^*) + \overline{K}(\underline{\partial}h(x^*)+w')] .$$

Hence,

$$0 \in [\underline{\partial}f(x^*) + w + \overline{K}(\underline{\partial}h(x^*)+w')] ,$$

i.e.,

$$[\underline{\partial}f(X^*) + w] \cap [-\overline{K}(\underline{\partial}h(x^*)+w')] \neq \emptyset ,$$

and this is the same as (7.5). The theorem has been proved. ■

<u>COROLLARY</u>. Let $h(x^*) = 0$. Assume that the function $h(x)$ is subdifferentiable and satisfies the nondegeneracy condition at the point x^*. Then,

$$-\overline{\partial}f(x^*) \subset \underline{\partial}f(x^*) - \Gamma^{+}(x^*) .$$

This follows from (7.16) and the fact that

$$\overline{\partial}h(x^*) = \{0\} , \qquad \overline{K}(\underline{\partial}h(x^*)) = -\Gamma^+(x^*) ,$$

where $\Gamma(x^*)$ is the cone of feasible directions of the set Ω at the point x^*.

Necessary conditions for a maximum can be established in a similar way.

THEOREM 7.3. Let a function $f(x)$ be Lipschitzian in a neighborhood of a point $x^{**} \in \Omega$. Moreover, if $h(x^{**}) = 0$, assume that the function $h(x)$ is also Lipschitzian in a neighborhood of the point x^{**} and that the nondegeneracy condition (6.3) is satisfied at the point x^*. Then, for the function $f(x)$ to attain its maximum on Ω at x^{**}, it is necessary that

$$-\underline{\partial}f(x^{**}) \subset \overline{\partial}f(x^{**}) \qquad \text{if} \quad h(x^{**}) < 0 , \tag{7.17}$$

$$-\underline{\partial}f(x^{**}) \subset \bigcap_{w \in \overline{\partial}h(x^{**})} [\overline{\partial}f(x^{**}) - \overline{K}(\underline{\partial}h(x^{**})+w)] \tag{7.18}$$

$$\text{if} \quad h(x^{**}) = 0 .$$

PROBLEM 7.1. Prove Theorem 7.3 and show that if $h(x)$ is a convex function, then (7.18) implies the condition

$$-\underline{\partial}f(x^{**}) \subset \overline{\partial}f(x^{**}) + \Gamma^+(x^{**}) .$$

2. Let $x_0 \in \Omega$. Assume that a function $f(x)$ is Lipschitzian in a neighborhood of the point x_0. If $h(x_0) = 0$, then we also assume that the function $h(x)$ is Lipschitzian in a neighborhood of x_0 and that the nondegeneracy condition (6.3) is satisfied at x_0.

DEFINITION. A point $x_0 \subset \Omega$, at which relation (7.4) is satisfied if $h(x_0) < 0$ and relation (7.5) is satisfied if $h(x_0) = 0$, is called an inf-*stationary point* of the function $f(x)$ on the set Ω. A point $x_0 \subset \Omega$, for which condition (7.17) is satisfied if $h(x_0) < 0$ and

(7.18) is satisfied if $h(x_0)=0$, called a sup-*stationary point* of the function $f(x)$ on the set Ω.

We consider in more detail the case where $h(x_0)=0$.

Suppose that the point x_0 is not inf-stationary, i.e., that condition (7.16) (or, equivalently, (7.5)) is not satisfied. Let

$$\rho(x_0) = \max_{\substack{w\in\overline{\partial}f(x_0)\\ w'\in\overline{\partial}h(x_0)}} \min_{\substack{z\in[\underline{\partial}f(x_0)+w]\\ z'\in\overline{K}(\underline{\partial}h(x_0)+w')}} \|z+z'\| \equiv \max_{\substack{w\in\overline{\partial}f(x_0)\\ w'\in\overline{\partial}h(x_0)}} d(w,w')$$

$$= d(w_0,w_0') , \tag{7.19}$$

where

$$d(w,w') = \min_{\substack{z\in[\underline{\partial}f(x_0)+w]\\ z'\in\overline{K}(\underline{\partial}h(x_0)+w')}} \|z+z'\| .$$

Since (7.5) does not hold, we have $\rho(x_0)>0$.

Let $d(w_0,w_0')=\|z_0+z'\|$ (i.e., $\|z_0+z_0'\|=\rho(x_0)$) and define

$$g_0 = -(z_0+z_0')\|z_0+z_0'\|^{-1} .$$

<u>LEMMA 7.2</u>. If $h(x_0)=0$ and the nondegeneracy condition (6.3) is satisfied, then the direction g_0 is the steepest descent direction of the function $f(x)$ at a point x_0 with respect to the set Ω and the value $\rho(x_0)=\|z_0+z_0'\|$ is the rate of steepest descent, i.e.,

$$\frac{\partial f(x_0)}{\partial g_0} = \min_{\substack{g\in\Gamma(x_0)\\ \|g\|=1}} \frac{\partial f(x_0)}{\partial g} = -\rho(x_0) . \tag{7.20}$$

<u>Proof</u>. As in the proof of Lemma 6.4 of Chapter 1, we can show that

$$g_0 \in [-K^{+}(\underline{\partial}h(x_0)+w_0')] ,$$

and then $g_0\in\Gamma(x_0)$ (see (6.4) and (6.11)). It only remains to prove (7.20). Assume that (7.20) is not satisfied. Then, there exists a $g_1\in\Gamma(x_0)$, $\|g_1\|=1$, such that

$$\frac{\partial f(x_0)}{\partial g_1} < -\rho(x_0) . \tag{7.21}$$

Since

$$\frac{\partial f(x_0)}{\partial g_1} = \min_{w \in \overline{\partial} f(x_0)} \; \max_{v \in [\underline{\partial} f(x_0) + w]} (v, g_1) ,$$

then we can find a vector $w_1 \in \overline{\partial} f(x_0)$ such that

$$\frac{\partial f(x_0)}{\partial g_1} = \max_{v \in [\underline{\partial} f(x_0) + w_1]} (v, g_1) . \tag{7.22}$$

On the other hand, since $g_1 \in \Gamma(x_0)$, then there exists a $w_1' \in \overline{\partial} h(x_0)$ such that

$$g_1 \in [-K^+(\underline{\partial} h(x_0) + w_1')]$$

(see (6.4) and (6.11)). Let us evaluate the quantity

$$d(w_1, w_1') = \min_{\substack{z \in [\underline{\partial} f(x_0) + w_1] \\ z \in \overline{K}(\underline{\partial} h(x_0) + w_1') .}} \|z + z'\| = \|z_1 + z_1'\| .$$

As in the proof of Lemma 6.4 of Chapter 1, we can show that

$$-\|z_1 + z_1'\| = \min_{g \in [-K^+(\underline{\partial} h(x_0) + w_1')]} \; \max_{v \in [\underline{\partial} f(x_0) + w_1]} (v, g)$$

$$\le \max_{v \in [\underline{\partial} f(x_0) + w_1]} (v, g_1) = \frac{\partial f(x_0)}{\partial g_1} .$$

From this and (7.21), we have

$$d(w_1, w_1') \ge -\frac{\partial f(x_0)}{\partial g_1} > \rho(x_0) ,$$

which contradicts (7.19). The lemma has been proved. ∎

REMARK. A steepest descent direction is not necessarily unique.

PROBLEM 7.2. Prove the result analogous to Lemma 7.2 for the case of the steepest ascent.

EXAMPLE 1. Let $x = (x^{(1)}, x^{(2)}) \in E_2$, $x_0 = (0,0)$, $f(x) = |x^{(1)}| - |x^{(2)}| + x^{(2)}$, $h(x) = -0.5|x^{(1)}| - x^{(2)}$.

Then,

$$f(x) = f_1(x) + f_2(x) + f_3(x) , \qquad h(x) = h_1(x) + h_2(x) ,$$

where $f_1(x) = |x^{(1)}|$, $f_2(x) = -|x^{(2)}|$, $f_3(x) = x^{(2)}$,

$$h_1(x) = -0.5|x^{(1)}|, \qquad h_2(x) = -x^{(2)}.$$

Construct the following set:

$$\Omega = \{x \in E_2 \mid h(x) \le 0\}$$

(Figure 38).

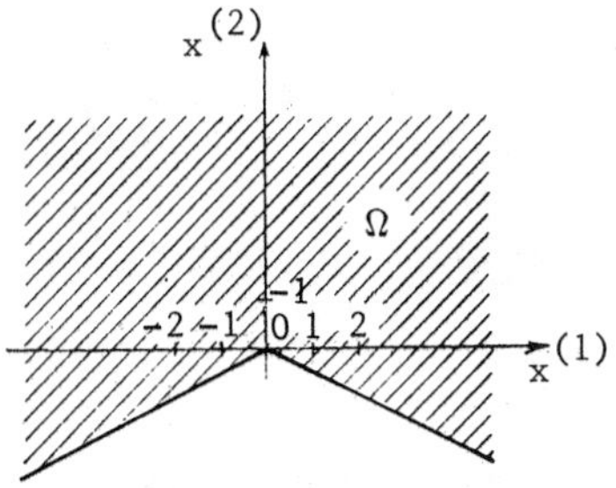

Figure 38

Since

$$\underline{\partial} f_1(x_0) = \text{co}\,\{A_1;A_2\}, \qquad \overline{\partial} f_1(x_0) = \{0\},$$

$$\underline{\partial} f_2(x_0) = \{0\}, \qquad \overline{\partial} f_2(x_0) = \text{co}\,\{B_1;B_2\},$$

$$\underline{\partial} f_3(x_0) = \{B_1\}, \qquad \overline{\partial} f_3(x_0) = \{0\},$$

$$\underline{\partial} h_1(x_0) = \{0\}, \qquad \overline{\partial} h_1(x_0) = \text{co}\,\{0.5A_1;0.5A_1\},$$

$$\underline{\partial} h_2(x_0) = \{B_2\}, \qquad \overline{\partial} h_2(x_0) = \{0\},$$

where $A_1 = (1,0)$, $A_2 = (-1,0)$, $B_1 = (0,1)$, $B_2 = (0,-1)$ (see Examples 1 - 4 in Section 3), then

$$\underline{\partial} f(x_0) = \text{co}\,\{A_1,A_2\} + \{B_1\}, \qquad \overline{\partial} f(x_0) = \text{co}\,\{B_1,B_2\},$$

$$\underline{\partial} h(x_0) = \{B_2\}, \qquad \overline{\partial} h(x_0) = \text{co}\,\{0.5A_1;0.5A_2\}.$$

First of all, let us verify that the nondegeneracy condition is satisfied at the point x_0. We have

$$\frac{\partial h(x_0)}{\partial g} = (B_2,g) + \min\,\{(0.5A_1;g),\ (0.5A_2;g)\}.$$

Find a $g \in E_2$, $\|g\| = 1$, such that the relation $\frac{\partial h(x_0)}{\partial g} = 0$ holds.

Clearly,

$$\frac{\partial h(x_0)}{\partial g_1} = \frac{\partial h(x_0)}{\partial g_2} = 0 ,$$

where

$$g_1 = \left(\frac{2}{\sqrt{5}}, \frac{-1}{\sqrt{5}}\right), \qquad g_2 = \left(\frac{-2}{\sqrt{5}}, \frac{-1}{\sqrt{5}}\right) .$$

We note that

$$\gamma(x_0) = \beta_1 \cup \beta_2 \cup \beta_3 ,$$

where

$$\beta_1 = \{g \in E_2 \mid g = \lambda(-2,z),\ \lambda > 0,\ z \in (-1,0]\} ,$$

$$\beta_2 = \{g \in E_2 \mid g = \lambda(y,1),\ \lambda > 0,\ y \in [-1,1]\} ,$$

$$\beta_3 = \{g \in E_2 \mid g = \lambda(2,z),\ \lambda > 0,\ z \in (-1,0]\} .$$

Clearly,

$$\gamma_1(x_0) = \gamma(x_0) \cup \{g = \lambda g_1 \mid \lambda \geq 0\} \cup \{g = \lambda g_2 \mid \lambda \geq 0\}$$
$$= \overline{\gamma(x_0)}$$

(see Figure 39), i.e., the nondegeneracy condition (6.3) is satisfied at the point x_0.

From Figure 40 it is clear that

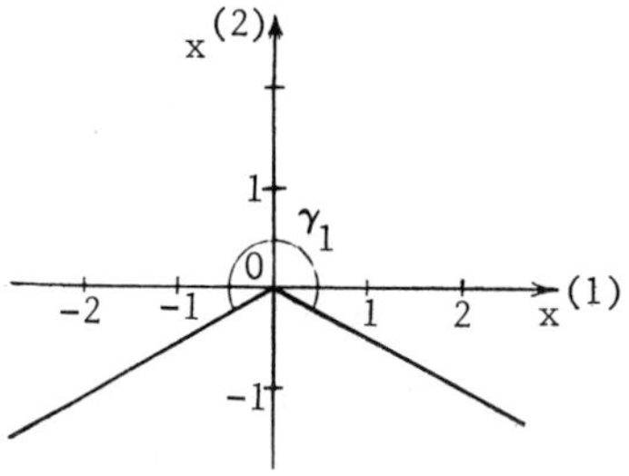

Figure 39

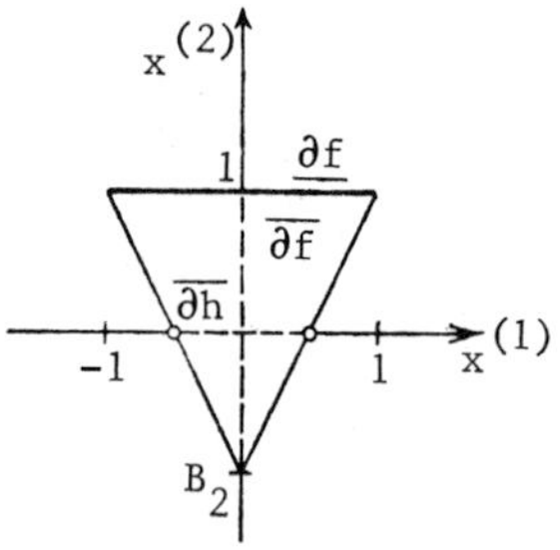

Figure 40

$$H \equiv \bigcap_{w \in \overline{\partial}h(x_0)} [\underline{\partial}f(x_0) + \overline{K}(\underline{\partial}h(x_0)+w] = \text{co}\ \{(-1,1),\ (1,1),\ (0,-1)\}.$$

Hence, $-\overline{\partial}f(x_0) \subset H$, i.e., condition (7.16) is satisfied.

Therefore, the point x_0 is an inf-stationary point of the function $f(x) = |x^{(1)}| - |x^{(2)}| + x^{(2)}$ on the set $\Omega = \{x \mid -0.5x^{(1)} - x^{(2)} \leq 0\}$.

In fact, the point x_0 is a minimum point of the function $f(x)$ on the set Ω.

EXAMPLE 2. Let again $x=(x^{(1)},x^{(2)})$, $x_0=(0,0)$ and the function $f(x)$ is the same as in Example 1.

Take the set $\Omega=\{x\in E_2 \mid h(x)\le 0\}$ (Figure 41), where $h(x) = -|x^{(1)}| - x^{(2)}$. We have

$$\underline{\partial}f(x_0) = \text{co }\{A_1,A_2\} + \{B_1\}, \qquad \overline{\partial}f(x_0) = \text{co }\{B_1,B_2\}$$

(see Example 1). It is easy to see that

$$\underline{\partial}h(x_0) = \{B_2\}, \qquad \overline{\partial}h(x_0) = \text{co }\{A_1,A_2\}.$$

As in Example 1, we can verify the nondegeneracy condition at the point x_0. The set H is shown in Figure 42:

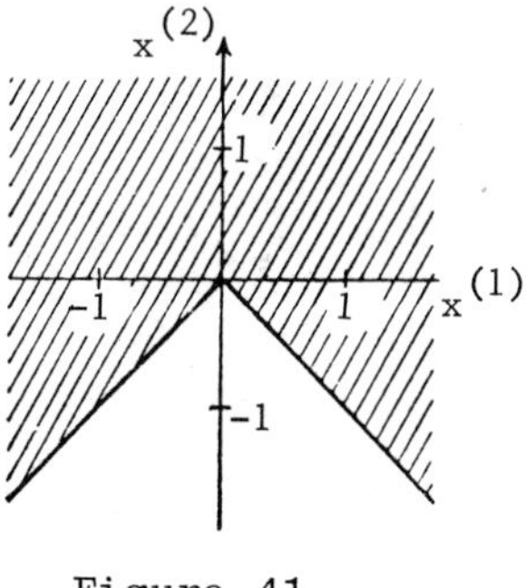

Figure 41

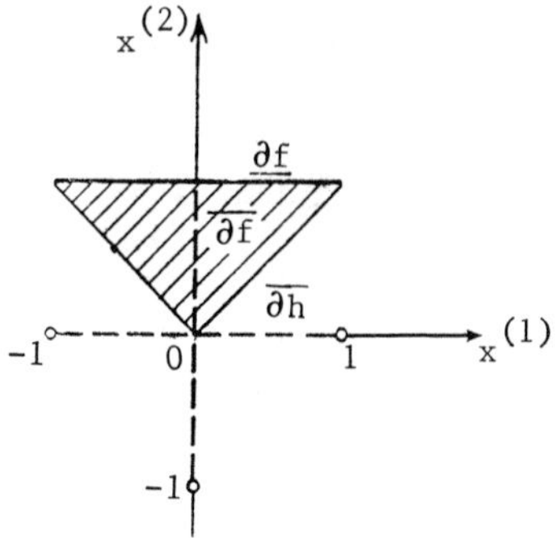

Figure 42

$$H \equiv \bigcap_{w\in\overline{\partial}h(x_0)} [\underline{\partial}f(x_0) + \overline{K}(\underline{\partial}h(x_0)+w)] = \text{co }\{(-1,1),\ (1,1),\ (0,0)\}.$$

It is clear that

$$-\overline{\partial}f(x_0) = -\text{co }\{B_1,B_2\} = \text{co }\{B_1,B_2\} \not\subset H.$$

Now, let us find steepest descent directions. For $w\in\overline{\partial}f(x_0)$, $w'\in\overline{\partial}h(x_0)$, we have

$$\underline{\partial}f(x_0) + w = \text{co }\{A_1+B_1+w,\ A_2+B_1+w\} \equiv U_w,$$

$$-\overline{K}(\underline{\partial}h(x_0)+w') = \{z=\lambda(-B_2-w')=\lambda(B_1-w') \mid \lambda\ge 0\} \equiv B_{w'}.$$

The sets U_w are segments parallel to the segment $\text{co}\{A_1,A_2\}$ and fill the square with the vertices at $(-1,0)$, $(-1,2)$, $(1,2)$ and $(1,0)$. The sets $B_{w'}$ are rays with vertices at the origin which pass through the points of the segment connecting the points $(-1,1)$ and $(1,1)$. Let us find the value $\rho(x_0)$ (see (7.19)):

$$\rho(x_0) = \max_{\substack{w\in\overline{\partial}f(x_0)\\ w'\in\overline{\partial}f(x_0)}} d(w,w') = d(w_0,w_0') = d(w_1,w_1') ,$$

where

$$w_0 = (0,1) , \quad w_0' = A_1 , \quad w_1 = w_0 = (0,1) , \quad w_1' = A_2$$

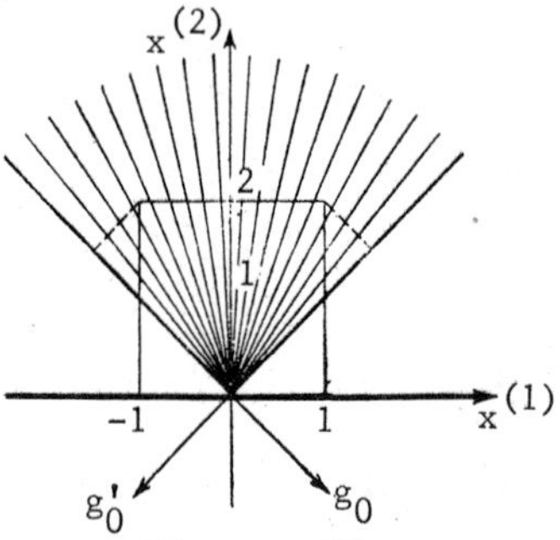

Figure 43

We have two directions of steepest descent (Figure 43):

$$g_0 = \left(\frac{\sqrt{2}}{2}, -\frac{\sqrt{2}}{2}\right) , \quad g_0' = \left(-\frac{\sqrt{2}}{2}, -\frac{\sqrt{2}}{2}\right) .$$

In addition,

$$\frac{\partial f(x_0)}{\partial g_0} = \max_{v\in\underline{\partial}f(x_0)} (v,g_0) + \min_{w\in\overline{\partial}f(x_0)} (w,g_0) = -\frac{\sqrt{2}}{2} .$$

Analogously, $\dfrac{\partial f(x_0)}{\partial g_0'} = -\dfrac{\sqrt{2}}{2}$. It is obvious that

$$\frac{\partial f(x_0)}{\partial g_0} = -\rho(x_0) = -\frac{\sqrt{2}}{2} .$$

3. Let us derive sufficient conditions for a local minimum of a function $f(x)$ on a set Ω. Let $x_0 \in \Omega$ be an inf-stationary point of the function $f(x)$ on the set Ω. Theorem 5.3 establishes sufficient conditions for a minimum for $h(x_0) < 0$. Now, we shall consider the case where $h(x_0) = 0$. Assume that the nondegeneracy condition (6.3) is satisfied at a point x_0 and that the functions $f(x)$ and $h(x)$ are Lipschitzian in a neighborhood of x_0. According to Theorem 7.1, we have (7.5) for arbitrary $w \in \bar{\partial} f(x_0)$ and $w' \in \bar{\partial} h(x_0)$, i.e.,

$$0 \in [\underline{\partial} f(x_0) + w + \bar{K}(\underline{\partial} h(x_0) + w')] \equiv L(w,w') . \tag{7.23}$$

Let us denote by $r(w,w')$ the radius of the maximum sphere centered at the origin and inscribed in $L(w,w')$. By virtue of (7.23), $r(w,w') \geq 0$. Let

$$r = \min_{w \in \bar{\partial} f(x_0),\, w' \in \bar{\partial} h(x_0)} r(w,w') .$$

Recall that

$$\begin{aligned} f(x_0+\alpha g) &= f(x_0) + \alpha \frac{\partial f(x_0)}{\partial g} + o(\alpha,g) , \\ h(x_0+\alpha g) &= h(x_0) + \alpha \frac{\partial h(x_0)}{\partial g} + o_1(\alpha,g) . \end{aligned} \tag{7.24}$$

THEOREM 7.4. Assume that, in (7.24), $\frac{o(\alpha,g)}{\alpha} \to 0$ and $\frac{o_1(\alpha,g)}{\alpha} \to 0$ are satisfied uniformly with respect to $g \in E_n$, $\|g\| = 1$. If $r > 0$, then the point x_0 is a local minimum and, moreover,

$$\min_{g \in \Gamma(x_0),\, \|g\|=1} \frac{\partial f(x_0)}{\partial g} = r .$$

Proof. Is obvious.

EXAMPLE 3. Let

$$x = (x^{(1)},x^{(2)}) \in E_2 , \qquad x_0 = (0,0) ,$$

$$f(x) = |x^{(1)}| - 0.5|x^{(2)}| + x^{(2)} ,$$

and let the set Ω be given by

$$\Omega = \{x \in E_2 \mid h(x) \leq 0\} ,$$

where $h(x) = -0.25|x^{(1)}| - x^{(2)}$.

As in Examples 1 and 2, we obtain that

$$\underline{\partial}f(x_0) = \text{co}\ \{A_1,A_2\} + \{B_1\} = \text{co}\ \{(1;1),\ (-1;1)\}\ ,$$

$$\overline{\partial}f(x_0) = \text{co}\ \{0.5B_1;0.5B_2\} = \text{co}\ \{(0;0.5),\ (0;-0.5)\}\ ,$$

$$\underline{\partial}h(x_0) = \{B_2\}\ ,$$

$$\overline{\partial}h(x_0) = \text{co}\ \{0.25A_1;0.25A_2\} = \text{co}\ \{(0.25;0),\ (-0.25);0)\}$$

(Figure 44).

Take $w \in \overline{\partial}f(x_0)$ and $w' \in \overline{\partial}h(x_0)$. Then

$$\underline{\partial}f(x_0) + w = \text{co}\ \{A_1+B_1+w,\ A_2+B_1+w\} \equiv U_w\ ,$$

$$K(\underline{\partial}h(x_0) + w') = \{z = \lambda(B_2+w')\ |\ \lambda \geq 0\} \equiv B_{w'}\ .$$

The set $L(w,w')$ (see (7.23)) is shown in Figure 45 for $w = (0,0.25)$, $w' = (-0.125,0)$. The rays ℓ_1, ℓ_2, ℓ_3 are parallel.

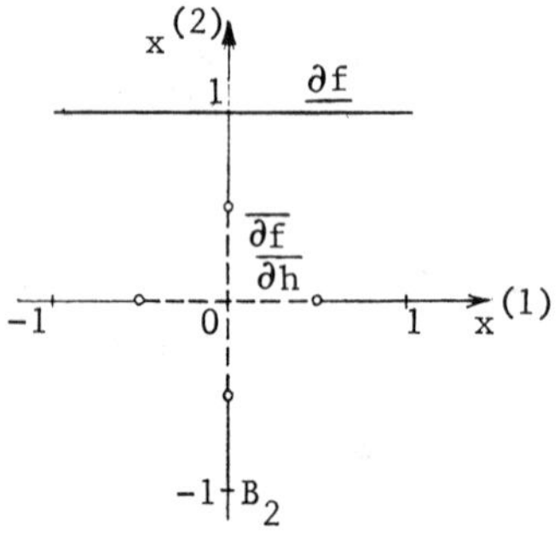

Figure 44

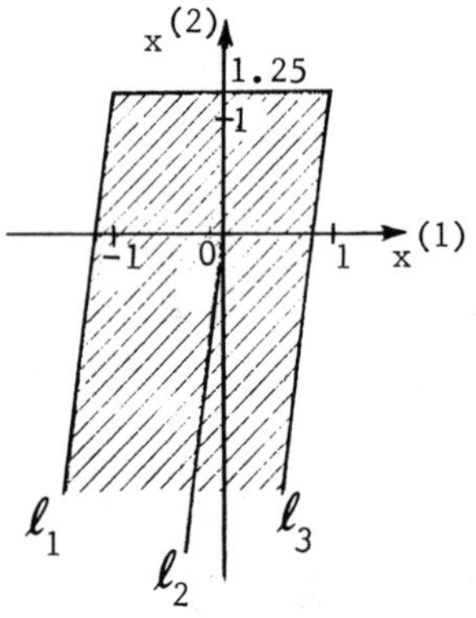

Figure 45

Clearly,

$$0 \in L(w,w') \qquad \forall w \in \overline{\partial}f(x_0), \quad \forall w' \in \overline{\partial}h(x_0)\ .$$

It is easy to calculate

$$r = \min_{w\in\overline{\partial}f(x_0),w'\in\overline{\partial}h(x_0)} r(w,w')\ .$$

Here, the minimum is attained for $w_0 = (0,0.5)$, $w_0' = (0.25,0)$ and for $w_1 = (0,0.5)$, $w_1' = (-0.25,0)$. Computing $r = r(w_0,w_0')$, we find that $r = 34^{-1}\sqrt{425} \approx 0.6063$.

8. THE DISTANCE FUNCTION FROM A POINT TO A SET

In this section, we shall study the directional differentiability of a function which represents the Euclidean distance from a point to a closed set $\Omega \subset E_n$. In Subsection 1, we shall consider the case where Ω is a convex set and in Subsection 2 the case where Ω is given in terms of a quasidifferentiable function. Such functions are of independent interest; they are usual in solutions of practical problems.

1. Let $\Omega \subset E_n$ be a closed convex set. Define the function

$$f(x) = \min_{y \in \Omega} \|x-y\| = \|x - y(x)\| .$$

It is clear that if $x \in \Omega$, then $f(x) = 0$. First of all, we note that $f(x)$ is a convex function on E_n. Indeed, we can show this.

Take points $x_1 \in E_n$ and $x_2 \in E_n$. Let

$$f(x_1) = \|x_1 - y_1\| , \qquad f(x_2) = \|x_2 - y_2\| ,$$

where $y_1 \in \Omega$, $y_2 \in \Omega$. For $\alpha \in [0,1]$, due to the convexity of Ω, we have

$$y_\alpha = \alpha y_1 + (1-\alpha)y_2 \in \Omega .$$

Hence,

$$\begin{aligned} f(\alpha x_1+(1-\alpha)x_2) &\le \|\alpha x_1+(1-\alpha)x_2-y_\alpha\| = \|\alpha(x_1-y_1)+(1-\alpha)(x_2-y_2)\| \\ &\le \alpha\|x_1-y_1\| + (1-\alpha)\|x_2-y_2\| = \alpha f(x_1) + (1-\alpha)f(x_2), \end{aligned}$$

which implies that $f(x)$ is a convex function.

Our aim is to find the subdifferential of this function.

Let $x_0 \in E_n$. If $x_0 \in \operatorname{int} \Omega$, then

$$\frac{\partial f(x_0)}{\partial g} \equiv \lim_{\alpha\to+0} \alpha^{-1}[f(x_0+\alpha g) - f(x_0)] = 0 \qquad \forall g \in E_n .$$

Therefore,

$$\partial f(x_0) = \{0\} . \tag{8.1}$$

If $x_0 \notin \Omega$, then

$$f(x_0) = \min_{y\in\Omega} \|y-x_0\| = \min_{y\in\Omega} \sqrt{(y-x_0)^2} = \min_{y\in\Omega} \phi(x_0,y) .$$

Here we show that the function $\phi(x,y) = \sqrt{(y-x)^2}$ is continuously differentiable in x in a neighborhood of the point x_0.

Since

$$\min_{y\in\Omega} \phi(x,y) = -\max_{y\in\Omega} (-\phi(x,y)) ,$$

then all the conditions of Theorem 14.1 of Chapter 1 are satisfied and, hence,

$$\frac{\partial f(x_0)}{\partial g} = \min_{y\in Q(x_0)} (\phi'_x(x_0,y), g) ,$$

where $Q(x_0) = \{y \in \Omega \mid \phi(x_0,y) = f(x_0)\}$. In our case, the set $Q(x_0)$ consists of a unique point $y(x_0)$. Hence

$$\frac{\partial f(x_0)}{\partial g} = (v,g) , \tag{8.2}$$

where

$$v = \frac{1}{\sqrt{(y(x_0)-x_0)^2}} (x_0 - y(x_0)) = \frac{x_0 - y(x_0)}{f(x_0)} , \tag{8.3}$$

i.e., the function $f(x)$ is continuously differentiable in a neighborhood of the point x_0.

Now, it remains only to consider the case where x_0 is a boundary point of the set Ω. We have

$$\frac{\partial f(x_0)}{\partial g} = \lim_{\alpha\to+0} H(\alpha) ,$$

where $H(\alpha) = \alpha^{-1}[f(x_0+\alpha g) - f(x_0)]$. Since $x_0 \in \Omega$, then $f(x_0) = 0$, i.e., $$H(\alpha) = \alpha^{-1} f(x_0+\alpha g) = \alpha^{-1} \min_{y\in\Omega} \|x_0+\alpha g-y\| .$$

For any $y \in \Omega$, we have

$$\|x_0+\alpha g-y\| = \|x_0 + \alpha g - (x_0+(y-x_0))\| = \alpha\| g - \alpha^{-1}(y-x_0)\| .$$

Since $\alpha > 0$, then $v(y,\alpha) = \alpha^{-1}(y-x_0) \in \gamma(x_0) = \{v = \lambda(x-x_0) | x\in\Omega\}$.

Hence,

$$\|x_0+\alpha g-y\| = \alpha\|g - v(y,\alpha)\| \geq \alpha \min_{v\in\Gamma(x_0)} \|g-v\| \equiv \alpha P ,$$

where $P = \min_{v\in\Gamma(x_0)} \|g-v\|$, and $\Gamma(x_0) = \overline{\gamma}(x_0)$ is the cone of feasible directions of the set Ω at the point x_0. Since P does not depend on y, then $\min_{y\in\Omega} \|x_0+\alpha g-y\| \geq \alpha P$, i.e.,

$$H(\alpha) \geq P . \tag{8.4}$$

Now, take an arbitrary $v \in \gamma(x_0)$. For sufficiently small $\alpha > 0$, we have $x_0+\alpha v \in \Omega$. Hence,

$$f(x_0+\alpha g) = \min_{y\in\Omega} \|x_0+\alpha g-y\| \leq \|x_0+\alpha g-x_0-\alpha v\| = \alpha\| g - v\| .$$

From this, we get

$$H(\alpha) \leq \|g - v\| .$$

Since this inequality is valid for any $v \in \gamma(x_0)$, then

$$H(\alpha) \leq \inf_{v\in\gamma(x_0)} \|g-v\| = \min_{v\in\Gamma(x_0)} \|g-v\| = P . \tag{8.5}$$

P does not depend on α; therefore from (8.4) and (8.5) we have

$$\frac{\partial f(x_0)}{\partial g} = P = \min_{v\in\Gamma(x_0)} \|g - v\| .$$

According to Corollary 3 of Lemma 6.4 in Chapter 1, we have

$$\min_{v\in\Gamma(x_0)} \|g - v\| = \max_{\substack{w\in[-\Gamma^+(x_0)] \\ \|w\|=1}} (w,g)$$

(by exchanging the roles of Γ and Γ^+), i.e.,

$$\frac{\partial f(x_0)}{\partial g} = \max_{\substack{w \in -\Gamma^+(x_0) \\ \|w\|=1}} (w,g) . \tag{8.6}$$

Thus,

$$\partial f(x_0) = \{w \in E_n \mid \|w\| \le 1,\ w \in [-\Gamma^+(x_0)]\} = (-\Gamma^+(x_0)) \cap S_1 ,$$

where

$$S_1 = \{w \in E_n \mid \|w\| \le 1\} .$$

Since we have $\Gamma^+(x_0) = \{0\}$ for $h(x_0) < 0$, then from (8.1), (8.2) and (8.6) we obtain the following lemma.

LEMMA 8.1. The function $f(x)$ is directionally differentiable on E_n and, in addition,

$$\frac{\partial f(x_0)}{\partial g} = \max_{v \in \partial f(x_0)} (v,g) , \tag{8.7}$$

where

$$f(x_0) = \begin{cases} \left\{\dfrac{x_0 - y(x_0)}{f(x_0)}\right\} & \text{if } x_0 \notin \Omega , \\ (-\Gamma^+(x_0) \cap S_1) & \text{if } x_0 \in \Omega . \end{cases} \tag{8.8}$$

PROBLEM 8.1. Find the subdifferential of the function

$$f(x) = \min_{y \in \Omega} \|x-y\|_m ,$$

where $\|x-y\|_m = \max_{i \in 1:n} \{|x^{(i)}-y^{(i)}|\}$, $x = (x^{(1)},\dots,x^{(n)})$, $y = (y^{(1)},\dots,y^{(n)})$, and $\Omega \subset E_n$ is a closed convex set.

2. Let $\Omega \in E_n$ be a closed (however, not necessarily convex) set.

Consider again the function

$$f(x) = \min_{y \in \Omega} \|x-y\| . \tag{8.9}$$

Let

$$Q(x_0) = \{y \in \Omega \mid \|x-y\| = f(x)\} . \tag{8.10}$$

It is clear that the set $Q(x)$ may consist of more than one point.

The function $f(x)$ is no longer necessarily convex. Let $x_0 \in E_n$. If $x_0 \in \operatorname{int}\Omega$, then

$$\frac{\partial f(x_0)}{\partial g} = 0 \qquad \forall g \in E_n . \tag{8.11}$$

If $x_0 \notin \Omega$, then $f(x_0) = \min_{y\in\Omega} \|y-x_0\|$. As in Subsection 1,

$$\frac{\partial f(x_0)}{\partial g} = \min_{y\in\Omega(x_0)} (\phi_x'(x_0,y),\ g) ,$$

where $\phi(x,y) = \sqrt{(y-x)^2}$. Indeed,

$$\frac{\partial f(x_0)}{\partial g} = \min_{v\in A} (v,g) , \tag{8.12}$$

where

$$A = \operatorname{co}\left\{\frac{x_0 - y}{f(x_0)} \;\middle|\; y \in Q(x_0)\right\} .$$

Now, let x_0 be a boundary point of the set Ω. Assume that the set Ω is given by

$$\Omega = \{x \in E_n \mid h(x) \le 0\} ,$$

where $h(x)$ is a quasidifferentiable function on E_n. Assume also that the function $h(x)$ is Lipschitzian in a neighborhood of the point x_0 and that the nondegeneracy condition (6.3) is satisfied at the point x_0.

Since x_0 is a boundary point of the set Ω, then $h(x_0) = 0$.

According to Lemma 6.1,

$$\Gamma(x_0) = \gamma_1(x_0) = \bar{\gamma}(x_0) ,$$

where

$$\gamma(x_0) = \left\{g \in E_n \;\middle|\; \frac{\partial h(x_0)}{\partial g} < 0\right\} . \tag{8.13}$$

For the same $H(\alpha)$ as defined in Subsection 1, we have

$$H(\alpha) = \alpha^{-1}[f(x_0+\alpha g) - f(x_0)] = \alpha^{-1} f(x_0+\alpha g) = \alpha^{-1} \min_{y\in\Omega} \|x_0+\alpha g-y\| .$$

Let a sequence $\{\alpha_k\}$ be such that

$$\alpha_k \to +0 , \qquad H(\alpha_k) \to \underline{\lim}_{\alpha\to+0} H(\alpha) \tag{8.14}$$

(such a sequence exists) and, moreover, we observe that

$$f(x_0+\alpha_k g) \equiv \min_{y\in\Omega} \|x_0+\alpha_k g-y\| = \|x_0+\alpha_k g-y_k\| = \alpha_k\left\|g-\frac{y_k-x_0}{\alpha_k}\right\|$$

$$= \alpha_k\left\|g-\frac{y_k-x_0}{\|y_k-x_0\|}\frac{\|y_k-x_0\|}{\alpha_k}\right\| = \alpha_k\|g-\lambda_k z_k\| , \qquad (8.15)$$

where

$$z_k = \frac{y_k-x_0}{\|y_k-x_0\|}, \qquad \lambda_k = \frac{\|y_k-x_0\|}{\alpha_k} .$$

Since

$$f(x_0+\alpha_k g) = \|x_0+\alpha_k g-y_k\| = \min_{y\in\Omega}\|x_0+\alpha g-y\| \le \|x_0+\alpha_k g-x_0\|$$

$$= \alpha_k \|g\| = \alpha_k ,$$

then it follows from (8.15) that $\alpha_k\|g-\lambda_k z_k\| \le \alpha_k$, i.e., $\|g-\lambda_k z_k\| \le 1$.

Therefore, since $\|z_k\| = 1$, we conclude that λ_k in (8.15) is bounded. We can assume that $\lambda_k \to \lambda \ge 0$, $z_k \to z_0$.

Hence, from (8.14) we have

$$\underline{\lim}_{\alpha\to+0} H(\alpha) = \lim_{k\to\infty} H(\alpha_k) = \|g-\lambda z_0\| .$$

Here, since $z_0 \in \Gamma(x_0)$, then $z_1 = \lambda z_0 \in \Gamma(x_0)$ and

$$\underline{\lim}_{\alpha\to+0} H(\alpha) = \|g-z_1\| \ge \min_{z\in\Gamma(x_0)} \|g-z\| \equiv P \qquad (8.16)$$

(it is implicitly assumed above that $y_k \not\equiv x_0$. For the case $y_k \equiv x_0$, relation (8.16) is obvious).

Now, let us take a sequence $\{\alpha_k\}$ such that

$$\alpha_k \to +0 , \qquad H(\alpha_k) \to \overline{\lim}_{\alpha\to+0} H(\alpha) . \qquad (8.17)$$

Fix an arbitrary $z\in\gamma(x_0)$. Then, it follows from (8.13) that for sufficiently small $\alpha>0$ we have $x_\alpha = x_0+\alpha z\in\Omega$; hence

$$f(x_0+\alpha_k g) = \min_{y\in\Omega}\|x_0+\alpha_k g-y\| \le \|x_0+\alpha_k g-x_{\alpha_k}\| = \alpha_k\|g-z\| .$$

From this and (8.17), we get

$$\overline{\lim_{\alpha\to+0}} H(\alpha) = \lim_{k\to\infty} H(\alpha_k) \leq \|g - z\| .$$

This inequality is valid for all $z \in \Gamma(x_0)$. Therefore,

$$\overline{\lim_{\alpha\to+0}} H(\alpha) \leq \inf_{z\in\gamma(\bar{x}_0)} \|g - z\| .$$

Since $\Gamma(x_0) = \bar{\gamma}(x_0)$, we get

$$\overline{\lim_{\alpha\to+0}} H(\alpha) \leq \min_{z\in\Gamma(x_0)} \|g - z\| \equiv P . \tag{8.18}$$

From (8.16) and (8.18), we can conclude that $\lim\limits_{\alpha\to+0} H(\alpha)$ exists and, moreover,

$$\frac{\partial f(x_0)}{\partial g} \equiv \lim_{\alpha\to+0} H(\alpha) = P \equiv \min_{z\in\Gamma(x_0)} \|g-z\| , \tag{8.19}$$

i.e., the function $f(x)$ is directionally differentiable at the point x_0. Since $\Gamma(x_0) = \gamma_1(x_0)$, then

$$\begin{aligned}
\Gamma(x_0) &= \left\{z \in E_n \;\middle|\; \max_{v\in\underline{\partial}h(x_0)} (v,z) + \min_{w\in\overline{\partial}h(x_0)} (w,z) \leq 0\right\} \\
&= \left\{z \in E_n \;\middle|\; \min_{w\in\overline{\partial}h(x_0)} \left[\max_{v\in[\underline{\partial}h(x_0)+w]} (v,z)\right] \leq 0\right\} \\
&= \bigcup_{w\in\overline{\partial}h(x_0)} \left\{\max_{v\in[\underline{\partial}h(x_0)+w]} (v,z) \leq 0\right\} \\
&= \bigcup_{w\in\overline{\partial}h(x_0)} [-K^+(\underline{\partial}h(x_0) + w] ,
\end{aligned}$$

where $K(A)$ is the conical hull of a set A and K^+ is the conjugate to a cone K. From (8.19),

$$\frac{\partial f(x_0)}{\partial g} = \min_{w\in\overline{\partial}h(x_0)} \min_{z\in[-K^+(\underline{\partial}h(x_0)+w)]} \|g - z\| .$$

According to Corollary 3 of Lemma 6.4 of Chapter 1,

$$\min_{z\in[-K^+(\underline{\partial}h(x_0)+w)]} ||g-z|| = \max_{\substack{v\in K(\partial h(x_0)+w)\\ ||\Gamma v||\le 1}} (v,g) .$$

Finally, we obtain

$$\frac{\partial f(x_0)}{\partial g} = \min_{w\in\overline{\partial}h(x_0)} \max_{\substack{v\in K(\partial h(x_0)+w)\\ ||\overline{v}||\le 1}} (v,g) . \tag{8.20}$$

From (8.11), (8.12) and (8.20) there follows

LEMMA 8.2. The function $f(x)$ defined by (8.9) is directionally differentiable at all $x_0 \in \text{int}\,\Omega$, at all $x_0 \notin \Omega$ and at those boundary points of the set Ω, in a neighborhood of which the function $h(x)$ is Lipschitzian and at which the nondegeneracy condition (6.3) is satisfied. Furthermore, the corresponding directional derivatives are calculated by using formulas (8.11), (8.12) and (8.20), respectively.

REMARK. From (8.11) and (8.12), we conclude that $f(x)$ is quasidifferentiable for $x_0 \in \text{int}\,\Omega$ and $x_0 \notin \Omega$. Formula (8.20), however, shows that the function $f(x)$ is no longer quasidifferentiable at the boundary points (i.e., at least, from (8.20), we are not able to confirm the quasidifferentiability of the function $f(x)$).

Now, we shall introduce an example which shows that the nondegeneracy condition at a boundary point x_0 is essential for the directional differentiability at the point x_0.

EXAMPLE 1. Let

$$x = (x^{(1)}, x^{(2)}) \in E_2 , \qquad \Omega = \{0\} \cup \bigcup_{k=0}^{\infty} D_k ,$$

where D_k is a convex hull of the points A_k, B_k, C_k:

$$A_k = (2^{-k}, 0) , \qquad B_k = (2^{1-k}, 0) , \qquad C_k = (2^{-k}, 2^{-k}) .$$

The set Ω is shown in Figure 46.

It is easy to see that every D_k (it is a triangle) can be described as

$$D_k = \{x \in E_2 \mid h_k(x) \le 0\} \quad ,$$

where

$$h_k(x) = \max\{-x^{(2)},\ 2^{-k}-x^{(1)},\ x^{(1)}+x^{(2)}-2^{-k+1}\} \quad .$$

Each of the functions $h_{k1}(x) = -x^{(2)}$, $h_{k2}(x) = 2^{-k} - x^{(1)}$, $h_{k3} = x^{(1)} + x^{(2)} - 2^{-k+1}$ is Lipschitzian with the same Lipschitz constant (say $L = 2$). Therefore $h_k(x)$ is also a Lipschitzian function with a constant $L = 2$.

Now it is clear that

$$\Omega = \{x \in E_2 \mid \bar{h}(x) \le 0\} \quad ,$$

where $\bar{h}(x) = \inf\limits_{k \in 1:\infty} h_k(x)$.

Since each of $h_k(x)$ is a Lipschitzian function with the same Lipschitzian constant $L = 2$, then $\bar{h}(x)$ is also a Lipschitzian function with a Lipschitzian constant $L=2$.

Unfortunately, it is not obvious that the function $\bar{h}$ is quasidifferentiable. But we can also describe Ω as follows:

$$\Omega = \{x \in E_n \mid h(x) \le 0\} \quad ,$$

where

$$h(x) = (x^{(1)2} + x^{(2)2})\, \bar{h}(x) \quad .$$

Let $x_0 = (0,0)$. Let us show that the function $h(x)$ is not only Lipschitzian but quasidifferentiable as well at x_0. First of all, note that $h(x_0) = 0$, i.e., $x_0 \in \Omega$. Take any $g = (g^{(1)}, g^{(2)}) \in E_2$. Then

$$h(x_0+\alpha g) - h(x_0) = \alpha^2 \|g\|^2 \inf_{k \in 1:\infty} h_k(x_0+\alpha g) \quad ,$$

where

$$h_k(x_0+\alpha g) = \max\{-\alpha g^{(2)},\ 2^{-k}-\alpha g^{(1)},\ \alpha(g^{(1)}+g^{(2)}) - 2^{-k+1}\} \quad .$$

Now it is clear that $h(x)$ is directionally differentiable at x_0 and

$$\frac{\partial h(x_0)}{\partial g} \equiv \lim_{\alpha \to +0} \frac{h(x_0+\alpha g)-h(x_0)}{\alpha} = \alpha \|g\|^2 \inf_{k \in 1:\infty} h_k(x_0+\alpha g)$$

$$= 0 \qquad \forall g \in E_2 \;, \qquad (8.21)$$

i.e., $h(x)$ is even quasidifferentiable with $Dh(x_0) = [\{0\},\{0\}]$. It means that at the point x_0, the nondegeneracy condition (6.3) is not satisfied, since $\gamma(x_0) = \emptyset$ and $\gamma_1(x_0) = E_2$.

Now let us consider the function $f(x) = \min_{y \in \Omega} \|x-y\|$. Since the nondegeneracy condition (6.3) is not satisfied at x_0, we cannot apply Lemma 8.2. Moreover, we shall demonstrate that this function is even not differentiable at x_0 in the direction $g_0 = (0,1)$.

It is clear that for the point x_0 the cone of feasible (in a broad sense) directions $\Gamma(x_0)$ is

$$\Gamma(x_0) = \{v = \lambda z \mid \lambda \geq 0, \ z = (1,\alpha), \ \alpha \in [0,1]\}$$

(see Figure 47).

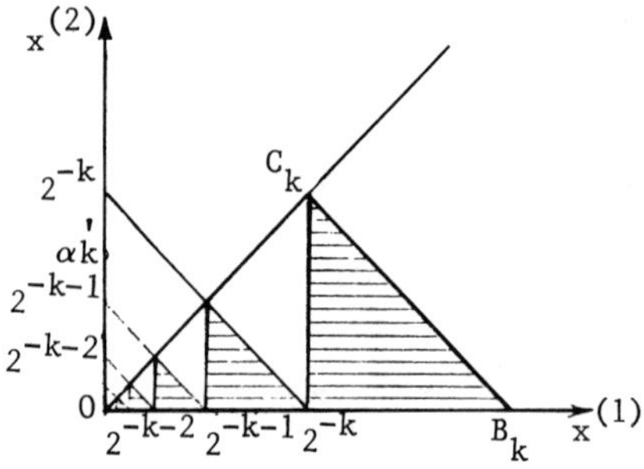

Figure 46

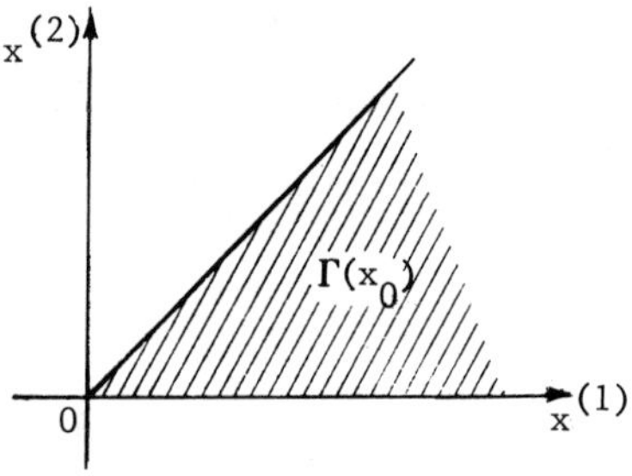

Figure 47

Consider the sequence of numbers $\{\alpha_k\}$, where $\alpha_k = 2^{-k}$, and the sequence of points $\{x_k\}$ where $x_k = x_0 + \alpha_k g_0 = (0,2^{-k})$. It is obvious that

$$f(x_k) \equiv \min_{y \in \Omega} \|x_k - y\| = \|x_k - C_{k-1}\|$$

$$= \sqrt{\left(\frac{1}{x^{k+1}}\right) + \left(\frac{1}{2^{k+1}} - \frac{1}{2^k}\right)^2} = \sqrt{2} \cdot 2^{-(k+1)} ,$$

i.e.,

$$H(\alpha_k) = \alpha_k^{-1}[f(x_0 + \alpha_k g_0) - f(x_0)]$$

$$= \sqrt{2} \cdot 2^k \cdot 2^{-(k+1)} = \frac{\sqrt{2}}{2} \qquad \forall k .$$

Now, let

$$\alpha'_k = \frac{1}{2}\left(\frac{1}{2^k}+\frac{1}{2^{k+1}}\right) = \frac{3}{2^{k+2}} .$$

For the points $x'_k = x_0 + \alpha'_k g_0 = (0, \frac{3}{2^{k+2}})$, we obtain

$$f(x'_k) = \min_{y\in\Omega} \|x'_k - y\| = \|x'_k - C_{k-2}\|$$

$$= \sqrt{\left(\frac{1}{2^{k+2}}\right)^2 + \left(\frac{1}{2^{k+2}} - \frac{3}{2^{k+2}}\right)^2} = \frac{\sqrt{5}}{2^{k+2}} ,$$

i.e.,

$$H(\alpha'_k) = (\alpha'_k)^{-1}[f(x_0+\alpha'_k g_0) - f(x_0)] = \sqrt{5}\cdot 3^{-1} \quad \forall k .$$

Thus, the limits of the function $H(\alpha)$ corresponding to the sequences $\{\alpha_k\}$ and $\{\alpha'_k\}$ do not coincide (they are, respectively, equal to $\sqrt{2}\cdot 2^{-1}$ and $\sqrt{5}\cdot 3^{-1}$), i.e., the function $f(x) = \min_{y\in\Omega} \|x-y\|$ at the point $x_0 = (0,0)$ is not differentiable in the direction g_0.

PROBLEM 8.2. Let $x = (x^{(1)}, x^{(2)}) \in E_2$. Consider the functions

$$r_1(x) = (x - x_1)^2 - 2 , \qquad r_2(x) = (x - x_2)^2 - 2 ,$$

$$r_3(x) = (x - x_3)^2 - 2 , \qquad r_4(x) = (x - x_4)^2 - 2 ,$$

where $x_1 = (1,1)$, $x_2 = (1,-1)$, $x_3 = (-1,1)$, $x_4 = (-1,-1)$. Let

$$h_1(x) = \max \{r_1(x), r_2(x)\} , \qquad h_2(x) = \max \{r_1(x), r_4(x)\} ,$$

$$h_3(x) = \max \{r_3(x), r_4(x)\} , \qquad h_4(x) = \max \{r_2(x), r_3(x)\} .$$

Define

$$h(x) = \min_{i\in 1:4} h_i(x) .$$

Here, we construct the set $\Omega = \{x\in E_2 \mid h(x)\le 0\}$ and the function $f(x) = \min_{y\in\Omega} \|x-y\|$. The set Ω is shown in Figure 48.

For the point $x_0 = (0,0)$, find $\frac{\partial f(x_0)}{\partial g}$ (it is necessary to use formula (8.20) and find the $\underline{\partial} h(x_0)$ and $\overline{\partial} h(x_0)$).

The set Ω is also known as Karlson-Pshenichnyj's propeller.

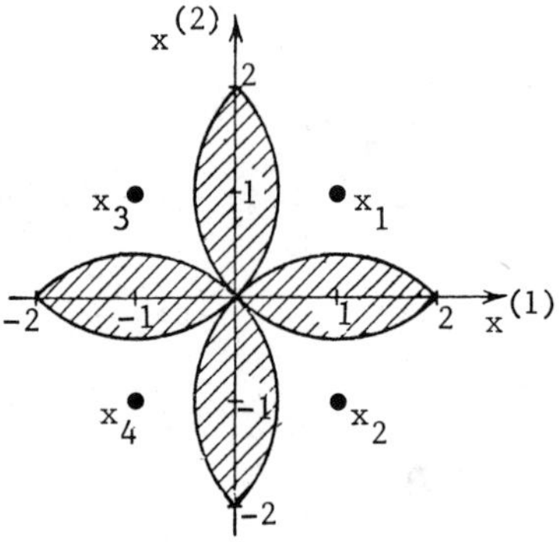

Figure 48

9. IMPLICIT FUNCTIONS

Let $f(z) = f(x,y)$ be a function of two variables, where $z = (x,y) \in E_2$, and let $z_0 = (x_0,y_0)$ be a solution of the equation $f(z) = 0$, i.e.,

$$f(x_0,y_0) = 0 \; . \tag{9.1}$$

Assume that the function $f(z)$ is quasidifferentiable in a neighborhood of the point z_0. Fix an arbitrary $g \in E_1$, $\|g\| = 1$ (in E_1 there exist only two such directions: i.e., $g_1 = +1$ and $g_2 = -1$), and consider the function $f(\alpha,y) = f(x_0+\alpha g,y)$, where $\alpha \geq 0$, $y \in E_1$.

The problem consists in determining the conditions under which there exist a number $\alpha_0 > 0$ and a continuous function $y(\alpha)$ given on $[0,\alpha_0]$, such that

$$F(\alpha,y(\alpha)) \equiv f(x_0+\alpha g,\; y(\alpha)) = 0 \quad \forall\, \alpha \in [0,\alpha_0] \; . \tag{9.2}$$

Since $f(z)$ is a quasidifferentiable function in a neighborhood of the point z_0, then

$$f(z_0+\alpha\eta) = f(z_0) + \alpha\Big[\max_{v \in \underline{\partial} f(z_0)} (v,\eta) + \min_{x \in \overline{\partial} f(z_0)} (w,\eta)\Big] + o(\alpha) \tag{9.3}$$

Here, $\eta = (g,q)$. Suppose that there exists a $q_0 \in E_1$ such that

$$\frac{\partial f(z_0)}{\partial \eta_0} \equiv \max_{v\in\underline{\partial} f(z_0)} (v,\eta_0) + \min_{w\in\overline{\partial} f(z_0)} (w,\eta_0) = 0 , \qquad (9.4)$$

where $\eta_0 = (g,q_0)$. Assume that the function $o(\alpha) = o(\alpha,q)$ in (9.3) is such that $\frac{o(\alpha,q)}{\alpha} \xrightarrow[\alpha\to+0]{} 0$ uniformly with respect to q in some neighborhood of the point $q_0 \in E_1$. Let $v = (v_1,v_2)$, $w = (w_1,w_2)$ and consider the function

$$h(q) = \max_{v\in\underline{\partial} f(z_0)} (v_1 g + v_2 q) + \min_{w\in\overline{\partial} f(z_0)} (w_1 g + w_2 q) . \qquad (9.5)$$

Considering (9.4), we have

$$h(q_0) = 0 . \qquad (9.6)$$

<u>THEOREM 9.1</u>. Let $h(q_0) = 0$. If the function $h(q)$ is strictly monotone in a neighborhood of the point q_0, then there exists a function $y(\alpha)$ which is given and continuous for $\alpha \in [0,\alpha_0]$, where $\alpha_0 > 0$, and satisfies (9.2). Here, $y(\alpha)$ has the right-hand derivative at the point $\alpha = 0$ and $y'_+(0) = q_0$.

<u>Proof</u>. Is analogous to that of the theorem for the smooth case and is left to the reader.

<u>REMARK 1</u>. The existence of an implicit function is established by considering function (9.5), the properties of which depend only on the quasidifferential of the function $f(z)$ at the point z_0. In particular, it can be seen that q_0 satisfying (9.6) is not necessarily unique.

<u>EXAMPLE 1</u>. Let $f(z) = |x| - |y|$, $z_0 = (0,0)$. It is obvious that $f(z_0) = 0$. This function was studied in Example 5 of Section 3.

Let us take $g = +1$. It is clear from Figure 18 that equation (9.6) has two solutions: $q_0 = -1$ and $\bar{q}_0 = -1$. It is also clear that the function $h(q)$ strictly decreases in a neighborhood of the point q_0 and strictly increases in a neighborhood of the point $\bar{q}_0$. Hence, the conditions of the theorem are satisfied, and two solutions of the equation $|\alpha| - |y| = 0$ (for $\alpha > 0$), $y(\alpha)$ and $\bar{y}(\alpha)$, exist and $y'_+(0) = -1$, $\bar{y}'_+(0) = -1$. Analogously, it is clear from Figure 13 that for $g = -1$ equation (9.6) has also two solutions: $q_0 = +1$ and $\bar{q}_0 = -1$ and the function $h(q)$ is strictly monotone in a neighborhood of these points.

EXAMPLE 2. Let $f(z) = ||x| + y|$, $z_0 = (0,0)$. Clearly, $f(z_0) = 0$. This function was considered in Example 10 of Section 3.

It is clear from Figure 23 that for $g = +1$ equation (9.6) has the solution $q_0 = -1$; however, the function $h(q)$ is not strictly monotone.

Nevertheless, the solution of the equation $||\alpha| + y| = 0$ exists and $y'_+(0) = -1$.

PROBLEM 9.1. Examine the case where $f(z)$ is a sub- or superdifferentiable function at a point z_0 where $f(z_0) = 0$.

REMARK 2. Consider the following system of equations:

$$f_i(z) \equiv f_i(x,y) = 0 \qquad \forall i \in I \equiv 1:n . \tag{9.7}$$

Here, $x \in E_m$, $y \in E_n$. Let a solution of system (9.7) $z_0 = (x_0, y_0)$ be known: $f_i(z_0) = 0 \ \forall i \in I$. Take $g \in E_m$ and consider the system:

$$f_i(x_0 + \alpha g, y) = 0 \qquad \forall i \in I .$$

We assume that the functions $f_i(z)$ are continuous and quasidifferentiable at the point z_0.

The problem is to discover whether there exist a number $\alpha_0 > 0$ and a vector-valued continuous function $y(\alpha)$ such that

$$y(0) = y_0 , \qquad f_i(x_0 + \alpha g, y(\alpha)) = 0$$
$$\forall \alpha \in [0, \alpha_0], \quad \forall i \in I . \tag{9.8}$$

From the quasidifferentiability of the functions $f_i(z)$, we have

$$f_i(x_0+\alpha g,\ y_0+\alpha q)$$

$$= f_i(x_0,y_0) + \alpha \left| \max_{v_i \in \underline{\partial} f_i(z_0)} ((v_{1i},g) + (v_{2i},q)) + \min_{w_i \in \overline{\partial} f_i(z_0)} ((w_{1i},g) + (w_{2i},g)) \right| + o_i(\alpha) \qquad \forall\, i \in I\ , \qquad (9.9)$$

where $v_i = [v_{1i},v_{2i}] \in E_m \times E_n$, $w_i = [w_{1i},w_{2i}] \in E_m \times E_n$.

Consider the system:

$$F_i(q) \equiv \max_{v_i \in \underline{\partial} f_i(z_0)} ((v_{1i},g) + (v_{2i},g)) + \min_{w_i \in \overline{\partial} f_i(z_0)} ((w_{1i},g) + (w_{2i},g)) = 0 \qquad \forall\, i \in I\ . \qquad (9.10)$$

Here, the vector $g \in E_n$ is unknown.

It follows from (9.9) that the problem of existence of an implicit function of several variables is reduced to that of studying the properties of the solutions of system (9.10). These systems are referred to as quasilinear, in the terminology used by Bellman and Kalaba.

Chapter 3

MINIMIZATION ON THE ENTIRE SPACE

1. NECESSARY AND SUFFICIENT CONDITIONS FOR A MINIMUM OF A CONVEX FUNCTION ON E_n

Let $f(x)$ be a convex function on E_n. Recall that in order that a convex function $f(x)$ attain its minimum value on E_n at a point x^*, it is necessary and sufficient that

$$0 \in \partial f(x^*)$$

(Section 1.6).

Let $\varepsilon \geq 0$. A point $x_\varepsilon \in E_n$ is said to be an ε-*stationary* point of the function $f(x)$ if

$$0 \in \partial_\varepsilon f(x_\varepsilon)$$

or, equivalently,

$$0 \leq f(x_\varepsilon) - f^* \leq \varepsilon , \tag{1.1}$$

where $f^* = \inf_{x \in E_n} f(x)$.

Recall also that if $0 \notin \partial_\varepsilon f(x_0)$ and

$$\frac{\partial_\varepsilon f(x_0)}{\partial g_0} = \max_{v \in \partial_\varepsilon f(x_0)} (v, g_0) < 0 ,$$

where $g_0 \in E_n$, then

$$f(x_0) - \inf_{\alpha > 0} f(x_0 + \alpha g_0) > \varepsilon \tag{1.2}$$

(see Subsection 1.3.8). Introduce the following notation:

$$\psi_\varepsilon(x) = \min_{\|g\|=1} \max_{v \in \partial_\varepsilon f(x)} (v, g) ;$$

$$\rho_\varepsilon(x) = \min_{v \in [\partial_\varepsilon f(x)]_{fr}} \|v\| ,$$

$$\rho(x) = \rho_0(x) , \qquad d_\varepsilon(x) = \min_{v \in \partial_\varepsilon f(x)} \|v\| , \qquad d(x) = d_0(x) .$$

It is clear that if $d_\varepsilon(x) > 0$, then $\rho_\varepsilon(x) = d_\varepsilon(x)$ and that if $d_\varepsilon(x) = 0$, then $\rho_\varepsilon(x)$ is the radius of the largest sphere centered at the origin and included in $\partial_\varepsilon f(x)$.

The following lemma establishes the relation between $\psi_\varepsilon(x)$, $\rho_\varepsilon(x)$ and $d_\varepsilon(x)$.

<u>LEMMA 1.1</u>. We have the following relation:

$$\psi_\varepsilon(x) = \begin{cases} -\rho_\varepsilon(x) = -d_\varepsilon(x) & \text{if } d_\varepsilon(x) > 0 , \\ \rho_\varepsilon(x) & \text{if } d_\varepsilon(x) = 0 . \end{cases}$$

<u>Proof</u>. Is obtained by applying Lemma 6.4 in Chapter 1 for $B = \partial_\varepsilon f(x)$, $\Gamma^+ = \{0\}$.

<u>LEMMA 1.2</u>. For any $x, y \in E_n$ and $\varepsilon \geq 0$, we have the inequality

$$f(y) - f(x) \geq \psi_\varepsilon(x) \|y-x\| - \varepsilon . \tag{1.3}$$

<u>Proof</u>. By the definition of $\partial_\varepsilon f(x)$ and $\psi_\varepsilon(x)$, we have

$$f(y) - f(x) \geq \max_{v \in \partial_\varepsilon f(x)} (v, y-x) - \varepsilon \geq \|y-x\| \psi_\varepsilon(x) - \varepsilon . \quad \blacksquare$$

<u>LEMMA 1.3</u>. Let $\varepsilon > 0$, x_ε be an ε-stationary point of a function $f(x)$ on E_n and $\psi_\varepsilon(x) > 0$.

Then, $f^* > -\infty$, the set $\mu^* = \{x \in E_n \mid f(x) = f^*\}$ is a nonempty convex compact set and

$$0 \leq f(x_\varepsilon) - f^* < \varepsilon \tag{1.4}$$

and, moreover, if $f^* < f(x_\varepsilon)$ then

$$(f(x_\varepsilon) - f^*)(\rho(x_\varepsilon))^{-1} \leq \|x^* - x_\varepsilon\| \leq (f^* - f(x_\varepsilon) + \varepsilon)(\rho_\varepsilon(x_\varepsilon))^{-1} . \tag{1.5}$$

Here x^* is a minimum point of the function $f(x)$ on E_n and $\rho(x) = \rho_0(x)$.

Proof. Since x_ε is an ε-stationary point of $f(x)$, then from (1.1) we have

$$0 \le f(x_\varepsilon) - f(x^*) < \varepsilon ; \qquad (1.6)$$

consequently, $f^* > -\infty$.

By virtue of (1.3), we have for any $x \in E_n$

$$f(x) - f(x_\varepsilon) \ge \psi_\varepsilon(x_\varepsilon) \|x-x_\varepsilon\| - \varepsilon .$$

Hence, considering the assumption that $\psi_\varepsilon(x_\varepsilon) > 0$,

$$\|x-x_\varepsilon\| \le (f(x) - f(x_\varepsilon) + \varepsilon)(\psi_\varepsilon(x_\varepsilon))^{-1} . \qquad (1.7)$$

From (1.7), in particular, it follows that the Lebesgue set $D(x_\varepsilon) = \{x \in E_n \mid f(x) \le f(x_\varepsilon)\}$ is bounded and, moreover,

$$D(x_\varepsilon) \subset S_\delta(x_\varepsilon) , \qquad \text{where} \quad \delta = \varepsilon(\psi_\varepsilon(x_\varepsilon))^{-1} .$$

In addition, the set $\mu^* \subset D(x_\varepsilon)$ is nonempty and bounded. The closedness and the convexity of the set μ^* are obvious. Now, assume that

$$f(x_\varepsilon) - f^* = \varepsilon . \qquad (1.8)$$

Letting $x = x^*$ in (1.7) and considering (1.8), we obtain $x_\varepsilon = x^*$; and this contradicts (1.8). Thus, equality (1.8) is impossible.

Hence, (1.4) follows from (1.6).

Furthermore, it follows from Lemma 1.1 that $\psi_\varepsilon(x_\varepsilon) = \rho_\varepsilon(x_\varepsilon)$ because $d_\varepsilon(x_\varepsilon) = 0$, and that $\psi(x_\varepsilon) = -\rho(x_\varepsilon)$ because $d(x_\varepsilon) > 0$. Now we obtain the second inequality of (1.5) by letting $x = x^*$ in (1.7). The first inequality of (1.5) follows from the fact that we have, according to (1.3),

$$f^* - f(x_\varepsilon) = f(x^*) - f(x_\varepsilon) \ge \psi(x_\varepsilon)\|x^*-x_\varepsilon\| = -\rho(x_\varepsilon)\|x^*-x_\varepsilon\| .$$ ∎

REMARK. The condition $f^* < f(x_\varepsilon)$ is necessary only for the validity of the left-hand side of (1.5).

2. MINIMIZATION OF A SMOOTH FUNCTION

The methods of nonsmooth optimization discussed below in methodological terms are closely related to methods for minimization of smooth functions.

Let a function $f(x)$ be continuously differentiable on E_n, but not necessarily convex. In order that $f(x)$ attain its minimum value on E_n at a point $x^* \in E_n$, it is necessary that

$$f'(x^*) = 0 . \tag{2.1}$$

This condition is also sufficient if $f(x)$ is a convex function on E_n.

The point x^* satisfying (2.1) is said to be a *stationary point* of the function $f(x)$ on E_n. If $f'(x_0) \neq 0$, the direction $g(x_0) = \frac{f'(x_0)}{\|f'(x_0)\|}$ is the steepest descent direction of the function $f(x)$ at the point x_0, i.e., $\frac{\partial f(x_0)}{\partial g(x_0)} = \min_{\|g\|=1} \frac{\partial f(x_0)}{\partial g}$.

Recall that, in this case, $\frac{\partial f(x_0)}{\partial g} = (f'(x_0), g)$.

A large number of optimization methods employ condition (2.1).

1. The Method of Continuous Descent.

Consider the system of ordinary differential equations:

$$\dot{x}(t) = -A(x(t))\, f'(x) , \tag{2.2}$$

$$x(0) = x_0 , \tag{2.3}$$

where $A(x)$ is an $n \times n$ matrix-valued function continuous on $D(x_0) = \{x \in E_n \mid f(x) \le f(x_0)\}$ and, moreover, there exists an $m > 0$ such that

$$(A(x)z, z) \ge m \|z\|^2 \quad \forall z \in E_n , \quad \forall x \in D(x_0) . \tag{2.4}$$

Under these assumptions, there exists a solution $x(t, x_0)$ of the system (2.2) - (2.3) on $[0,\infty)$. We also note that

$$x(t,x_0) \in D(x_0) \qquad \forall t \geq 0 . \tag{2.5}$$

LEMMA 2.1. If the set $D(x_0)$ is bounded, then all the accumulation points of the solution $x(t,x_0)$ are stationary points of $f(x)$ on E_n.

As discrete analogues of the continuous method (2.2), we have the method of steepest descent, the method of steepest descent with a constant step-size and Newton's method (see [16], [59], [61], [105], [114]).

2. The Method of Steepest Descent.

We choose an arbitrary $x_0 \in E_n$ as an initial approximation. Let a point $x_k \in E_n$ have already been found. If $f'(x_k) = 0$, then x_k is stationary and the process terminates. If $f'(x_k) \neq 0$, then we let $g_k = f'(x_k)$ and consider the ray

$$\{x \in E_n \mid x = x_{k\alpha} \equiv x_k - \alpha g_k, \ \alpha \geq 0\} .$$

Next, we find $\min_{\alpha\geq 0} f(x_{k\alpha}) = f(x_{k\alpha_k})$ (assume that $\inf_{\alpha\geq 0} f(x_{k\alpha})$ is attained) and let $x_{k+1} = x_{k\alpha_k}$. Thus, we obtain a sequence $\{x_k\}$. If this sequence consists of a finite number of points, then its final element is stationary by construction.

Otherwise, the following theorem is valid.

THEOREM 2.1. If the set $D(x_0)$ is bounded, then every accumulation point of the sequence $\{x_k\}$ is stationary.

It can be shown that under some quite natural assumptions the convergence rate of the method of steepest descent is geometric.

REMARK. As a g_k we can choose an arbitrary vector such that

$$(g_k, g'(x_k)) \geq \|g_k\| \cdot \|f'(x_k)\| \cdot \theta ,$$

where $g_k \neq 0$ and $\theta \in (0,1]$ is fixed.

3. The Method of Steepest Descent With a Constant Step-Size.

Let $f'(x)$ satisfy a Lipschitz condition on E_n: there exists an $L < \infty$ such that $\|f'(x) - f'(z)\| \le L\|x-z\|$ $\forall x,y \in E_n$.

Let us fix $\lambda \in (0,\frac{2}{L})$. Take an $x_0 \in E_n$. Let $x_k \in D(x_0)$ have already been found. If $f'(x_k) = 0$, then the point x_k is stationary and the process terminates. If $f'(x_k) \ne 0$, then let
$x_{k+1} = x_k - \lambda g'(x_k)$.

For the sequence $\{x_k\}$, Theorem 2.1 holds.

3. THE METHOD OF STEEPEST DESCENT

1. Let a function $f(x)$ be convex on E_n. In Section 1.8 we introduced the concept of an ε-derivative of a convex function $f(x)$ at a point x_0 in the direction g:

$$\frac{\partial_\varepsilon f(x_0)}{\partial g} = \max_{v \in \partial_\varepsilon f(x_0)} (v,g) .$$

If $0 \in \partial_\varepsilon f(x_0)$, then the point x_0 is ε-stationary and, in this case,

$$0 \le f(x_0) - f^* \le \varepsilon ,$$

where $f^* = \inf_{x \in E_n} f(x)$.

If $0 \notin \partial_\varepsilon f(x_0)$, then the direction $g_\varepsilon(x_0) = -v_\varepsilon(x_0)\|v_\varepsilon(x_0)\|^{-1}$, where $v_\varepsilon(x_0) \in \partial_\varepsilon f(x_0)$, $\|v_\varepsilon(x_0)\| = \min_{v \in \partial_\varepsilon f(x_0)} \|v\| = d_\varepsilon(x_0)$, is the ε-steepest descent direction, i.e.,

$$\frac{\partial f(x_0)}{\partial g_\varepsilon(x_0)} = \min_{\|g\|=1} \frac{\partial_\varepsilon f(x_0)}{\partial g} .$$

The direction $g(x_0)$ is unique.

In Section 1.8 we showed that if, for some $g \in E_n$, the inequality $\frac{\partial_\varepsilon f(x_0)}{\partial g} < 0$ holds, then

$$\inf_{\alpha > 0} f(x_0+\alpha g) < f(x_0) - \varepsilon , \qquad (3.1)$$

i.e., the function $f(x)$ can be decreased in the direction g not less than by ε.

2. **ε-Algorithm.** Let us fix $\varepsilon > 0$ and choose an arbitrary initial approximation $x_0 \in E_n$. Let the kth approximation $x_k \in E_n$ have already been found. If $0 \in \partial_\varepsilon f(x_k)$, then x_k is an ε-stationary point of the function $f(x)$ on E_n. If $0 \notin \partial_\varepsilon f(x_k)$, then let us find a vector $g_k \in E_n$, $\|g_k\| = 1$ such that

$$\frac{\partial_\varepsilon f(x_k)}{\partial g_k} = \max_{v \in \partial_\varepsilon f(x_k)} (v, g_k) < 0 . \tag{3.2}$$

After this, we find a point $x_{k+1} = x_k(\alpha_k)$ on the ray $\{x_k(\alpha) = x_k + \alpha g_k \mid \alpha \geq 0\}$ such that

$$f(x_{k+1}) \leq f(x_k) - \varepsilon . \tag{3.3}$$

The existence of such a point x_{k+1} follows from (3.1).

On the basis of (3.3), we conclude that either an ε-stationary point of the function $f(x)$ on E_n is obtained after a finite number of steps (this occurs if $f^* > -\infty$) or $f^* = -\infty$.

Let us discuss here how to find a vector g_k satisfying (3.2) in the case $0 \notin \partial_\varepsilon f(x_k)$. Clearly, we can take as g_k a vector $g_\varepsilon(x_k)$, i.e., the direction of ε-steepest descent of the function $f(x)$ at the point x_k. Indeed, if

$$0 \notin \partial_\varepsilon f(x_k) , \qquad v_\varepsilon \in \partial_\varepsilon f(x_k) , \qquad \|v_\varepsilon(x_k)\| = d_\varepsilon(x_k) ,$$

$$g_\varepsilon(x_k) = -v_\varepsilon(x_k) \, \|v_\varepsilon(x_k)\|^{-1} ,$$

then we have

$$\frac{\partial_\varepsilon f(x_k)}{\partial g_\varepsilon(x_k)} = -\|v_\varepsilon(x_k)\| < 0 .$$

If we take as g_k the vector $g_\varepsilon(x_k)$, i.e., the direction of ε-steepest descent of the function $f(x)$ at the point x_k, the resulting method is called the *method of ε-steepest descent.*

To find $v_\varepsilon(x_k)$, we need to solve the following problem:

Find a $v_\varepsilon(x_k) \in \partial_\varepsilon f(x_k)$ such that

$$\|v_\varepsilon(x_k)\| = \min_{v \in \partial_\varepsilon f(x_k)} \|v\| . \tag{3.4}$$

If in the process of solving the problem (2.4) we find a vector $v_k \in \partial_\varepsilon f(x_k)$ such that $v_k \neq 0$ and

$$(v, v_k) \geq \theta(v_k, v_k) \qquad \forall v \in \partial_\varepsilon f(x_k) , \tag{3.5}$$

where $\theta \in (0,1]$, then the vector $g_k = -v_k\|v_k\|^{-1}$ satisfies (3.2).

REMARK 1. If, instead of a vector g_k such that $\frac{\partial_\varepsilon f(x_k)}{\partial g_k} < 0$, we take a direction such that $\frac{\partial f(x_k)}{\partial g_k} < 0$, then this method does not necessarily give the convergence (a minimum point of the function $f(x)$ on E_n is not necessarily among all the accumulation points of the sequence $\{x_k\}$). This is explained by the fact that, in this case, we do not have an estimate of the type (3.1), i.e., the guaranteed decrease of the function $f(x)$ at each step.

3. In the presented scheme of the ε-algorithm for the construction of a point x_{k+1} satisfying (3.3), generally, it is necessary to make a one-dimensional minimization of the function $f(x_k+\alpha g_k)$ with respect to $\alpha \geq 0$, having to calculate the value of the function $f(x)$ at a sufficiently great number of points.

If $g_k = -v_k\|v_k\|^{-1}$, where v_k satisfies (3.5), then it is possible to choose a step-size from the point x_k in the direction g_k by calculating the value of the function $f(x)$ on the ray $\{x_k(\alpha) \mid \alpha \geq 0\}$ only once. Once can choose α_k, for example, in the following manner:

$$\alpha_k = \varepsilon\left[\frac{\partial_\varepsilon f(x_k)}{\partial g_k} - \frac{\partial f(x_k)}{\partial g_k}\right]^{-1} ;$$

then by virtue of Lemma 8.6 in Chapter 1, the inequalities

$$f(x_{k+1}) \le f(x_k) + \alpha_k \frac{\partial_\varepsilon f(x_k)}{\partial g_k} \qquad (3.6)$$

are satisfied for each k.

If the Lebesgue set $D(x_0) = \{x \in E_n \mid f(x) \le f(x_0)\}$ is bounded, then any accumulation point of the sequence $\{x_{k+1} = x_k + \alpha_k g_k\}$ is an ε-stationary point of the function $f(x)$ on E_n. To show this, let us assume that there exists a subsequence $\{x_{k_s}\}$ such that $x_{k_s} \to x^*$, however, assume that $0 \notin \partial_\varepsilon f(x^*)$ (hence, $d_\varepsilon(x^*) > 0$).

Then, for sufficiently large k_s, by virtue of the upper semi-continuity of $\partial_\varepsilon f(x)$, the inequality

$$\frac{\partial_\varepsilon f(x_{k_s})}{\partial g_{k_s}} \le -\theta \| v_{k_s} \| \le -\theta \| v_\varepsilon(x_{k_s}) \| \le -\frac{1}{2}\theta d_\varepsilon(x^*)$$

holds (see (3.5)).

Therefore, from (3.6) and the form of α_k, we have

$$f(x_{k_s+1}) \le f(x_{k_s}) - \frac{1}{2}\theta \varepsilon b^{-1} d_\varepsilon(x^*) \quad ,$$

where $b = \max_{x \in D(x_0)} \max_{v \in \partial_\varepsilon f(x)} \| v \|$.

Consequently, $f(x_k) \xrightarrow[k\to\infty]{} -\infty$ and this contradicts the boundedness of $f(x)$ on $D(x_0)$.

4. Now we shall show how to apply the ε-algorithm in order to find minimum points of a function $f(x)$ on E_n.

Let us arbitrarily choose a point $x_0 \in E_n$ and a number $\varepsilon_0 > 0$.

Applying the ε-algorithm of Subsection 2 with $\varepsilon = \varepsilon_0$ and the initial approximation $x_{00} = x_0$, we shall either discover that $f^* = -\infty$ or, after a finite number of steps t_0, find a point $x_1 = x_{0t_0}$ which is an ε-stationary point of the function $f(x)$ on E_n.

We consider now only the case where $f^* > -\infty$. Let $x_{10} = x_1$ and choose $\varepsilon_1 < \varepsilon_0$. Letting the ε-algorithm start with $\varepsilon = \varepsilon_1$ and the point x_{10}, after a finite number of steps t_1 we find a point $x_2 = x_{1t_1}$ which is an ε_1-stationary point of the function $f(x)$ on E_n.

Proceeding analogously, we construct a sequence $\{x_k\}$, each point of which is an ε_{k-1}-stationary point of the function $f(x)$ on E_n.

THEOREM 3.1. If $f^* > -\infty$, $\varepsilon_k \to 0$. Then,

$$\lim_{k\to\infty} f(x_k) = f^* , \tag{3.7}$$

and if the set $D(x_0)$ is bounded, then any accumulation point of the sequence $\{x_k\}$ is a minimum point of the function $f(x)$ on E_n.

Proof. Since for each k the point x_k is an ε_{k-1}-stationary point of the function $f(x)$ on E_n, then

$$0 \le f(x_k) - f^* \le \varepsilon_{k-1} . \tag{3.8}$$

From (3.8) and the fact that $\varepsilon_k \to 0$, (3.7) follows. If $D(x_0)$ is bounded, then the sequence $\{x_k\}$ has accumulation points as $x_k \in D(x_0)$ for each k. Let $x_{k_s} \to x^*$. Then, by virtue of (3.7), we have $f(x^*) = f^*$.

The theorem has been proved. ∎

5. We shall describe one more modification of the ε-algorithm for finding minimum points of the function $f(x)$ on the entire space.

Let us arbitrarily choose an initial approximation $x_0 \in E_n$.

If $0 \in \partial_\varepsilon f(x_0)$, then x_0 is a minimum point. Now let $0 \notin \partial f(x_0)$.

Find a number $\varepsilon_0 > 0$ such that $0 \in \partial_{\varepsilon_0} f(x_0)$, i.e., $f(x_0) - f^* \le \varepsilon_0$.

As an ε_0, we can take, for example, $f(x_0) - f^*$.

We note that if for any $\varepsilon > 0$ we have

$$0 \notin \partial_\varepsilon f(x_0) , \tag{3.9}$$

then $f^* = -\infty$. Indeed we can show this. The condition $0 \in \partial_\varepsilon f(x_0)$ is equivalent to the condition $f(x_0) - f^* \le \varepsilon$. Then, (3.9) is equivalent to the inequality $f(x_0) - f^* > \varepsilon$. Since this inequality is valid for all $\varepsilon > 0$, we obtain $f^* = -\infty$.

Now assume that $f^* > -\infty$.

We sort out the numerical sequence $\varepsilon_{0q} = 2^{-q}\varepsilon_0$, $q = 1,2,\ldots$, until the relation $0 \notin \partial_{\varepsilon_1} f(x_0)$ holds for some $q_0 \ge 1$ and $\varepsilon_1 = \varepsilon_{0q_0}$.

Here, clearly, $\varepsilon_1 \le 2^{-1}\varepsilon_0$ and

$$0 \in \partial_{2\varepsilon_1} f(x_0) , \qquad f(x_0) - f^* \le 2\varepsilon_1 . \tag{3.10}$$

Now find a vector $g_0 \in E_n$, $\|g_0\| = 1$, such that $\dfrac{\partial_{\varepsilon_1} f(x_0)}{\partial g_0} < 0$, and find on the ray $\{x_0(\alpha)=x_0+\alpha g_0 \mid \alpha \ge 0\}$ a point $x_1 = x_0(\alpha_0)$ such that

$$f(x_1) - f(x_0) \le -\varepsilon_1 . \tag{3.11}$$

According to (3.1), such a point x_1 necessarily exists. It follows from (3.10) and (3.11) that

$$f(x_1) \le f(x_0) - \varepsilon_1 \le f^* + \varepsilon_1 .$$

Consequently, $0 \in \partial_{\varepsilon_1} f(x)$ and the point x_1 is an ε_1-stationary point of the function $f(x)$ on E_n.

Continuing analogously, we construct a sequence $\{x_k\}$, which either is finite (in this case, a minimum point is found) or is such that every point x_k is an ε_k-stationary point of the function $f(x)$ on E_n, where $\varepsilon_k = 2^{-q_{k-1}}\varepsilon_{k-1}$ and

$$0 \le f(x_k) - f^* \le 2^{-q_{k-1}}\varepsilon_{k-1} \le 2^{-k}\varepsilon_0 \qquad \forall k . \qquad (3.12)$$

In this case, the following theorem holds.

THEOREM 3.2. Let $f^* > -\infty$. Then

$$\lim_{k\to\infty} f(x_k) = f^*$$

and the rate of convergence is geometric. Moreover, if the set $D(x_0)$ is bounded, then an arbitrary accumulation point of the sequence $\{x_k\}$ is a minimum point of the function $f(x)$ on E_n.

Proof. Is analogous to that of Theorem 3.1. The inequality (3.12) implies that the rate of convergence is geometric.

6. To find f^*, we can also proceed in the following way.

Let us arbitrarily choose an initial approximation $x_0 \in E_n$ and fix positive numbers a and b, $a < b$. Assume that the set $D(x_0)$ is bounded and denote its diameter by D. Let the kth approximation $x_k \in D(x_0)$ have already been found. Take an $\varepsilon_k > 0$ such that

$$D a d_{\varepsilon_k}(x_k) \le \varepsilon_k \le D b d_{\varepsilon_k}(x_k) . \qquad (3.13)$$

By the continuity of $\partial_\varepsilon f(x_k)$ in $\varepsilon > 0$ and the fact that $d_\varepsilon(x_k) \xrightarrow[\varepsilon\to\infty]{} 0$, such an ε_k exists.

We find a $g_k \in E_n$, $\|g_k\| = 1$, such that $\dfrac{\partial_{\varepsilon_k} f(x_k)}{\partial g_k} < 0$, and on the ray $\{x_k(\alpha)=x_k+\alpha g_k \mid \alpha\ge 0\}$ we find a point $x_{k+1} = x_k(\alpha_k)$ such that

$$f(x_{k+1}) \le f(x_k) - \varepsilon_k . \qquad (3.14)$$

Continuing analogously, we construct a sequence $\{x_k\}$. In this case, Theorem 3.2 also holds. Moreover, we have

$$f(x_{k+1}) - f^* \le (1 - a(1+b)^{-1})(f(x_k) - f^*) \qquad \forall k. \qquad (3.15)$$

Indeed, by virtue of Lemmas 1.1 and 1.2, we have

$$f(x_k) - f^* \leq \varepsilon_k + d_{\varepsilon_k}(x_k) \; ; \qquad (3.16)$$

and combining (3.13), (3.14) and (3.16), we obtain (3.15).

REMARK 2. If a function $f(x)$ is strongly convex or if it is strictly convex and the set $D(x_0)$ is bounded, then the sequences $\{x_k\}$ in Theorems 3.1 and 3.2 converge to a unique minimum point of the function $f(x)$ on E_n.

7. The modifications of the ε-algorithm considered in this subsection as well as the ε-algorithm itself are mainly of theoretical interest because at each step it is necessary to know all the corresponding sets $\partial_\varepsilon f(x)$. Unfortunately, a class of functions for which ε-subdifferentials are efficiently computable, is not sufficiently large.

Thus, on the one hand ε-subdifferentials are very convenient and have many important and useful properties (the continuity of $\partial_\varepsilon f(x)$, the guaranteed estimate (3.1) of decrease of the function by ε if $0 \notin \partial_\varepsilon f(x)$, the possibility of having the geometric rate of convergence of a sequence $\{f(x_k)\}$ to f^* with the strong convexity assumption on the function $f(x)$), which do not have analogues in the theory of nonlinear programming and minimax problems. On the other hand, ε-subdifferentials are almost noncomputable because the ε-subdifferential describes some global properties of the function that makes the methods stated in this subsection practically inapplicable.

EXAMPLE. Let $f(x) = (Ax,x) + (a,x)$, where A is an $n\times n$ symmetric positive definite matrix and $a \in E_n$. Then,

$$\partial_\varepsilon f(x) = \{2A(x+v)+a \mid (Av,v)\leq\varepsilon\} ,$$

i.e., in order to find $\partial_\varepsilon f(x)$, even for such a <<simple>> function as a quadratic function, it is necessary to know the entire set of the solutions of the nonlinear problem $(Av,v) \le \varepsilon$.

Furthermore, in our case, the method of steepest descent consists in constructing a sequence $\{x_k\}$ according to the formula

$$x_{k+1} = x_k - (f'(x_k), f'(x_k))(2Af'(x_k), f'(x_k))^{-1} f'(x_k) .$$

Here, only to find the direction of ε-steepest descent of the function $(Ax,x)+(a,x)$ in the ε-algorithm, we need at each step to solve the problem

$$\min_{(Av,v)\le\varepsilon} \|2A(x+v)+a\| .$$

As a result, an auxiliary problem which we need to solve while applying the method of ε-steepest descent is, at least formally, more complicated than the original minimization problem.

4. THE SUBGRADIENT METHOD FOR MINIMIZING A CONVEX FUNCTION

1. Subgradient Method. Let a function $f(x)$ be finite and convex on E_n. Let us choose an arbitrary point $x_0 \in E_n$ and a sequence of numbers $\{\lambda_k\}$ such that

$$\lambda_k \to +0 , \qquad \sum_{k=0}^{\infty} \lambda_k = +\infty . \tag{4.1}$$

Let $x_k \in E_n$ have already been found. Take an arbitrary $v_k \in \partial f(x_k)$. If $v_k = 0$, then $0 \in \partial f(x_k)$ and, hence, the point x_k is a minimum point of $f(x)$ on E_n, and the process terminates. Otherwise, we let

$$x_{k+1} = x_k - \lambda_k v_k \|v_k\|^{-1} \equiv x_k + z_k , \tag{4.2}$$

where

$$z_k = -\lambda_k v_k \|v_k\|^{-1} , \qquad \|z_k\| = \lambda_k .$$

Thus, we get a sequence $\{x_k\}$. If this sequence consists of a finite number of points, then, by construction, its final element is a minimum point of the function $f(x)$ on E_n.

Now we consider the case where the sequence $\{x_k\}$ is infinite.

As above, let

$$M^* = \{x \in E_n \mid f(x) = f^*\} ,$$

where

$$f^* = \inf_{x \in E_n} f(x) , \qquad \rho(x, M^*) = \min_{z \in M^*} \|z-x\|$$

(the set M^* can be empty).

THEOREM 4.1. If the set M^* is nonempty and bounded, then

$$\rho(x_k, M^*) \to 0 , \qquad f(x_k) \to f^* , \tag{4.3}$$

as $k \to \infty$.

Proof. We show first that

$$\varliminf_{k\to\infty} \rho(x_k, M^*) = 0 . \tag{4.4}$$

Suppose that this is not the case. Then, there exist an $a > 0$ and a $K_1 < \infty$ such that

$$\rho(x_k, M^*) \geq 2a > 0 \qquad \forall k \geq K_1. \tag{4.5}$$

Since the set M^* is bounded, then, according to Corollary 4 of Lemma 9.1 in Chapter 1, there exists a $c > 0$ such that

$$D_c \equiv \{x \mid f(x) \leq f^*+c\} \subset S_a(M^*) , \tag{4.6}$$

where

$$S_a(M^*) = \{x \mid \rho(x, M^*) \leq a\} .$$

There exists an $r > 0$ such that

$$S_r(\bar{x}) \equiv \{x \mid \|x-\bar{x}\| \leq r\} \subset D_c \qquad \forall \bar{x} \in M^* . \tag{4.7}$$

Let us fix an arbitrary $\bar{x} \in M^*$ (Figure 49). From the definition of $\partial f(x_k)$, we have

$$f(x) - f(x_k) \geq (v_k, x-x_k) \qquad \forall x \in E_n . \tag{4.8}$$

From (4.5) and (4.6), we obtain $f(x_k) > f^* + c \quad \forall k \geq K_1$ and, considering (4.8), we obtain

$$(v_k,\ x-x_k) < 0 \qquad \forall x \in D_c ,$$

i.e.,

$$(z_k,\ x_k-x) < 0 \qquad \forall x \in D_c . \qquad (4.9)$$

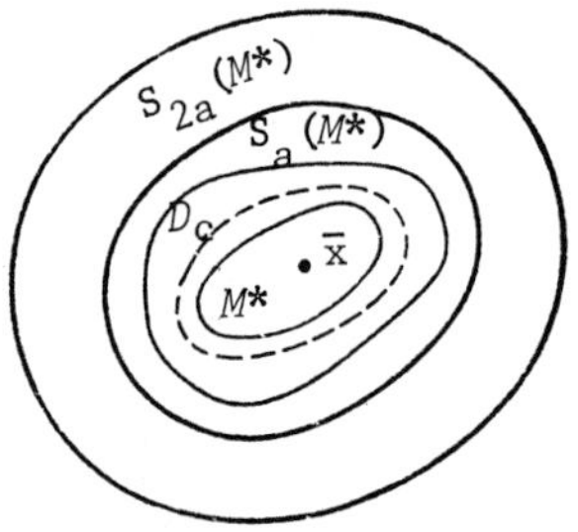

Figure 49

Take an $\tilde{x}_k = \bar{x} - rz_k \|z_k\|^{-1}$. It follows from (4.7) that $\tilde{x}_k \in D_c$ and, hence, from (4.9)

$$(z_k,\ x_k-\tilde{x}_k) = (z_k,\ x_k - \bar{x} + rz_k\|z_k\|^{-1}) < 0 ,$$

i.e.,

$$(z_k,\ x_k-\bar{x}) < -r\lambda_k .$$

Next,

$$\|x_{k+1}-\bar{x}\|^2 = \|x_k+z_k-\bar{x}\|^2 = \|x_k-\bar{x}\|^2 + 2(z_k,x_k-\bar{x}) + z_k^2$$

$$< \|x_k-\bar{x}\|^2 - 2r\lambda_k + \lambda_k^2 .$$

Since $\lambda_k \to +0$, then for sufficiently large k,

$$\|x_{k+1} - \bar{x}\|^2 < \|x_k - \bar{x}\|^2 - r\lambda_k .$$

Proceeding analogously, we obtain

$$\|x_{k+s} - \bar{x}\|^2 < \|x_k - \bar{x}\|^2 - r\sum_{i=0}^{s-1}\lambda_{k+i} . \qquad (4.10)$$

Hence, by virtue of (4.1) it follows that $\|x_k - \bar{x}\| \to -\infty$, which is impossible. This contradiction proves (4.4).

Now, let us fix a $\delta > 0$. As above, we can find an $r_\delta > 0$ which satisfies

$$\|x_{k+1} - \bar{x}\|^2 < \|x_k - \bar{x}\|^2 - r_\delta \lambda_k \qquad \forall \bar{x} \in M^* \tag{4.11}$$

for all sufficiently large k such that $\rho(x_k, M^*) \geq \delta$. From (4.11),

$$\rho^2(x_{k+1}, M^*) < \rho^2(x_k, M^*) - r_\delta \lambda_k . \tag{4.12}$$

On the other hand, we note that, by virtue of (4.4), there exist k such that

$$\rho(x_k, M^*) < \delta . \tag{4.13}$$

Since $x_{k+1} = x_k + z_k$, $\|z_k\| = \lambda_k$, then, considering (4.1), we can assume that $\rho(x_{k+1}, M^*) < 2\delta$ for k satisfying (4.13). Here, if $\rho(x_{k+1}, M^*) \geq \delta$, then, by virtue of (4.12),

$$\rho(x_{k+2}, M^*) < \rho(x_{k+1}, M^*) < 2\delta .$$

We proceed further analogously. Thus, if $\rho(x_{k+i}, M^*) < \delta$, then $\rho(x_{k+i+1}, M^*) < 2\delta$; if $\delta \leq \rho(x_{k+i}, M^*) < 2\delta$, then $\rho(x_{k+i+1}, M^*) < \rho(x_{k+i}, M^*) < 2\delta$. Finally, for all $i \geq 1$, we have $\rho(x_{k+i}, M^*) < 2\delta$.

By virtue of arbitrariness of δ, we obtain that $\rho(x_k, M^*) \to 0$ and then $f(x_k) \to f^*$, which completes the proof of the theorem. ■

<u>COROLLARY</u>. If $\operatorname{int} M^* \neq \emptyset$, then the sequence $\{x_k\}$ is finite, i.e., for some k we have $v_k = 0$.

<u>Proof</u>. Assume that this is not the case, i.e., $v_k \neq 0$ for all k. Then, the sequence $\{x_k\}$ is infinite. Take an arbitrary $\bar{x} \in \operatorname{int} M^*$. There exists an $r > 0$ such that $S_r(\bar{x}) \subset M^*$. Find the point $\tilde{x}_k = \bar{x} - r z_k \|z_k\|^{-1}$. It is clear that $\tilde{x}_k \in S_r(\bar{x}) \subset M^*$ and, hence, $f(\tilde{x}_k) = f^*$. By the definition of a subgradient,

$$f(\tilde{x}_k) - f(x_k) \geq (v_k, \tilde{x}_k - x_k) .$$

Since $f(x_k) \geq f^* = f^*(\tilde{x}_k)$, then we have

$$(v_k, \tilde{x}_k - x_k) \leq f(\tilde{x}_k) - f(x_k) \leq 0 ,$$

i.e.,

$$(v_k, \bar{x}-x_k) \leq r(v_k, z_k\|z_k\|^{-1}) .$$

Since $z_k = -\lambda_k v_k \|v_k\|^{-1}$, then we have $(z_k, x_k-\bar{x}) \leq -r\lambda_k$. Arguing as in proving Theorem 4.1, we obtain

$$\|x_{k+1} - \bar{x}\|^2 \leq \|x_k - \bar{x}\|^2 - r\lambda_k . \tag{4.14}$$

Indeed, $\|x_{k+s} - \bar{x}\| \xrightarrow[s\to\infty]{} -\infty$, and this is impossible. This contradiction proves the finiteness of the sequence $\{x_k\}$. ∎

REMARK 1. After obtaining inequality (4.14), the proof can be completed differently: it follows from (4.14) that the distance to the point $\bar{x}$ strictly decreases and, moreover, we conclude that after a finite number of steps, it is possible to find a point x_{k+s} such that $x_{k+s} \in \operatorname{int} M^*$. Incidentally, if $x \in \operatorname{int} M^*$, then $\partial f(x) = \{0\}$ (see Problem 6.2 in Chapter 1). Hence, $v_{k+s} = 0$ and the process terminates, i.e., the sequence $\{x_k\}$ is finite.

REMARK 2. The subgradient method introduced above is often called the *method of generalized gradient descent* (GGD-method).

2. ε-Subgradient Method. The set of ε-subgradients (the ε-subdifferential) is, generally speaking, richer than the set $\partial f(x)$.

Therefore, to find an element in $\partial_\varepsilon f(x)$ is perhaps simpler than to find an element in $\partial f(x)$. Therefore we can describe the following method for finding ε-stationary points.

Fix an $\varepsilon > 0$. Let a numerical sequence $\{\lambda_k\}$ satisfy (4.1) and take an arbitrary $x \in E_n$. Assume that $x_k \in E_n$ has already been found.

Take an arbitrary $v_k \in \partial_\varepsilon f(x_k)$.

If $v_k = 0$, then $0 \in \partial_\varepsilon f(x_k)$ and the point x_k is an

ε-stationary point of the function $f(x)$ on E_n. In this case, as shown in Subsection 1.8.3, the relation $0 \le f(x_k) - f^* \le \varepsilon$ holds, where $f^* = \inf_{x \in E_n} f(x)$.

If $v_k \neq 0$, then we define

$$x_{k+1} = x_k - \lambda_k v_k \|v_k\|^{-1} \equiv x_k + z_k . \tag{4.15}$$

THEOREM 4.2. If the set M^* is nonempty and bounded, then

$$\varliminf_{k\to\infty} \rho(x_k, D_\varepsilon) = 0 , \qquad \rho(x_k, S_{a_\varepsilon}(M^*)) \xrightarrow[k\to\infty]{} 0 , \tag{4.16}$$

where

$$D_\varepsilon = \{x \mid f(x) \le f^*+\varepsilon\} , \qquad a_\varepsilon = \max_{x \in Q_\varepsilon} (x, M^*) ,$$

$$Q_\varepsilon = \{x \mid f(x) = f^*+\varepsilon\} .$$

Proof. First of all we show that

$$\varliminf_{k\to\infty} \rho(x_k, D_\varepsilon) = 0 . \tag{4.17}$$

If we assume that this is not the case, then there exist an $a > 0$ and a $K_1 < \infty$ such that

$$\rho(x_k, D_\varepsilon) \ge 2a > 0 \qquad \forall k > K_1 . \tag{4.18}$$

Since M^* is a bounded set, then the set D_ε is also bounded and, hence, there exists a $c > 0$ such that

$$D_{c+\varepsilon} = \{x \mid f(x) \le f^*+c+\varepsilon\} \subset S_a(D_\varepsilon) . \tag{4.19}$$

Hence, of course, $D_c \subset S_a(D_\varepsilon)$. We note that there exists an $r > 0$ such that

$$S_r(\bar{x}) = \{x \mid \|x-\bar{x}\| \le r\} \subset D_c \qquad \forall \bar{x} \in M^* . \tag{4.20}$$

Let us fix an arbitrary $\bar{x} \in M^*$. By the definition of $\partial_\varepsilon f(x_k)$, we have

$$f(x) - f(x_k) \ge (v_k, x-x_k) - \varepsilon \qquad \forall x \in E_n . \tag{4.21}$$

It follows from (4.18) and (4.19) that $f(x) > f^* + \varepsilon + c$ and, hence, from (4.21)

$$(v_k,\ x-x_k) \le f(x) - f(x_k) + \varepsilon < f(x) - f^* - c\ . \quad (4.22)$$

For $x \in D_c$, we have $f(x) \le f^* + c$. Now, from (4.22) we obtain

$$(v_k,\ x-x_k) < 0 \qquad \forall x \in D_c\ . \quad (4.23)$$

Let us take $\tilde{x}_k = \bar{x} - rz_k\|z_k\|^{-1}$. Then, from (4.20), $\tilde{x}_k \in D_c$. Indeed, from (4.23) with $x = \tilde{x}_k$ and (4.15)

$$(z_k,\ x_k-\tilde{x}_k) < 0\ .$$

Hence

$$(z_k,\ x_k-\bar{x}) < -r\lambda_k\ .$$

Using the same arguments as in proving Theorem 4.1, we establish that $\|x_k-\bar{x}\| \to -\infty$, and this is impossible.

Thus, (4.17) has been proved.

It was actually shown above that for any $\delta > 0$ the inequality

$$\rho(x_{k+1}, M^*) < \rho(x_k, M^*)$$

holds for sufficiently large k such that $\rho(x_k, D_\varepsilon) > \delta$. Now using the same arguments as in proving the second half of Theorem 4.1, we conclude that $\rho(x_k, S_{a_\varepsilon}(M^*)) \to 0$. ■

Thus, there exists a subsequence $\{k_s\}$ such that $x_{k_s} \to x^*$, where $f(x^*) \le f^* + \varepsilon$. Furthermore, for any $\delta > 0$ there exists a $K_\delta < \infty$ such that

$$\rho(x_k, M^*) \le a_\varepsilon + \delta \qquad \forall k \ge K_\delta\ .$$

It has already been noticed that $a_\varepsilon \xrightarrow[\varepsilon\to+0]{} 0$ (see Corollary 3 of Lemma 9.1 in Chapter 1).

3. ε_k-Subgradient Method. We shall describe yet another iterative method in which ε_k-subgradients are used at each step and where $e_k \xrightarrow[k\to\infty]{} 0$.

Let again a sequence $\{\lambda_k\}$ satisfy condition (4.1). Choose a sequence $\{\varepsilon_k\}$ such that

$$\varepsilon_k \to +0 , \tag{4.24}$$

and take an arbitrary point $x_0 \in E_n$. Let a point $x_k \in E_n$ have already been found. Choose an arbitrary $v_k \in \partial_{\varepsilon_k} f(x_k)$ and define $x_{k+1} = x_k + z_k$, where

$$z_k = \begin{cases} 0 & \text{if } v_k = 0 , \\ -\lambda_k v_k \|v_k\|^{-1} & \text{if } v_k \neq 0 . \end{cases} \tag{4.25}$$

THEOREM 4.3. If the set M^* is nonempty and bounded, then

$$\rho(x_k, M^*) \to 0 , \qquad f(x_k) \to f^* , \tag{4.26}$$

as $k \to \infty$.

Proof. First we establish that

$$\varliminf_{k\to\infty} \rho(x_k, M^*) = 0 . \tag{4.27}$$

Assume that this is not the case, then there exist an $a > 0$ and a $K_1 < \infty$ such that

$$\rho(x_k, M^*) \geq 2a > 0 \qquad \forall k \geq K_1 . \tag{4.28}$$

Without loss of generality, we can assume that $v_k \neq 0$ because if there exist infinitely many k such that $v_k = 0$, then we can find a sequence $\{x_{k_s}\}$ for which we have $\rho(x_{k_s}, M^*) \to 0$. (Use the boundedness of D_{ε_0}.)

As in the proof of Theorem 4.1, there exists a $c > 0$ such that

$$D_c = \{x \mid f(x) \leq f^* + c\} \subset S_a(M^*) .$$

Here, we can assume that

$$\inf_{x \notin S_{2a}(M^*)} f(x) - f^* \equiv d > c > 0. \tag{4.29}$$

We note that there exists an $r > 0$ such that

$$S_r(\bar{x}) = \{x \mid \|x - \bar{x}\| \leq r\} \subset D_c \qquad \forall \bar{x} \in M^* . \tag{4.30}$$

Let us fix an arbitrary $\bar{x} \in M^*$. Then,

$$f(x) - f(x_k) \geq (v_k, x - x_k) - \varepsilon_k \qquad \forall x \in E_n . \tag{4.31}$$

From (4.28) and (4.30), we have $f(x_k) > f^* + d$ and, hence, from (4.31) we have

$$(v_k, x - x_k) \le f(x) - f(x_k) + \varepsilon_k < f^* + c - f^* - d + \varepsilon_k < 0$$

for sufficiently large k such that $\varepsilon_k < d - c$. Now, using the same arguments as in proving Theorem 4.1, we obtain (4.26).

4. Subgradient Methods With Almost Complete Relaxation. None of the methods described in this section are relaxation methods, that is the value of the function does not necessarily decrease as the point x_k goes to the point x_{k+1}. A remarkable feature of subgradient methods is their extraordinary simplicity, because all we have to do is to find some subgradient (or ε-subgradient) at each step. However, a serious drawback of these methods is a very slow rate of convergence. To overcome this, we can use various techniques, for example, the following modification of the subgradient method. As before, let a sequence $\{\lambda_k\}$ satisfy condition (4.1). We construct the sequence $\{x_k\}$ as follows:

$$x_{k+1} = x_k - \alpha_k v_k \|v_k\|^{-1} \equiv x_k + \alpha_k g_k ,$$

where $\alpha_k \in [\beta\lambda_k, \gamma\lambda_k]$, the numbers $\beta > 0$ and $\gamma > \beta$ are fixed. The choice α_k from the interval $[\beta\lambda_k, \gamma\lambda_k]$ can be arbitrary, e.g.,

1) α_k is an arbitrary number in $[\beta\lambda_k, \gamma\lambda_k]$;

2) $f(x_k + \alpha_k g_k) = \min_{\alpha \in [\beta\lambda_k, \gamma\lambda_k]} f(x_k + \alpha g_k)$;

3) $f(x_k + \alpha_k g_k) = \min_{i \in 1:N} f(x_k + \xi_{ik} g_k)$,

where

$$\xi_{ik} = \beta_i \lambda_k , \qquad \beta \le \beta_1 , \qquad \beta_i = 2^{i-1}\beta_1, \ldots,$$

$$\beta_N = 2^{N-1}\beta_1 \le \gamma .$$

If, using 2 or 3, we take a sufficiently small β and a sufficiently large γ, e.g.: $\beta = 10^{-10}$, $\gamma = 10^{10}$, the gradient method becomes a method with almost complete relaxation.

In this case, the convergence is guaranteed because, as before, the sequence $\{\alpha_k\}$ satisfies the condition

$$\alpha_k \to +0 , \qquad \sum_{k=0}^{\infty} \alpha_k = +\infty .$$

Analogous procedure is applicable to ε-subgradient and ε_k-subgradient methods.

5. Subgradient Method With a Constant Step-Size. Assume that we choose an arbitrary $x_0 \in E_n$ and $\lambda > 0$. Suppose x_k has already been constructed. Find an arbitrary $v_k \in \partial f(x_k)$ and define

$$x_{k+1} = x_k - \lambda v_k \| v_k \|^{-1} . \qquad (4.32)$$

If $v_k = 0$, the process terminates because x_k is a minimum point.

As a result, we obtain a sequence $\{x_k\}$. If this sequence is finite, then, by construction, its final element is a minimum point. Now, let the sequence $\{x_k\}$ be infinite and the set M^* be bounded. Define

$$r = 2^{-1}\lambda , \qquad S_r(M^*) = \{x \mid \rho(x, M^*) \le r\} ,$$

$$\max_{x \in S_r(M^*)} f(x) = f^* + c_r .$$

We note that the maximum here is attained on the boundary (see Lemma 9.7 in Chapter 1) and that $c_r > 0$,

$$D_{c_r} = \{x \mid f(x) \le f^* + c_r\} , \qquad r_1 = \max_{x \in D_{c_r}} \rho(x, M^*) .$$

<u>THEOREM 4.4</u>. If the sequence $\{x_k\}$ is infinite, then

$$\rho(x_k, S_{\lambda + r_1}(M^*)) \xrightarrow[k\to\infty]{} 0 ,$$

and, moreover,

$$\underline{\lim_{k\to\infty}} \ (x_k, D_{c_r}) = 0 \ . \tag{4.33}$$

Proof. First we establish (4.33). If we assume that (4.33) does not hold, then there exist an $a > 0$ and a $K_1 > \infty$ such that

$$\rho(x_k, D_{c_r}) \geq a \qquad \forall k \geq K_1 \ . \tag{4.34}$$

Here we note that there exist a $c' > c_r$ and an $r' > r$ such that

$$D_{c'} \equiv \{x \mid f(x) \leq f^* + c'\} \subset S_a(D_{c_r}) \ , \tag{4.35}$$

$$S_{r'}(M^*) \subset D_{c'} \ .$$

(Figure 50).

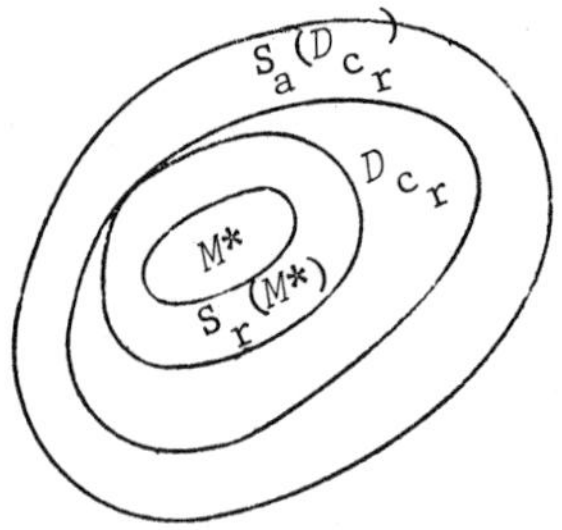

Figure 50

Take an arbitrary $\bar{x} \in M^*$. It is clear that

$$S_{r'}(\bar{x}) = \{x \mid \|x-\bar{x}\| \leq r'\} \subset D_{c'} \ . \tag{4.36}$$

From the definition of $\partial f(x_k)$, we have

$$f(x) - f(x_k) \geq (v_k, x-x_k) \qquad \forall x \in E_n \ . \tag{4.37}$$

Since it follows from (4.34) and (4.35) that $f(x_k) > f^* + c'$, then (4.37) implies

$$(v_k, x-x_k) < 0 \qquad \forall x \in D_{c'} \ , \tag{4.38}$$

for $k \geq K_1$. Let $\tilde{x}_k = \bar{x} + r' v_k \|v_k\|^{-1}$. Then it is clear that $\tilde{x}_k \in D_{c'}$ and, hence, from (4.38),

$$(-\lambda v_k \|v_k\|^{-1}, x_k - \tilde{x}_k) < 0 \ .$$

Therefore,

$$(-\lambda v_k\|v_k\|^{-1},\ x_k-\bar{x}) + (\lambda v_k\|v_k\|^{-1},\ r' v_k\|v_k\|^{-1}) < 0 ,$$

i.e.,

$$(-\lambda v_k\|v_k\|^{-1},\ x_k-\bar{x}) < -\lambda r' .$$

We obtain

$$\begin{aligned}\|x_{k+1} - \bar{x}\|^2 &= \|x_k - \lambda v_k\|v_k\|^{-1} - \bar{x}\|^2 \\ &= \|x_k-\bar{x}\|^2 + 2(-\lambda v_k\|v_k\|^{-1},\ x_k-\bar{x}) + \lambda^2 \\ &< \|x_k - \bar{x}\|^2 + \lambda[-2r' + \lambda] . \qquad (4.39)\end{aligned}$$

Since $r' > r = \frac{\lambda}{2}$, then it follows from (4.39) that $\|x_k-\bar{x}\| \to -\infty$, which is impossible. This contradiction proves (4.33).

Now, take an arbitrary $\delta > 0$. As above, there exist $K_\delta < \infty$ such that inequality (4.39) holds for $k > K_\delta$ which satisfy $\rho(x_k, D_{c_r}) > \delta$. From (4.33), there exist k such that $\rho(x_k, D_{c_r}) \le \delta$. Since $x_{k+1} = x_k - \lambda v_k\|v_k\|^{-1}$, then $\|x_{k+1} - x_k\| = \lambda$ and, hence, $x_{k+1} \in S_{\lambda+\delta}(D_{c_r})$. Here, if $\rho(x_{k+1}, D_{c_r}) > \delta$, then, by virtue of (4.39), we obtain

$$\rho(x_{k+2}, M^*) < \rho(x_{k+1}, M^*) \le \lambda + r_1 + \delta$$

for sufficiently large k (for $k \ge K_\delta$). If $\rho(x_{k+1}, D_{c_r}) \le \delta$, then we again obtain $\rho(x_{k+2}, D_{c_r}) \le \lambda + \delta$, i.e., $x_{k+2} \in S_{\lambda+\delta}(D_{c_r})$.

Let us continue analogously. Then, by the arbitrariness of $\delta > 0$, we conclude that

$$\rho(x_k,\ S_{\lambda+r_1}(M^*)) \xrightarrow[k\to\infty]{} 0 .$$

The theorem has been proved. ■

<u>REMARK 3</u>. From formula (5.26) in Chapter 1, it follows that

$$f(x+z) = f(x) + \int_0^1 \max_{v\in\partial f(x+\tau z)} (v,z)\, d\tau .$$

Hence, if $\|z\| \le r$, then $c_r \le rB + \frac{\lambda B}{2}$, where

$$B = \max_{x \in S_r(M^*)} \max_{v \in \partial f(x)} \|v\| .$$

It is clear that choosing an appropriate λ we can obtain an arbitrarily small c_r.

REMARK 4. From the practical point of view, it is convenient to use the following modification of the method described above.

Let $A > 0$ be fixed. Let us construct a sequence $\{x_k\}$ by the formula:

$$x_{k+1} = \begin{cases} x_k - \lambda v_k \|v_k\|^{-1} & \text{if } \|v_k\| > A , \\ x_k - \lambda v_k & \text{if } \|v_k\| \leq A . \end{cases} \tag{4.40}$$

For the case of smooth functions (see Section 2), this method assures the convergence of the sequence $\{x_k\}$ to a minimum point.

PROBLEM 5.1. Investigate the ε-subgradient method with a constant step-size (the sequence $\{x_k\}$ is constructed by (4.32) or (4.40) and there $v_k \in \partial_\varepsilon f(x_k)$).

6. We shall consider again the subgradient method introduced in Subsection 1. Let a sequence $\{\lambda_k\}$ satisfy (4.1) and, moreover,

$$\sum_{k=0}^{\infty} \lambda_k^2 = A < \infty . \tag{4.41}$$

THEOREM 4.5. If the set M^* is nonempty, then the sequence $\{x_k\}$ converges to a point in the set M^*.

Proof. Take an arbitrary $x^* \in M^*$. Similarly as in Subsection 1, we have

$$\|x_{k+1} - x^*\|^2 = \|x_k - x^*\|^2 + \frac{2\lambda_k}{\|v_k\|} (v_k, x^* - x_k) + \lambda_k^2 . \tag{4.42}$$

Since $v_k \in \partial f(x_k)$ and $x^* \in M^*$, then

$$0 \geq f(x^*) - f(x_k) \geq (v_k, x^* - x_k) . \tag{4.43}$$

Hence

$$(v_k, x^* - x_k) \leq 0 \qquad \forall k . \tag{4.44}$$

Considering this, we obtain from (4.42) that

$$\|x_k - x^*\|^2 \le \|x_0 - x^*\|^2 + \sum_{i=0}^{k-1} \lambda_i^2 .$$

Therefore, the boundedness of the sequence $\{x_k\}$ follows from (4.41). Indeed, the sequence $\{v_k\}$ is also bounded, i.e., there exists an $L < \infty$ such that

$$\|v_k\| \le L \qquad \forall k . \quad (4.45)$$

We show now that there exists a subsequence $\{x_{k_s}\}$ such that

$$a_{k_s} = (v_{k_s}, x^* - x_{k_s}) \to 0 . \quad (4.46)$$

Let us assume that this is not the case. Then, it follows from (4.44) that we can find a $b > 0$ and a $K < \infty$ such that

$$a_k \le -b \qquad \forall k > K . \quad (4.47)$$

From (4.42), (4.45), (4.47) and (4.1) we obtain $\|x_k - x^*\|^2 \to -\infty$, and this is impossible. Thus, there exists a subsequence $\{x_{k_s}\}$ which satisfies (4.46).

It follows from (4.43) and (4.46) that

$$f(x_{k_s}) \to f(x^*) . \quad (4.48)$$

Since the sequence $\{x_k\}$ is bounded, then, without loss of generality, we can assume that $x_{k_s} \to \bar{x}$. From (4.48), we can conclude that $\bar{x} \in M^*$. Now we show that the sequence $\{x_k\}$ converges to the point $\bar{x}$. In (4.42) and (4.44) we can take the point $\bar{x}$ instead of x^*.

For any $\varepsilon > 0$, there exists a $K_1 < \infty$ such that

$$\|x_{k_s} - \bar{x}\|^2 < \frac{\varepsilon}{2} , \qquad \sum_{i=k_s}^{\infty} \lambda_i^2 < \frac{\varepsilon}{2} \qquad \forall k_s > K_1 .$$

For such k_s, we obtain from (4.42) and (4.44)

$$\|x_{k_s+N} - \bar{x}\|^2 \leq \|x_{k_s} - \bar{x}\|^2 + \sum_{i=k_s}^{k_s+N-1} \lambda_i^2 < \varepsilon \qquad \forall N \geq 1 ,$$

and this implies that $x_k \to \bar{x}$.

REMARK 5. Thus, under the additional condition (4.41), the sequence $\{x_k\}$ constructed in the subgradient method has a unique accumulation point and that point belongs to the set M^*: the set M^* can be unbounded. As a sequence $\{\lambda_k\}$ satisfying (4.1) and (4.41), for example, we can take the sequence $\{\frac{1}{k}\}$. Condition (4.41) can be employed also for the methods in Subsections 2 and 3.

5. THE MULTISTEP SUBGRADIENT METHOD

For the minimization of smooth functions, multistep methods are extremely efficient. To find the successive $(k+1)^{st}$ approximation these methods utilize the information about the value of the function and its derivatives not only at the kth step but also at one or several previous steps. For example, the conjugate-gradient method belongs to such methods. We can expect that multistep methods will also be efficient for solving nonsmooth problems. In problems for smooth functions, gradients at close points are not very different from each other, by virtue of the continuous differentiability of the function; however, in problems for convex nonsmooth functions it is difficult to evaluate on the basis of some subgradient at a certain point the entire subdifferential, even at the same point. Hence, knowing the subgradients at some neighboring points may give more information about the subdifferential and, therefore, about directions of descent. We proceed to describe the multistep method.

1. Let $f(x)$ be a finite convex function on E_n. In Section 1.5 we showed that

$$f(x) = \max_{z \in E_n} [f(z) + (v(z), x-z)] , \quad (5.1)$$

where $v(z)$ is an arbitrary fixed element of $\partial f(z)$. Let the p-tuple of the points $\tau = \{y_1,\ldots,y_p\}$, $y_i \in E_n$ be given, where $p \ge 1$.

Define

$$\phi(x) = \max_{y \in \tau} [f(y) + (v(y), x-y)] .$$

Then,

$$\phi(x_0+z) = \max_{y \in \tau} [f(y) + (v(y), x_0-y) + (v(y), z)] .$$

Find the value

$$\min_{\|z\|=\alpha} \phi(x_0+z) = \phi(x_0 + z_0(\alpha)) .$$

The curve $\{x \mid x=x_0+z_0(\alpha), \alpha \ge 0\}$ is said to be an *approximating curve* of the function $f(x)$ constructed at the point x_0 with respect to the set τ. If $p=1$ and $\tau = \{x_0\}$, then the approximating curve is the ray $\{x \mid x=x_0-\alpha v(x_0), \alpha \ge 0\}$. We introduce the following method of successive approximation. Let us choose a number $N > n+2$ and a nonincreasing sequence of numbers $\{\lambda_k\}$ such that

$$\lambda_k \to +0 , \quad \sum_{k=0}^{\infty} \lambda_k = +\infty . \quad (5.2)$$

Take an arbitrary point $x_0 \in E_n$ and fix an arbitrary $A > f(x_0)$.

Let $\sigma_0 = \{x_0\}$. Suppose that the point $x_k \in E_n$ and the set σ_k have already been found, where

$$\sigma_k = \{x_{j_{1k}}, \ldots, x_{j_{s_k k}}\} , \quad j_{1k} < j_{2k} < \cdots < j_{s_k k} ,$$

$$j_{1k} \ge k - 1 - N , \quad |\sigma_k| = s_k \le n+1 . \quad (5.3)$$

Here, we recall that $|\sigma|$ denotes the number of elements of the set σ.

Take an arbitrary $v(x_k) \in \partial f(x_k)$. If $v(x_k) = 0$, then x_k is a minimum point of the function $f(x)$ on E_n and the process terminates.

Now we consider the case $v(x_k) \neq 0$. If $f(x_k) > A$, then we let

$$\sigma_k' = \{x_k\} \tag{5.4}$$

and if $f(x_k) \leq A$, we take

$$\sigma_k' = \sigma_k \cup \{x_k\} . \tag{5.5}$$

By virtue of (5.3),

$$|\sigma_k'| \leq n + 2 . \tag{5.6}$$

Here we find the value

$$\min_{\|z\| \leq \lambda_k} \phi_k(z) = \phi_k(z_k) , \tag{5.7}$$

where

$$\phi_k(z) = \max_{y \in \sigma_k'} B(x_k+z, y) , \tag{5.8}$$

$$B(x,y) = f(y) + (x-y, v(y)) .$$

We note that in the case (5.4) we have

$$z_k = -\lambda_k v_k \|v_k\|^{-1} , \tag{5.9}$$

where $v_k = v(x_k)$. Introduce the sets

$$R_k = \{y \in \sigma_k' \mid B(x_k+z_k, y) = \phi_k(z_k)\} , \qquad L_k = \text{co} \bigcup_{y \in R_k} v(y) .$$

Since $R_k \subset \sigma_k'$, then from (5.6) we have $|R_k| \leq n+2$. Due to the necessary and sufficient condition for a minimum of the function $\phi_k(z)$ on the set $\{z \mid \|z\|^2 \leq \lambda_k^2\}$ (see Section 1.6), we obtain:

1. If $\|z_k\| < \lambda_k$, then

$$0 \in L_k . \tag{5.10}$$

If $z_k = 0$, then x_k is a minimum point and the process terminates.

2. If $\|z_k\| = \lambda_k$, then there exists a $\gamma_k \geq 0$ such that

$$-2\gamma_k z_k \in L_k \quad . \tag{5.11}$$

By Carathéodory's theorem (Theorem 1.1 in Chapter 1), there exists a set $R_k' \subset R_k$ such that $||R_k'|| \le n+1$ and

$$0 \in \text{co} \bigcup_{y \in R_k'} v(y) \equiv L_k' \qquad \text{in the case (5.10)} ,$$

$$-2\gamma_k z_k \in L_k' \qquad \text{in the case (5.11)}.$$

In particular, if $|R_k| \le n+2$, then we can take $R_k' = R_k$.

Here, define

$$x_{k+1} = x_k + z_k \ ,$$

$$\sigma_{k+1} = \begin{cases} R_k' & \text{if } x_{k-N-1} \notin R_k' \ , \\ R_k' \setminus \{x_{k-N-1}\} & \text{if } x_{k-n-1} \in R_k' \ . \end{cases}$$

Thus, the basis σ_{k+1} is some subset of the set of points $\{x_k, \ldots, x_{k-N}\}$.

Let

$$f^* = \inf_{x \in E_n} f(x) \ , \qquad M^* = \{z \in E_n \mid f(z) = f^*\} \ ,$$

$$\rho(x, M^*) = \min_{z \in M^*} ||z-x|| \ .$$

<u>THEOREM 5.1</u>. If the set M^* is nonempty and bounded, then

$$\rho(x_k, M^*) \to 0 \ , \qquad f(x_k) \to f^* \ . \tag{5.12}$$

<u>Proof</u>. From the boundedness of M^* we obtain the boundedness of the set

$$D_{A_1} = \{x \in E_n \mid f(x) \le f^* + A_1\} \ , \qquad \text{where } A_1 = A - f^*$$

(see Section 1.9). Moreover (see Lemmas 9.1 and 9.2 in Chapter 1), there exist a $b > 0$ and a $B < \infty$ such that

$$\begin{aligned} ||v|| &\ge b & \forall v \in \partial f(x), \quad x \notin D_{A_1} \ , \\ ||v|| &= B & \forall v \in \partial f(x), \quad x \in D_{A_1} \ . \end{aligned} \tag{5.13}$$

First we shall prove that

$$\lim_{k\to\infty} \rho(x_k, M^*) = 0 . \tag{5.14}$$

Here we assume that this is not the case and, next, prove that there exists a $d > 0$ such that

$$\min_{v\in L_k} \|v\| = \|v_k\| \geq d > 0 \qquad \forall k . \tag{5.15}$$

To do this, we assume that this is not the case. Then, there exists a subsequence of indices $\{k_s\}$ such that

$$\min_{v\in L_{k_s}} \|v\| = \|v_{k_s}\| \xrightarrow[s\to\infty]{} 0 , \tag{5.16}$$

where $v_{k_s} \in L_{k_s}$.

Then, without loss of generality, we can assume that $x_{k_s} \to x^* \in D_{A_1}$ because if this is not the case, then, by virtue of (5.13), we obtain a contradiction to (5.16).

Since $|R'_{k_s}| \leq n + 2$, then we can also assume that for all k_s

$$|R'_{k_s}| = m , \qquad v_{k_s} = \sum_{i\in 1:m} \alpha_{k_s i} v_{k_s i}$$

and

$$v_{k_s i} \to v_i^* , \qquad y_{k_s i} \to y_i^* , \qquad \alpha_{k_s i} \to \alpha_i^* \qquad \forall i \in 1:m ,$$

where $m \leq N + 2$, $\sum_{i\in 1:m} \alpha_{k_s i} = 1$.

It follows from the upper semicontinuity of the mapping $\partial f(x)$ that $v_i^* \in \partial f(y_i^*)$. Since $|R'_{k_s}| = m$, then

$$\phi_{k_s}(z_{k_s}) \equiv \max_{y\in\sigma'_k} B(x_{k_s+1}, y) = f(x_{k_s}) + (v_{k_s i}, x_{k_s} - y_{k_s i}) + (v_{k_s i}, z_{k_s})$$

$$\forall i \in 1:m . \tag{5.17}$$

Since $x_{k_s} \in \sigma'_{k_s}$, then it follows from the definition of $\phi_{k_s}(z)$ that

$$\phi_{k_s}(x_{k_s+1}) \geq f(x_{k_s}) + (v(x_{k_s}), z_{k_s}) . \tag{5.18}$$

Moreover,

$$f(x_{k_s}+z_{k_s}) \geq \phi_{k_s}(z_{k_s}) , \qquad (5.19)$$

because

$$f(x+z) = \max_{y\in E_n} B(x+z, y) \geq \max_{y\in\sigma_k} B(x+z, y) = \phi_k(x+z) \quad \forall x,z \in E_n , \quad \forall k .$$

Taking the limit in (5.17) - (5.19) as $k_s \to \infty$, we obtain

$$f(x^*) \geq \max_{i\in 1:m} [f(y_i^*) + (v_i^*, x^* - y_i^*)] \leq f(x^*) ,$$

i.e.,

$$v(x^*) = \max_{i\in 1:m} [f(y_i^*) + (v_i^*, x^* - y_i^*)] . \qquad (5.20)$$

Here, it follows from (5.16) that

$$0 \in \text{co}\,\{v_i^* \mid i\in 1:m\} , \qquad (5.21)$$

and from (5.17) that

$$f(x^*) = f(y_i^*) + (v_i^*, x^* - y_i^*) \quad \forall i \in 1:m . \qquad (5.22)$$

Relations (5.21) and (5.22) imply that the function

$$\phi^*(x) = \max_{i\in 1:m} [f(y_i^*) + (v_i^*, x - y_i^*)]$$

attains its minimum value on E_n at the point x^* and the minimum value is equal to $f(x^*)$ by virtue of (5.20). On the other hand,

$$f(x) = \max_{x\in E_n} [f(y) + (v(y),x-y)] \geq \phi^*(x) ,$$

where $v(y) \in \partial f(x)$. Hence,

$$f^* \equiv \inf_{x\in E_n} f(x) \geq \min_{x\in E_n} \phi^*(x) = f(x^*) ,$$

which implies (5.14).

Consequently, if there is a subsequence of indices $\{k_s\}$ which satisfies (5.16), then (5.14) holds as well.

Thus, for some $d > 0$, relation (5.15) holds. By hypothesis, (5.14) does not hold, i.e., there exists an $a > 0$ such that

$$\rho(x_k, M^*) \geq 2a \qquad \forall k \tag{5.23}$$

From (5.11) and (5.15), it follows that

$$z_k = -\alpha_k \sum_{y \in R_k} \alpha'_k(y) v(y) , \qquad \sum_{y \in R_k} \alpha'_k(y) = 1 ,$$
$$\alpha_k > 0 , \qquad \alpha'_k(y) \geq 0 , \qquad y \in R_k . \tag{5.24}$$

Since $v_k \in L_k$, $\|v_k\| \geq d$ and $\|z_k\| = \lambda_k$, then $\alpha_k \leq d^{-1}\lambda_k$.

Hence,

$$\max_{y \in R_k} \alpha_k(y) \xrightarrow[k\to\infty]{} 0 , \tag{5.25}$$

where $\alpha_k(y) = \alpha_k \alpha'_k(y)$.

By virtue of the boundedness of M^*, there exists a $c > 0$ such that

$$D_c = \{x \mid f(x) \leq f^*+c\} \subset S_a(M^*) ,$$

where

$$S_a(M^*) = \{x \mid \rho(x, M^*) \leq a\} .$$

Take an arbitrary $\bar{x} \in M^*$. It is clear that $\bar{x} \in \operatorname{int} D_c$, i.e., there exists an $r > 0$ such that

$$S_r(\bar{x}) \subset D_c .$$

From the definition of $\partial f(x_k)$ we have

$$f(x) - f(x_k) \geq (v(x_k), x-x_k) \qquad \forall x \in E_n . \tag{5.26}$$

Since, by virtue of (5.23) we have $f(x_k) > f^* + c$, then from (5.26)

$$(v(x_k), x-x_k) < 0 \qquad \forall x \in D_c . \tag{5.27}$$

If $f(x_k) > A$, then, by construction, $z_k = -\lambda_k v(x_k) \cdot \|v(x_k)\|^{-1}$ and, hence, from (5.27)

$$(z_k, x_k-x) < 0 \qquad \forall x \in D_c . \tag{5.28}$$

Now we consider the case $f(x_k) \leq A$. Take a $y \in \sigma'_k$. From (5.13) we have $\|v(y)\| \leq B$. There exists an $i \geq k - N$ such that $y = x_i$.

Since

$$(v(y),\ y-x_k) = (v(y),\ x_i - x_k)$$
$$= (v(y),\ (x_i-x_{i+1}) + (x_{i+1}-x_{i+2}) + \cdots + (x_{k-1}-x_k))\ ,$$

then

$$(v(y),\ y-x_k) \le B \sum_{j=i+1}^{k} \|x_{j-1} - x_j\| \le B\,N\,\lambda_{kN}\ ,$$

where $\lambda_{kN} = \max_{i\in(k-N):k} \lambda_i$. We note that $\lambda_{kN} \to 0$. From this and (5.27), we have for $y \in \sigma_k'$

$$(v(y),\ x-x_k) = (v(y),\ x-y) + (v(y),\ y-x_k) < b\,N\,\lambda_{kN} \qquad \forall x \in D_c\ . \tag{5.29}$$

From (5.24) and (5.29),

$$(z_k,\ x_k-x) = \sum_{y\in\sigma_k} \alpha_k(y)(v(y),\ x-x_k) < B\,N\,\lambda_{kN}\alpha_k \equiv \lambda_{kN}\gamma_k \qquad \forall x \in D_c\ . \tag{5.30}$$

Here, $\gamma_k = BN\alpha_k$. From (5.25), we obtain $\gamma_k \to 0$. The inequality (5.30) has been established in the case $f(x_k) \le A$. In the case $f(x_k) > A$, (5.28) holds. Hence, we can assume that inequality (5.30) holds for all k.

Take an $x = \bar{x} - rz_k\|z_k\|^{-1}$. Then, from (5.30)

$$(z_k,\ x_k - \bar{x} + rz_k\|z_k\|^{-1}) < \gamma_k\lambda_{kN}\ ,$$

i.e.,

$$(z_k,\ x_k - \bar{x}) < [-r\lambda_k + \gamma_k\lambda_{kN}]\ .$$

On the other hand,

$$\|x_{k+1}-\bar{x}\|^2 = \|x_k + z_k - \bar{x}\|^2 = \|x_k - \bar{x}\|^2 + 2(z_k,\ x_k-\bar{x}) + z_k^2$$
$$< \|x_k-\bar{x}\|^2 + [-2r\lambda_k + 2_k\gamma_{kN}] + \lambda_k^2\ .$$

Since $\lambda_k \to +0$ and $\gamma_k \to 0$, then for sufficiently large k we get

$$\|x_{k+1} - \bar{x}\|^2 < \|x_k - \bar{x}\|^2 - r\lambda_k\ .$$

Continuing analogously yields

$$\|x_{k+s} - \bar{x}\|^2 < \|x_k - \bar{x}\|^2 - r\sum_{i=0}^{s-1}\lambda_{k+i} .$$

By virtue of (5.2), $\|x_k - \bar{x}\| \to -\infty$, which is impossible. Thus, (5.14) has been established. Take an arbitrary $\delta > 0$. Arguing as above, we conclude that there exists an $r_\delta > 0$ such that for all $\bar{x} \in M^*$, the inequality

$$\|x_{k+1} - \bar{x}\|^2 < \|x_k - \bar{x}\|^2 - r_\delta\lambda_k$$

holds for all sufficiently large k satisfying $\rho(x_k, M^*) \geq \delta$.

Hence,

$$\rho^2(x_{k+1}, M^*) < \rho^2(x_k, M^*) - r_\delta\lambda_k . \tag{5.31}$$

By virtue of (5.14), there exists a k such that $\rho(x_k, M^*) < \delta$. Since $x_{k+1} = x_k + z_k$, $\|z_k\| \leq \lambda_k$, then we can assume that $\rho(x_{k+1}, M^*) < 2\delta$. If $\rho(x_{k+1}, M^*) \geq \delta$, then, by virtue of (5.31),

$$\rho(x_{k+2}, M^*) < \rho(x_{k+1}, M^*) < 2\delta .$$

Let us continue in a similar way. Thus, if $\rho(x_{k+i}, M^*) < \delta$, then $\rho(x_{k+i+1}, M^*) < 2\delta$; if $\delta \leq \rho(x_{k+i}, M^*) < 2\delta$, then

$$\rho(x_{k+i+1}, M^*) < \rho(x_{k+i}, M^*) < 2\delta .$$

Consequently, we obtain that $\rho(x_{k+i}, M^*) < 2\delta$ for $i > 1$. Hence by the arbitrariness of δ, we conclude that $\rho(x_k, M^*) \to 0$, and then $f(x_k) \to f^*$, which completes the proof of the theorem. ∎

<u>COROLLARY</u>. As in the proof of Corollary of Theorem 4.1, it is possible to show that if $\operatorname{int} M^* \neq \emptyset$, then the sequence $\{x_k\}$ is finite

<u>REMARK</u>. The problem (5.7) is a linear programming problem if we take in (5.7) the m-norm: $\|z\| = \max_{i \in 1:n} |z^{(i)}|$ instead of the Euclidean norm. Here, $z = (z^{(1)}, \ldots, z^{(n)})$.

2. Multistep ε-Subgradient Method. As in Subsection 4.2, in the method described above we can choose $v(x_k) \in \partial_\varepsilon f(x_k)$, where $\varepsilon > 0$ is fixed. As a result, we obtain a sequence $\{x_k\}$. If this sequence is finite, then, by construction, its final point is ε-stationary because the process terminates only if $v(x_k) = 0$, i.e., $0 \in \partial_\varepsilon f(x_k)$.

If the sequence $\{x_k\}$ is infinite, then the following theorem holds.

THEOREM 5.2. If the set M^* is nonempty and bounded, then

$$\underline{\lim}_{k\to\infty} \rho(x_k, D_\varepsilon) \to 0 , \qquad \rho(x_k, S_{a_\varepsilon}(M^*)) \xrightarrow[k\to\infty]{} 0 ,$$

where

$$D_\varepsilon = \{x \in E_n \mid f(x) \le f^*+\varepsilon\} ,$$

$$a_\varepsilon = \max_{x\in Q_\varepsilon} \rho(x, M^*) ,$$

$$Q_\varepsilon = \{x \mid f(x) = f^*+\varepsilon\} .$$

Proof. Is analogous to that of Theorem 4.2.

6. THE RELAXATION SUBGRADIENT METHOD

The method stated in this section is methodologically close, on the one hand, to the method of ε-steepest descent for the minimization of an arbitrary convex function $f(x)$ and, on the other hand, to the conjugate-gradient method for the minimization of a continuously differentiable function.

In the method of ε-steepest descent, it is necessary to know the whole set $\partial_\varepsilon f(x)$ in order to find a direction of descent; however, in the relaxation-subgradient method, some finite number of points in $\partial_\varepsilon f(x)$ approximating this set are used for constructing a direction of descent.

1. The Conceptual Algorithm. We suppose that a function $f(x)$ is strictly convex on E_n and that the value of the function $f(x)$ and its subgradient can be calculated and a minimum of the function $f(x)$ on the ray can be found.

Let $d(x) = \min_{v \in \partial f(x)} \|v\|$. First, we introduce the relaxation-subgradient method for finding points $x_0 \in E_n$ such that $d(x_0) \le \varepsilon$, where $\varepsilon > 0$, and next we show how to apply it for finding minimum points of the function $f(x)$ on E_n.

Let us fix positive numbers ε_0, δ and an integer $m_0 \ge n$.

Take an arbitrary initial approximation $x_0 \in E_n$. Assume that the set $D(x_0) = \{x \in E_n \mid f(x) \le f(x_0)\}$ is bounded.

Suppose the kth approximation $x_k \in E_n$ has already been found. The transition to the $(k+1)^{st}$ approximation x_{k+1}, which is called the kth cycle, is realized in the following way.

Let $x_{k0} = x_k$ and compute an arbitrary vector $v_{k0} \in \partial f(x_{k0})$. If $\|\bar{v}_{k,-1}\| \le \varepsilon_0$, where $\bar{v}_{k,-1} = v_{k0}$, then we let $t_k = 0$, $x_{k+1} = x_{kt_k}$, and the kth cycle terminates (we denote by t_k the number of steps in the kth cycle). If $\|\bar{v}_{k,-1}\| > \varepsilon_0$, then let us find on the ray $\{x_{k0}(\alpha) = x_{k0} - \alpha v_{k0} \mid \alpha \ge 0\}$ a point $x_{k1} = x_{k0}(\alpha_{k0})$ such that

$$f(x_{k1}) = \min_{\alpha \ge 0} f(x_{k0}(\alpha)) .$$

We shall pay attention to the fact that the case where $x_{k+1} = k_{x0}$ is possible. This occurs if $-v_{k0}$ is not a direction of descent of $f(x)$ at x_{k0}.

If $\|x_{k1} - x_{k0}\| \ge \delta$ or $f(x_{k0} - f(x_{k1}) \ge \delta$, then we let $t_k = 1$, $x_{k+1} = x_{kt_k}$ and the kth cycle terminates. Otherwise, let us find a vector $\bar{v}_{k0} \in \partial f(x_{k1})$ such that

$$(v_{k0}, \bar{v}_{k0}) \le 0 .$$

Such a vector $\bar{v}_{k0} \in \partial f(x_{k1})$ can always be found, because, due to the necessary condition for a minimum of the function $f(x_{k0}(\alpha))$ with respect to α, we should have

$$\frac{\partial f(x_{k1})}{\partial f(-v_{k0})} = \max_{v \in \partial f(x_{k1})} (v, -v_{k0}) \geq 0 .$$

If $\|\bar{v}_{k0}\| \leq \varepsilon_0$, then we let $t_k = 1$, $x_{k+1} = x_{kt_k}$ and the kth cycle terminates. If $\|\bar{v}_{k0}\| > \varepsilon_0$, then let us find a vector v_{k1} such that

$$v_{k1} = \beta_{k0}\bar{v}_{k0} + (1 - \beta_{k0})v_{k0} , \qquad \beta_{k0} \in (0,1) ,$$

and

$$\|v_{k1}\| = \min_{\beta \in [0,1]} \|\beta\bar{v}_{k0} + (1-\beta)v_{k0}\| ,$$

i.e., v_{k1} is the nearest point to the origin in the segment $[v_{k0}, \bar{v}_{k0}]$.

It is easy to see that

$$v_{k1} = v_{k0} - (v_{k0}, \bar{v}_{k0} - v_{k0}) \|\bar{v}_{k0} - v_{k0}\|^{-2}(\bar{v}_{k0} - v_{k0}) .$$

Now, let us find on the ray $\{x_{k1}(\alpha) = x_{k1} - \alpha v_{k1} \mid \alpha \geq 0\}$ a point $x_{k2} = x_{k1}(\alpha_{k1})$ such that

$$f(x_{k2}) = \min_{\alpha \geq 0} f(x_{k1}(\alpha)) .$$

We continue analogously until in a number of steps t_k at least one of the following conditions is satisfied:

(a) $t_k = m_0$;

(b) $\|x_{k,t_k-1} - x_{kt_k}\| \geq \delta$; (6.1)

(c) $f(x_{k,t_k-1}) - f(x_{kt_k}) \geq \delta$; (6.2)

(d) $\|\bar{v}_{k,t_k-1}\| \leq \varepsilon_0$.

Now, we take a point x_{kt_k} as the $(k+1)^{st}$ approximation x_{k+1}. The kth cycle terminates and, moreover, if condition (d) is satisfied, then the process of successive approximations terminates because $d(x_{k+1}) \leq \varepsilon_0$.

We note that, by construction,

$$(v_{kt}, \bar{v}_{kt}) \leq 0 \qquad \forall t \in 0:\ell_k , \tag{6.3}$$

where $\ell_k = t_k - 1$.

The proof of convergence of the relaxation-subgradient method is based on inequality (6.3) and the following lemma.

LEMMA 6.1. Let a function $f(x)$ be strictly convex on E_n, the set $D(x_0)$ be bounded and sequence $\{x_k\}_{k=0}^{\infty}$ be such that

$$f(x_{k+1}) = \min_{\alpha \in [0,1]} f(x_k + \alpha(x_{k+1} - x_k)) . \tag{6.4}$$

Then

$$\lim_{k\to\infty} \|x_{k+1} - x_k\| = 0 .$$

Proof. Assume that this is not the case. Then, there exist a number $h > 0$ and a sequence $\{x_{k_s}\}$ such that

$$\|x_{k_s+1} - x_{k_s}\| \geq h \qquad \forall s .$$

Since $f(x_{k+1}) \leq f(x_k)$, then $x_k \in D(x_0)$ for each k. By virtue of the boundedness of $D(x_0)$, without loss of generality, it is possible to assume that $x_{k_s} \to \bar{x}$, $x_{k_s+1} \to \bar{\bar{x}}$ and, here, $\|\bar{x} - \bar{\bar{x}}\| \geq h$.

Since $f(x_{k+1}) \leq f(x_k)$ and $f(x)$ is bounded from below we have

$$f(\bar{x}) = f(\bar{\bar{x}}) . \tag{6.5}$$

It follows from the fact that $f(x_{k_s}) \to \bar{f} = f(\bar{x})$, $f(x_{k_s+1}) \to \bar{f} = f(\bar{\bar{x}})$ (since $\{f(x_k)\}$ is a decreasing sequence bounded from below and therefore there exists $\lim \{f(x_k)\}$, and all partial limits must be the same).

The function $f(x)$ is convex, therefore

$$f(2^{-1}\bar{x} + 2^{-1}\bar{\bar{x}}) < f(\bar{x}) .$$

By virtue of (6.4), we have

$$f(x_{k_s+1}) = \min_{\alpha\in[0,1]} f(x_{k_s} + \alpha(x_{k_s+1} - x_{k_s})) \le f(2^{-1}x_{k_s} + 2^{-1}x_{k_s+1}) .$$

Taking the limit in this inequality, we obtain

$$f(\bar{\bar{x}}) \le f(2^{-1}\bar{x} + 2^{-1}\bar{\bar{x}}) < f(\bar{x}) ,$$

and this contradicts (6.5). The lemma has been proved. ∎

We note that Lemma 6.1 also holds for a continuous strictly quasiconvex function $f(x)$.

Now, we shall go into the problem of convergence of the relaxation-subgradient method. If the sequence $\{x_k\}$ is finite, then, by construction, we have $d(x_k) \le \varepsilon_0$ for the obtained point x_k.

Otherwise, the following theorem holds.

THEOREM 6.1. Let a function $f(x)$ be strictly convex on E_n and the set $D(x_0)$ be bounded. Then, if x^* is an accumulation point of the sequence $\{x_k\}$, we have

$$d(x^*) \le \max\{\varepsilon_0, bq^{m_0-1}\} \equiv d_0 ,$$

where

$$b = \max_{x\in D(x_0)} \max_{v\in\partial f(x)} \|v\| , \qquad q = [1 + (\varepsilon_0 b^{-1})^2]^{-\frac{1}{2}} .$$

In particular, if $m_0 - 1 > [\ln q]^{-1} \ln(\varepsilon_0 b^{-1})$, then $d(x^*) \le \varepsilon_0$.

Proof. The existence of accumulation points of the sequence $\{x_k\}$ follows from the boundedness of $D(x_0)$ and the fact that $x_k \in D(x_0)$. Suppose that the assertion of the theorem is not true, i.e., assume that $x_{k_s} \to x^*$ but $d(x^*) = \bar{d} > d_0$.

Since the function $f(x)$ is strictly convex, the set $D(x_0)$ is bounded and (6.4) is satisfied at each step, then, due to Lemma 6.1, for sufficiently large k the termination of the k-step is not possible because of (6.1), or due to the uniform continuity of the function $f(x)$ on $D(x_0)$ and because of (6.2). Hence, for sufficiently large k, say for $k \ge K$, we have

$$t_k = m_0 . \tag{6.6}$$

Let $\varepsilon = 2^{-1}(\bar{d} - d_0)$. Since $d(x^*) = \bar{d} > d_0$, then, by virtue of the upper semicontinuity of the point-to-set mapping $\partial f(x)$, there exists a $\delta > 0$ such that

$$\partial f(x) \subset S_\varepsilon(\partial f(x^*)) \qquad \forall x \in S_\delta(x^*) , \qquad (6.7)$$

i.e., for x sufficiently close to x^*, all the sets $\partial f(x)$ are included in some convex neighborhood of the set $\partial f(x^*)$ and the distance from the origin to this neighborhood is greater than d_0 (in Figure 51, the neighborhood is shown by a dotted line).

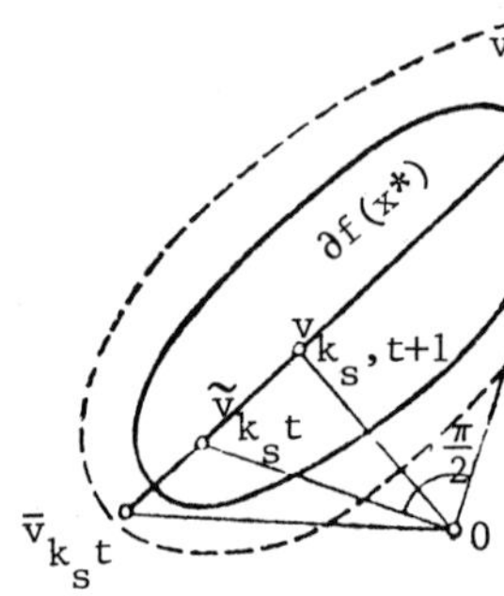

Figure 51

Since the neighborhood is convex, then, together with arbitrary two points in this neighborhood, the whole segment joining these points clearly belongs to this neighborhood.

Thus, it follows from (6.6) and (6.7) that for sufficiently large k_s we have

$$\|v_{k_s t}\| > d_0 \geq \varepsilon_0 \qquad \forall t \in 0:(m_0-1) . \qquad (6.8)$$

From (6.8) and the definition of the number b, we obtain

$$\|v_{k_s, t+1}\| = \|v_{k_s t}\| \frac{\|\tilde{v}_{k_s t}\|}{\sqrt{\|\tilde{v}_{k_s t}\|^2 + \|v_{k_s t}\|^2}} \leq \|v_{k_s t}\| \left\{\sqrt{1 + (\varepsilon_0 b^{-1})^2}\right\}^{-1}$$

(see Figure 51), where $\tilde{v}_{k_s t}$ is a point lying on the segment

$[\bar{v}_{k_s t}, v_{k_s t}]$ such that the angle between $\tilde{v}_{k_s t}$ and $v_{k_s t}$ is equal to $\pi/2$.

Hence,

$$\|v_{k_s, t+1}\| \le \|v_{k_s 0}\| q^{t+1} \le bq^{t+1} \qquad \forall t \in 0:(m_0-1) .$$

From this and (6.6), we conclude that

$$\|v_{k_s, m_0-1}\| = \|v_{k_s \ell_{k_s}}\| \le bq^{m_0-1} ; \qquad (6.9)$$

in particular, if $m_0 > m$, then $\|v_{k_s, m_0-1}\| \le \varepsilon_0$, where $m = 1 + \ln(\varepsilon_0 b^{-1})(\ln q)^{-1}$. However, (6.9) contradicts (6.8). The theorem has been proved. ■

<u>REMARK 1</u>. Conditions (6.1) and (6.2) are verified in order to avoid excessive calculations at the start of the process of successive approximations.

<u>REMARK 2</u>. If it turns out that $x_{kt} = x_{k0}$ for all $t \in 1:t_k$ and $t_k = m_0$, then, as follows from the proof of Theorem 6.1, $d(x_k) \le d_0$, and in this case the process of successive approximation can be terminated.

The kth cycle is called *complete* if $t_k = m_0$. It follows from Lemma 6.1 that, for sufficiently large k, $t_k = m_0$, i.e., all the cycles are complete and, consequently, within the kth cycle one constructs points $x_{k0}, x_{k1}, \ldots, x_{km_0}$, subgradients $\bar{v}_{k,-1}, \bar{v}_{k0}, \ldots, \bar{v}_{k,m_0-2}$; $\bar{v}_{kt} \in \partial f(x_{k,t+1})$ and vectors $v_{k1}, v_{k2}, \ldots, v_{k,m_0-1}$ such that for each $t \in 0:(m_0-2)$, the vector $v_{k,t+1}$ is the point in the segment $[v_{kt}, \bar{v}_{kt}]$ nearest to the origin; here, $(v_{kt}, \bar{v}_{kt}) \le 0$ and for each $t \in 0:(m_0-1)$, the equality $f(x_{k,t+1}) = \min_{\alpha \ge 0} f(x_{kt}(\alpha))$ should hold.

Using the relaxation-subgradient method, we construct a sequence $\{x_k\}$. For its arbitrary accumulation point x^*, due to

Lemma 6.1, we have $d(x^*) \le d_0$ and, due to Lemma 1.2,

$$0 \le f(x^*) - f^* \le Dd_0 , \tag{6.10}$$

where D is the diameter of the set $D(x_0)$ and $f^* = \inf_{x \in E_n} f(x)$.

It follows from (6.10) that for any $\bar{d} > d_0$ after a finite number of steps we get a point x_k such that

$$0 \le f(x_k) - f^* \le D\bar{d} .$$

This method can also be used for finding minimum points of a function $f(x)$ on E_n. In that case, conditions (a) and (d) are to be replaced by

(a) $t_k = m_k$,

(d) $\|v_{kt_k}\| \le \varepsilon_k$,

and, here, the transition to the $(k+1)^{st}$ cycle takes place after any of conditions (a) - (d) is satisfied.

THEOREM 6.2. Let a function $f(x)$ be strictly convex, the set $D(x_0)$ be bounded and $\varepsilon_k \to 0$, $m_k \to \infty$.

Then, any accumulation point of the sequence $\{x_k\}$ is a minimum point of the function $f(x)$ on E_n.

Proof. Suppose that this is not the case, i.e., assume that $x_{k_s} \to x^*$ and that, for some $d_0 > 0$, $d(x^*) > d_0$.

Since the mapping $\partial f(x)$ is upper semicontinuous, then for sufficiently large k_s all the sets $\partial f(x_{k_s})$ are included in some convex neighborhood of the set $\partial f(x^*)$, which is distant from the origin by more than d_0.

Now we consider k_s such that $\varepsilon_k < d_0$, $t_{k_s} = m_{k_s}$ (see the analogous part in the proof of Theorem 6.1) and

$$m_{k_s} > 1 + (\ln q)^{-1} \ln (d_0 b^{-1}) ,$$

where

$$b = \max_{x \in D(x_0)} \max_{v \in \partial f(x)} \|v\| , \qquad q = (1 + d_0 b^{-1})^{-\frac{1}{2}} .$$

Then, for sufficiently large k_s, on the one hand, we have

$$\|v_{k_s t}\| > d_0 \qquad \forall t \in 0:(m_{k_s}-1) ,$$

and, on the other hand, we have

$$\|v_{k_s, m_{k_s}-1}\| \le bq^{m_{k_s}-1} \le d_0$$

(see the proof of Theorem 6.1). Thus, we obtain a contradiction.

The theorem has been proved. ■

2. The Relationship Between the Relaxation-Subgradient Method and the Conjugate-Gradient Method. We suppose that the cycles of the relaxation-subgradient method are of the same length, $(m \ge n)$, i.e., conditions (b), (c), (d) are not verified (this occurs if $\varepsilon_0 = 0$ and $\delta = \infty$).

The relaxation-subgradient method can be expressed in concise form:

$$x_0 \in E_n , \qquad v_0 \in \partial f(x_0) , \qquad \beta_0 = 1 , \qquad p_0 = v_0 ,$$

$$x_{k+1} = x_k - \alpha_k p_k , \qquad f(x_{k+1}) = \min_{\alpha \ge 0} f(x_k - \alpha p_k) ,$$

$$p_k = p_{k-1} + \beta_k (v_k - p_{k-1}) , \qquad v_k \in \partial f(x_k) ,$$

$$(v_k, p_{k-1}) \le 0 , \tag{6.11}$$

$$\beta_k = \begin{cases} 1 & \text{if } k \text{ is a multiple of } m, \\ -(p_{k-1}, v_k - p_{k-1}) \| v_k - p_{k-1} \|^{-2} & \text{otherwise} . \end{cases}$$

It is easy to see that such a description of the relaxation-subgradient method is formally similar to the conjugate-gradient method.

In fact, these two methods are closely related, i.e., the

relaxation-subgradient method becomes the conjugate-gradient method, if it is applied for the minimization of smooth functions.

The conjugate-gradient methods with renewal, as well as the relaxation-subgradient method are of the cyclic character (the length of the cycle -- m' $(m' \geq n)$ -- is the number of steps after which the descent in the direction of the antigradient of the function $f(x)$ is carried out).

Recall that the *Fletcher-Reeves* conjugate-gradient *method* for minimization of a continuously differentiable function $f(x)$ on E_n consists in the construction of a sequence $\{x'_k\}$ according to the following rules:

$$x'_0 \in E_n , \qquad v'_0 = f'(x'_0) , \qquad \beta'_0 = 0 , \qquad p'_0 = v'_0 ,$$

$$x'_{k+1} = x'_k - \alpha'_k p'_k , \qquad f(x'_{k+1}) = \min_{\alpha \geq 0} f(x'_k - \alpha p'_k) ,$$

$$p'_k = v'_k + \beta'_k p'_{k-1} ,$$

$$v'_k = f'(x'_k) = \left(\frac{\partial f(x'_k)}{\partial x^{(1)}} , \dots , \frac{\partial f(x'_k)}{\partial x^{(n)}} \right) ,$$

$$b_k = \begin{cases} 0 & \text{if } k \text{ is a multiple of } m' , \\ \|v'_k\|^2 \cdot \|v'_{k-1}\|^2 & \text{otherwise} . \end{cases}$$

<u>THEOREM 6.3.</u> Let a function $f(x)$ be strictly convex and continuously differentiable on E_n and the set $D(x_0) = \{x \in E_n \mid f(x) \leq f(x_0)\}$ be bounded. Then, if the relaxation-subgradient method and Fletcher Reeves conjugate-gradient method are initiated from the same point $(x_0 = x'_0)$ and cycles in both methods are of the same length $(m = m')$ then for any k we have

$$x_k = x'_k , \tag{6. 12}$$

$$p_k = \beta_k p'_k , \tag{6.13}$$

i.e., both methods generate the same sequence of successive approximations.

Proof. Is derived by induction. For $k=0$, equalities (6.12), (6.13) hold by hypothesis of the Theorem and by construction, because we have $\partial f(x)=\{f'(x)\}$ for a convex differentiable function $f(x)$. Here, assuming that equalities (6.12), (6.13) hold for any $k\in 0:\ell$, where $\ell\geq 0$, we show that they hold also for $k=\ell+1$.

Since $x_\ell = x'_\ell$ and $p_\ell = \beta_\ell p'_\ell$, then

$$\{x_\ell-\alpha p_\ell \mid \alpha\geq 0\} = \{x'_\ell-\alpha p'_\ell \mid \alpha\geq 0\}$$

and, therefore, $x_{\ell+1}=x'_{\ell+1}$, since by construction the points $x_{\ell+1}$ and $x'_{\ell+1}$ give the minima of the function $f(x)$ on the rays $\{x_\ell-\alpha p_\ell \mid \alpha\geq 0\}$ and $\{x'_\ell-\alpha p'_\ell \mid \alpha\geq 0\}$, respectively.

If $\ell+1$ is a multiple of m, then $\beta_{\ell+1}=1$, $\beta'_{\ell+1}=0$ and $p_{\ell+1}=\beta_{\ell+1}p'_{\ell+1}=f'(x_{\ell+1})$.

Now consider the case where $\ell+1$ is not a multiple of m. If the function $f(x)$ is convex and smooth, then instead of (6.11), we have the equality

$$(v_k,\ p_{k-1}) = 0 \qquad \forall\, k \tag{6.14}$$

and, moreover, if k is not a multiple of m, then

$$p_k = \|p_k\|^2\,\|v_k\|^{-2}(v_k+\|v_k\|^2\|p_{k-1}\|^{-2}p_{k-1}) . \tag{6.15}$$

We show this. Since

$$\begin{aligned} p_k &= \beta_k v_k + (1-\beta_k)p_{k-1} , \\ (p_k,p_{k-1}) &= (p_k,v_k) = \|p_k\|^2 , \end{aligned} \tag{6.16}$$

then by virtue of (6.14) we have

$$(p_k,v_k) = \beta_k(v_k,v_k) + (1-\beta_k)(p_{k-1},v_k) = \beta_k\|v_k\|^2 ,$$

$$(p_k,p_{k-1}) = \beta_k(v_k,p_{k-1}) + (1-\beta_k)(p_{k-1},p_{k-1}) = (1-\beta_k)\|p_{k-1}\|^2 ,$$

yielding

$$\beta_k = \|p_k\|^2 \|v_k\|^{-2}, \qquad 1 - \beta_k = \|p_k\|^2 \|p_{k-1}\|^{-2}.$$

Now, (6.15) obviously follows from (6.16) and the representations of the numbers β_k and $1-\beta_k$.

Using equality (6.15) for $k = \ell+1$ and the relations

$$p_\ell = \|p_\ell\|^2 \|v_\ell\|^{-2} p'_\ell, \qquad v_{\ell+1} = v'_{\ell+1}, \qquad v_\ell = v'_\ell,$$

yields

$$\begin{aligned} p_{\ell+1} &= \|p_{\ell+1}\|^2 \|v_{\ell+1}\|^{-2}(v_{\ell+1} + \|v_{\ell+1}\|^2 \|p_\ell\|^{-2} p_\ell) \\ &= \|p_{\ell+1}\|^2 \|v_{\ell+1}\|^{-2}(v_{\ell+1} + \|v_{\ell+1}\|^2 \|p_\ell\|^{-2} \|p_\ell\|^2 \|v_\ell\|^{-2} p'_\ell) \\ &= \beta_{\ell+1}(v'_{\ell+1} + \|v'_{\ell+1}\|^2 \|v'_\ell\|^{-2} p'_\ell) = \beta_{\ell+1} p'_{\ell+1}. \end{aligned}$$

Thus, $p_{\ell+1} = \beta_{\ell+1} p'_{\ell+1}$ (Figure 52). The theorem has been proved. ∎

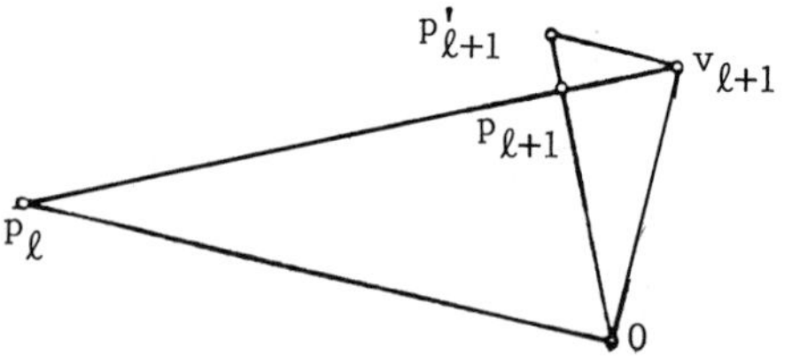

Figure 52

<u>REMARK 3</u>. It is useful to interrupt the cycle when conditions equivalent to (6.1), (6.2) are satisfied in the case of the conjugate-gradient method as well, because the gradient method is quite efficient far from the minimum.

3. In Subsection 6.1, we considered the relaxation-subgradient method with complete relaxation, i.e., a method in which the minimum in the search direction is found exactly at each step.

It is clear that, in most cases, it is impossible to implement complete relaxation--to do so, an auxiliary infinite iterative procedure is required. Hence, our immediate aim is to construct such a modification of the relaxation-subgradient method that requires no exact solution of one-dimensional minimization problems. In addition, we shall also get rid of the strict-convexity constraint imposed on an objective function $f(x)$.

In proving the convergence of the relaxation-subgradient method, Lemma 6.1 plays an essential role: hence, our first priority is to obtain an analogue of Lemma 6.1.

LEMMA 6.2. Let a function $f(x)$ be defined on E_n and a sequence $\{x_k\}$ be such that

$$x_{k+1} = x_k + \alpha_k g_k , \qquad f(x_{k+1}) \le f(x_k) - \delta_2 \alpha_k \|v_k\| ,$$
$$g_k = -v_k \|v_k\|^{-1} , \qquad \|v_k\| \ge \varepsilon_0 ,$$

where $\varepsilon_0 > 0$, $\delta_2 > 0$ are fixed.

If the sequence $\{f(x_k)\}$ is bounded from below, then the sequence $\{x_k\}$ converges.

Proof. By hypothesis, we have

$$f(x_{k+1}) \le f(x_0) - \delta_2 \varepsilon_0 \sum_{t=1}^{k} \|x_{t-1} - x_t\|$$

and

$$\sum_{t=1}^{k} \|x_{t-1} - x_t\| \le (\varepsilon_0 \delta_2)^{-1} (f(x_0) - \inf_k f(x_k)) < \infty \quad \forall k .$$

This implies that the sequence $\{x_k\}$ converges.

Thus, it is shown how to choose a step-size such that the norm of the difference between two consecutive approximations tends to zero, and, moreover, the sequence $\{x_k\}$ converges. ∎

Another important requirement which guarantees the convergence of the relaxation-subgradient method is to obtain vectors v_{kt}, $\bar{v}_{kt}$ satisfying (6.3), i.e., such vectors that the angle between them is not less than the right angle. Even if we construct the sequence $\{x_k\}$ as required in Lemma 6.2, the point x_{k+1} is not necessarily near the point at which the function $f(x)$ attains the minimum on the ray $\{x_k+\alpha g_k \mid \alpha\geq 0\}$. Hence, one needs to weaken condition (6.3).

LEMMA 6.3. Let $f(x)$ be a convex function on E_n. If for some $\alpha > 0$ and $\delta_1 > 0$

$$f(x+\alpha g) \geq f(x) - \delta_1\alpha\|v\| , \tag{6.17}$$

where $g = -v\|v\|^{-1}$, $v \in E_n$, then

$$(v,\bar{v}) \leq \delta_1\|v\|^2 \qquad \forall\bar{v} \in \partial f(x+\alpha g) . \tag{6.18}$$

Proof. In fact, for any $\bar{v} \in \partial f(x+\alpha g)$, considering (6.17), we have

$$f(x) \geq f(x+\alpha g) - \alpha(g,\bar{v}) \geq f(x) - \delta_1\alpha\|v\| - \alpha(g,\bar{v}) ,$$

i.e., $-(g,\bar{v}) \leq \delta_1\|v\|$, yielding (6.18). ∎

REMARK 4. It is proved quite similarly that if $f(x+\alpha g) \geq f(x)$, then $(v,\bar{v}) \leq 0 \quad \forall\bar{v} \in \partial f(x+\alpha g)$.

Let us proceed to describe the relaxation-subgradient method for the minimization of a convex (not necessarily strictly convex) function $f(x)$ on E_n, where an auxiliary problem of one-dimensional minimization can be solved approximately.

Let us fix positive numbers $a, \varepsilon_0, \delta, \delta_1, \delta_2$, where $0 < \delta_2 < \delta_1 <$

Choose an arbitrary initial approximation $x_0 \in E_n$, a number $\nu_0 > 0$ and an integer $m_0 > 1$.

Let the kth approximation $x_k \in E_n$ be given. We shall describe the kth cycle in this method, i.e., the process of transitio from x_k to x_{k+1}.

Take a number $\nu_k > 0$ and an integer $m_k > 1$ which is the maximum length of the kth cycle.

Let $x_{k0} = x_k$ and calculate an arbitrary subgradient

$$v_{k0} \in \partial f(x_{k0}).$$

If $\|v_k\| \le \varepsilon_0$, then we set $t_k = 0$, $x_{k+1} = x_{kt_k}$, and the kth cycle terminates. Otherwise, we find points $x_{k1} = x_{k0}(\alpha_{k0})$ and $\bar{x}_{k1} = x_{k0}(\bar{\alpha}_{k0})$ on the ray $\{x_{k0}(\alpha) = x_{k0} + \alpha g_{k0} \mid \alpha \ge 0\}$, where $g_{k0} = -v_{k0} / \|v_{k0}\|^{-1}$, such that

$$f(x_{k1}) \le f(x_{k0}) - \delta_2 \alpha_{k0} \|v_{k0}\| \} , \qquad (6.19)$$

$$f(\bar{x}_{k1}) \ge \min \{f(x_{k1}), f(x_{k0}) - \delta_1 \bar{\alpha}_{k0} \|v_{k0}\| \} , \qquad (6.20)$$

where

$$\bar{\alpha}_{k0} \ge \alpha_{k0} , \qquad \nu_k \le \bar{\alpha}_{k0} \le \nu_k + (a+1)\alpha_{k0} \qquad (6.21)$$

(Figure 53).

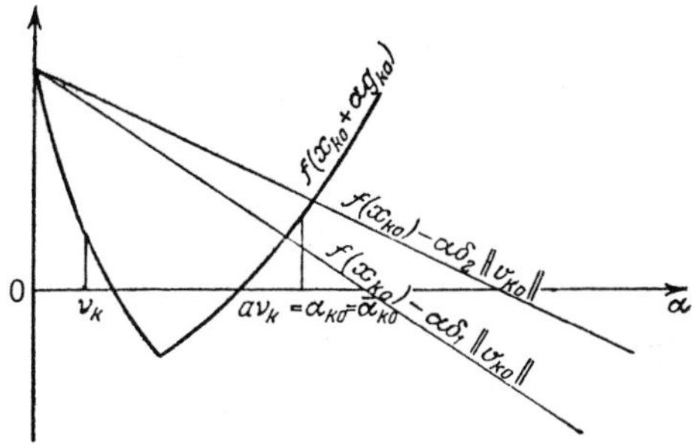

Figure 53

Here again, let us pay attention to the fact that it can happen that $x_{k1} = x_{k0}$. We also note that the case where $\bar{x}_{k1} = x_{k1}$ is not excluded.

If $\|x_{k0} - x_{k1}\| \ge \delta$ or $f(x_{k0}) - f(x_{k1}) \ge \delta$, then we suppose that $t_k = 1$, $x_{k+1} = x_{kt_k}$, and there the kth cycle terminates. Otherwise, we compute an arbitrary (this is the essential difference from the method with complete relaxation) subgradient $\bar{v}_{k0} \in \partial f(\bar{x}_{k1})$.

By virtue of (6.20) and Lemma 6.3, the inequality $(v_{k0}, \bar{v}_{k0}) \le \delta_1 \|v_{k0}\|$ should hold. Moreover, if the value $f(x_{k1})$ is the smallest one in (6.20), then, due to Remark 4, $(v_{k0}, \bar{v}_{k0}) \le 0$.

Now, let us find a vector v_{k1} which is the nearest point to the origin in the segment $[v_{k0},\bar{v}_{k0}]$, i.e.,

$$v_{k1} = \beta_{k0}\bar{v}_{k0} + (1-\beta_{k0})v_{k0} , \qquad \beta_{k0} \in [0,1] ,$$

$$\|v_{k1}\| = \min_{\beta\in[0,1]} \|\beta\bar{v}_{k0} + (1-\beta)v_{k0}\| .$$

If $\|v_{k1}\| \le \varepsilon_0$, then we assume that $t_k = 1$, $x_{k+1} = x_{kt_k}$, and the kth cycle terminates. If $\|v_{k1}\| > \varepsilon_0$, then, with the point x_{k1} and the direction $g_{k1} = -v_{k1}\|v_{k1}\|^{-1}$, we carry out the second step of the kth cycle, which is done like all other steps of the kth cycle, according to the same rules as for the first step. As a result, we shall perform the steps where $t_k \le m_k$, and at least one of the following conditions is satisfied:

(a) $t_k = m_k$;

(b) $\|x_{k,t_k-1} - x_{kt_k}\| \ge \delta$;

(c) $f(x_{k,t_k-1}) - f(x_{kt_k}) \ge \delta$;

(d) $\|v_{kt_k}\| \le \varepsilon_0$.

After that, we let $x_{k+1} = x_{kt_k}$, and the kth cycle terminates; in this case, if $x_{kt} = x_{k0}$ $\forall t \in 0:t_k$ and the cycle is interrupted owing to condition (d), then the whole process of successive approximations terminates: a point x_k such that $d(x_k) \le \|v_{kt_k}\| \le \varepsilon_0$ is obtained.

Let us note that by construction we have

$$(v_{kt},\bar{v}_{kt}) \le \delta_1\|v_{kt}\|^2 \qquad \forall t \tag{6.22}$$

due to Lemma 6.3.

<u>THEOREM 6.4</u>. Let

$$\nu_k \to +0 , \qquad m_k \to \infty. \tag{6.23}$$

Then, if the sequence $\{f(x_k)\}$ is bounded from below, the sequence $\{x_k\}$ converges to a point x^* such that

$$d(x^*) = \min_{v \in \partial f(x^*)} \|v\| \leq \varepsilon_0 . \tag{6.24}$$

Proof. Since the sequence $\{f(x_k)\}$ is bounded from below, then according to Lemma 6.2

$$x_k \to x^* , \quad x_{kt} \to x^* \quad \forall t \in 0:t_k . \tag{6.25}$$

It follows from (6.23) and (6.25) that for sufficiently large k, say for $k \geq K$, the termination of the kth cycle is impossible if conditions (b), (c) are satisfied.

According to (6.21), we have for all $t \in 0:(t_k-1)$

$$\|\bar{x}_{k,t+1} - x_{kt}\| \leq \nu_k + a\|x_{k,t+1} - x_{kt}\| ,$$

implying, by (6.23) and (6.25), that

$$\bar{x}_{kt} \to x^* \quad \forall t \in 1:t_k , \tag{6.26}$$

since $\|\bar{x}_{kt}-x^*\| \leq \|\bar{x}_{kt}-x_{kt}\| + \|x_{kt}-x^*\|$.

Thus, all the points x_{kt}, $t \in 0:t_k$, $\bar{x}_{kt}$, $t \in 1:t_k$, $\forall k$ belong to some compact set, which we denote by X. Since the mapping $\partial f(x)$ is bounded on X, then

$$b = \max_{x \in X} \max_{v \in \partial f(x)} \|v\| < \infty.$$

Consequently,

$$\|\bar{v}_{kt}\| \leq b,$$

$$\|v_{kt}\| \leq b \qquad \forall t, \quad \forall k . \tag{6.27}$$

Let us take here an integer $m \geq 1+(\ln \tilde{q})^{-1} \cdot \ln(\varepsilon_0 b^{-1})$, where $\tilde{q}^2 = 1-(1-\delta_1)^2[1-2\delta_1+(b\varepsilon_0^{-1})^2]^{-1}$. Without loss of generality, we suppose that $m_k > m$ for all $k \geq K$.

Now, we shall show that in this case $t_k \leq m < m_k$.

Assume that $t_k > m$. Then, considering (6.22), (6.27) and the fact that $\|v_{kt}\| > \varepsilon_0$ for all $t \in 0:(t_k-1)$, we have

$$\|v_{k,t+1}\|^2 = \min_{\beta\in[0,1]} \|\beta\bar{v}_{kt} + (1-\beta)v_{kt}\|^2$$
$$= \|v_{kt}\|^2 \min_{\beta\in[0,1]} \left[\beta^2 \frac{\|\bar{v}_{kt}\|^2}{\|v_{kt}\|^2} + 2\beta(1-\beta)\frac{(v_{kt},\bar{v}_{kt})}{\|v_{kt}\|^2} + (1-\beta)^2\right]$$
$$\le \|v_{kt}\|^2 \min_{\beta\in[0,1]} [\beta^2(b\varepsilon_0^{-1})^2 + 2\beta(1-\beta)\delta_1 + (1-\beta)^2]$$
$$= \|v_{kt}\|^2\left(1 - \frac{(1-\delta_1)^2}{(b\varepsilon_0^{-1})^2 - 2\delta_1 + 1}\right) = \|v_{kt}\|^2\tilde{q}^2 .$$

It follows from this that

$$\|v_{kt}\| \le \|v_{k0}\|\tilde{q}^t \le b\tilde{q}^t \qquad \forall t \in 0:(t_k-1) ,$$

and, therefore, $\|v_{k,m-1}\| \le b\tilde{q}^{m-1} \le \varepsilon_0$. On the other hand, we have $\|v_{kt}\| > \varepsilon_0$ $\forall t \in 0:(t_k-1)$. Consequently, $t_k \le m$.

Thus, if the sequence $\{f(x_k)\}$ is bounded from below, then all the cycles are finite and the length of each of them does not exceed the number m; for all $k \ge K$, the kth cycle is terminated only if condition (d) is satisfied:

$$\|v_{kt_k}\| \le \varepsilon_0 . \tag{6.28}$$

Now, assume that (6.24) does not hold. Then,

$$\varepsilon_0 < \min_{v\in\partial f(x^*)} \|v\| \equiv \bar{\varepsilon} .$$

Since the mapping $f(x)$ is upper semicontinuous, then there exists a $\delta > 0$ such that

$$\partial f(x) \subset S_\varepsilon(\partial f(x^*)) \equiv D \qquad \forall x \in S_\delta(x^*),$$

where $\varepsilon = 2^{-1}(\bar{\varepsilon} - \varepsilon_0)$.

Thus,

$$\|v\| > \varepsilon_0 \qquad \forall v \in D .$$

However, by virtue of (6.25) and (6.26), we have $v_{kt_k} \in D$ for sufficiently large k and, hence, $\|v_{kt_k}\| > \varepsilon_0$, which contradicts (6.28). The theorem has been proved. ∎

This method can also be applied to finding minimum points of the the function $f(x)$ on E_n. To do this, instead of condition (d) $\|v_{kt_k}\| \le \varepsilon_0$, we need to verify condition (d) $\|v_{kt_k}\| \le \varepsilon_k$, as $\varepsilon_k \to 0$.

In this case we have the following theorem.

THEOREM 6.5. Let

$$\varepsilon_k \to 0 , \qquad \nu_k \to 0 , \qquad m_k \to \infty .$$

Then, either

$$f(x_k) \xrightarrow[k\to\infty]{} -\infty ,$$

or for some k and the corresponding t

$$f(x_{kt}+\alpha g_{kt}) \xrightarrow[\alpha\to\infty]{} -\infty ,$$

or an arbitrary accumulation point of the sequence $\{x_k\}$ is a minimum point of the function $f(x)$ on E_n.

PROBLEM 6.1. Prove Theorem 6.5.

REMARK 5. In the method with complete relaxation, a vector $\bar{v}_{kt} \in \partial f(x_{k,t+1})$ such that $(v_{kt},\bar{v}_{kt}) \le 0$ is computed at the point $x_{k,t+1}$, and such a subgradient exists. On the other hand, in the method described above, an arbitrary $\bar{v}_{kt} \in \partial f(\bar{x}_{k,t+1})$ is computed at the point $\bar{x}_{k,t+1}$, where $(v_{kt},\bar{v}_{kt}) \le \delta_1\|v_{kt}\|^2$. Since a convex function is differentiable almost everywhere, i.e., $\partial f(x)$ is a singleton, i.e., it consists of a unique point -- the gradient of the function $f(x)$ -- almost everywhere, then it can be seen that it is of little advantage to have <<arbitrary $v \in \partial f(x)$>> rather than <<$v \in \partial f(x)$ exists>>. However, if the function $f(x)$ is nonsmooth, then during the line-search we may hit precisely the set of measure zero, where $f(x)$ is a convex compact set with the nonempty interior and, hence, in <<$v \in \partial f(x)$ exists>> it is indeed necessary to initiate a search for <<what exists>>. The easy implementation adds to the superiority of this method over the method with complete relaxation.

7. THE RELAXATION ε-SUBGRADIENT METHOD

In Section 6, it was assumed that the value of an objective function $f(x)$ and its subgradient can be calculated exactly. However, this cannot be done on a computer because, first of all, a computer operates only rational numbers and, second, for the computation of the objective function and, a fortiori, its subgradients, we need an infinite iterative process. As an example of such a function, we have the maximum function $f(x) = \max_{y \in G} \phi(x,y)$ (see Section 1.17).

Hence, it is important to construct a modification of the relaxation-subgradient method, which allows approximate calculations of the objective function $f(x)$ and its subgradients, i.e., we can use some function $f_\varepsilon(x)$ such that

$$f(x) - \varepsilon \le f_\varepsilon(x) \le f(x) + \varepsilon , \tag{7.1}$$

instead of the objective function $f(x)$, and ε-subgradients $v_\varepsilon \in \partial_\varepsilon f(x)$ instead of subgradients $v \in \partial f(x)$.

We shall be constructing this modification in the following way.

1. First, let us obtain an analogue of Lemma 6.3 for the case under consideration. Take positive numbers δ_1, c, ν, ε, ε_0 and δ_2, where $\delta_1 \in (0,1)$, $\delta_2 < \delta_1$, $\varepsilon \le (c+2)^{-1}(\delta_1-\delta_2)\varepsilon_0\nu$.

Let $x_1 = x_0+\alpha_0 g_0$, where $g_0 = -v_0\|v_0\|^{-1}$, $v_0 \in E_n$.

<u>LEMMA 7.1</u>. If $\|v_0\| \ge \varepsilon_0$, $\alpha_0 \ge \nu$ and

$$\mu \equiv f_\varepsilon(x_1) - [f_\varepsilon(x_0) - \delta_2\alpha_0\|v_0\|] \ge 0 , \tag{7.2}$$

then

$$(v_0,\bar{v}_0) \le \delta_1\|v_0\|^2 \quad \forall \bar{v}_0 \in \partial_{c\varepsilon+\mu} f(x_1) . \tag{7.3}$$

Proof. By (7.2), we have

$$f_\varepsilon(x_1) = f_\varepsilon(x_0) - \delta_1\alpha_0\|v_0\| + (\delta_1-\delta_2)\alpha_0\|v_0\| + \mu$$
$$\geq f_\varepsilon(x_0) - \delta_1\alpha_0\|v_0\| + (c+2)\varepsilon + \mu ,$$

which together with (7.1) yield

$$f(x_1) \geq f_\varepsilon(x_1) - \varepsilon \geq f_\varepsilon(x_0) - \delta_1\alpha_0\|v_0\| + (c+1)\varepsilon + \mu$$
$$\geq f(x_0) - \delta_1\alpha_0\|v_0\| + c\varepsilon + \mu .$$

Now, the validity of (7.3) follows from the fact that for any $\bar{v}_0 \in \partial_{c\varepsilon+\mu} f(x_1)$ the inequality

$$f(x_0) \geq f(x_1) - \alpha_0(g_0,\bar{v}_0) - (c\varepsilon+\mu)$$

is satisfied. ∎

Lemma 7.1 shows what accuracy of the computation of values of the function $f(x)$ is required in order that (7.3) follow from (7.2).

2. Let us proceed to describe the relaxation ε-subgradient method.

Fix positive numbers $a, c, \nu, \varepsilon, \delta, \varepsilon_0, \delta_i;\ i \in 1:3$, such that $\delta_1 \in (0,1)$, $0 < \delta_3 < \delta_2 < \delta_1$, $\varepsilon \leq (c+2)^{-1}(\delta_1-\delta_2)\varepsilon_0\nu$. Choose an arbitrary initial approximation $x_0 \in E_n$ and an integer m_0. Let the kth approximation $x_k \in E_n$ be already found. The transition to the $(k+1)^{st}$ approximation x_{k+1} (the kth cycle) is accomplished in the following way.

Let an integer m_k be the maximum length of the kth cycle.

Let $x_{k0} = x_k$ and calculate the value of $f_\varepsilon(x_{k0})$ and an arbitrary $c\varepsilon$-subgradient $v_{k0} \in \partial_{c\varepsilon} f(x_{k0})$. If $\|v_{k0}\| \leq \varepsilon_0$, then we set $t_k = 0$, $x_{k+1} = x_{kt_k}$, and the kth cycle terminates. If $\|v_{k0}\| > \varepsilon_0$, then on the ray $\{x_{k0}(\alpha) = x_{k0}+\alpha g_{k0} \mid \alpha \geq 0\}$, where $g_{k0} = -v_{k0}\|v_{k0}\|^{-1}$, let us find points $x_{k1} = x_{k0}(\alpha_{k0})$ and $\bar{x}_{k1} = x_{k0}(\bar{\alpha}_{k0})$ such that (for $t = 0$)

$$\bar{\alpha}_{kt} \geq \alpha_{kt}, \qquad \nu \leq \bar{\alpha}_{kt} \leq \nu + a\alpha_{kt}; \tag{7.4}$$

here, either $\alpha_{kt} = 0$ or $\alpha_{kt} \geq \nu$, and

$$f_\varepsilon(x_{k,t+1}) \leq f_\varepsilon(x_{kt}) - \delta_3 \alpha_{kt} \| v_{kt} \| , \tag{7.5}$$

$$f_\varepsilon(\bar{x}_{k,t+1}) - \min\{f_\varepsilon(x_{k,t+1}),\ f_\varepsilon(x_{kt}) - \delta_2 \bar{\alpha}_{kt} \| v_{kt} \|\} \geq 0 . \tag{7.6}$$

If (7.5) is satisfied for any $\alpha_{kt} \geq 0$, then $f_\varepsilon(x_{kt}(\alpha)) \xrightarrow[\alpha\to\infty]{} -\infty$, therefore, $f(x_{kt}(\alpha)) \xrightarrow[\alpha\to\infty]{} -\infty$, and the process of successive approximations terminates.

If $\|x_{k0} - x_{k1}\| \geq \delta$ or $f_\varepsilon(x_{k0}) - f_\varepsilon(x_{k1}) \geq \delta$, then we set $t_k = 1$, $x_{k+1} = x_{kt_k}$, i.e., the kth cycle terminates. Otherwise, let us calculate an arbitrary $c\varepsilon$-subgradient $\bar{v}_{k0} \in \delta_{c\varepsilon} f(\bar{x}_{k1})$. Since $\|v_{k0}\| > \varepsilon_0$, $\nu \leq \bar{\alpha}_{k0}$, $\varepsilon \leq (c+2)^{-1}(\delta_1 - \delta_2)\varepsilon_0 \nu$ and (7.6) is satisfied, then, by Lemma 7.1,

$$(v_{k0}, \bar{v}_{k0}) \leq \delta_1 \|v_{k0}\|^2 .$$

After this, we find a vector v_{k1} which is nearest to the origin in the segment $[v_{k0}, \bar{v}_{k0}]$.

If $\|v_{k1}\| \leq \varepsilon_0$, then we set $t_k = 1$, $x_{k+1} = x_{kt_k}$, and the kth cycle terminates. If $\|v_{k1}\| > \varepsilon_0$, then on the ray $\{x_{k1}(\alpha) = x_{k1} + \alpha g_1 \mid \alpha > 0\}$, where $g_{k1} = -v_{k1}\|v_{k1}\|^{-1}$, let us find points $x_{k2} = x_{k1}(\alpha_{k1})$ and $\bar{x}_{k2} = x_{k1}(\bar{\alpha}_{k1})$ such that (7.4) - (7.6) are satisfied for $t = 1$.

We continue in a similar way until in a number of steps t_k at least one of the following conditions is satisfied:

(a) $t_k = m_k$;

(b) $\|x_{k,t_k-1} - x_{kt_k}\| \geq \delta$;

(c) $f_\varepsilon(x_{k,t_k-1}) - f_\varepsilon(x_{kt_k}) \geq \delta$;

(d) $\|v_{kt_k}\| \leq \varepsilon_0$.

Now, we let $x_{k+1} = x_{kt_k}$, and the kth cycle terminates. In this case, if the kth cycle is interrupted condition (d) is satisfied and $x_{kt} = x_{k0}$ for all $t \in 1:t_k$, then the whole process of successive approximations also terminates.

As a result, a sequence $\{x_k\}$ has been constructed.

Let $\bar{v}_{k,-1} = v_{k0}$.

$$\ell_k = \begin{cases} t_k - 1 & \text{if the } k^{th} \text{ cycle is interrupted if condition (d) is satisfied ,} \\ t_k - 2 & \text{otherwise .} \end{cases}$$

We note that, by construction, due to Lemma 7.1,

$$(v_{kt}, \bar{v}_{kt}) \le \delta_1 \| v_{kt} \|^2 \qquad \forall t \in 0:\ell_k \, .$$

THEOREM 7.1. Let

$$m_k \to \infty \, . \tag{7.7}$$

Then, either

$$f(x_k) \xrightarrow[k\to\infty]{} -\infty \, ,$$

or for some k and the corresponding t

$$f(x_{kt} + \alpha g_{kt}) \xrightarrow[\alpha\to\infty]{} -\infty \, ,$$

or the process of successive approximation terminates at a point x_k such that

$$f(z) - f(x_k) \ge -\varepsilon_0 \| z - x_k \| - (c+2)\varepsilon - (1+\delta_2)\nu b_0 \qquad \forall z \in E_n \, , \tag{7.8}$$

where $b_0 = \max\limits_{t \in -1:\ell_k} \| v_{kt} \|$.

Proof. Let us first consider the case where the sequence $\{x_k\}$ is infinite. Then, we shall show that there exists a subsequence $\{x_{k_s}\}$ such that

$$x_{k_s+1} \ne x_{k_s} \qquad \forall s \, . \tag{7.9}$$

Suppose that this is not the case. Then, starting from some K, we should have

$$x_k = x_{k0} = \cdots = x_{kt_k} = x_{k+1} \tag{7.10}$$

for all $k \ge K$ and, therefore,

$$\varepsilon_0 < \|v_{kt}\| \le b \qquad \forall\, t \in 1:t_k ,$$
$$\|\bar{v}_{kt}\| \le b \qquad \forall\, t \in -1:\ell_k , \tag{7.11}$$

where $b \le \max_{v \in \partial_{c\varepsilon} f(x_k)} \|v\| < \infty$.

Since the sequence $\{x_k\}$ is infinite, then it follows from (7.10) and (7.11) that $t_k = m_k$ for any $k \ge K$.

By virtue of (7.7), without loss of generality, we can suppose that $m_k > m$, where the integer m is such that

$$m > 1 + \ln^{-1}\tilde{q} \cdot \ln(\varepsilon_0 b^{-1}) ,$$
$$\tilde{q} = [1 - (1-\delta_1)^2(1 - 2\delta_1 + (b\varepsilon_0^{-1})^2)^{-1}]^{\frac{1}{2}} .$$

Then, as in proving Theorem 6.4, we show that $\|v_{k,m-1}\| \le \varepsilon_0$, and this contradicts (7.11).

Thus, if the sequence $\{x_k\}$ is infinite, then there exists a subsequence $\{x_{k_s}\}$, for which (7.9) holds. Then, from (7.4) and (7.5), we conclude that

$$f_\varepsilon(x_{k_s+1}) \le f_\varepsilon(x_{k_s}) - \delta_3 \nu \varepsilon_0 \qquad \forall\, s .$$

It follows from this inequality and the inequality $f_\varepsilon(x_{k+1}) \le f_\varepsilon(x_k)$, that $f_\varepsilon(x_k) \xrightarrow[k\to\infty]{} -\infty$, and this implies by virtue of (7.1) that $f(x_k) \xrightarrow[k\to\infty]{} -\infty$.

Let us consider the case where the sequence $\{x_k\}$ is finite. Assume that the process terminates at some point $x_{k+1} = x_k$.

Then, (7.10) holds and $\|v_{kt_k}\| \le \varepsilon_0$.

Since $\bar{v}_{k,-1} = v_{k0} \in \partial_{c\varepsilon} f(x_{k0})$, then

$$f(z) - f(x_k) \ge (\bar{v}_{k,-1}, z - x_k) - c\varepsilon \qquad \forall z \in E_n . \tag{7.12}$$

According to (7.4) and (7.10), we have

$$x_k - \bar{x}_{kt} = -\nu g_{k,t-1} = \nu v_{k,t-1}\|v_{k,t-1}\|^{-1} \qquad \forall t \in 1:t_k . \tag{7.13}$$

Since, for any $t \in 1:t_k$ and any $z \in E_n$,

$$f(z) - f(\bar{x}_{kt}) \geq (\bar{v}_{k,t-1}, z-\bar{x}_{kt}) - c\varepsilon$$
$$= (\bar{v}_{k,t-1}, z-x_k) - c\varepsilon + (\bar{v}_{k,t-1}, x_k - \bar{x}_{kt}) ,$$

then, considering (7.13) and the estimate $\|\bar{v}_{k,t-1}\| \leq b_k$, we obtain

$$f(z) - f(\bar{x}_{kt}) \geq (\bar{v}_{k,t-1}, z-x_k) - c\varepsilon - \nu b_k$$
$$\forall z \in E_n, \quad \forall t \in 1:t_k . \qquad (7.14)$$

By virtue of (7.10), (7.6) and (7.13), we get

$$f(\bar{x}_{kt}) - f(x_k) \geq -2\varepsilon - \delta_2 \nu \|v_{k,t-1}\| \geq -2\varepsilon - \delta_2 \nu b_k \qquad \forall t \in 1:t_k ,$$

which together with (7.14) and (7.12) imply that

$$f(z) - f(x_k) \geq (\bar{v}_{k,t-1}, z-x_k) - (c+2)\varepsilon - (1+\delta_2)\nu b_k$$
$$\forall z \in E_n, \quad \forall t \in 0:t_k . \qquad (7.15)$$

It is clear that $v_{kt_k} \in \text{co } \{v_{k,t-1} \mid t \in 0:t_k\}$, i.e.,

$$v_{kt_k} = \sum_{t=-1}^{\ell_k} \lambda_t \bar{v}_{kt} , \qquad \sum_{t=-1}^{\ell_k} \lambda_t = 1 , \qquad \lambda_t \geq 0 \;\; \forall t \in -1:\ell_k . \; (7.16)$$

Now, multiply each of inequalities (7.15) by the corresponding coefficient λ_t in representation (7.16) of the vector v_{kt_k} and sum up the obtained $t_k + 1$ inequalities. Then,

$$f(z) - f(x_k) \geq (v_{kt_k}, z-x_k) - (c+2)\varepsilon - (1+\delta_2)\nu b_k$$
$$\geq -\|v_{kt_k}\| \cdot \|z-x_k\| - (c+2)\varepsilon - (1+\delta_2)\nu b_k \qquad \forall z \in E_n .$$

Since $\|v_{kt_k}\| \leq \varepsilon_0$, we have (7.8). The theorem has been proved. ■

3. Theorem 7.1 and, in particular, estimate (7.8) illustrate how we can apply the method described in Subsection 7.2, which we called the ε-algorithm, to find minimum points of $f(x)$ on E_n.

Take sequences of positive numbers $\{\varepsilon_{0k}\}$, $\{\varepsilon_k\}$ and $\{\nu_k\}$ which tend to zero. Choose an initial approximation $\tilde{x}_0 \in E_n$ arbitrarily.

Let us apply the ε-algorithm for $\varepsilon_0 = \varepsilon_{01}$, $\varepsilon = \varepsilon_1$, $\nu = \nu_1$, starting from $x_0 = \tilde{x}_0$. As a result either we establish that $f^* = \inf_{x \in E_n} f(x) = -\infty$ or we find a point x_k satisfying (7.8). Let $\tilde{x}_1 = x_k$. We apply the ε-algorithm for $\varepsilon_0 = \varepsilon_{02}$, $\varepsilon = \varepsilon_2$, $\nu = \nu_2$, starting from $x_0 = \tilde{x}_1$, and either we establish that $f^* = -\infty$ or we find a point x_k satisfying (7.8), here again, for variables ε_0, ε, ν. Let $\tilde{x}_2 = x_k$.

Continuing further analogously, either we establish that $f^* = -\infty$ or we obtain an infinite sequence $\{\tilde{x}_k\}$, each point of which satisfies, by virtue of (7.8), the relation

$$f(z) - f(\tilde{x}_k) \geq -\varepsilon_{0k}\|z-\tilde{x}_k\| - (c+2)\varepsilon_k - (1+\delta_2)\nu_k b_{k-1} \qquad \forall z \in E_n.$$

Hence, it is clear that if the set $D(\tilde{x}_0) = \{x \in E_n \mid f(x) \leq f(\tilde{x}_0)\}$ is bounded, then $f(\tilde{x}_k) \to f^*$, and moreover, any accumulation point of the sequence $\{\tilde{x}_k\}$ is a minimum point of the function $f(x)$ on E_n.

4. In all of the variants of the relaxation-subgradient method considered above, the direction of descent was determined by solving the simplest problem of quadratic programming, that is finding the point nearest to the origin in the segment $[v_{kt}, \bar{v}_{kt}]$. If the dimension of the space E_n is not too large and a computer can be employed, then the efficiency of the relaxation-subgradient method will significantly improve, if we apply a more complicated auxiliary problem to choose the direction of descent, in other words, if we find the point which is nearest to the origin, in a polyhedron spanned on the set $H_{kt} = \{\bar{v}_{k,-1}, \bar{v}_{k0}, \ldots, \bar{v}_{k,t-1}\}$, or on some part of it.

In this case, the vectors $v_{k,t+1}$, $t \in 1:t_k$, are constructed in the following way: we choose some subset $\bar{H}_{kt}$ of the set H_{kt} and find a vector $v_{k,t+1} \in L_{kt}$, where $L_{kt} = \text{co}\,\{v_{kt}, \bar{v}_{kt}, \bar{H}_{kt}\}$, such that

$$\|v_{k,t+1}\| = \min_{v \in L_{kt}} \|v\| . \tag{7.17}$$

In this case the relaxation-subgradient method becomes similar to the method of ε-steepest descent, in methodological sense as well as in terms of efficiency.

Since $[v_{ky}, \bar{v}_{kt}] \subset L_{kt}$, then all the theorems proved in Sections 3.6 and 3.7 remain valid if we find a vector $v_{k,t+1}$ from condition (7.17).

Knowing the dimension of the original problem of the minimization of a function $f(x)$ on E_n, it is possible to estimate the dimension of the quadratic programming problem (7.17), for a particular computer. The more vectors of H_{kt} are included in $\bar{H}_{kt}$, the more exactly the set L_{kt} approximates the set $\partial_\varepsilon f(x_{k,t+1})$ and the more considerably the function $f(x)$ is decreased when one moves from the point $x_{k,t+1}$ in the direction

$$g_{k,t+1} = -v_{k,t+1}\|v_{k,t+1}\|^{-1} .$$

5. To conclude this section, let us discuss the auxiliary problem of line-search for the points $x_{k,t+1} = x_{kt}(\alpha_{kt})$ and $\bar{x}_{k,t+1} = x_{kt}(\bar{\alpha}_{kt})$ satisfying (7.4) - (7.6). Obviously, this can be done in many ways. Here we show how this is done by using the golden-section method which is usually applied for one-dimensional minimization on a segment. We shall also describe a modification of the golden-section method for the search on a ray.

Let

$$x(\alpha) = x_{kt}(\alpha) , \qquad x = x_{kt} , \qquad x_+ = x_{k,t+1} ,$$

$$\bar{x}_+ = \bar{x}_{k,t+1}, \qquad f(\alpha) = f_\varepsilon(x(\alpha)),$$

$$\xi_s = f(\alpha_s) - f(\alpha_{s-1}) + \delta_3\alpha_s\|v_{kt}\|,$$

$$\mu_s = f(\alpha_s) - \min\{f(\alpha_{s-1}), f(\alpha_0) - \delta_2\alpha_s\|v_{kt}\|\}.$$

Let

$$\alpha_0 = 0, \qquad \alpha_1 = \nu, \qquad q = 2^{-1}(1+\sqrt{5}) > 1.$$

Compute $f(\alpha_0,)$, $f(\alpha_1)$ and μ_1. If $\mu_1 \geq 0$, then we let $x_+ = x$, $\bar{x}_+ = x(\alpha_1)$; otherwise, we compute $f(\alpha_1)$, μ_2 and ξ_2, where $\alpha_2 = \nu(1+q) = \nu\sum_{i=0}^{1} q^i$.

Let $f(\alpha_s)$, μ_s and ξ_s be already found, where $\alpha_s = \nu\sum_{i=0}^{s-1} q^i$.

If $\mu_s < 0$, then we let $\alpha_{s+1} = \nu\sum_{i=0}^{s} q^i$ and compute $f(\alpha_{s+1})$, μ_{s+1}, ξ_{s+1}.

Let $\mu_s \geq 0$. If $\xi_{s-1} \leq 0$ and $\alpha_s \leq \nu + a\alpha_{s-1}$ or $\xi_s \leq 0$, then we let $\bar{x}_+ = x(\alpha_s)$,

$$x_+ = \begin{cases} \bar{x}_+ & \text{if } \xi_s \leq 0, \\ x(\alpha_{s-1}) & \text{if } \xi_s > 0,\ \xi_{s-1} \leq 0,\ \alpha_s \leq \nu + av_{s-1}. \end{cases}$$

Otherwise (here, as before, we consider the case $\mu_s \geq 0$), we continue the search on the segment $[\alpha_{s-2}, \alpha_s]$ using the golden-section method. We note that α_{s-1} is the golden section of the segment $[\alpha_{s-2}, \alpha_s]$. We show this next. It is well known that the golden section of a segment is a division of this segment into parts such that the ratio of the length of the shorter part to the length of the longer part is equal to the ratio of the length of the longer part to the length of the whole segment. Hence, we have

$$\alpha_{s-2} = \nu\sum_{i=0}^{s-3} q^i, \qquad \alpha_{s-1} = \nu\sum_{i=1}^{s-2} q^i, \qquad \alpha_s = \nu\sum_{i=0}^{s-1} q^i,$$

$$\frac{\alpha_s - \alpha_{s-1}}{\alpha_s - \alpha_{s-2}} = \frac{q^{s-1}}{q^{s-1} + q^{s-2}} = \frac{q}{1+q},$$

$$\frac{\alpha_{s-1} - \alpha_{s-2}}{\alpha_s - \alpha_{s-1}} = \frac{q^{s-2}}{q^{s-1}} = \frac{1}{q},$$

and one of the solutions of the equation $\frac{q}{1+q} = \frac{1}{q}$ is $2^{-1}(1+\sqrt{5})$, which proves that α_{s-1} is the golden section of $[\alpha_{s-2}, \alpha_s]$.

On the segment $[\alpha_{s-2}, \alpha_s]$, the search is continued in the following way. Let $\alpha_{s+1} = \alpha_s$, $\alpha_s = \alpha_{s-2} + \alpha_{s+1} - \alpha_{s-1}$ and compute $f(\alpha_s)$, μ_s and ξ_s.

The case $\mu_s \geq 0$ is considered similarly.

Let $\mu_s < 0$. Then, if $\alpha_{s+1} \leq \nu + a\alpha_s$, we let $x_+ = x(\alpha_s)$, $\bar{x}_+ = x(\alpha_{s+1})$; otherwise, we let $x_+ = x(\alpha_s)$, $\bar{x}_+ = x(\alpha_{s+1})$ and continue the search on the segment $[\alpha_{s-2}, \alpha_s]$.

It is clear that such a process is either finite, or
$f(\alpha_s) \xrightarrow[\alpha_s \to \infty]{} -\infty$.

<u>REMARK 1</u>. If the kth cycle is interrupted if condition (a) is satisfied and $x_{kt} = x_k$ $\forall t \in 0:t_k$, then this indeed implies that the number m_k is inappropriate. Then one needs to take a larger m_k and continue the kth cycle.

<u>REMARK 2</u>. From the computational point of view, the case where $\bar{x}_{kt} = x_{kt}$ is most convenient, since the point x_{kt} is taken as the consecutive point and the subgradient $\bar{v}_{k,t-1}$ is computed also at the point x_{kt}; there is no need to store superfluous points in the computer.

If $\alpha_{k,t-1} \geq \nu$, then it is always possible to obtain $\bar{x}_{kt} = x_{kt}$, after a finite number of computations of the function. However, it is not worthwhile to make much effort specially for that.

8. THE KELLEY METHOD

1. The Kelley Method. Let $f(x)$ be a finite convex function on E_n. Then

$$f(x) = \max_{z \in E_n} [f(z) + (v(z), x-z)] \equiv \max_{z \in E_n} B(z,x) \qquad (8.1)$$

(see (5.11) in Chapter 1), where $v(z) \in \partial f(z)$ (for each point z, one arbitrary vector of $\partial f(z)$ is chosen).

Our problem consists in finding

$$\min_{x \in \Omega} f(x) ,$$

where $\Omega \subset E_n$ is a bounded closed set. Choose an arbitrary collection σ_0 of a finite number of points in E_n. Let the set σ_k be already found.

Define the function

$$f_k(x) = \max_{z \in \sigma_k} B(z,x)$$

and find

$$\min_{x \in \Omega} f_k(x) = f_k(x_k) . \tag{8.2}$$

If

$$f_k(x_k) = f(x_k) , \tag{8.3}$$

then the point x_k is a minimum point of the function $f(x)$ on Ω.

Indeed, $f_k(x) \le f(x)$ and, hence,

$$f_k(x_k) = \min_{x \in \Omega} f_k(x) \le \min_{x \in \Omega} f(x) \equiv f^* \le f(x_k) . \tag{8.4}$$

Therefore, it follows from (8.3) that $f^* = f(x_k)$.

If $f_k(x_k) < f(x_k)$, then we let $\sigma_{k+1} = \sigma_k \cup \{x_k\}$ and proceed analogously. Note that $x_k \in \Omega$,

$$\begin{aligned} f_{k+1}(x_{k+1}) &\ge f_k(x_k) , \\ B(x_k, x_k) &= f(x_k) . \end{aligned} \tag{8.5}$$

As a consequence, we construct a sequence of sets $\{\sigma_k\}$ and a sequence of points $\{x_k\}$.

<u>THEOREM 8.1</u>. Every accumulation point of the sequence $\{x_k\}$ is a minimum point of the function $f(x)$ on Ω.

<u>Proof</u>. The existence of accumulation points follows from the boundedness and the closedness of the set Ω. Let $x_{k_s} \to x^* \in \Omega$.

It is necessary to show that

$$f(x^*) = f^* . \tag{8.6}$$

At first, we note that $f(x_{k_s}) \to f(x^*)$ due to the continuity of the function $f(x)$. Let us fix an arbitrary k_{s_0}. Then, for $k_s > k_{s_0}$, by virtue of (8.4),

$$f^* \geq f_{k_s}(x_{k_s}) \geq B(x_{k_{s_0}}, x_{k_s}) = f(x_{k_{s_0}}) + (v_{k_{s_0}}, x_{k_s} - x_{k_{s_0}}) . \tag{8.7}$$

Considering the boundedness of $v(z)$ on Ω (the mapping $\partial f(z)$ is bounded on any bounded set) and taking the limit under the condition $k_s > k_{s_0}$ as $k_s \to \infty$, we obtain from (8.7)

$$f^* \geq f(x^*) . \tag{8.8}$$

It is clear that

$$f(x^*) \geq \inf_{x \in \Omega} f(x) = f^* . \tag{8.9}$$

Note that (8.7) - (8.9) imply that $f_{k_s}(x_{k_s}) \to f^*$.

We obtain (8.6) from (8.8) and (8.9). The theorem has been proved. ■

<u>REMARK 1</u>. If $f(x_k) - f_k(x_k) \leq \varepsilon$, then from the inequality $f_k(x) \leq f(x)$ we have

$$\min_{x \in \Omega} f_k(x) = f_k(x_k) \leq \min_{x \in \Omega} f(x) = f^* .$$

Hence, $f(x_k) - \varepsilon \leq f^*$, i.e., x_k is an ε-stationary point of the function $f(x)$ on Ω.

<u>REMARK 2</u>. Now we consider the problem of the minimization of the function $f(x)$ on the entire space E_n. Assume that the set

$$M^* = \{x \in E_n \mid f(x) = \inf_{x \in E_n} f(x)\}$$

is nonempty and bounded. Choose a compact set $\Omega \subset E_n$ such that $M^* \subset \operatorname{int} \Omega$ (it is in our interest to choose a set Ω as simple as possible, for example, take a set Ω given by linear inequalities).

Let us apply the Kelly method. It follows from Theorem 8.1 that

$$\min_{x \in \Omega} f_k(x) = \min_{x \in E_n} f_k(x)$$

for large k, i.e., the set Ω plays an auxiliary role and is used only at the start of the process. If the set Ω is a priori unknown (however, it is known that M^* is bounded), then one can take an arbitrary set $\Omega_0 \subset E_n$ and carry out some steps; if, at some step k_0, it turns out that

$$x_{k_0} \in \text{int}\, \Omega_0 , \qquad (8.10)$$

then for $k > k_0$ we can solve the problem

$$\min_{x \in E_n} f_k(x)$$

instead of problem (8.2), by virtue of (8.4). If the inclusion (8.10) does not hold until much later, then one can try to take a new $\Omega_0' \supset \Omega_0$.

<u>REMARK 3</u>. Let

$$R_k = \{z \in \sigma_k \mid f_k(x_k) = B(z, x_k)\} .$$

It is easy to observe that in this case problem (8.2) is equivalent to the problem

$$\min_{x \in \Omega} \bar{f}_k(x) ,$$

where

$$\bar{f}_k(x) = \max_{z \in R_k} B(z,x) .$$

Assume that $f(x)$ is a strongly convex function and let x^* be a minimum point of the function $f(x)$ on Ω. Due to Theorem 8.1, the sequence $\{x_k\}$ converges to the point x^* (for a strongly convex function, there exists a unique minimum point). We show that an arbitrary point $\bar{x} \neq x^*$ can be encountered in R_k only for a finite number k. To show this, assume that this is not the case.

Let $\bar{x} \in R_{k_s}$, $k_s \to \infty$. We have

$$B(\bar{x}, x_{k_s}) = f(\bar{x}) + (v(\bar{x}), x_{k_s} - \bar{x}) = f_{k_s}(x_{k_s}) .$$

Note that $x_{k_s} \to x^*$. It follows from the proof of Theorem 8.1 (see (8.7) - (8.9)) that $f_{k_s}(x_{k_s}) \to f(x^*)$. Therefore, the last equality implies that

$$f(\bar{x}) + (v(\bar{x}), x^* - \bar{x}) = f(x^*) .$$

However, the function $f(x)$ is strongly convex and hence there exists an $m > 0$ such that

$$f(x^*) = f(\bar{x} + (x^* - \bar{x})) \geq f(\bar{x}) + (v(\bar{x}), x^* - \bar{x}) + m\|x^* - \bar{x}\|^2$$

(see (9.17) in Chapter 1). Since $\bar{x} \neq x^*$, then from this we have $f(x^*) > f(\bar{x}) + (v(\bar{x}, x^* - \bar{x})$. Therefore, we obtain a contradiction to the above-given equality.

Thus, every point $x_k \neq x^*$ is <<essential>> only a finite number of times.

This fact may be, perhaps, used in the practical realization of the Kelley method.

2. The Kelley ε-Method. For each $z \in E_n$, we choose a certain vector $v(z) \in \partial_\varepsilon f(z)$ and let

$$F(x) = \sup_{z \in E_n} [f(z) + (v(z), x - z)] . \tag{8.11}$$

By the definition of $\partial_\varepsilon f(z)$, we have

$$f(x) - f(z) \geq (v(z), x - z) - \varepsilon \quad \forall x \in E_n .$$

Hence

$$f(x) \leq \sup_{z \in E_n} [f(z) + (v(z), x - z)] \equiv F(x) \leq f(x) + \varepsilon . \tag{8.12}$$

The function $F(x)$ is also convex as a supremum of convex functions.

Let $\Omega \subset E_n$ be an arbitrary bounded convex set. Our problem consists in finding $\min_{x\in\Omega} f(x) \equiv f^*$. Recall that if $x^* \in \Omega$ and $f^* \le f(x^*) \le f^* + \varepsilon$, then the point x^* is called an ε-stationary point of the function $f(x)$ on Ω.

Note also that in this case $x^* \in \operatorname{int} \Omega$, then

$$0 \in \partial_\varepsilon f(x^*)$$

(see Lemma 8.1 in Chapter 1). Recall that the set $\bigcup_{z\in\Omega} \partial_\varepsilon f(z)$ is bounded.

Now we describe one generalization of the Kelley method for finding ε-stationary points. Let σ_0 be an arbitrary finite collection of points from E_n and let σ_k have already been found. Define the function

$$f_k(x) = \max_{z\in\sigma_k} B(z,x) ,$$

where

$$B(z,x) = f(z) + (v(z), x-z) , \qquad v(z) \in \partial_\varepsilon f(z) ,$$

and find

$$f_k(x_k) = \min_{x\in\Omega} f_k(x) . \tag{8.13}$$

If $f_k(x_k) \ge f(x_k)$, then from (8.12),

$$f^* \le f(x_k) \le f_k(x_k) = \min_{x\in\Omega} f_k(x) \le \min_{x\in\Omega} f(x) + \varepsilon = f^*+\varepsilon. \tag{8.14}$$

Hence, $f^* \le f(x_k) \le f^*+\varepsilon$, i.e., x_k is an ε-stationary point of the function $f(x)$ on Ω and the process terminates.

If $f_k(x_k) < f(x_k)$, then we let $\sigma_{k+1} = \sigma_k \cup \{x_k\}$, and we continue analogously. As a result, we get a sequence of sets $\{\sigma_k\}$ and points $\{x_k\}$ such that

$$x_k \in \Omega , \qquad f_k(x_k) \le f_{k+1}(x_{k+1}) . \tag{8.15}$$

We note also that if for some k and $\mu > 0$ we have

$$f(x_k) - f_k(x_k) \le \mu , \tag{8.16}$$

then, similarly to (8.14), we obtain

$$f^* \leq f(x_k) \leq f^* + \mu + \varepsilon \quad , \tag{8.17}$$

and this implies that x_k is an $(\varepsilon+\mu)$-stationary point of the function $f(x)$ on Ω.

THEOREM 8.2. Every accumulation point of the sequence $\{x_k\}$ is an ε-stationary point of the function $f(x)$ on Ω.

Proof. The existence of accumulation points follows from the boundedness of the set Ω. Let $x_{k_s} \to \tilde{x} \in \Omega$. Since

$$f_{k_s}(x_{k_s}) \geq B(x_{k_{s_0}}, x_{k_s}) = f(x_{k_{s_0}}) + (v(x_{k_{s_0}}), x_{k_s} - x_{k_{s_0}})$$

for $k_{s_0} < k_s$, then, taking the limit under the condition $k_s > k_{s_0}$ as $k_s, k_{s_0} \to \infty$ and considering the boundedness of v_{k_s} and (8.14), we obtain

$$f^* + \varepsilon \geq f(\tilde{x}) \geq f^* \tag{8.18}$$

and therefore, $\tilde{x}$ is an ε-stationary point of $f(x)$ on Ω. ∎

REMARK 4. If the set Ω is given by linear inequalities, then problem (8.13) is a linear programming problem.

REMARK 5. For $\varepsilon = 0$, we obtain the <<pure>> Kelley method.

REMARK 6. In Subsections 8.1, 8.2, the convexity of the set Ω was not assumed. The only requirement is the solvability of the problem of type (8.2).

3. A Modification of the Kelley Method With Accelerated Convergence.

Let f be a convex function defined and finite on the n-dimensional Euclidean space E_n, and let $\Omega \subset E_n$ be a compact convex set. It is required to find a point $x^* \in \Omega$ such that

$$f(x^*) = \min_{x \in \Omega} f(x) .$$

Choose $v(u) \in \partial f(u)$ and let us introduce the function

$$F(x,u,\varepsilon) = f(u) + (\tfrac{1}{2}+\varepsilon)(v(u),\ x-u) \quad . \tag{8.19}$$

Take an arbitrary point $x_0 \in \Omega$ and put $\sigma_0 = \{x_0\}$. Let $\sigma_K = \{x_0, x_1, \ldots, x_k\}$ have been found. Let us choose $\varepsilon_k = \varepsilon(x_k) \geq 0$ such that

$$F(x^*,\ x_k,\ \varepsilon_k) \leq f(x^*) \quad . \tag{8.20}$$

Such an ε_k exists for any k since for $\varepsilon = \frac{1}{2}$ it follows from the definition of $\partial f(x_k)$ that

$$f(x^*) \geq f(x_k) + (v(x_k),\ x^* - x_k) \ .$$

Therefore we can assume that $0 \leq \varepsilon_k \leq \frac{1}{2}$. Next, let us introduce the function

$$\phi_k(x) = \max_{i \in 0:k} F(x,\ x_i,\ \varepsilon_i) \tag{8.21}$$

and find

$$x_{k+1} = \arg\min\ \{\phi_k(x) \mid x \in \Omega\} \quad . \tag{8.22}$$

Now take $\sigma_{k+1} = \sigma_k \cup \{x_{k+1}\}$ and continue in the same manner.

THEOREM 8.3. If for some k

$$\phi_k(x_{k+1}) = f(x_{k+1}) \ ,$$

then x_{k+1} is a minimum point of f on Ω. Otherwise any limit point of the sequence $\{x_k\}$ is a minimum point of the function f on Ω.

Proof. The first part of the Theorem is obvious:

$$f(x_{k+1}) \geq f(x^*) \geq \phi_k(x^*) \geq \phi_k(x_{k+1}) = f(x_{k+1}) \ ,$$

which implies

$$f(x^*) = f(x_{k+1})$$

(the inequality $f(x^*) \geq \phi_k(x^*)$ follows from (8.20) and (8.21)).

To prove the rest of the Theorem, assume the opposite. Then there exists a subsequence $\{x_{k_s}\}$ such that $x_{k_s} \to \bar{x}$, $k_s \to \infty$ and

$$f(\bar{x}) > f(x^*) \quad . \tag{8.23}$$

By construction,

$$\phi_{k_s}(x_{k_{s+1}}) = \max_{i\in 0:x_{k_{s+1}}} F(x_{k_{s+1}}, x_i, \varepsilon_i) \geq F(x_{k_{s+1}}, x_{k_s}, \varepsilon_{k_s})$$

$$= f(x_{k_s}) + (\tfrac{1}{2}+\varepsilon_{k_s})(v(x_{k_s}), (x_{k_{s+1}} - x_{k_s}))$$

$$\xrightarrow[k_{s\to\infty}]{} f(\bar{x}) .$$

On the other hand, since $\phi_k(x) \leq \phi_{k+\ell}(x)\ \forall \ell > 0,\ \forall x$, then

$$\phi_{k_s}(x_{k_{s+1}}) \leq \phi_{k_{s+1}-1}(x_{k_{s+1}}) \leq \phi_{k_{s+1}-1}(x^*) \leq f(x^*) ,$$

i.e., $f(\bar{x}) \leq f(x^*)$, which contradicts (8.23).

<u>REMARK 1</u>. If $\varepsilon_k = \frac{1}{2}\ \forall k \in 0:\infty$, the method becomes the Kelley method.

<u>REMARK 2</u>. Let f be a quadratic function

$$f(x) = (Ax,x) + (b,x) ,$$

where A is an $n\times n$ positive definite matrix, $b \in E_n$, $\Omega \subset E_n$ is a convex set. If $x^* = \arg\min\{f(x) \mid x\in\Omega\} \in \operatorname{int}\Omega$, then by the necessary condition for a minimum,

$$f'(x^*) = 0 . \tag{8.24}$$

Therefore

$$f(x) + \tfrac{1}{2}(f'(x), x^*-x) - f(x^*)$$

$$= (Ax,x) + (b,x) + \tfrac{1}{2}(2Ax+b, x^*-x) - (Ax^*,x^*) - (b,x^*)$$

$$= (Ax^*, x-x^*) + \tfrac{1}{2}(b, x-x^*) = (Ax^*+\tfrac{1}{2}b, x-x^*) .$$

Since $f'(x^*) = 2Ax^*+b$, then from (8.24) we have

$$f(x) + \tfrac{1}{2}(f'(x), x^*-x) - f(x^*) = (f'(x^*), x-x^*) = 0 \qquad \forall x \in \Omega,$$

i.e.,

$$F(x^*, x, 0) \leq f(x^*) \qquad \forall x \in \Omega \tag{8.25}$$

and in (8.20) we can choose $\varepsilon_k = 0$.

Thus, for a quadratic convex function we can always take $\varepsilon_k = 0\ \forall k$.

THEOREM 8.4. If f is a strongly convex twice continuously differentiable function, then there exists a sequence $\{\varepsilon_k\}$ satisfying condition (8.20) such that $\varepsilon_k \to 0$ as $k \to \infty$.

Proof. Since f is twice continuously differentiable, then the matrix of the second derivatives is strictly positive definite. Let $x^* = \arg\min \{f(x) \mid x \in \Omega\}$. Assume that $x^* \in \operatorname{int} \Omega$. Since f is strongly convex, then there exists $\mu > 0$ such that

$$f(x^*) \geq f(x) + (f'(x), x^*-x) + \mu\|x^*-x\|^2 \qquad \forall x \in E_n ,$$

implying

$$(f'(x), x^*-x) \leq -\mu\|x^*-u\|^2 . \tag{8.26}$$

Let us introduce the function

$$f_1(x) = f(x^*) + (f'(x^*), x-x^*) + \tfrac{1}{2}(f''(x^*)(x-x^*), x-x^*) .$$

Clearly

$$f(x) = f(x^* + (x-x^*)) = f_1(x) + o(\|x-x^*\|^2) , \tag{8.27}$$

where

$$o(\|x-x^*\|^2) = \tfrac{1}{2}(f''(x^* + \theta(x)(x-x^*))) - (f''(x^*)(x-x^*), (x-x^*)) ,$$

$$\theta = \theta(x) \in (0,1)$$

and

$$f_1(x^*) = f(x^*) . \tag{8.28}$$

Since f_1 is a quadratic function, then it follows from (8.25) that

$$f_1(x) + \tfrac{1}{2}(f_1'(x), x^*-x) = f_1(x^*) \qquad \forall x \in \Omega . \tag{8.29}$$

From (8.27) we obtain

$$f_1(x) = f(x) + o(\|x-x^*\|^2) .$$

Therefore

$$f_1'(x) = f'(x) + o(\|x-x^*\|)$$

and (8.28) and (8.29) imply

$$\begin{aligned} f(x) + o(\|x-x^*\|^2) + \tfrac{1}{2}((f'(x),x^*-x) + (o(\|x-x^*\|), x^*-x)) &\leq f_1(x^*) \\ &= f(x^*) , \end{aligned}$$

or

$$f(x) + \tfrac{1}{2}(f'(x),\ x^*-x) + o(\|x-x^*\|^2) \leq f(x^*) . \qquad (8.30)$$

Since

$$o(\|x-x^*\|^2) = \alpha(x)\|x - x^*\|^2 , \qquad \text{where} \quad \alpha(x) \xrightarrow[x\to x^*]{} 0 ,$$

then it follows from (8.26) that

$$\frac{|\alpha(x)|}{\mu}(f'(x),\ x^*-x) \leq -|\alpha(x)|\ \|x - x^*\|^2 .$$

Moreover,

$$\frac{|\alpha(x)|}{\mu}(f'(x),\ x^*-x) \leq \alpha(x)\ \|x - x^*\|^2 .$$

Hence (8.30) implies

$$f(x) + (\tfrac{1}{2} + \varepsilon(x))(f'(x),\ x^*-x) \leq f(x^*) , \qquad (8.31)$$

where

$$\varepsilon(x) = \frac{|\alpha(x)|}{\mu} \xrightarrow[x\to x^*]{} 0 . \qquad (8.32)$$

Thus, if in the method described above (see (5)) we choose $\varepsilon_k = \varepsilon(x_k)$, then

(1) $x_k \to x^*$ (since (8.31) implies (8.20),

(2) $\varepsilon_k \to 0$ (due to (8.32)).

REMARK 3. This subsection is based on the results obtained by V.N. Tarasov and N.K. Popova.

9. MINIMIZATION OF A SUPREMUM-TYPE FUNCTION

Let X be a linear normed space, $G \subset E_p$ be a bounded set and let $\Omega \subset X$.

A continuous function $\phi(x,y)$ is given on $\Omega \times G$ and, in addition,

$$-\infty < \inf_{x\in\Omega} \sup_{y\in G} \phi(x,y) \equiv f^* < +\infty .$$

Let

$$f(x) = \sup_{y\in G} \phi(x,y) . \qquad (9.1)$$

Our problem consists in finding

$$\inf_{x\in\Omega} f(x) .$$

Assume that there exists a finite set $\sigma_0 \in G$ such that

$$\inf_{x\in\Omega} \max_{y\in\sigma_0} \phi(x,y) > -\infty . \tag{9.2}$$

Let the function $\phi(x,y)$ satisfy a Lipschitz condition with respect to y, i.e., there exists an $L < \infty$ such that

$$|\phi(x,y') - \phi(x,y'')| \le L\|y' - y''\| \qquad \forall x \in \Omega, \quad y',y'' \in G. \tag{9.3}$$

LEMMA 9.1. If $\{x_k\}$ $(x_k \in \Omega\ \forall k)$ is an arbitrary sequence and a sequence $\{y_k\}$ $(y_k \in \Omega\ \forall k)$ is such that

$$\phi(x_k, y_{k+1}) \ge \max_{0\le i\le k} \phi(x_k,y_i) , \tag{9.4}$$

then

$$\lim_{k\to\infty} [\phi(x_k,y_{k+1}) - \max_{0\le i\le k} \phi(x_k,y_i)] = 0 . \tag{9.5}$$

Proof. Assume that (9.5) is not valid; then it follows from (9.4) that there exist a sequence of indices $\{k_s\}$ and a number $\varepsilon_0 > 0$ such that

$$\phi(x_{k_s}, y_{k_s+1}) - \max_{0\le i\le k_s} (x_{k_s},y_i) \ge \varepsilon_0 .$$

Hence, considering (9.3) and (9.4),

$$\varepsilon_0 \le \phi(x_{k_s}, y_{k_s+1}) - \phi(x_{k_s},y_i) \le L\|y_{k_s+1} - y_i\| \qquad \forall i \in 0:k_s,$$

i.e.,

$$\|y_{k_s+1} - y_i\| \ge L^{-1}\varepsilon_0 \equiv \varepsilon \qquad \forall i \in 0:k_s ,$$

which contradicts the boundedness of the set G (every bounded set in a finite-dimensional space is covered by a finite number of spheres with the given radius ε). The lemma has been proved. ∎

Choose and fix $\varepsilon \ge 0$, $\delta \ge 0$.

Take an arbitrary set $\sigma_0 \in G$ satisfying (9.2). Let $\sigma_k \in G$ have already been found. Construct the function

$$\phi_k(x) = \max_{y\in\sigma_k} \phi(x,y)$$

and find a point $x_k \in \Omega$ such that

$$\phi_k(x_k) \le \inf_{y \in \Omega} \phi_k(x) + \delta , \tag{9.6}$$

i.e., we find the infimum of the function $\phi_k(x)$ with accuracy of δ.

Now we find a point $y_{k+1} \in G$ such that

$$\phi(x_k, y_{k+1}) \ge f(x_k) - \varepsilon , \tag{9.7}$$

$$\phi(x_k, y_{k+1}) \ge \phi_k(x_k) , \tag{9.8}$$

i.e., we solve problem (9.1) with accuracy of ε.

We let $\sigma_{k+1} = \sigma_k \cup \{y_{k+1}\}$ and continue analogously.

<u>THEOREM 9.1</u>. We have the following relations:

$$0 \le \underline{\lim}_{k\to\infty} f(x_k) - f^* \le \varepsilon + \delta ,$$

$$0 \le \overline{\lim}_{k\to\infty} f(x_k) - f^* \le \varepsilon + \delta .$$

<u>Proof</u>. It follows from (9.6) and (9.7) that

$$f^* \le f(x_k) \le \phi(x_k, y_{k+1}) + \varepsilon ,$$

$$f^* = \inf_{y \in \Omega} \sup_{y \in G} \phi(x,y) \ge \inf_{x \in \Omega} \max_{y \in \sigma_k} \phi(x,y) = \inf_{x \in \Omega} \phi_k(x)$$

$$\ge \phi_k(x_k) - \delta .$$

Hence,

$$0 \le f(x_k) - f^* \le \phi(x_k, y_{k+1}) + \varepsilon - f^*$$

$$\le \phi(x_k, y_{k+1}) - \phi_k(x_k) + \varepsilon + \delta . \tag{9.9}$$

It follows from Lemma 9.1 that $\phi(x_k, y_{k+1}) - \phi_k(x_k) \to 0$.

Therefore, (9.9) implies that for any $\mu > 0$ there exists a $K < \infty$ such that

$$0 \le f(x_k) - f^* \le \varepsilon + \delta + \mu \qquad \forall k > K . \qquad \blacksquare$$

<u>COROLLARY</u>. If, instead of (9.6) and (9.7), the points x_k and y_{k+1} satisfy the inequalities

$$\phi_k(x_k) \le \inf_{x\in\Omega} \phi_k(x) + \delta_k ,$$

$$\phi(x_k, y_{k+1}) \ge f(x_k) - \varepsilon_k ,$$

where $\delta_k \to +0$, $\varepsilon_k \to +0$, then

$$\lim_{k\to\infty} f(x_k) = f^*$$

and, moreover, any accumulation point of the sequence $\{x_k\}$, if such point exists, is a minimum point of the function $f(x)$ on the set Ω.

REMARK 1. Above, the function $\phi(x,y)$ was not assumed to be convex in x and the set Ω was not necessarily convex, either.

REMARK 2. Since any convex function is a maximum function (formula (5.11) in Chapter 1), this method can also be applied to finding a minimum of a convex function. As a result, the Kelley method is obtained (see Section 3.8).

10. MINIMIZATION OF A CONVEX MAXIMUM-TYPE FUNCTION AND THE EXTREMUM-BASIS METHOD

1. Let

$$f(x) = \max_{y\in G} \phi(x,y) ,$$

where G is a compact set, the function $\phi(x,y)$ is continuous in x and y on $E_n \times G$ together with $\phi'_x(x,y)$, and is strongly convex in x with a strong convexity constant $m > 0$, for each fixed $y \in G$, i.e

$$\phi(x+\Delta, y) \ge \phi(x,y) + (\phi'_x(x,y), \Delta) + m\|\Delta\|^2 \qquad (10.1)$$

$$\forall [x,y] \in E_n \times G , \qquad \forall \Delta \in E_n .$$

Take any $x_0 \in E_n$. Let

$$b = \min_{y\in G} \phi(x_0, y) , \qquad K_1 = \max_{y\in G} \|\phi'_x(x_0,y)\| .$$

From (4.1), it is clear that

$$\phi(x,y) \geq b - K_1\|x-x_0\| + m\|x-x_0\|^2 \qquad \forall [x,y] \in E_n \times G .$$

Hence,

$$\phi(x,y) > f(x_0) , \tag{10.2}$$

if $\|x-x_0\| > z$, where

$$z = [\tfrac{1}{2}K_1 + (\tfrac{1}{4}K_1^2 + m(f(x_0) - b))^{\frac{1}{2}}]m^{-1} .$$

Let $K = \max_{x \in S_z(x_0), y \in G} \|\phi_x'(x,y)\|$. Our problem consists in finding $\min_{x \in E_n} f(x)$.

In Section 3.1, it was shown that for a point $x^* \in E_n$ to be a minimum point of the function $f(x)$, it is necessary and sufficient that

$$0 \in L(x^*) , \tag{10.3}$$

where

$$L(x) = \text{co } H(x) , \qquad H(x) = \{\phi_x'(x,y) \mid y \in R(x)\} ,$$

$$R(x) = \{y \in G \mid \phi(x,y) = f(x)\} .$$

To each point $x_0 \in E_n$, there corresponds a set $L(x_0)$. If $0 \notin L(x_0)$, then the direction

$$g(x_0) = -z(x_0)\|z(x_0)\|^{-1} ,$$

where $z(x_0) = \min_{z \in L(x_0)} \|z\|$, is the steepest descent direction of the function $f(x)$ at the point x_0. The ability of computing the steepest descent direction, or some directions approximating it, makes it possible to develop numerical minimization methods (see Section 3.3). In that case, however, it is necessary to consider all the points of the set $R(x)$ and to minimize the function $f(x)$ on the ray (in the direction of descent). In practice, the set G, if it includes an infinite number of points, is discretized (i.e., it is replaced by a finite set of points); however, for more exact approximation of $f(x)$, it is necessary to take into account more points of the set G. It is advantageous to develop a minimization method of the function $f(x)$, to implement which one can do with comparatively small number of points of the set G.

Let us describe first <<a conceptual algorithm>> in the terminology of E. Polak [96]. Choose an arbitrary collection of $n+2$ points of the set G:

$$\sigma_0 = \{y_{01},\dots,y_{0,n+2}\}, \qquad y_{0i} \in G \quad \forall i \in 1:(n+2).$$

The collection σ_0 is called a basis.

Let points $x_0,\dots,x_{k-1}$ and a basis

$$\sigma_k = \{y_{k1},\dots,y_{k,n+2}\}, \qquad y_{ki} \in G \quad \forall i \in 1:(n+2)$$

have already been constructed. Define the function

$$f_k(x) = \max_{i\in 1:(n+2)} \phi(x,y_{ki})$$

and find a point x_k such that

$$f_k(x_k) = \min_{x\in E_n} f_k(x) \quad . \tag{10.4}$$

The solution of problem (10.4) is simpler than the minimization of the function $f(x)$, because to find the function $f_k(x)$ it is necessary to calculate the values of the function $f(x,y)$ only at $n+2$ points.

Due to the necessary condition for a minimum of the function $f_k(x)$, we have

$$0 \in L_k , \tag{10.5}$$

where

$$L_k = \text{co } H_k , \qquad H_k = \{\phi'_x(x_k,y_{ki}) \mid i \in R_k\} ,$$

$$R_k = \{i \mid \phi(x_k,y_{ki}) = f_k(x_k)\} \quad .$$

Since, by Carathéodory's theorem (Theorem 1.1 in Chapter 1), an arbitrary point of the set L_k can be represented as a convex combination of $n+1$ or fewer points of the set H_k, then there exists at least one index $i_k \in 1:(n+2)$, which we can omit, i.e., either $i_k \in R_k$ or we can form the origin in (10.5) without the vector $\phi'_x(x_k, y_{ki_k})$. Let $\bar{y}_k \in G$ be such that

$$\phi(x_k, \bar{y}_k) = f(x_k) = \max_{y \in G} \phi(x_k, y) , \qquad (10.6)$$

i.e., $\bar{y}_k \in R(x_k)$. Now we construct a new basis $\sigma_{k+1} = \{y_{k+1,i}, \ldots, y_{k+1,n+2}\}$ as follows:

$$y_{k+i,i} = \begin{cases} y_{ki} , & i \notin i_k , \\ \bar{y}_{ki} , & i = i_k . \end{cases}$$

The basis σ_{k+1} is different by one point from the basis σ_k and includes $n+2$ points. If $f(x_k) = f_k(x_k)$, then x_k is a minimum point of the function $f(x)$, because, in this case, $\phi(x_k, y_{ki}) = f(x_k)$, i.e., $y_{ki} \in R(x_k)$ and, by virtue of (10.5), the necessary and sufficient condition (10.3) is satisfied. If $f(x_k) > f_k(x_k)$, then, starting from the basis σ_{k+1}, we construct a function $f_{k+1}(x)$ and continue analogously.

As a result, we have a sequence of points $\{x_k\}$. If this sequence is finite, then its final point is a minimum point of the function $f(x)$. If the sequence $\{x_k\}$ is infinite, then the following theorem holds.

<u>THEOREM 10.1</u>. The sequence of points $\{x_k\}$ converges to a minimum point of the function $f(x)$.

<u>Proof</u>. First of all, we note that the sequence $\{x_k\}$ is bounded because it follows from (10.2) that $f_k(x) > f(x_0) \ge f_k(x_0)$, provided $\|x - x_0\| > z$. Hence,

$$\|x_k - x_0\| \le z \qquad (10.7)$$

since $f_k(x_k) = \min_{x \in E_n} f_k(x) \le f_k(x_0)$.

Define $\Delta_k = x_{k+1} - x_k$. We have from (10.1) that

$$\phi(x_{k+1},y_{ki}) = \phi(x_k+\Delta_k,y_{ki})$$

$$\geq \phi(x_k,y_{ki}) + (\phi'_x(x_k,y_{ki}),\Delta_k) + m\|\Delta_k\|^2 . \quad (10.8)$$

For $i \in R_k$, we have $\phi(x_k,y_{ki}) = f_k(x_k)$.

By virtue of (10.5), there exists $y_{ki} \in \sigma_k$ such that

$$(\phi'_x(x_k,y_{ki}), \Delta_k) \geq 0 . \quad (10.9)$$

Hence, from (10.8), we have for such an i,

$$\phi(x_{k+1},y_{ki}) \geq f_k(x_k) + m\Delta_k^2 ,$$

i.e.,

$$f_{k+1}(x_{k+1}) \equiv \max_i \phi(x_{k+1},y_{k+1,i}) \geq f_k(x_k) + m\Delta_k^2 . \quad (10.10)$$

Then it follows that

$$\Delta_k \xrightarrow[k\to\infty]{} 0 . \quad (10.11)$$

To show this, we assume that this is not the case, i.e., we assume that there exist a number $a > 0$ and a subsequence $\{k_s\}$ such that $\|\Delta_{k_s}\| \geq a$. Then, from (10.10) we have $f_k(x_k) \to +\infty$; on the other hand, we have $f(x_k) \geq f_k(x_k)$; hence, $f(x_k) \to +\infty$, which is impossible since $\{x_k\}$ is a bounded sequence and $f(x)$ is a continuous function.

Thus, (10.11) has been proved.

From (10.6), we have

$$f(x_k) = f_{k+1}(x_k) . \quad (10.12)$$

Since

$$\phi(x_{k+1},y) = \phi(x_k+\Delta_k,y) = \phi(x_k,y) + (\phi'_x(x_k+\xi_k\Delta_k,y), \Delta_k) ,$$

where $\xi_k \in (0,1)$, then

$$|f(x_{k+1}) - f(x_k)| \leq K\|\Delta_k\| ,$$

$$|f_{k+1}(x_{k+1}) - f_{k+1}(x_k)| = K\|\Delta_k\| ,$$

and from (10.12) we obtain

$$\begin{aligned}|f_{k+1}(x_{k+1})-f(x_{k+1})| &\le |f_{k+1}(x_{k+1})-f_{k+1}(x_k)| + |f_{k+1}(x_k)-f(x_{k+1})| \\ &= |f_{k+1}(x_{k+1})-f_{k+1}(x_k)| + |f(x_k)-f(x_{k+1})| \\ &\le 2K\|\Delta_k\| \ . \end{aligned} \tag{10.13}$$

From (10.5), we have $0 \in \text{co}\{\phi_x'(x_{k+1},y_{k+1,i}) \mid i\in R_k\}$, where $R_k = \{i \mid \phi(x_{k+1},y_{k+1,i})=f_{k+1}(x_{k+1})\}$. Hence, from (10.13),

$$0 \in L_{\varepsilon_k}(x_{k+1}) \quad ,$$

where

$$\varepsilon_k = 2K\|\Delta_k\| \ , \qquad L_\varepsilon(x) = \text{co } H_\varepsilon(x) \ ,$$

$$H_\varepsilon(x) = \{\phi_x'(x,y) \mid y\in G:\ f(x)-\phi(x,y) \le \varepsilon\} \ .$$

Indeed x_{k+1} is an ε_k-stationary point of the function $f(x)$ on E_n (see Section 3.1).

Since $\varepsilon_k \to 0$, then the sequence of ε_k-stationary points converges to a unique minimum point of the function $f(x)$ on E_n. ∎

Above, we assumed that the function $\phi(x,y)$ is strongly convex with respect to x. This condition is essential, as the following example illustrates.

<u>EXAMPLE 1</u>. Let $x\in E_1$, $G = 1:4$, and

$$\phi(x,1) \equiv 1 \ , \qquad \phi(x,2) = x^2 \ , \qquad \phi(x,3) = 2-x \ , \qquad \phi(x,4) = 2+x \ .$$

Also let

$$f(x) = \max_{y\in G} \phi(x,y) = \max_{i=1:4} \phi(x,i) \ .$$

Since $n = 1$, then the basis should consist of three points.

Let $\sigma_0 = \{1,2,4\}$. It is clear that

$$\min_{x\in E_1} f_0(x) = \min_{x\in E_1} \max \{1, x^2, 2+x\} = f_0(-1) = 1 \ ,$$

i.e., $x_0 = -1$. In this case,

$$R_0 = \{1,2,4\} \ , \qquad H_0 = \{\phi_x'(-1,1),\ \phi_x'(-1,2),\ \phi_x'(-1,4)\} = \{0,-2,1\} \ .$$

As i_0 we can take $i_0 = 4$. Since

$$f(x_0) = \max_{y \in 1:4} \phi(x_0, y) = \phi(x_0, 3) = 3 ,$$

then $\bar{y}_0 = 3$, hence, $\sigma_1 = \{1,2,3\}$.

Now we have $f_1(x) = \max \{1, x^2, x-2\}$, and, therefore, $\min_{x \in E_1} f_1(x) = f_1(+1) = 1$, i.e., $x_1 = 1$. In this case, $R_0 = \{1,2,3\}$, $H_0 = \{0,2,-1\}$.

We can let $i_1 = 3$. Since $f(x_1) = \phi(x_1, 4) = 3$, then $\bar{y}_1 = 4$. Again, we have $\sigma_2 = \{1,2,4\} = \sigma_0$. Similarly, we obtain

$$\sigma_3 = \sigma_1, \ldots, \sigma_{2k} = \sigma_0, \sigma_{2k+1} = \sigma_1 .$$

Thus, the cycling occurs.

It is obvious that $\min f(x) = f(0) = 2$.

The cycling occurs because the strong convexity of the function $\phi(x,y)$ is insufficient (the functions $\phi(x,1)$, $\phi(x,3)$, $\phi(x,4)$ are not strongly convex).

2. Generally speaking, the auxiliary problem (10.4) in the above-mentioned method cannot be solved in a finite number of steps, therefore, the method in Subsection 9.1 is not implementable.

In the sequel, we shall describe three more methods, two of which enable us to implement the successive approximation methods. The proof of convergence is given in [20], [33].

First, we shall describe the ε-method. Choose a sequence $\{\varepsilon_k\}$ such that

$$\varepsilon_k \to +0 , \qquad \sum_{k=0}^{\infty} \varepsilon_k < \infty . \tag{10.14}$$

Let us take an arbitrary point $x_0 \in E_n$ and an arbitrary $(n+2)$-point basis

$$\sigma_0 = \{y_{01}, \ldots, y_{0,n+2}\} , \qquad y_{0i} \in G \quad \forall i \in 1:(n+2)$$

Let a point $x_k \in E_n$ and the basis

$$\sigma_k = \{y_{k1}, \ldots, y_{k,n+2}\}, \qquad y_{ki} \in G \quad \forall i \in 1:(n+2)$$

have already been constructed.

Let

$$f_k(x) = \max_{i \in 1:(n+2)} \phi(x, y_{ki}) \tag{10.15}$$

and find a point x_{k+1} such that

$$0 \in L_{k\varepsilon_k}, \tag{10.16}$$

where

$$L_{k\varepsilon_k} = \text{co}\, H_{k\varepsilon_k}, \qquad H_{k\varepsilon_k} = \{\phi'_x(x_{k+1}, y_{ki}) \mid i \in R_{k\varepsilon_k}\},$$

$$R_{k\varepsilon_k} = \{i \in 1:(n+2) \mid \phi(x_{k+1}, y_{ki}) \geq f_k(x_{k+1}) - \varepsilon_k\}.$$

The point x_{k+1} is an ε_k-stationary point of the function $f_k(x)$.

Now we construct a new basis σ_{k+1} in the same way as in Subsection 10.1.

If $f(x_{k+1}) = f_k(x_{k+1})$ and, moreover, (10.5) is valid, then the point x_{k+1} is a minimum point of the function $f(x)$ and the process terminates.

Otherwise, starting from the point x_{k+1} and the basis σ_{k+1}, we construct the function $f_{k+1}(x)$ and find an ε_{k+1}-stationary point of the function. If the sequence of points $\{x_k\}$ constructed in such a way is finite, then its terminal point is a minimum point of the function $f(x)$. Otherwise, the following theorem holds.

THEOREM 10.2. The sequence of points $\{x_k\}$ converges to a minimum point of the function $f(x)$.

3. Unfortunately, the ε-method described in Subsection 9.2 is still not convenient to use, because the finding of ε-stationary points involves in many cases an infinite procedure.

The ρ-method which follows does not have this drawback.

Let a sequence $\{\rho_k\}$ be such that

$$\rho_k \to +0 , \qquad \sum_{k=0}^{\infty} \rho_k < \infty . \tag{10.17}$$

Suppose a point x_k and a basis σ_k have already been found and a function $f_k(x)$ has been defined (see (10.15)). Let us find a point x_{k+1} such that

$$\min_{z \in L_k} \| z \| \le \rho_k .$$

Here, L_k is the set defined after relation (10.5). We construct a new basis σ_{k+1} in the same way as in Subsection 10.1. As a result, we get a sequence $\{x_k\}$.

THEOREM 10.3. The sequence of points $\{x_k\}$ converges to a minimum point of the function $f(x)$.

4. To conclude, we shall describe the (ε,ρ)-method, which is a combination of the ε-method and the ρ-method.

Let sequences $\{\varepsilon_k\}$ and $\{\rho_k\}$ satisfy relations (10.14) and (10.17). As x_{k+1}, we choose a point such that

$$\min_{z \in L_{k\varepsilon_k}} \| z \| \le \rho_k , \tag{10.18}$$

where $L_{k\varepsilon_k}$ is the set defined after the inclusion (10.16). The change of the basis is performed as above. The finding of a point x_{k+1} satisfying inequality (10.18) requires a finite number of steps.

For the sequence thus obtained, the convergence theorem also holds.

REMARK 1. The methods described can be generalized to the case where the minimization is performed on the set given by inequalities ([20], [33]).

REMARK 2. Since any convex function can be represented as a maximum function of its supports (i.e., any convex function is the maximum of a linear function; and any convex set is representable as the intersection of half-planes), the extremum-basis method can be used to solve the problem of minimizing a convex function on a convex set; in this case, it is necessary to solve a linear programming problem of small dimension on each step. To avoid possible cycling (because linear functions are not strongly convex), special methods need to be developed (see [20]).

Thus, unlike the Kelley method (see Section 3.8), the extremum-basis method does not entail any increase in the number of constraints for the auxiliary linear programming problem.

11. A NUMERICAL METHOD FOR MINIMIZING QUASIDIFFERENTIABLE FUNCTIONS

In Chapter 2 we studied properties of quasidifferentiable functions. It was shown how to verify necessary conditions for an extremum and how to find steepest descent directions. But, of course, one cannot apply these directions in a straight way since "a jamming effect" is possible. It requires special precautions to avoid these difficulties.

Here we describe an algorithm for minimizing a certain class of quasidifferentiable functions, i.e., that produced by smooth compositions of max-type functions. The main feature of the algorithm is that at each step it is necessary to consider a bundle of auxiliary directions and points, of which only one can be chosen for the next step. This requirement seems to arise from the intrinsic nature of nondifferentiable functions.

Let

$$f(x) = F(x, y_1(x), \dots, y_m(x)) , \tag{11.1}$$

where

$$x \in E_n , \qquad y_i(x) \equiv \max_{j \in I_i} \phi_{ij}(x) , \qquad I_i \equiv 1:N_i ,$$

and functions $F(x,y_1,\ldots,y_m)$ and $\phi_{ij}(x)$ are continuously differentiable on E_{n+m} and E_n, respectively.

Take any $g \in E_n$. Then for $\alpha \geq 0$ we have

$$y_i(x+\alpha g) = y_i(x) + \alpha \frac{\partial y_i(x)}{\partial g} + o_i(\alpha,g) ,$$

where

$$\frac{\partial y_i(x)}{\partial g} \equiv \lim_{\alpha \to +0} \frac{y_i(x+\alpha g) - y_i(x)}{\alpha} = \max_{j \in R_i(x)} (\phi'_{ij}(x), g) ,$$

$$\phi'_{ij}(x) = \frac{\partial \phi_{ij}(x)}{\partial x} ,$$

$$R_i(x) = \{j \in I_i \mid \phi_{ij}(x) = y_i(x)\} ,$$

$$\frac{o_i(\alpha,g)}{\alpha} \xrightarrow[\alpha \to +0]{} 0 . \tag{11.2}$$

This leads to

$$f(x+\alpha g) = f(x) + \alpha \left[\left(\frac{\partial F(y(x))}{\partial x}, g \right) + \sum_{i \in I} \frac{\partial F(y(x))}{\partial y_i} \frac{\partial y_i(x)}{\partial g} \right] + o(\alpha,g) , \tag{11.3}$$

where

$$I \equiv 1:m , \qquad y(x) \equiv (x, y_1(x), \ldots, y_m(x)) ,$$

and

$$\frac{o(\alpha,g)}{\alpha} \xrightarrow[\alpha \to +0]{} 0 . \tag{11.4}$$

It is clear that convergence in (11.2) and (11.4) is uniform with respect to $g \in S_1 \equiv \{g \in E_n \mid \|g\| = 1\}$. Let

$$I_+(x) = \left\{ i \in I \mid \frac{\partial F(y(x))}{\partial y_i} > 0 \right\} ,$$

$$I_-(x) = \left\{ i \in I \mid \frac{\partial F(y(x))}{\partial y_i} < 0 \right\} .$$

Then from (11.3) we have

$$f(x+\alpha g) = f(x) + \alpha\left|\left(\frac{\partial F(y(x))}{\partial x}, g\right) + \sum_{i\in I_+(x)} \max_{j\in R_i(x)}\left(\frac{\partial F(y(x))}{\partial y_i}\phi'_{ij}(x), g\right)\right.$$

$$\left. + \sum_{i\in I_-(x)} \min_{j\in R_i(x)}\left(\frac{\partial F(y(x))}{\partial y_i}\phi'_{ij}(x), g\right)\right| + o(\alpha, g) . \qquad (11.5)$$

It follows from (11.5) that f is quasidifferentiable and

$$Df(x) = [\underline{\partial} f(x), \overline{\partial} f(x)] ,$$

where

$$\underline{\partial} f(x) = \text{co } A(x) , \qquad \overline{\partial} f(x) = \text{co } B(x) ,$$

$$A(x) = \left\{v\in E_n \mid v = \frac{\partial F(y(x))}{\partial x} + \sum_{i\in I_+(x)} \frac{\partial F(y(x))}{\partial y_i}\phi'_{ij}(x), \quad j\in R_i(x)\right\} ,$$

$$B(x) = \left\{w\in E_n \mid w = \sum_{i\in I_-(x)} \frac{\partial F(y(x))}{\partial y_i}\phi'_{ij}(x), \quad j\in R_i(x)\right\} .$$

The following lemmas can be derived from the above necessary and sufficient conditions (see (5.1) and (5.10) of Chapter 2).

<u>LEMMA 11.1</u>. For any set of coefficients

$$\{\lambda_{ij} \mid i\in I_-(x^*), \ j\in R_i(x^*), \ \lambda_{ij}\ge 0, \ \sum_{j\in R_i(x^*)}\lambda_{ij} = 1\}$$

there exists another set of coefficients

$$\{\lambda_{ij} \mid i\in I_+(x^*), \ j\in R_i(x^*), \ \lambda_{ij}\ge 0, \ \sum_{j\in R_i(x^*)}\lambda_{ij} = 1\}$$

such that

$$\frac{\partial F(y(x^*))}{\partial x_i} + \sum_{i\in I}\frac{\partial F(y(x^*))}{\partial y_i}\sum_{j\in R_i(x^*)}\lambda_{ij}\phi'_{ij}(x^*) = 0 . \qquad (11.6)$$

(If $\frac{\partial F(y(x^*))}{\partial y_i} = 0$ put $\lambda_{ij} = 0 \ \forall j\in R_i(x^*)$.) Condition (11.6) is a multipliers rule -- note the difference between it and the Lagrange multipliers rule for mathematical programming.

It follows from (11.6) that x^* is a stationary point of the smooth function

$$F_\lambda(x) = F\left(x, \sum_{j\in R_1(x^*)}\lambda_{1j}\phi_{1j}(x), \ldots, \sum_{j\in R_m(x^*)}\lambda_{mj}\phi_{mj}(x)\right) ,$$

and if $\underline{\partial} f(x^*)$ consists of more than one point, then the set $\{\lambda_{ij}\}$ is not unique. (Of course, it may not be unique even if $\underline{\partial} f(x^*)$ is a singleton.)

LEMMA 11.2. If for any $w \in B(x^*)$ there exist sets

$$\{v_i \mid i \in 1:(n+1)\} \qquad \text{and} \qquad \left\{\alpha_i \mid \alpha_i > 0, \ \sum_{i=1}^{n+1} \alpha_i = 1\right\}$$

such that the vectors $\{v_i\}$ form a simplex (i.e., vectors $\{v_i - v_{n+1} \mid i \in 1:n\}$ are linearly independent) and $w = \sum_{i=1}^{n+1} \alpha_i v_i$, then x^* is a local minimum point of f on E_n.

We shall now introduce the following sets, where $\varepsilon \geq 0$, $\mu \geq 0$:

$$R_{i\varepsilon}(x) = \{j \in I_i \mid \phi_{ij}(x) \geq y_i(x) - \varepsilon\} ,$$

$$\underline{\partial}_\varepsilon f(x) = \text{co}\left\{v \in E_n \mid v = \frac{\partial F(y(x))}{\partial x} + \sum_{i \in I_+(x)} \frac{\partial F(y(x))}{\partial y_i} \phi'_{ij}(x), \ j \in R_{i\varepsilon}(x)\right\} ,$$

$$B_\mu(x) = \left\{w \in E_n \mid w = \sum_{i \in I_-(x)} \frac{\partial F(y(x))}{\partial y_i} \phi'_{ij}(x), \ j \in R_{i\mu}(x)\right\} .$$

Let f be defined by (11.1). A point $x^* \in E_n$ will be called an ε-*inf-stationary point* of f on E_n if

$$-\overline{\partial} f(x^*) \subset \underline{\partial}_\varepsilon f(x^*) .$$

We shall describe an algorithm for finding an ε-inf-stationary point, with $\varepsilon > 0$ and $\mu > 0$ fixed.

Choose an arbitrary $x_0 \in E_n$. Suppose that x_k has been found. If

$$-\overline{\partial} f(x_k) \subset \underline{\partial}_\varepsilon f(x_k) , \tag{11.7}$$

then x_k is an ε-inf-stationary point and the process terminates. If, on the other hand, (11.7) is not satisfied, then for every $w \in B_\mu(x_k)$

$$\min_{v \in \underline{\partial}_\varepsilon f(x_k)} \|w+v\| = \|w + v_k(w)\| .$$

If $w + v_k(w) \neq 0$, then let $g_k(w) = -\frac{w+v_k(w)}{\|w+v_k(w)\|}$ and compute

$$\min_{\alpha \geq 0} f(x_k + \alpha g_k(w)) = f(x_k + \alpha_k(w) g_k(w)) \quad . \tag{11.8}$$

If $w + v_k(w) = 0$, then take $\alpha_k(w) = 0$ and find

$$\min_{w \in B_\mu(x_k)} f(x_k + \alpha_k(w) g_k(w)) = f(x_k + \alpha_k(w_k) g_k(w_k)) \quad .$$

We then set

$$x_{k+1} = x_k + \alpha_k(w_k) g_k(w_k) \quad .$$

It is clear that

$$f(x_{k+1}) < f(x_k) \quad . \tag{11.9}$$

By repeating this procedure we obtain a sequence of points $\{x_k\}$. If it is a finite sequence (i.e., consists of a finite number of points), then its final element is an ε-inf-stationary point by construction. Otherwise the following result holds.

<u>THEOREM 11.1</u>. If the set $D(x_0) = \{x \in E_n \mid f(x) \leq f(x_0)\}$ is bounded, then any limit point of the sequence $\{x_k\}$ is an ε-inf-stationary point of f on E_n.

<u>Proof</u>. The existence of limit points follows from the boundedness of $D(x_0)$. Let x^* be a limit point of $\{x_k\}$, i.e., $x^* = \lim_{k_s \to \infty} x_{k_s}$. It is clear that

$$x^* \in D(x_0) \quad .$$

Assume that x^* is not an ε-inf-stationary point. Then there exists a $w^* \in B_0(x^*)$ such that

$$\min_{v \in \underline{\partial}_\varepsilon f(x^*)} \|w^* + v\| = a > 0 \quad . \tag{11.10}$$

We shall denote by w_k^* the point in $B\mu(x_k)$ which is nearest to w^* and by $\rho(x_k^*)$ the distance of w_k^* from w^*. It is obvious that $\rho(x_{k_s}^*) \xrightarrow[k_s \to \infty]{} 0$. It may also be seen that the mapping $\underline{\partial}_\varepsilon f(x)$ is upper semicontinuous. From (11.10) and the above statements it follows that there exists a $K \to \infty$ such that

$$\min_{v \in \underline{\partial}_\varepsilon f(x_{k_s})} \|w^*_{k_s} + v\| \equiv \|w^*_{k_s} + v(w^*_{k_s})\| = a_{k_s} \geq \frac{a}{2} \qquad \forall k_s > K \,. \quad (11.11)$$

Now we have

$$\begin{aligned} f(x_{k_s} + \alpha g_{k_s}) &= f(x^* + (x_{k_s} - x^* + \alpha g_{k_s})) \\ &= f(x^*) + \frac{\partial f(x^*)}{\partial [x_{k_s} - x^* + \alpha g_{k_s}]} + o(\|x_{k_s} - x^* + \alpha g_{k_s}\|), \quad (11.12) \end{aligned}$$

where

$$g_{k_s} \equiv g_{k_s}(w^*_{k_s}) = -\frac{w^*_{k_s} + v_{k_s}(w^*_{k_s})}{\|w^*_{k_s} + v_{k_s}(w^*_{k_s})\|} \,,$$

$$\begin{aligned} &\frac{\partial f(x^*)}{\partial [x_{k_s} - x^* + \alpha g_{k_s}]} \\ &= \sum_{i \in I_+(x^*)} \max_{j \in R_i(x^*)} \left(\frac{\partial F(y(x^*))}{\partial x} + \frac{\partial F(y(x^*))}{\partial y_i} \phi'_{ij}(x^*),\ x_{k_s} - x^* + \alpha g_{k_s} \right) \\ &\quad + \sum_{i \in I_-(x^*)} \min_{j \in R_i(x^*)} \left(\frac{\partial F(y(x^*))}{\partial y_i} \phi'_{ij}(x^*),\ x_{k_s} - x^* + \alpha g_{k_s} \right). \quad (11.13) \end{aligned}$$

Since

$$\max_{i \in I} a_i + \min_{i \in I} b_i \leq \max_{i \in I} [a_i + b_i] \leq \max_{i \in I} a_i + \max_{i \in I} b_i \,,$$

$$\min_{i \in I} a_i + \min_{i \in I} b_i \leq \min_{i \in I} [a_i + b_i] \leq \min_{i \in I} a_i + \max_{i \in I} b_i \,,$$

it follows from (11.13) that

$$\begin{aligned} &\frac{\partial f(x^*)}{\partial [x_{k_s} - x^* + \alpha g_{k_s}]} \\ &= \alpha \left[\left(\frac{\partial F(y(x^*))}{\partial x}, g_{k_s} \right) + \sum_{i \in I_+(x^*)} \max_{j \in R_i(x^*)} \left(\frac{\partial F(y(x^*))}{\partial y_i} \phi'_{ij}(x^*), g_{k_s} \right) \right. \\ &\quad \left. + \sum_{i \in I_-(x^*)} \min_{j \in R_i(x^*)} \left(\frac{\partial F(y(x^*))}{\partial y_i} \phi'_{ij}(x^*), g_{k_s} \right) \right] + \sum_{i \in I} \beta_i(\alpha, x_{k_s} - x^*) \\ &= \alpha \frac{\partial f(x^*)}{\partial g_{k_s}} + \sum_{i \in I} \beta_i(\alpha, x_{k_s} - x^*) \,, \quad (11.14) \end{aligned}$$

where

$$\beta_i(\alpha, x_{k_s}-x^*) \in \left[\min_{j\in R_i(x^*)} \left(\frac{\partial F(y(x^*))}{\partial x} + \frac{\partial F(y(x^*))}{\partial y_i} \phi'_{ij}(x^*),\ x_{k_s}-x^* \right), \max_{j\in R_i(x^*)} \left(\frac{\partial F(y(x^*))}{\partial x} + \frac{\partial F(y(x^*))}{\partial y_i} \phi'_{ij}(x^*),\ x_{k_s}-x^* \right) \right] .$$

It is clear that $\beta_i(\alpha, x_{k_s}-x^*) \underset{k_s\to\infty}{\Longrightarrow} 0$ uniformly with respect to α.

From (11.11) it also follows that for k_s sufficiently large,

$$\frac{\partial f(x^*)}{\partial g_{k_s}} \le -\frac{a}{4} .$$

From (11.14) and (11.12) we conclude that there exist values of $\alpha_0 > 0$ and k_s such that

$$f(x_{k_s}+\alpha_0 g_{k_s}) < f(x^*) .$$

But this is impossible since

$$f(x_{k_s+1}) = \min_{w\in B_\mu(x_{k_s})} f(x_{k_s}+\alpha_{k_s}(w) g_{k_s}(w)) \le f(x_{k_s}+\alpha_{k_s}(w^*_{k_s}) g_{k_s}(w^*_{k_s}))$$

$$= \min_{\alpha>0} f(x_{k_s}+\alpha g_{k_s}(w^*_{k_s})) \le f(x_{k_s}+\alpha_0 g_{k_s}) < f(x^*) .$$

This contradicts (11.9) and the fact that $f(x_k) \underset{k\to\infty}{\Longrightarrow} f(x^*)$. ∎

<u>REMARK</u>. Now it is clear that an analogous method can be applied to minimize other classes of nonsmooth functions, e.g.,

$$f(x) = \max_{i\in I} \min_{j\in J_i} \phi_{ij}(x) .$$

It is also possible to extend the method to the constrained case (see, e.g., [6*]).

Chapter 4

CONSTRAINED MINIMIZATION

1. NECESSARY AND SUFFICIENT CONDITIONS FOR A MINIMUM OF A CONVEX FUNCTION ON A CONVEX SET

1. Let $S \subset E_n$ be an open set, $\Omega \subset S$ be a closed convex set and let a convex function $f(x)$ be defined on Ω.

The problem consists in finding a point $x^* \in \Omega$ such that

$$f^* \equiv f(x^*) = \inf_{x \in \Omega} f(x) \quad . \tag{1.1}$$

In the investigation of problem (1.1), necessary and sufficient conditions for a minimum of the function $f(x)$ on the set Ω play an important role. Let us recall them (see Sections 1.6 and 1.10).

Let $\gamma(x_0) = \{v = \lambda(z-x_0) \mid \lambda>0, z\in\Omega\}$.

The set $\Gamma(x_0) = \overline{\gamma(x_0)}$ is said to be a *cone of feasible directions* of the set Ω at the point x_0. Denote by $\Gamma^+(x_0)$ the conjugate cone to a cone $\Gamma(x_0)$. The set $\gamma(x_0)$ is a convex cone, $\Gamma(x_0)$ and $\Gamma^+(x_0)$ are closed convex cones and, moreover, $0 \in \gamma(x_0) \cap \Gamma^+(x_0)$ (see Section 1.3).

We recall here the following theorem proved in Section 1.6.

<u>THEOREM 1.1</u>. For the function $f(x)$ to attain its minimum value on Ω at a point $x^* \in \Omega$, it is necessary and sufficient that

$$\partial f(x^*) \cap \Gamma^+(x^*) \neq \emptyset \quad . \tag{1.2}$$

If we use the concept of a conditional subdifferential of the function $f(x)$ on the set Ω at the point x_0:

$$\partial^{\Omega} f(x_0) = \{v \in E_n \mid f(z)-f(x_0) \geq (v, z-x_0) \quad \forall z \in \Omega\} ,$$

then, considering the relation (see Lemma 10.2 in Chapter 1)

$$\partial^{\Omega} f(x_0) = \partial f(x_0) - \Gamma^{+}(x_0) , \tag{1.3}$$

condition (1.2) can be written in the following form:

$$0 \in \partial^{\Omega} f(x^*) . \tag{1.4}$$

Let $x_0 \in \Omega$ and let

$$\phi(x_0) = \min_{\|g\|=1} \frac{\partial^{\Omega} f(x_0)}{\partial g} , \qquad \psi(x_0) = \min_{\|g\|=1, g \in \Gamma(x_0)} \frac{\partial f(x_0)}{\partial g} ,$$

$$d(x_0) = \min_{v \in \partial^{\Omega} f(x_0)} \|v\| , \qquad \rho(x_0) = \min_{v \in [\partial^{\Omega} f(x_0)]_{fr}} \|v\| .$$

If $0 \in \partial^{\Omega} f(x_0)$, then $\rho(x_0)$ is the radius of the largest sphere centered at the origin and included in $\partial^{\Omega} f(x_0)$: if $0 \notin \partial^{\Omega} f(x_0)$, then $\rho(x_0) = \min_{v \in \partial^{\Omega} f(x_0)} \|v\| = d(x_0)$, i.e., $\rho(x_0)$ is the distance from the origin to the set $\partial^{\Omega} f(x_0)$.

Condition (1.2) is equivalent to the inequality

$$\psi(x^*) \geq 0$$

(see Lemma 6.3 in Chapter 1).

<u>LEMMA 1.1</u>. Let $x_0 \in \Omega$. We have the following relations:

$$\psi(x_0) = \phi(x_0) , \tag{1.5}$$

$$\psi(x_0) = \begin{cases} -\rho(x_0) = -d(x_0) & \text{if } d(x_0) > 0 , \\ \rho(x_0) & \text{if } d(x_0) = 0 . \end{cases} \tag{1.6}$$

<u>Proof</u>. Applying Lemmas 6.2 and 6.4 of Chapter 1 for the sets $B = \partial f(x_0)$ and $\Gamma^{+} = \Gamma^{+}(x_0)$, we have proved the lemma. ∎

Take an $x_0 \in \Omega$. If $\psi(x_0) < 0$, $\|v(x_0)\| = \rho(x_0)$ and $v(x_0) \in \partial^{\Omega} f(x_0)$, then the vector $g_0 \equiv g(x_0) = -v(x_0)\|v(x_0)\|^{-1}$ is the steepest descent direction of the function $f(x)$ at the point x_0 with respect to the set Ω and it follows from (1.5) that

$$-\rho(x_0) = \frac{\partial f(x_0)}{\partial g_0} = \min_{\|g\|=1, g\in\Gamma(x_0)} \frac{\partial f(x_0)}{\partial g} = \min_{\|g\|=1} \frac{\partial^{\Omega} f(x_0)}{\partial g}$$

(see Section 1.6).

Let

$$T_\eta(x) = \{v \in -\Gamma^+(x) \mid \|v\| = \eta\} ,$$

$$L_\eta(x) = \text{co}\,\{\partial f(x) \cup T_\eta(x)\} ,$$

where $\eta > 0$ is an arbitrary fixed number.

We note that if $\Gamma^+(x) = 0$, then $L_\eta(x) = \partial f(x)$.

<u>LEMMA 1.2</u>. For condition (1.2) to be satisfied, it is necessary that

$$0 \in L_\eta(x^*) \tag{1.7}$$

and, moreover, this condition is also sufficient if the set Ω has interior points.

<u>Proof</u>. Necessity. Let (1.2) be satisfied. Then, there exist a $v \in \partial f(x^*)$ and a $w \in \Gamma^+(x^*)$ such that $v = w$. If $w = 0$, then $v = 0$ and, hence, (1.7) is satisfied. If $w \neq 0$, then $g = -\eta w\|w\|^{-1} \in T_\eta(x^*)$ and since $v - w = 0$, then

$$\eta v\|w\|^{-1} + g = 0 . \tag{1.8}$$

From (1.8) we obtain for $\alpha = \eta\|w\|^{-1}(1+\eta\|w\|^{-1})^{-1}$ that $v_\alpha = \alpha v + (1-\alpha)g = 0$.

Since $\alpha \in (0,1]$, $v \in \partial f(x^*)$ and $g \in T_\eta(x^*)$, then $v_\alpha \in L_\eta(x^*)$ and, hence, (1.7) is satisfied.

<u>Sufficiency</u>. If $T_\eta(x^*) \neq \emptyset$, then it follows from (1.7) that $0 \in \partial f(x^*)$. On the other hand, $0 \in \Gamma^+(x^*)$ and, therefore, (1.2) is satisfied.

Now, let $T_\eta(x^*) \neq \emptyset$. It follows from (1.7) that there exists $v \in \partial f(x^*)$, $g \in \text{co}\, T_\eta(x^*)$ and a number $\alpha \in [0,1]$ such that

$$\alpha v + (1-\alpha)g = 0 . \tag{1.9}$$

Since the set Ω has interior points, then $0 \notin \text{co}\, T_\eta(x^*)$. Hence, in representation (1.9) we should have $\alpha > 0$. Then, by virtue of (1.9), we get

$$v = (\alpha-1)\alpha^{-1}g .$$

However, we have $(\alpha-1)\alpha^{-1}g$ $(= v) \in \Gamma^+(x^*)$ because $(\alpha-1)\alpha^{-1} < 0$. Therefore, (1.2) is satisfied in this case, too.

Lemma 1.2 has been proved. ∎

Let $x_0 \in \Omega$ and $\psi(x_0) < 0$. Then,

$$f(x_0) - f(x) \leq d(x_0)\|x-x_0\| \qquad \forall x \in \Omega . \tag{1.10}$$

Define $d_\eta(x_0) = \min\limits_{v \in L_\eta(x_0)} \|v\|$.

To estimate $f(x_0) - f^*$ in terms of $d_\eta(x_0)$, we establish the relation between $d(x_0)$ and $d_\eta(x_0)$.

Let $A, B \in E_n$ be convex compact sets and let

$$L = \text{co}\,\{A \cup B\} , \qquad \tilde{d} = \min_{v \in L} \|v\| ,$$

$$\tilde{\rho} = \min_{v \in L_{fr}} \|v\| , \qquad \tilde{\psi} = \min_{\|g\|=1} \max_{v \in L} (v,g) .$$

Moreover, let $0 \notin A$ and $\Gamma^+ = K(A)$, where $K(A) = \{\lambda v \mid \lambda \geq 0,\ v \in A\}$. By virtue of Lemma 3.5 in Chapter 1, the set Γ^+ is a closed convex cone.

<u>LEMMA 1.3.</u> Let

$$\|v\| \geq a > 0 \quad \forall v \in A, \quad \|v\| \leq b \quad \forall v \in B. \tag{1.11}$$

Then

$$\tilde{d} \leq d \leq (1 + 2ba^{-1})\tilde{d} , \tag{1.12}$$

where

$$d = \min_{v \in B - \Gamma^+} \|v\| .$$

Proof. Take an arbitrary $\lambda \in (0,1)$. If $\tilde{d} > \lambda a$, then, considering $0 \in \Gamma^+$, $B \subset B - \Gamma^+$ and (1.11), we have

$$d = \min_{v \in B-\Gamma^+} \|v\| \leq \min_{v \in B} \|v\| \leq b \leq b(\lambda a)^{-1}\tilde{d} ,$$

i.e.,

$$d \leq b(\lambda a)^{-1}\tilde{d} . \tag{1.13}$$

Now we consider the case $\tilde{d} \leq \lambda a$. Let $v_0 \in L$ and $\|v_0\| = \tilde{d}$. Then, the vector v_0 can be represented in the form

$$v_0 = \alpha v_1 + (1-\alpha)v_2 , \qquad \alpha \in [0,1], \quad v_1 \in A, \quad v_2 \in B .$$

By virtue of (1.11), we have

$$\lambda a \geq \tilde{d} = \|\alpha v_1 + (1-\alpha)v_2\| \geq \alpha\|v_1\| - (1-\alpha)\|v_2\| \geq \alpha a - (1-\alpha)b ,$$

and from this

$$\alpha \leq (\lambda a+b)(a+b)^{-1} < 1 ,$$

because $\lambda \in (0,1)$.

Since $\alpha \in [0,1)$, then $\alpha(1-\alpha)^{-1}v_1 \in -\Gamma^+$ and hence

$$\tilde{d} = \|\alpha v_1 + (1-\alpha)v_2\| = (1-\alpha)\|\alpha(1-\alpha)^{-1}v_1 + v_2\|$$

$$\geq (1-\alpha)d \geq a(1-\lambda)(a+b)^{-1}d .$$

Therefore, if $\tilde{d} \leq \lambda a$, then

$$d \leq (a+b)[(1-\lambda)a]^{-1}\tilde{d} . \tag{1.14}$$

From (1.13) and (1.14), we obtain

$$d \leq \min_{\lambda \in (0,1)} \max \{(a+b)[(1-\lambda)a]^{-1}, b(\lambda a)^{-1}\}\tilde{d} = (1 + 2ba^{-1})\tilde{d} .$$

Now prove the first inequality of (1.12).

Let $d = \|v_1+v_2\|$, $v_1 \in -\Gamma^+$, $v_2 \in B$. Since $\Gamma^+ = -K(A)$, then there exists a $w_1 \in A$ and a $\lambda_1 > 0$ such that $v_1 = \lambda_1 w_1$. Then,

$$d = (1+\lambda_1)\|\lambda_1(1+\lambda_1)^{-1}w_1 + (1+\lambda_1)^{-1}v_2\| \geq (1+\lambda_1)\tilde{d} \geq \tilde{d} .$$

Hence, Lemma 1.3 has been proved. ■

Now assume that the set Ω is bounded (hence, $b \equiv \max_{x\in\Omega} \max_{v\in\partial f(x)} \|v\| < \infty$).

Let D be the diameter of the set Ω, r be the radius of the largest sphere included in Ω and, moreover, let $r > 0$.

Then, for any point $x_0 \in \Omega$, the conjugate cone $\Gamma^+(x_0)$ is included in the solid angle $2\pi - \alpha$, where $\sin(\frac{\alpha}{2}) = D^{-1}r$, and hence

$$\min_{x\in\Omega} \min_{v\in \mathrm{co}\, T_\eta(x)} \|v\| \geq \eta r D^{-1}$$

(Figures 54 and 55).

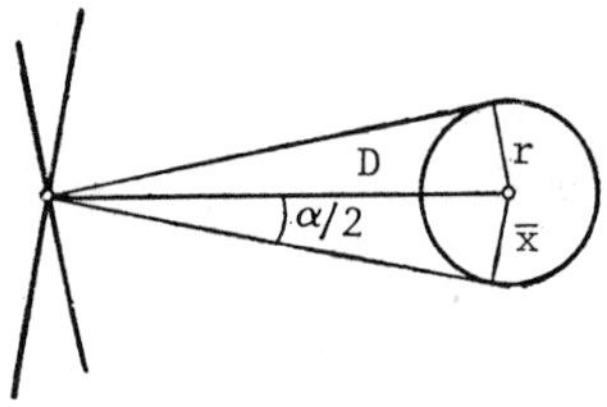

Figure 54

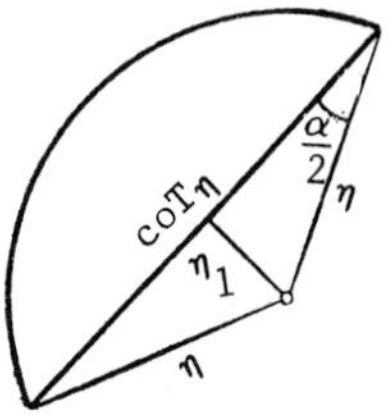

Figure 55

Finally, from (1.10) and Lemma 1.3 we have

$$0 \leq f(x_0) - f^* \leq D(1 + 2bD(\eta r)^{-1}) d_\eta(x_0) . \tag{1.15}$$

2. We shall consider the case where the set Ω is given by

$$\Omega = \{x \in E_n \mid h(x) \leq 0\} ,$$

where $h(x)$ is a convex function on E_n.

Here, we assume that the Slater condition is satisfied, i.e., $h(\bar{x}) < 0$ for some $\bar{x} \in E_n$. Let $x_0 \in \Omega$. In this case,

$$\Gamma^+(x_0) = \begin{cases} \{0\} & \text{if } h(x_0) < 0 , \\ -K(\partial h(x_0)) & \text{if } h(x_0) = 0 , \end{cases}$$

(see Section 1.12), where $K(A) = \{w = \lambda v \mid \lambda \geq 0, v \in A\}$.

In particular, if $h(x) = \max_{j\in I} h_j(x)$, $J = 1:m$, and $h_j(x)$ are continuously differentiable convex functions on E_n, then

$$\partial h(x_0) = \text{co}\,\{h_j'(x_0) \mid j \in Q(x_0)\} ,$$

where $Q(x_0) = \{j \in J \mid h_j(x_0) = h(x_0)\}$. Hence,

$$\Gamma^+(x_0) = \begin{cases} \{0\} & \text{if } h(x_0) < 0 , \\ \{v = -\sum_{j \in Q(x_0)} \lambda_j h_j'(x_0) \mid \lambda_j \geq 0 \ \forall \lambda_j \in Q(x_0)\} & \text{if } h(x_0) = 0 . \end{cases}$$

For any $x_0 \in \Omega$, we introduce the set

$$L(x_0) = \begin{cases} \partial f(x_0) & \text{if } h(x_0) < 0 , \\ \text{co}\,\{\partial f(x_0) \cup \partial h(x_0)\} & \text{if } h(x_0) = 0 . \end{cases}$$

THEOREM 1.2. Let the Slater constraint qualification be satisfied.

For the function $f(x)$ to attain its minimum value on the set Ω at a point $x^* \in \Omega$, it is necessary and sufficient that

$$0 \in L(x^*) . \tag{1.16}$$

Proof. Necessity. We need to prove that (1.16) follows from (1.2). Let (1.2) be satisfied. This implies that we can find a $v \in \partial f(x^*)$ and a $w \in \Gamma^+(x^*)$ such that

$$v - w = 0 . \tag{1.17}$$

If $h(x^*) < 0$, then $\Gamma^+(x^*) = \{0\}$, i.e., $w = 0$ and from (1.17) we have $v = 0$, i.e., $0 \in \partial f(x^*) \subset L(x^*)$.

Now let $h(x^*) = 0$. Since the Slater constraint qualification is satisfied, then $\Gamma^+(x^*) = -K(\partial h(x^*))$. Hence, there exist a $\lambda \geq 0$ and a $z \in \partial h(x^*)$ such that $w = -\lambda z$, and from (1.17) we obtain $v + \lambda z = 0$.

Consequently, $(1+\lambda)^{-1} v + \lambda(1+\lambda)^{-1} z = 0$. On the other hand, clearly, $(1+\lambda)^{-1} v + \lambda(1+\lambda)^{-1} z \in L(x^*)$: indeed $0 \in L(x^*)$ and this gives us the proof.

Sufficiency. It is necessary to show that (1.2) follows from (1.16). If $h(x^*) < 0$, then this is obvious.

Now let $h(x^*) = 0$. The inclusion (1.16) implies that there exist $v \in \partial f(x^*)$, $w \in \partial h(x^*)$ and a number $\alpha \in [0,1]$ such that

$$\alpha v + (1-\alpha)w = 0 . \tag{1.18}$$

The number α in (1.18) cannot be equal to 0 because, due to the Slater constraint qualification, $0 \notin \partial h(x^*)$.

Dividing (1.18) by α, we obtain

$$v = (\alpha-1)\alpha^{-1}w .$$

On the other hand, $(\alpha-1)\alpha^{-1}w \in \Gamma^+(x^*)$ and, therefore, $\partial f(x^*) \cap \Gamma^+(x^*) \neq \emptyset$. Theorem 1.2 has been proved. ■

Let $x_0 \in \Omega$ and define

$$\tilde{\psi}(x_0) = \min_{\|g\|=1} \max_{v \in L(x_0)} (v,g) , \qquad \tilde{d}(x_0) = \min_{v \in L(x_0)} \|v\| ,$$

$$\tilde{\rho}(x_0) = \min_{v \in [L(x_0)]_{fr}} \|v\| .$$

If $0 \in L(x_0)$, then $\tilde{\rho}(x_0)$ is the radius of the maximum sphere centered at 0 and included in $L(x_0)$; and if $0 \in L(x_0)$, then $\tilde{\rho}(x_0) = \tilde{d}(x_0)$, i.e., $\tilde{\rho}(x_0)$ is the distance from the origin to the set $L(x_0)$.

Applying Lemmas 6.2 - 6.4 in Chapter 1, we obtain for the sets $B = L(x_0)$ and $\Gamma^+ = 0$ that (1.16) is equivalent to the condition that $\tilde{\psi}(x^*) \geq 0$, and also, that

$$\tilde{\psi}(x_0) = \begin{cases} -\tilde{\rho}(x_0) = -\tilde{d}(x_0) & \text{if } \tilde{d}(x_0) > 0 , \\ \tilde{\rho}(x_0) & \text{if } \tilde{d}(x_0) = 0 . \end{cases}$$

Let $\tilde{\psi}(x_0) < 0$ and let $\tilde{v}(x_0) \in L(x_0)$, $\|\tilde{v}(x_0)\| = \tilde{\rho}(x_0)$. The vector $\tilde{g} = \tilde{g}(x_0) = -\tilde{v}(x_0)\|\tilde{v}(x_0)\|^{-1}$ is said to be the *direction of quasisteepest descent of the function* $f(x)$ *on the set* Ω *at the point* x_0 [36].

Let $h(x_0)=0$ and $\tilde{d}(x_0)>0$. We shall pay a special attention to the fact that if $g(x_0)$ and $\tilde{g}(x_0)$ are the directions of steepest and quasisteepest descents of the function $f(x)$ on the set at the point x_0, then it may happen that

$$x_0 + \alpha g(x_0) \notin \Omega \qquad \forall\, \alpha > 0 .$$

However, there always exists a number $\alpha_0 > 0$ such that

$$x_0 + \alpha\tilde{g}(x_0) \in \Omega \qquad \forall\, \alpha \in [0,\alpha_0] .$$

This is explained by the fact that $\dfrac{\partial h(x_0)}{\partial g(x_0)} \le 0$, and $\dfrac{\partial h(x_0)}{\partial \tilde{g}(x_0)} < 0$.

Thus, $\tilde{g}(x_0)$ is a direction of descent of the function $f(x)$ at the point x_0 and, moreover, in this direction it is possible to leave the point x_0, remaining in the set Ω. This is clear from the following. We have the expansion

$$f(x_0) + \alpha\tilde{g}(x_0)) = f(x_0) + \alpha\frac{\partial f(x_0)}{\partial \tilde{g}(x_0)} + o(\alpha) , \qquad \frac{o(\alpha)}{\alpha} \xrightarrow[\alpha\to+0]{} 0 .$$

For sufficiently small α, we have $\dfrac{o(\alpha)}{\alpha} \le \frac{1}{2}\tilde{\rho}(x_0)$.

Hence, for such α

$$f(x_0 + \alpha\tilde{g}(x_0)) \le f(x_0) - \tfrac{1}{2}\alpha\tilde{\rho}(x_0) .$$

Analogously, we can show that for sufficiently small α

$$h(x_0 + \alpha\tilde{g}(x_0)) \le -\tfrac{1}{2}\alpha\tilde{\rho}(x_0) .$$

We also note that $\tilde{\rho}(x_0) \le \rho(x_0)$. Indeed, let us show this.

Let

$$\rho(x_0) = \|v_0 + \lambda_0 w_0\| ,$$

where

$$v_0 \in \partial f(x_0) , \qquad w_0 \in \partial h(x_0) , \qquad \lambda_0 \ge 0 .$$

If $\lambda_0 = 0$, then $\tilde{\rho}(x_0) = \rho(x_0)$. If $\lambda_0 > 0$, then

$$\rho(x_0) = (1+\lambda_0)\|v_0(1+\lambda_0)^{-1} + \lambda_0 w_0(1+\lambda_0)^{-1}\| \ge (1+\lambda_0)\tilde{\rho}(x_0) .$$

Hence we have $\tilde{\rho}(x_0) \le \rho(x_0)$, and, in turn, $\frac{\partial f(x_0)}{\partial g(x_0)} \le \frac{\partial f(x_0)}{\partial \tilde{g}(x_0)}$. Moreover, if $\tilde{\rho}(x_0) = \rho(x_0)$, then $g(x_0) = \tilde{g}(x_0)$ and if $\tilde{\rho}(x_0) < \rho(x_0)$, then $g(x_0) \ne \tilde{g}(x_0)$ and $\frac{\partial f(x_0)}{\partial g(x_0)} < \frac{\partial f(x_0)}{\partial \tilde{g}(x_0)}$.

Now we assume that

$$\max_{v\in\partial f(x_0)} \|v\| \le b , \qquad \min_{v\in\partial h(x_0)} \|v\| \ge a .$$

Then it follows from (1.10) and Lemma 1.3 that

$$0 \le f(x_0) - f(x) \le \|x-x_0\| (1+2a^{-1}b)\tilde{d}(x_0) \qquad \forall x \in \Omega. \quad (1.19)$$

PROBLEM 1.1. Let $\Omega_1 = \{x\in E_n \mid h_1(x)\le 0\}$, where $h_1(x) = \max_{j\in J_1} h_{1j}(x)$, $h_{1j}(x) = (A_j, x) + b_j$, $\|A_j\| > 0 \quad \forall j \in J_1 \equiv 1:N$.

Prove that for the function $f(x)$ to attain its minimum value on the set $\Omega \cap \Omega_1$ at a point $x^* \in \Omega \cap \Omega_1$, it is necessary and sufficient that

$$L(x^*) \cap \Gamma_1^+(x^*) \ne \emptyset ,$$

where

$$L(x) = \begin{cases} \partial f(x) & \text{if } h(x) < 0 , \\ \text{co } \{\partial f(x) \cup \partial h(x)\} & \text{if } h(x) = 0 , \end{cases}$$

$$\Gamma_1^+(x) = \begin{cases} \{0\} & \text{if } h_1(x) < 0 , \\ \{v = -\sum_{j\in Q_1(x)} \lambda_j A_j \mid \lambda_j \ge 0 \ \forall j \in Q_1(x)\} & \text{if } h_1(x) = 0 , \end{cases}$$

$$Q_1(x) = \{j \in J_1 \mid h_{1j}(x) = h_1(x)\} .$$

2. ε-STATIONARY POINTS

Let a convex function $f(x)$ be given on some open convex set including a closed convex set Ω.

1. Take an $x_0 \in \Omega$ and an $\varepsilon \ge 0$. Let

$$\psi_\varepsilon^\Omega(x_0) = \min_{\|g\|=1} \sup_{v\in\partial_\varepsilon^\Omega f(x_0)} (v,g) ,$$

$$d_\varepsilon^\Omega(x_0) = \min_{v\in\partial_\varepsilon^\Omega f(x_0)} \|v\| ,$$

$$\rho_\varepsilon^\Omega(x_0) = \min_{v\in[\partial_\varepsilon^\Omega f(x_0)]_{fr}} \|v\| , \qquad \rho^\Omega(x_0) = \rho_0^\Omega(x_0) ,$$

$$d^\Omega(x_0) = d_0^\Omega(x_0) , \qquad \psi^\Omega(x_0) = \psi_0^\Omega(x_0) ,$$

$$f^* = \inf_{x\in\Omega} f(x) .$$

It is clear that if $0 \notin \partial_\varepsilon^\Omega f(x_0)$, then $\rho_\varepsilon^\Omega(x_0) = d_\varepsilon^\Omega(x_0)$ and if $0 \in \partial_\varepsilon^\Omega f(x_0)$, then $\rho_\varepsilon^\Omega(x_0)$ is the radius of the largest sphere centered at the origin and included in $\partial_\varepsilon^\Omega f(x_0)$.

Applying Corollary 2 of Lemma 6.4 in Chapter 1 for the set $B = \partial_\varepsilon^\Omega f(x_0)$, we get

$$\psi_\varepsilon^\Omega(x_0) = \begin{cases} -\rho_\varepsilon^\Omega(x_0) = -d_\varepsilon^\Omega(x_0) & \text{if } d_\varepsilon^\Omega(x_0) > 0 , \\ \rho_\varepsilon^\Omega(x_0) & \text{if } d_\varepsilon^\Omega(x_0) = 0 . \end{cases} \tag{2.1}$$

The point $x_0 \in \Omega$ at which $\psi_\varepsilon^\Omega(x_0) \geq 0$ or, equivalently, $0 \leq f(x_0) - f^* \leq \varepsilon$ is called an *ε-stationary point* of the function $f(x)$ on the set Ω.

The following lemmas are obvious generalizations of Lemmas 1.2 and 1.3 in Chapter 3.

<u>LEMMA 2.1</u>. For any $x_0 \in \Omega$, we have the estimate

$$f(y) - f(x_0) \geq \|y - x_0\| \psi_\varepsilon^\Omega(x_0) - \varepsilon \qquad \forall\, y \in \Omega . \tag{2.2}$$

<u>LEMMA 2.2</u>. Let $x_\varepsilon \in \Omega$, $\varepsilon > 0$ and $\psi_\varepsilon^\Omega(x_\varepsilon) > 0$. Then, $f^* > -\infty$, the set $D^* = \{x\in\Omega \mid f(x)=f^*\}$ is a nonempty convex compact set and

$$0 \leq f(x_\varepsilon) - f^* < \varepsilon .$$

Moreover, if $f(x_\varepsilon) > f^*$, then

$$\frac{f(x_\varepsilon) - f^*}{\rho^\Omega(x_\varepsilon)} \le \|x_\varepsilon - x^*\| \le \frac{f^* - f(x_\varepsilon) + \varepsilon}{\rho_\varepsilon^\Omega(x_\varepsilon)} , \tag{2.3}$$

where x^* is a minimum point of the function $f(x)$ on the set Ω.

Let a function $f(x)$ be strongly convex on Ω and let m be its strong convexity constant. Following the arguments in Lemma 9.6 and its Corollary in Chapter 1, we can show that for any $x_0 \in \Omega$

$$f(x_0) - f(x^*) \le (4m)^{-1}(d_\varepsilon^\Omega(x_0) + 2\sqrt{\varepsilon m}\,)^2 , \tag{2.4}$$

$$\|x_0 - x^*\| \le (2m)^{-1}(d_\varepsilon^\Omega(x_0) + 2\sqrt{\varepsilon m}\,) , \tag{2.5}$$

$$f(x_0) - f(x^*) \le (4m)^{-1}[d^\Omega(x_0)]^2 , \tag{2.6}$$

$$\|x_0 - x^*\| \le (2m)^{-1}d^\Omega(x_0) . \tag{2.7}$$

2. Let $f(x)$ and $h(x)$ be convex functions on E_n. Let us fix an $x_0 \in E_n$. Assume that there exists a point $\bar{x} \in E_n$ such that

$$h(\bar{x}) \le h(x_0) - a , \tag{2.8}$$

where $a > 0$.

<u>LEMMA 2.3</u>. If, for some $\varepsilon \ge 0$ and $\mu \in [0,\alpha)$, the inclusion

$$0 \in \operatorname{co} \{\partial_\varepsilon f(x_0) \cup \partial_\mu h(x_0)\} \tag{2.9}$$

holds, then

$$f(x_0) \le f^*_{\mu-h(x_0)} + \varepsilon , \tag{2.10}$$

where

$$f^*_{\mu-h(x_0)} = \inf_{x\in\Omega_{\mu-h(x_0)}} f(x) ,$$

$$\Omega_{\mu-h(x_0)} = \{x \in E_n \mid h(x) \le h(x_0)-\mu\} .$$

<u>Proof</u>. From (2.9) follows the existence of $v_0 \in \partial_\varepsilon f(x_0)$, $w_0 \in \partial_\varepsilon h(x_0)$ and $\alpha_1 \in [0,1]$ such that

$$\alpha_1 v_0 + (1-\alpha_1)w_0 = 0 . \tag{2.11}$$

Here, $\alpha_1 > 0$ because otherwise we have $w_0 = 0$, which is impossible by virtue of (2.8) (if $w_0 = 0$, then this contradicts (2.8)).

Hence, (2.11) can be rewritten in the form

$$v_0 + \alpha_0 w_0 = 0 , \tag{2.12}$$

where

$$\alpha_0 = (1-\alpha_1)\alpha_1^{-1} .$$

Due to the definition of $\partial_\varepsilon f(x_0)$ and $\partial_\mu f(x_0)$, we have

$$f(x) - f(x_0) \geq (v_0, x-x_0) - \varepsilon \qquad \forall\, x \in E_n ,$$

$$h(x) - h(x_0) \geq (w_0, x-x_0) - \mu \qquad \forall\, x \in E_n ,$$

which together with (2.12) imply

$$f(x) - f(x_0) + \alpha_0(h(x) - h(x_0)) \geq (v_0+\alpha_0 w_0,\ x-x_0) - (\varepsilon + \alpha_0\mu)$$

$$= -(\varepsilon + \alpha_0\mu) \qquad \forall x \in E_n . \tag{2.13}$$

For $x \in \Omega_{\mu-h(x_0)}$, we have $h(x) \leq h(x_0) - \mu$. Then, from (2.13)

$$f(x) - f(x_0) \geq -\varepsilon \qquad \forall x \in \Omega_{\mu-h(x_0)} ,$$

and this implies that

$$f(x) \geq f^*_{\mu-h(x_0)} + \varepsilon .$$ ∎

<u>COROLLARY</u>. Assume that there exists a point $\tilde{x} \in E_n$ such that $f(\tilde{x}) \leq f(x_0) - b$ and $b > \varepsilon$. If (2.9) holds, then

$$h(x_0) \leq h^*_{\varepsilon-f(x_0)} + \mu ,$$

where

$$f^*_{\varepsilon-f(x_0)} = \inf \{h(x) \mid f(x) \leq f(x_0)-\varepsilon\} .$$

3. THE CONDITIONAL-GRADIENT METHOD

In this section we assume that $\Omega \in E_n$ is an arbitrary convex compact set.

1. We shall first obtain necessary and sufficient conditions for a minimum of a convex function $f(x)$ on the set Ω and use them in the conditional-gradient method.

Introduce the function

$$\Phi_\varepsilon(x) = \min_{z\in\Omega} \max_{v\in\partial_\varepsilon f(x)} (v, z-x) ,$$

where $\varepsilon \geq 0$ is an arbitrary number, and let

$$\Phi(x) = \Phi_0(x) .$$

Here, we note that

$$\Phi_\varepsilon(x) \leq \varepsilon \qquad \forall x \in \Omega \quad \forall \varepsilon \geq 0 . \tag{3.1}$$

THEOREM 3.1. For the function $f(x)$ to attain its minimum value on the set Ω at a point $x^* \in \Omega$, it is necessary and sufficient that

$$\Phi(x^*) = \min_{z\in\Omega} \max_{v\in\partial_\varepsilon f(x^*)} (v, z-x^*) = 0 .$$

Proof. Necessity. Assume that this is not the case.

Then, there exists a point $z^* \in \Omega$ such that

$$\max_{v\in\partial f(x^*)} (v, z^*-x^*) < 0$$

(because $\Phi(z) \leq 0 \ \forall z \in \Omega$ from (3.1)).

Here, let $g = z^* - x^*$ and, by virtue of the upper semicontinuit of the mapping $\partial_\varepsilon f(x)$ with respect to ε on $[0,\infty)$, we can find an $\varepsilon_0 > 0$ such that

$$\frac{\partial_{\varepsilon_0} f(x^*)}{\partial g} = \max_{v\in\partial_{\varepsilon_0} f(x^*)} (v,g) < 0 .$$

Then, according to (8.19) in Chapter 1,

$$f(x^*) - \inf_{\alpha>0} f(x^*+\alpha g) > \varepsilon_0 .$$

Consequently, for sufficiently small $\alpha \in (0,1]$ we have

$$x^* + \alpha g \in \Omega , \qquad f(x^*+\alpha g) < f(x^*) .$$

This contradicts the assumption that x^* is a minimum point of the function $f(x)$ on the set Ω.

Sufficiency. Let $\Phi(x^*) = 0$. According to the definition of $\partial f(x^*)$, we have for any $z \in \Omega$

$$f(z) \geq f(x^*) + \max_{v \in \partial f(x^*)} (v, z-x^*) \geq f(x^*) + \Phi(x^*) = f(x^*) ,$$

which proves the sufficiency.

Let

$$f^* = \min_{x \in \Omega} f(x) .$$

<u>LEMMA 3.1</u>. For any $x_0 \in \Omega$ and any $\varepsilon \geq 0$, we have the estimate

$$0 \leq f(x_0) - f^* \leq \varepsilon - \Phi_\varepsilon(x_0) . \quad (3.2)$$

<u>Proof</u>. From the definition of $\partial_\varepsilon f(x_0)$, we have

$$f(z) \geq f(x_0) + \max_{v \in \partial_\varepsilon f(x_0)} (v, z-x_0) - \varepsilon ,$$

$$\min_{z \in \Omega} f(z) \geq f(x_0) + \min_{z \in \Omega} \max_{v \in \partial_\varepsilon f(x_0)} (v, z-x_0) - \varepsilon ,$$

and the latter inequality implies (3.2). ■

<u>COROLLARY</u>. If $\Phi_\varepsilon(x_0) = 0$, then from (3.2)

$$0 \leq f(x_0) - f^* \leq \varepsilon . \quad (3.3)$$

2. Now we shall show how to use the conditional-gradient method for finding points, at which (3.3) is satisfied, i.e., ε-stationary points of the function $f(x)$ on the set Ω.

Fix a number $\varepsilon > 0$. We choose an initial approximation $x_0 \in \Omega$ arbitrarily.

Suppose a point $x_k \in \Omega$ has already been found. If $\Phi_\varepsilon(x_k) = 0$, then (3.3) is satisfied at the point x_k and the process terminates.

If $\Phi_\varepsilon(x_k) = \max_{v \in \partial_\varepsilon f(x_k)} (v, z_k - x_k) < 0$, then, as the $(k+1)^{st}$

approximation we take a point $x_{k+1} = x_k + \alpha_k(z_k - x_k)$ such that

$$f(x_{k+1}) = \min_{\alpha\in[0,1]} f(x_k + \alpha(z_k - x_k)) .$$

Note that $x_{k+1} \in \Omega$ and $f(x_{k+1}) < f(x_k)$.

The process of successive approximations will continue analogously.

If the sequence $\{x_k\}$ is finite, then its final element is ε-stationary by construction. If the sequence $\{x_k\}$ is infinite, then the following theorem holds.

THEOREM 3.2. At any accumulation point x^* of the sequence $\{x_k\}$, $\Phi_\varepsilon(x^*) = 0$ and

$$0 \le f(x^*) - f^* \le \varepsilon .$$

Proof. Assume that this is not the case, i.e., $x_{k_s} \to x^*$ but $\Phi_\varepsilon(x^*) = -a < 0$. Then, by virtue of the upper semicontinuity of the mapping $\partial_\varepsilon f(x)$ with respect to x, the relation

$$\frac{\partial_\varepsilon f(x_{k_s})}{\partial g_{k_s}} = \max_{v\in\partial_\varepsilon f(x_{k_s})} (v, g_{k_s}) = \Phi_\varepsilon(x_{k_s}) \le -\tfrac{1}{2}a < 0 \qquad (3.4)$$

holds for sufficiently large k_s, where $g_{k_s} = z_{k_s} - x_{k_s}$.

The following two cases are possible:

1. $\min_{\alpha\ge 0} f(x_k(\alpha)) = \min_{\alpha\in[0,1]} f(x_k(\alpha))$;
2. $\inf_{\alpha\ge 0} f(x_k(\alpha)) < \min_{\alpha\in[0,1]} f(x_k(\alpha))$

(here, $x_k(\alpha) = x_k + \alpha(z_k - x_k)$).

For case 1, according to (8.19) in Chapter 1, we have

$$f(x_{k+1}) \le f(x_k) - \varepsilon .$$

Consider case 2. Due to Lemma 8.6 in Chapter 1, for all

$$\alpha \in \left[0,\ \varepsilon\left(\frac{\partial_\varepsilon f(x_k)}{\partial g_k} - \frac{\partial f(x_k)}{\partial g_k}\right)^{-1}\right]$$

we have

$$f(x_k + \alpha g_k) \le f(x_k) + \alpha \frac{\partial_\varepsilon f(x_k)}{\partial g_k} , \qquad (3.5)$$

where $g_k = z_k - x_k$.

A fortiori, (3.5) is satisfied for

$$\alpha = -\varepsilon \left(\frac{\partial f(x_k)}{\partial g_k} \right)^{-1} = -\varepsilon(\Phi(x_k))^{-1} .$$

Therefore, for case 2 we have

$$f(x_{k+1}) \le f(x_k) + \min \{1, -\varepsilon(\Phi(x_k))^{-1}\} \Phi_\varepsilon(x_k) .$$

Let $b = \max_{x \in \Omega} |\Phi(x)|$. The number b is finite because the mapping $\partial f(x)$ is bounded on Ω. Then

$$f(x_{k+1}) \le f(x_k) + \beta \Phi_\varepsilon(x_k) ,$$

where

$$\beta = \min \{1, \varepsilon b^{-1}\} .$$

Thus, for both cases the inequality

$$f(x_{k+1}) \le f(x_k) + \max \{-\varepsilon, \beta\Phi_\varepsilon(x_k)\} \qquad \forall k \qquad (3.6)$$

holds. From (3.4) and (3.6), we conclude that for sufficiently large k_s

$$f(x_{k_s+1}) \le f(x_{k_s}) - \min \{\varepsilon, \tfrac{1}{2}\beta a\} ,$$

and hence, it follows that $f(x_k) \to -\infty$; and this contradicts the boundedness of the function $f(x)$ on the set Ω. ∎

We note that it is appropriate to apply the conditional-gradient method when the set Ω has sufficiently simple structure, for example, when it is given by linear inequalities and it is possible to efficiently calculate the ε-sbudifferential of the function $f(x)$.

REMARK 1. Instead of finding a point x_{k+1} from the condition for minimum of the function $f(x)$ on the interval $[x_k, z_k]$, we can proceed in the following manner: for $i = 0,1,2,\ldots$, check the inequality

$$f(x_k + 2^{-i}(z_k - x_k)) \le f(x_k) - 2^{-i}\Phi_\varepsilon(x_k) . \qquad (3.7)$$

Let (3.7) have been satisfied for the first time for $i = i_k$. Then we obtain

$$x_{k+1} = x_k + 2^{-i_k}(z_k - x_k) \equiv x_k + \alpha_k(z_k - x_k) .$$

Due to Lemma 8.6 in Chapter 1, the number i_k is finite, moreover,

$$\alpha_k \ge \min\{1, \varepsilon[2(\Phi_\varepsilon(x_k) - \Phi(x_k))]^{-1}\} .$$

PROBLEM 3.1. Prove Theorem 3.2 for the above-described line-search modification of the conditional-gradient method.

3. We shall next describe a modification of the conditional-gradient method for finding minimum points of the function $f(x)$ on the set Ω.

Fix some $\varepsilon_0 > 0$ and $\delta_0 > 0$. As an initial approximation, we take an arbitrary point $x_0 \equiv x_{00} \in \Omega$. Applying the conditional-gradient method with $\varepsilon = \varepsilon_0$, after a finite number of steps we find a point $x_{0t_0} \in \Omega$ such that

$$\Phi_{\varepsilon_0}(x_{0t_0}) \ge -\delta_0$$

(this follows from Theorem 3.2).

Let

$$\varepsilon_1 = \frac{\varepsilon_0}{2} , \qquad \delta_1 = \frac{\delta_0}{2} , \qquad x_1 = x_{10} = x_{0t_0} .$$

After taking x_1 as an initial approximation, we again apply the conditional-gradient method, here, however, with $\varepsilon = \varepsilon_1$. After a finite number of steps, we get a point x_{1t_1} such that $\Phi_{\varepsilon_1}(x_{1t_1}) \ge -\delta_1$.

Continuing this process, we obtain a sequence $\{x_k\}$ which satisfies

$$\Phi_{\varepsilon_k}(x_{k+1}) \ge -\delta_k , \qquad (3.8)$$

where

$$\varepsilon_k = \frac{\varepsilon_0}{2^k}, \qquad \delta_k = \frac{\delta_0}{2^k}, \qquad k = 0,1,2,\dots .$$

Clearly, $f(x_{k+1}) \le f(x_k)$ and $x_k \in \Omega$ for all k.

THEOREM 3.3. Any accumulation point of the sequence $\{x_k\}$ is a minimum point of the function $f(x)$ on the set Ω.

Proof. Assume that this is not the case, i.e., assume that $x_{k_s} \to x^*$ but $\Phi(x^*) = -a < 0$. By virtue of the upper semicontinuity of the mapping $\partial_\varepsilon f(x)$ with respect to x and ε on $E_n \times [0,\infty)$, we can find an $\varepsilon > 0$ such that, for sufficiently large k_s,

$$\Phi_\varepsilon(x_{k_s}) \le \tfrac{1}{2}a . \tag{3.9}$$

Now we consider k_s which satisfy (3.9) and, in addition,

$$\varepsilon_{k_s} \le \varepsilon , \qquad \delta_{k_s} < \frac{a}{2} .$$

Then, it follows from (3.9) that

$$\Phi_{\varepsilon_{k_s}}(x_{k_s+1}) \le -\tfrac{1}{2}a ,$$

which contradicts (3.8). ∎

REMARK 2. The boundedness of the set Ω is not necessary if we use the function

$$\tilde{\Phi}_\varepsilon(x) = \min_{z\in\Omega,\|z-x\|\le 1} \max_{v\in\partial_\varepsilon f(x)} (v, z-x)$$

instead of the function $\Phi_\varepsilon(x)$.

4. THE METHOD OF STEEPEST DESCENT FOR THE MINIMIZATION OF CONVEX FUNCTIONS

1. In this section we shall consider the methods related to the direction of steepest (or nearly steepest) descent. First, continuous methods for the minimization or methods of differential descent will be discussed. Though continuous methods for the minimization cannot be implemented on a computer, however, as they are,

they can be used to construct discrete algorithms (one of them will be given below in Section 4.3).

Let a function $f(x)$ be finite and convex on E_n. Then, as proved in Lemma 5.5 of Chapter 1, the function $f(x)$ is Lipschitzian on any convex bounded set $G \subset E_n$.

Let a vector-valued function $f(x)$ be continuously differentiable on $[0,T]$, where $T > 0$. Then,

$$x(t) = x_0 + \int_0^t \dot{x}(\tau)\, d\tau ,$$

where $x_0 = x(0)$, $\dot{x}(\tau)$ is a vector-valued function continuous on $[0,T]$.

From (5.26) in Chapter 1, we have for $t \in [0,T]$

$$f(x(t)) = f(x_0) + \int_0^t \frac{\partial f(x(\tau))}{\partial \dot{x}(\tau)}\, d\tau \quad . \tag{4.1}$$

Here, the integral is interpreted in the Lebesgue sense.

Let $\Omega \subset E_n$ be a closed convex set. As is established in Theorem 11.2 of Chapter 1, the mapping $\partial_\varepsilon^\Omega f(x)$ is continuous in the Kakutani sense with respect to ε and x on $(0,\infty) \times \Omega$.

Denote by $v_\varepsilon(x)$ a point of the set $\partial_\varepsilon^\Omega f(x)$ nearest to the origin, i.e.,

$$\|v_\varepsilon(x)\| = \min_{v \in \partial_\varepsilon^\Omega f(x)} \|v\| \ . \tag{4.2}$$

<u>LEMMA 4.1</u>. A vector-valued function $v_\varepsilon(x)$ is continuous with respect to ε and x on the set $(0,\infty) \times \Omega$.

<u>Proof</u>. Let $\varepsilon_0 > 0$, $x_0 \in \Omega$. Since

$$\partial_\varepsilon f(x) \subset \partial_\varepsilon^\Omega f(x)$$

for any $\varepsilon > 0$ and $x \in \Omega$, then

$$0 \le \min_{v \in \partial_\varepsilon^\Omega f(x)} \|v\| \le \min_{v \in \partial_\varepsilon f(x)} \|v\| \ .$$

The function $h(x,\varepsilon) = \min_{v\in\partial_\varepsilon f(x)} \|v\|$ is bounded in the neighborhood of the point $[\varepsilon_0, x_0]$. Hence, the function

$$H(x,\varepsilon) = \min_{v\in\partial_\varepsilon^\Omega f(x)} \|v\|$$

is also bounded for $x \in S_\delta(x_0) \cap \Omega$, $\varepsilon \in S_\delta(\varepsilon_0) = \{\varepsilon>0 \mid |\varepsilon-\varepsilon_0| \le \delta\}$, $\delta > 0$. Denote $v_{\varepsilon_0}(x_0) = v_0$. Let $\varepsilon_k \to \varepsilon_0$, $x_k \to x_0$.

By virtue of the lower semicontinuity of the mapping $\partial_\varepsilon^\Omega f(x)$, there exists a sequence $\{v_k\}$ such that $v_k \in \partial_{\varepsilon_k}^\Omega f(x_k)$, $v_k \to v_0$. Therefore

$$\|v_{\varepsilon_k}(x_k)\| = \min_{v\in\partial_{\varepsilon_k}^\Omega f(x_k)} \|v\| \le \|v_k\| . \tag{4.3}$$

The sequence $\{v_{\varepsilon_k}(x_k)\}$ is bounded. Hence, we can assume that

$$v_{\varepsilon_k}(x_k) \to \bar{v} . \tag{4.4}$$

Since the mapping $\partial_\varepsilon^\Omega f(x)$ is upper semicontinuous, then $\bar{v} \in \partial_{\varepsilon_0}^\Omega f(x_0)$. It follows from (4.2) - (4.4) that

$$\|\bar{v}\| \le \|v_0\| . \tag{4.5}$$

Hence, from (4.2) and (4.5) we conclude that $\bar{v} = v_0$.

The lemma has been proved. ∎

Consider the following system of differential equations

$$\begin{aligned} \dot{x}(t) &= -v_{\varepsilon(t)}(x(t)) , \\ x(0) &= x_0 \in \Omega , \end{aligned} \tag{4.6}$$

where $\varepsilon(t)$ is a continuous function on $[0,\infty)$ and, in addition, $\varepsilon(t) \xrightarrow[t\to\infty]{} +0$.

Due to Lemma 4.1, the vector function $v_\varepsilon(x)$ is continuous with respect to ε and x on the set $(0,\infty) \times \Omega$. Due to the Peano theorem, the solution $x(t,x_0)$ of the system (4.6) exists and $x(t,x_0)$ is a continuously differentiable function on $[0,\infty)$. Denote

$$\|v_\varepsilon(x)\| = \min_{v\in\partial_\varepsilon^\Omega f(x)} \|v\| .$$

Let $x \in \Omega$. If $0 \notin \partial_\varepsilon^\Omega f(x)$, then

$$\frac{\partial_\varepsilon^\Omega f(x)}{\partial \dot{x}} = \sup_{v\in\partial_\varepsilon^\Omega f(x)} (v,\dot{x}) = - \min_{v\in\partial_\varepsilon^\Omega f(x)} (v,-\dot{x})$$

$$= - \min_{v\in\partial_\varepsilon^\Omega f(x)} (v, v_\varepsilon(x)) = -\|v_\varepsilon(x)\|^2 < 0 .$$

From this, it is clear that if $x \in \Omega$ is not an ε-stationary point of the function f on the set Ω, then

1. $\dot{x} = -v_\varepsilon(x) \in \gamma(x) \equiv \{v=\lambda(z-x) \mid \lambda>0,\ z\in\Omega\}$, and $g_\varepsilon = -v_\varepsilon(x)\|v_\varepsilon(x)\|^{-1}$ is a direction of ε-steepest descent of the function f at the point $x \in \Omega$ (with respect to the set Ω). Hence, there exists a $T = T(x) > 0$ such that

$$x + tg_\varepsilon \in \Omega \qquad \forall t \in [0,T] ;$$

2. we have the inequality

$$\inf_{\alpha>0,\, x+\alpha g_\varepsilon\in\Omega} f(x+\alpha g_\varepsilon) < f(x) - \varepsilon , \tag{4.8}$$

i.e., it is possible to reduce the value of the function f by more than ε, without leaving the set Ω.

If for some point $\bar{x} \in \Omega$ it happens that $v_\varepsilon(\bar{x}) = 0$, i.e., $0 \in \partial_\varepsilon^\Omega f(\bar{x})$, then $\bar{x}$ is an ε-stationary point of the function f on the set Ω.

Define

$$D(x_0) = \{x \in \Omega \mid f(x) \le f(x_0)\} .$$

<u>THEOREM 4.1</u>. If the set $D(x_0)$ is bounded, then any accumulation point of the solution $x(t,x_0)$ of the system (4.6) is a minimum point of the function $f(x)$ on the set Ω.

<u>Proof</u>. Since

$$x(t) \equiv x(t,x_0) = x_0 + \int_0^t \dot{x}(\tau)\, d\tau ,$$

then from (4.1) and (4.7) we have

$$f(x(t)) = f(x_0) + \int_0^t \frac{\partial f(x(\tau))}{\partial \dot{x}(\tau)} d\tau \le f(x_0) + \int_0^t \frac{\partial^{\Omega}_{\varepsilon(\tau)} f(x(\tau))}{\partial \dot{x}(\tau)} d\tau \qquad (4.9)$$

because

$$\frac{\partial f(x(\tau))}{\partial \dot{x}(\tau)} = \max_{v \in \partial f(x(\tau))} (v, \dot{x}(\tau)) \le \sup_{v \in \partial^{\Omega}_{\varepsilon(\tau)} f(x(\tau))} (v, \dot{x}(\tau)) .$$

Similarly to (4.7), it follows that

$$\sup_{v \in \partial^{\Omega}_{\varepsilon(\tau)} f(x(\tau))} (v, \dot{x}(\tau)) = -\|v_{\varepsilon(\tau)}(x(\tau))\|^2 . \qquad (4.10)$$

Hence,

$$f(x(t)) \le f(x_0) - \int_0^t \|v_{\varepsilon(\tau)}(x(\tau))\|^2 d\tau .$$

From this, it is clear that $f(x(t)) \le f(x_0)$ $\forall t \ge 0$, i.e.,

$$x(t) \in D(x_0) \qquad \forall t \ge 0 .$$

By virtue of the boundedness of the set $D(x_0)$, accumulation points of the solution $x(t)$ exist. Let x^* be an accumulation point. This implies that there exists a sequence $\{t_k\}$ such that

$$t_k \to \infty , \qquad x(t_k) \to x^* .$$

Clearly, $x^* \in \Omega$. We need to show that x^* is a minimum point of the function $f(x)$ on Ω, i.e., $0 \in \partial^{\Omega}_{\varepsilon} f(x^*)$. Assume that this is not the case, i.e., $\|v_0(x^*)\| = a > 0$. From the upper semicontinuity of the mapping $\partial^{\Omega}_{\varepsilon} f(x)$, it follows that there exist an $\varepsilon_0 > 0$ and a $\delta > 0$ such that $\|v_{\varepsilon}(x)\| \ge \frac{a}{2}$ $\forall \varepsilon \in [0, \varepsilon_0]$, $\forall x \in S_{\delta}(x^*)$. Therefore, there exists a $K_1 < \infty$ such that

$$\|v_{\varepsilon(t_k)}(x(t_k))\| \ge \frac{a}{2} \qquad \forall k \ge K_1 .$$

Moreover, due to the continuity of $\varepsilon(t)$ and $x(t)$, there exist a number $\Delta > 0$ and an integer $K_2 \ge K_1$ such that

$$\|v_{\varepsilon(t)}(x(t))\| \ge \frac{a}{4} \qquad \forall\, t \in [t_k-\Delta, t_k+\Delta], \quad \forall\, k \ge K_2 .$$

Without loss of generality, we can assume that

$$t_{k+1} - t_k > 2\Delta \qquad \forall\, k \ge K_2 .$$

From (4.9), we have

$$f(x(t)) \le f(x_0) - \int_0^t \|v_{\varepsilon(\tau)}(x(\tau))\|^2 \, d\tau \le f(x_0) - \frac{a^2 \Delta m(t)}{8} , \tag{4.11}$$

where $m(t)$ is an integer such that $t_{K_2+m(t)} \le t \le t_{K_2+m(t)+1}$. Clearly, $m(t) \xrightarrow[t\to\infty]{} \infty$. Hence, it follows from (4.11) that $f(x(t)) \xrightarrow[t\to\infty]{} -\infty$, and this contradicts the boundedness of the continuous function $f(x)$ on the compact set $D(x_0)$. The theorem has been proved. ■

2. Let us consider the system

$$\begin{aligned} \dot{x}(t) &= -g_\varepsilon(x(t)) , \\ x(0) &= x_0 \in \Omega , \end{aligned} \tag{4.12}$$

where

$$g_\varepsilon(x(t)) = \begin{cases} v_\varepsilon(x)\|v_\varepsilon(x)\|^{-1} & \text{if } v_\varepsilon(x) \ne 0 , \\ 0 & \text{if } v_\varepsilon(x) = 0 , \end{cases}$$

and $\varepsilon > 0$ is fixed. The function on the right-hand side of the system (4.12) is continuous while $v_\varepsilon(x) \ne 0$ and, due to the Peano theorem, a solution of the system exists. As above, we can prove that $x(t) \in D(x_0)$ $\forall t > 0$ and that any accumulation point of the solution is ε-stationary. Moreover, the following theorem holds.

THEOREM 4.2. If the set $D(x_0)$ is bounded and the function $f(x)$ is strongly convex on E_n, then there exists a $T < \infty$ such that $v_\varepsilon(x(T)) = 0$, i.e., the system arrives at an ε-stationary point within a finite time.

Proof. Let $f(x)$ be a strongly convex function on E_n with a strong convexity constant $m > 0$. Due to Corollary 1 of Lemma 17.1 in Chapter 1, we have

$$\frac{\partial f(x)}{\partial g} \le \frac{\partial_\varepsilon f(x)}{\partial g} - 2\|g\|\sqrt{\varepsilon m} \quad . \tag{4.13}$$

Since $\partial_\varepsilon f(x) \subset \partial_\varepsilon^\Omega f(x)$, then from (4.13)

$$\frac{\partial f(x)}{\partial g} \le \frac{\partial_\varepsilon^\Omega f(x)}{\partial g} - 2\|g\|\sqrt{\varepsilon m} \quad . \tag{4.14}$$

If $v_\varepsilon(x(t)) \ne 0$ (from the definition of $g_\varepsilon(x(t))$ and the system (4.12), this implies $v_\varepsilon(x(\tau)) \ne 0$ for all $\tau \in [0,t]$), then we have from (4.9), (4.10) and (4.14)

$$f(x(t)) \le f(x_0) - \int_0^t \|v_{\varepsilon(\tau)}(x(\tau))\|^2 \, d\tau - 2t\sqrt{\varepsilon m} \ . \tag{4.15}$$

Assume that the assertion of the theorem is not true: then $v_\varepsilon(x(t)) \ne 0$ for any $t > 0$ and $f(x(t)) \to -\infty$ by virtue of (4.15), which contradicts the boundedness of the set $D(x_0)$.

The theorem has been proved. ■

REMARK 1. From (4.12), it follows that if $v_\varepsilon(x(T)) = 0$, then $g_\varepsilon(x(t)) = 0$ for $t \ge T$, and hence, $x(t) \equiv x(T)$ for all $t \ge T$.

REMARK 2. In the system (4.6), it is also possible to normalize the right-hand side.

3. Now we describe the following discrete method for the minimization of a convex function $f(x)$ on a closed convex set Ω.

Take an arbitrary $x_0 \in \Omega$. If $0 \in \partial^\Omega f(x_0)$, then x_0 is a minimum point of the function $f(x)$ on the set Ω and the process terminates.

If $0 \notin \partial^\Omega f(x_0)$, then there exists a number $\varepsilon_0 > 0$ such that $0 \in \partial_{\varepsilon_0}^\Omega f(x_0)$. If $0 \notin \partial_\varepsilon^\Omega f(x_0)$ for any $\varepsilon > 0$, then $f^* \equiv \inf_{x\in\Omega} f(x) = -\infty$.

In the sequel, we assume that $f^* > -\infty$.

Considering the numerical sequence $\varepsilon_{0q} = 2^{-q}$, $q = 1,2,\ldots$, we first find a $q_0 \geq 1$ such that $0 \notin \partial^{\Omega}_{\varepsilon_1} f(x_0)$, $0 \in \partial^{\Omega}_{2\varepsilon_1} f(x_0)$, where $\varepsilon_1 = \varepsilon_{0q_0} = \dfrac{\varepsilon_0}{2^{q_0}}$. Then, $\varepsilon_1 \leq \frac{1}{2}\varepsilon_0$ and

$$f(x_0) - f^* \leq 2\varepsilon_1 . \tag{4.16}$$

Now we find a vector $g_0 \in \gamma(x_0)$, $\|g_0\| = 1$, such that $\dfrac{\partial^{\Omega}_{\varepsilon_1} f(x_0)}{\partial g_0} < 0$, and on the set $\{x_0(\alpha)=x_0+\alpha g_0 \mid \alpha \geq 0\} \cap \Omega$ we find a point $x_1 = x_0(\alpha_0)$ such that

$$f(x_1) - f(x_0) \leq -\varepsilon_1 . \tag{4.17}$$

As (4.8) implies, such a point can always be found. From (4.16) and (4.17), we have $f(x_1) \leq f(x_0) - \varepsilon_1 \leq f^* + \varepsilon_1 = f^* + \dfrac{\varepsilon_0}{2^{q_0}}$.

Hence, $0 \in \partial^{\Omega}_{\varepsilon_1} f(x_1)$, i.e., the point x_1 is an ε_1-stationary point of the function $f(x)$ on the set Ω.

Continuing analogously, we get a sequence $\{x_k\}$, which is either finite or such that each point of it is an ε_k-stationary point of the function $f(x)$ on Ω, where $\varepsilon_{k+1} = \dfrac{\varepsilon_k}{2^{q_k}}$, $q_k \geq 1$, i.e.,

$$f(x_{k+1}) - f^* \leq \frac{\varepsilon_k}{2^{q_k}} \leq \frac{\varepsilon_0}{2^k} \qquad \forall\, k \tag{4.18}$$

implying the following theorem.

THEOREM 4.3. Let $f^* > -\infty$. Then, $\lim_{k\to\infty} f(x_k) = f^*$, and the rate of convergence of $f(x)$ to f^* is geometric. Moreover, if the set $M^* = \{x \in \Omega \mid f(x)=f^*\}$ is bounded, then any accumulation point of the sequence $\{x_k\}$ is a minimum point of the function $f(x)$ on the set Ω.

5. THE $(\mathcal{E},\mu)$-SUBGRADIENT METHOD IN THE PRESENCE OF CONSTRAINTS

1. Let $G \subset E_n$ be a closed convex set having interior points. Take an arbitrary point $x_0 \in E_n$. Let us find a point

$y_0 \equiv y(x_0)$ such that

$$\min_{y\in G} \|y-x_0\| = \|y(x_0) - x_0\| . \tag{5.1}$$

The point $y_0 \equiv y(x_0)$ (it is unique for any $x_0 \in E_n$ because the norm in (5.1) is Euclidean) is called the projection of the point x_0 onto G.

We note that if for some $x_0 \in G$ and $v_0 \in E_n$

$$y(x_0 + v_0) = x_0 ,$$

then

$$(x-x_0, v_0) \le 0 \qquad \forall x \in G .$$

Hence, $v_1 = -v_0 \in \Gamma^+(x_0)$. From this, in particular, it is clear that if $v_0 \ne 0$, then x_0 is a boundary point of the set G (because for an interior point x_0 we have $\Gamma^+(x_0) = \{0\}$).

2. Let convex functions $f(x)$ and $h(x)$ be given on E_n. Define

$$\Omega_1 = \{x \in E_n \mid h(x) \le 0\} , \qquad \Omega = G \cap \Omega_1 .$$

The problem consists in finding the value $\min_{x\in\Omega} f(x) \equiv f^*$.

Take numbers $\varepsilon > 0$, $\mu > 0$ and a sequence $\{\lambda_k\}$ such that

$$\lambda_k \xrightarrow[k\to\infty]{} +0 , \qquad \sum_{k=0}^{\infty} \lambda_k = \infty . \tag{5.2}$$

Assume that there exists a point $\bar{x} \in \text{int } G$ such that

$$h(\bar{x}) < -\mu . \tag{5.3}$$

In particular, this implies that $\text{int } \Omega_\mu \ne \emptyset$, where $\Omega_\mu = \{x\in G \mid h(x) \le -\mu\}$. We also assume that $f^* > -\infty$.

Let us introduce the following method of successive approximations.

Take an arbitrary $x_0 \in G$. Suppose $x_k \in G$ has already been found.

If $h(x_k) \le 0$, then let us take an arbitrary $v_k \in \partial_\varepsilon f(x_k)$.

If $v_k = 0$, then

$$0 \le f(x_k) - \inf_{x \in E_n} f(x) \le \varepsilon , \qquad (5.4)$$

and in this case the process terminates.

If $h(x_k) > 0$, then let us take an arbitrary $w_k \in \partial_\mu f(x_k)$.

By virtue of (5.3), we have $w_k \ne 0$. Let

$$g_k = \begin{cases} -v_k \|v_k\|^{-1} & \text{if } h(x_k) \le 0 , \\ -w_k \|w_k\|^{-1} & \text{if } h(x_k) > 0 , \end{cases} \qquad (5.5)$$

and find

$$x_{k+1} = y(x_k + \lambda_k g_k) \qquad (5.6)$$

(see (5.1)). We note that if $G = E_n$ or $x_k + \lambda_k g_k \in G$, then $x_{k+1} = x_k + \lambda_k g_k$.

Let us continue analogously. As a result, we obtain a sequence $\{x_k\}$. If the sequence is finite, then, by construction, its final element x_k belongs to Ω and satisfies (5.4). Now we consider the case where the sequence $\{x_k\}$ contains an infinite number of points. We define the set

$$D_c = \{x \in \Omega \mid f(x) \le f_\mu^* + \varepsilon + c\} , \qquad (5.7)$$

where $f^* = \inf_{x \in \Omega_\mu} f(x)$.

Since $f^* > -\infty$, the set D_c is nonempty for any $c \ge 0$.

THEOREM 5.1. For any $c \ge 0$, the sequence $\{x_k\}$ contains an infinite number of points belonging to the set D_c.

Proof. Assume that this is not the case. Then, there exist a $c > 0$ and a $K < \infty$ such that

$$x_k \notin D_c \qquad \forall k \ge K . \qquad (5.8)$$

By virtue of (5.3), we have $\operatorname{int} \{x \in \Omega_\mu \mid f(x) \le f^* + c\} \ne \emptyset$. Hence, there exist a point $\tilde{x} \in \Omega_\mu$ and a number $r > 0$ such that

$$x \in G, \qquad f(x) \le f_\mu^* + c , \qquad h(x) \le -\mu \qquad \forall x \in S_r(\tilde{x}) . \qquad (5.9)$$

If $h(x_k) \le 0$, then, according to (5.5), $g_k = -v_k\|v_k\|^{-1}$, where $v_k \in \partial_\varepsilon f(x_k)$. Therefore,

$$f(x) - f(x_k) \ge (v_k, x-x_k) - \varepsilon \qquad \forall x \in E_n. \quad (5.10)$$

Since $h(x_k) \le 0$ (i.e., $x_k \in \Omega$), then it follows from (5.7) and (5.8) that $f(x_k) > f^* + \varepsilon + c$.

Hence, from (5.9) and (5.10) we have

$$(v_k, x-x_k) \le f(x) + \varepsilon - f(x_k) < f^*_\mu + c + \varepsilon - (f^*_\mu + \varepsilon + c) = 0 \qquad \forall x \in S_r(\tilde{x}) .$$

Therefore,

$$(g_k, x_k-x) < 0 \qquad \forall x \in S_r(\tilde{x}) . \quad (5.11)$$

If $h(x_k) > 0$, then, according to (5.5), $g_k = -w_k\|w_k\|^{-1}$, where $w_k \in \partial_\mu h(x_k)$. Then

$$h(x) - h(x_k) \ge (w_k, x-x_k) - \mu \qquad \forall x \in E_n$$

and, considering (5.9),

$$(w_k, x-x_k) \le h(x) + \mu - h(x_k) < -\mu + \mu - 0 = 0 \qquad \forall x \in S_r(\tilde{x}) .$$

Hence, (5.11) holds again.

Take a point $x_{1k} = \tilde{x} - rg_k$. Clearly, $x_{1k} \in S_r(\tilde{x})$. It follows from (5.11) that $(g_k, x_k - x_{1k}) < 0$, i.e.,

$$(g_k, x_k-\tilde{x}) < -r \qquad \forall k . \quad (5.12)$$

By virtue of (5.6) and (5.1),

$$\|x_{k+1} - \tilde{x}\|^2 = \|y(x_k + \lambda_k g_k) - \tilde{x}\|^2 \le \|x_k + \lambda_k g_k - \tilde{x}\|^2$$
$$\le \|x_k - \tilde{x}\|^2 + 2\lambda_k(g_k, x_k-\tilde{x}) + \lambda_k^2 .$$

This and (5.12) imply that for sufficiently large k

$$\|x_{k+1} - \tilde{x}\|^2 \le \|x_k - \tilde{x}\|^2 - r\lambda_k .$$

Then, from (5.2) we obtain $\|x_k-\tilde{x}\| \xrightarrow[k\to\infty]{} -\infty$, which is impossible. The theorem has been proved. ∎

COROLLARY 1. There exists a subsequence $\{x_{k_s}\}$ such that

$$x_{k_s} \in G, \qquad h(x_{k_s}) \le 0, \qquad \varlimsup_{k_s \to \infty} f(x_{k_s}) \le f^*_\mu + \varepsilon .$$

Proof is given by letting c to $+0$ in (5.7).

COROLLARY 2. Assume that the set

$$D^* = \{x \in \Omega_\mu \mid f(x) \le f^*_\mu\}$$

is bounded. Then,

$$\rho(x_k, S_a(D^*)) \xrightarrow[k\to\infty]{} 0 ,$$

where

$$a = \max_{x \in D} \rho(x, D^*) , \qquad D = \{x \in \Omega \mid f(x) \le f^* + \varepsilon\} .$$

Proof. We follow the same arguments as those in proving the corresponding part of Theorem 4.1 in Chapter 3.

6. THE SUBGRADIENT METHOD WITH A CONSTANT STEP-SIZE

Let $f(x)$ and $h(x)$ be convex functions on E_n and let $\Omega = \{x \in E_n \mid h(x) \le 0\}$. Assume that for some $\mu_0 > 0$

$$\Omega_{\mu_0} \ne \emptyset , \tag{6.1}$$

where $\Omega_{\mu_0} = \{x \in E_n \mid h(x) \le -\mu_0\}$.

Consider the following method of successive approximations for the minimization of the function $f(x)$ on the set Ω.

Let us fix the numbers $\varepsilon \ge 0$, $\lambda > 0$, $\beta > 0$ and $\mu \in [0, \mu_0)$. Take an arbitrary $x_0 \in E_n$. Suppose $x_k \in E_n$ has already been obtained.

If $h(x_k) \le \beta$, then take an arbitrary $v_k \in \partial_\varepsilon f(x_k)$. If $v_k = 0$ and $h(x_k) \le 0$, then the process terminates and, furthermore, the point $x_k \in \Omega$ is an ε-stationary point of the function $f(x)$ on E_n.

If $h(x_k) > \beta$, then we take an arbitrary $w_k \in \partial_\mu h(x_k)$. Since $\mu < \mu_0$, then, by virtue of (6.1) and Lemma 9.2 in Chapter 1, $w_k \ne 0$.

Let

$$g_k = \begin{cases} -v_k \|v_k\|^{-1} & \text{if } h(x_k) \le \beta , \\ -w_k \|w_k\|^{-1} & \text{if } h(x_k) > \beta , \end{cases} \tag{6.2}$$

and take

$$x_{k+1} = x_k + \lambda g_k . \tag{6.3}$$

We continue analogously. As a result, we get a sequence $\{x_k\}$. If this sequence is finite, then by construction, its final element belongs to the set Ω and is an ε-stationary point of the function $f(x)$ on E_n. We consider the case where the sequence $\{x_k\}$ contains an infinite number of points.

Let

$$d_\mu^* = \inf_{x\in\Omega_\mu} f(x) , \qquad D^* = \{x\in\Omega_\mu \mid f(x) = f_\mu^*\} ,$$

$$d_{\lambda/2} = \sup_{x\in S_{\lambda/2}(\Omega_\mu)} h(x) + \mu, \qquad b_\gamma = \inf_{x\notin S_\gamma(\Omega)} h(x) ,$$

$$D_c = \{x\in\Omega_{-b\gamma} \mid f(x)\le c\} , \qquad c_\lambda = \sup_{x\in S_{\lambda/2}(D^*)} f(x) ,$$

$$r_1 = \sup_{x\in D_{\varepsilon+c_\lambda}} \rho(x, D^*) , \qquad \Omega_c = \{x = E_n \mid h(x)\le c\} .$$

We assume that the set Ω is bounded. Then, $\{c_\lambda, b_\gamma, d_{\lambda/2}\} < \infty$.

THEOREM 6.1. Let $d_{\lambda/2} < \beta \le b_\gamma$. Then,

$$\lim_{k\to\infty} \rho(x_k, D_{\varepsilon+c_\lambda}) = 0 , \tag{6.4}$$

$$\lim_{k\to\infty} \rho(x_k, S_{r_1+\lambda}(D^*)) = 0 . \tag{6.5}$$

Proof. First we prove (6.4). We assume that this is not the case. This implies that we can find an $a > 0$ and a $K < \infty$ such that

$$x_k \notin S_a(D_{\varepsilon+c_\lambda}) \qquad \forall\, k > K . \tag{6.6}$$

It is clear that there exists a $c > c_\lambda$ such that

$$D_{\varepsilon+c} \subset S_a(D_{\varepsilon+c_\lambda}) . \tag{6.7}$$

It is easy to see that

$$S_{\lambda/2}(D^*) \subset D_c . \tag{6.8}$$

Choose an $r > \lambda/2$ such that

$$S_r(\bar{x}) \subset D_c \cap S_r(\Omega_\mu) \quad , \tag{6.9}$$

$$d_r = \max_{m \in S_r(\Omega_\mu)} h(x) + \mu \leq \beta \quad , \tag{6.10}$$

where $\bar{x} \in D^*$. From (6.8) and the condition that $d_{\lambda/2} < \beta$, we can always find such an r.

If $h(x_k) \leq \beta$, then, by virtue of (6.2), $g_k = -v_k \|v_k\|^{-1}$, where $v_k \in \partial_\varepsilon f(x_k)$. Hence

$$f(x) - f(x_k) \geq (v_k, x-x_k) - \varepsilon \qquad \forall x \in E_n \ . \tag{6.11}$$

If $h(x_k) > \beta$, then, by virtue of (6.2), $g_k = w_k \|w_k\|^{-1}$, where $w_k \in \partial_\mu h(x_k)$. Hence

$$h(x) - h(x_k) \geq (w_k, x-x_k) - \mu \qquad \forall x \in E_n \ . \tag{6.12}$$

Let $x_k \in \Omega_{-b_\gamma}$. Then, according to (6.6) and (6.7), $f(x_k) > \varepsilon + c$ and if $h(x_k) \leq \beta$, then, noting (6.9) we conclude that $f(x) - f(x_k) > -\varepsilon$ for any $x \in S_r(\bar{x})$. Hence, from (6.11) we have $(v_k, x-x_k) < 0$ and

$$(g_k, x_k - x) < 0 \qquad \forall x \in S_r(\bar{x}), \quad \forall k \geq K \ . \tag{6.13}$$

If $x_k \in \Omega_{-b_\gamma}$ and $h(x_k) > \beta$, then, noting (6.10), we get $h(x) - h(x_k) < d_r - \mu - \beta \leq -\mu$ for any $x \in S_r(\bar{x})$. Hence, according to (6.12), $(w_k, x-x_k) < 0$ and, again, (6.13) holds.

Finally, if $x_k \notin \Omega_{-b_\gamma}$, then $h(x_k) > b_\lambda \geq \beta$ and, again, (6.13) holds.

Thus, (6.13) holds in each case.

Let $\bar{x}_k = \bar{x} - rg_k$. It is clear that $\bar{x}_k \in S_r(\bar{x})$. Hence, from (6.13) we have $(g_k, x_k - \bar{x}_k) < 0$, i.e.,

$$(g_k, x_k - \bar{x}) < -r \qquad \forall k \geq K \ . \tag{6.14}$$

According to (6.3),

$$\|x_{k+1}-\bar{x}\|^2 = \|x_k+\lambda g_k-\bar{x}\|^2 = \|x_k-\bar{x}\|^2 + 2\lambda(g_k, x_k-\bar{x}) + \lambda^2 .$$

From this and (6.14), it follows that

$$\|x_{k+1}-\bar{x}\|^2 < \|x_k-\bar{x}\|^2 - 2\lambda r + \lambda^2 . \tag{6.15}$$

By virtue of (6.15), $\|x_k-\bar{x}\| \xrightarrow[k\to\infty]{} -\infty$, and this is impossible. This contradiction proves (6.4).

Now we show that (6.5) holds. Take an arbitrary $\delta > 0$. According to (6.4), there exists an infinite number of k such that

$$\rho(x_k, D_{\varepsilon+c_\lambda}) \le \delta .$$

For such k, we have

$$x_{k+1} \in S_{\lambda+\delta}(D_{\varepsilon+c_\lambda}) \subset S_{r_1+\lambda+\delta}(D^*)$$

since $\|x_{k+1}-x_k\| = \lambda$. If in this case we have $\rho(x_{k+1}, D_{\varepsilon+c_\lambda}) > \delta$, then for k larger than some K_δ, (6.15) is satisfied and hence

$$x_{k+2} \in S_{r_1+\lambda+\delta}(D^*) . \tag{6.16}$$

If $\rho(x_{k+1}, D_{\varepsilon+c_\lambda}) < \delta$, then we have again (6.16) because $\|x_{k+2}-x_{k+1}\| = \lambda$.

Arguing analogously, we conclude that

$$x_k \in S_{r_1+\lambda+\delta}(D^*) \qquad \forall\, x \ge K_\delta .$$

Hence, by virtue of the arbitrariness of δ, we get (6.5). ∎

PROBLEM 6.1. Let $G \subset E_n$ be a closed convex set. Describe a subgradient method with a constant step-size and a projection onto the set G for the minimization of the function $f(x)$ on the set $G \cap \{x \in E_n \mid h(x) \le 0\}$.

Prove Theorem 6.1 for such a method.

7. THE MODIFIED ($\mathcal{E}, \mu$)-SUBGRADIENT METHOD IN THE PRESENCE OF CONSTRAINTS

Let us fix numbers $\varepsilon \geq 0$, $d_1 > 0$, $\mu \in [0,\gamma)$, $d_2 \in (0,\gamma-\mu)$ and take a sequence of positive numbers $\{\lambda_k\}$ such that

$$h(\bar{x}) = -\gamma < 0 . \tag{7.1}$$

Let $f(x)$ and $h(x)$ be convex functions on E_n. The problem consists in finding $f^* = \inf_{x \in \Omega} f(x)$, where $\Omega = \{x \in E_n \mid h(x) \leq 0\}$.

Assume that there exist an $\bar{x} \in E_n$ and a $\gamma > 0$ such that

$$\lambda_k \to 0 , \qquad \sum_{k=0}^{\infty} \lambda_k = \infty . \tag{7.2}$$

As x_0, let us take an arbitrary point in E_n. Suppose a point $x_k \in E_n$ has already been found.

If $h(x_k) \leq d_1$, then find an arbitrary $v_k \in \partial_\varepsilon f(x_k)$ and if $v_k = 0$ and $h(x_k) \leq 0$, then the process terminates because $x_k \in \Omega$ and x_k is an ε-stationary point of the function $f(x)$ on E_n (a fortiori on Ω).

If $h(x_k) \geq -d_2$, then find an arbitrary $w_k \in \partial_\mu h(x_k)$. By virtue of (7.1) and the condition $d_2 + \mu < \gamma$, we should have $w_k \neq 0$.

Let

$$z_k = \begin{cases} -\lambda_k v_k \|v_k\|^{-1} & \text{if } h(x_k) < -d_2 , \\ -\lambda_k w_k \|w_k\|^{-1} & \text{if } h(x_k) > d_1 . \end{cases}$$

If $-d_2 \leq h(x_k) \leq d_1$, then we define z_k as follows:

$$\|z_k\| \leq \lambda_k , \qquad \min_{\|z\| \leq \lambda_k} \Phi_k(z) = \Phi_k(z_k) , \tag{7.3}$$

where $\Phi_k(z) = \max \{(v_k, z), h(x_k) + (w_k, z)\}$.

We note that if $v_k = 0$, then $z_k = -\lambda_k w_k \|w_k\|^{-1}$.

If $\|z_k\| < \lambda_k$ and $h(x_k) \leq 0$, then the process terminates. In this case, $\Phi_k(z_k) = \min_{z \in E_n} \Phi_k(z)$ and, due to the necessary condition for a minimum of the function $\Phi_k(z)$ on E_n, we have $0 \in \mathrm{co}\{v_k, w_k\}$.

Then, $0 \in \text{co}\,\{\partial_\varepsilon f(x_k) \cup \partial_\mu h(x_k)\}$ and, by virtue of Lemma 2.3,

$$f(x_k) \leq f^*_\mu + \varepsilon \quad . \tag{7.4}$$

It is clear that if $\|z_k\| < \lambda_k$ and $h(x_k) > 0$, then $z_k \neq 0$.

Now we let

$$x_{k+1} = x_k + z_k \; .$$

We continue analogously. As a result, we get a sequence $\{x_k\}$. If this is finite, then, by construction, its final element belongs to the set Ω and either it is an ε-stationary point of the function $f(x)$ on E_n or it satisfies (7.4). Now let us consider the case where the sequence $\{x_k\}$ contains an infinite number of points. For any $c_1 \geq 0$, $c_2 \in [0,d_2]$ we let

$$\Omega_{\mu+c_2} = \{x \in E_n \mid h(x) \leq -(\mu+c_2)\} \quad ,$$

$$\Omega_{-c_2} = \{x \in E_n \mid h(x) \leq c_2\} \quad ,$$

$$f^*_{\mu+c_2} = \inf_{x \in \Omega_{\mu+c_2}} f(x) \; ,$$

$$\mathcal{D}_{c_1,c_2} = \{x \in \Omega_{-c_2} \mid f(x) \leq f^*_{\mu+c_2} + \varepsilon + c_1\} \; ,$$

$$\mathcal{D}_{c_2} = \{x \in \Omega_{-c_2} \mid f(x) \leq f^*_{\mu+c_2} + \varepsilon\} = \mathcal{D}_{0,c_2} \; .$$

Assume that $f^* > -\infty$; then the sets $\mathcal{D}_{c_1,c_2}$ and $\mathcal{D}_{c_2}$ are nonempty. We note that $\text{int}\,\Omega_{\mu+c_2} \neq \emptyset$ since $c_2 \leq d_2$.

Assume, moreover, that

$$\sup_{x \in Q} \; \sup_{v \in \partial_\varepsilon f(x) \cup \partial_\mu h(x)} \|v\| < \infty \; , \tag{7.6}$$

where $Q = \{x \in E_n \mid -d_2 \leq h(x) \leq d_1\}$.

<u>THEOREM 7.1</u>. If there exists a subsequence of indices $\{k_s\}$ such that $\|z_{k_s}\| < \lambda_{k_s}$, then, for any $c_2 \in (0,d_2]$, an infinite number of points of the sequence $\{x_k\}$ belongs to the set $\mathcal{D}_{c_2}$.

Otherwise, for any $c_1 > 0$, $c_2 \in (0,d_2]$, an infinite number of points of the sequence $\{x_k\}$ belongs to the set D_{c_1,c_2}.

Proof. Suppose that there exists a subsequence of indices $\{k_s\}$ such that $\|z_{k_s}\| < \lambda_{k_s}$. Then, it follows from (7.3) that

$$\Phi_{k_s}(z_{k_s}) = \min_{z \in E_n} \Phi_{k_s}(z)$$

and, due to the necessary condition for a minimum of the function $\Phi_{k_s}(z)$ on E_n, we have

$$0 \in \text{co}\,\{v_{k_s}, w_{k_s}\} \subset \text{co}\,\{\partial_\varepsilon f(x_{k_s}) \cup \partial_\mu h(x_{k_s})\} .$$

Then, by virtue of Lemma 2.3,

$$f(x_{k_s}) \le f^*_{\mu-h(x_{k_s})} + \varepsilon . \tag{7.7}$$

On the basis of (7.2), (7.6) and

we conclude that

$$h(x_{k_s}) \xrightarrow[s\to\infty]{} 0 .$$

Hence, for sufficiently large k_s,

$$-h(x_{k_s}) \le c_2 , \tag{7.8}$$

$$\Omega_{\mu+c_2} \subset \Omega_{\mu-h(x_{k_s})} , \qquad f^*_{\mu-h(x_{k_s})} \in f^*_{\mu+c_2} . \tag{7.9}$$

From (7.7) and (7.9), we obtain

$$f(x_{k_s}) \le f^*_{\mu+c_2} + \varepsilon . \tag{7.10}$$

Thus, for sufficiently large k_s, (7.8) and (7.10) are satisfied and the first assertion of the theory has been proved.

Next, without loss of generality, we can assume that $\|z_k\| = \lambda_k$ for all k. Then, by virtue of the necessary condition for a minimum of the function $\Phi_k(z)$ on the set $\{z \in E_n \mid \|z\| \le \lambda_k\}$, we have for each k

$$z_k = \lambda_k g_k , \qquad g_k = -(\alpha_k v_k + \beta_k w_k) , \tag{7.11}$$
$$\|g_k\| = 1 , \qquad \alpha_k \geq 0 , \qquad \beta_k \geq 0 .$$

Suppose that the assertion of the theorem is not true, i.e., there exist $c_1 > 0$, $c_2 \in (0,d_2]$ and $K < \infty$ such that

$$x_k \notin D_{c_1,c_2} \qquad \forall k \geq K . \tag{7.12}$$

We can suppose that $c_2 \leq d_1$. Let us consider the set

$$D'_{c_1,c_2} = \{x \in \Omega_{\mu+c_2} \mid f(x) \leq f^*_{\mu+c_2} + c_1\} .$$

Since $c_2 \leq d_2$ and $d_2+\mu < \gamma$, then, of course, $c_2+\mu < \gamma$. Hence, $\operatorname{int} D'_{c_1,c_2} \neq \emptyset$ and there exist a number $r > 0$ and a point $\tilde{x} \in \Omega_{\mu+c_2}$ such that $S_r(\tilde{x}) \subset D'_{c_1,c_2}$. This implies that

$$h(x) \leq -(\mu+c_2) , \qquad f(x) \leq f^*_{\mu+c_2} + c_1 \qquad \forall x \in S_r(\tilde{x}) . \tag{7.13}$$

If $h(x_k) \leq c_2$, then, by virtue of (7.12) and (7.13), we have $f(x) - f(x_k) < f^*_{\mu+c_2} + c_1 - f'_{\mu+c_2}+\varepsilon+c_1) = -\varepsilon$ for any $x \in S_r(\tilde{x})$.

Hence,

$$(v,x-x_k) < 0 \qquad \forall x \in S_r(\tilde{x}), \quad \forall k \geq K, \quad \forall v \in \partial_\varepsilon f(x_k). \tag{7.14}$$

If $h(x_k) > -c_2$, then, by virtue of (7.13), we have $h(x) - h(x_k) < -(\mu+c_2) - (-c_2) = -\mu$ for any $x \in S_r(\tilde{x})$. Hence,

$$(w,x-x_k) < 0 \qquad \forall x \in S_r(\tilde{x}), \quad \forall k \geq K, \quad \forall w \in \partial_\mu h(x_k). \tag{7.15}$$

If $h(x_k) < -c_2$, then it is easy to see that we should have $z_k = -\lambda_k v_k \|v_k\|^{-1}$, where $v_k \in \partial_\varepsilon f(x_k)$, for sufficiently large k, i.e., for $k \geq K_1 \geq K$. Then, from (7.14) we have $(v_k,x-x_k) < 0$ and from (7.11)

$$(g_k,x_k-x) < 0 \qquad \forall k \in S_r(\tilde{x}) . \tag{7.16}$$

If $h(x_k) > c_2$, then we should have $z_k = -\lambda_k w_k \|w_k\|^{-1} = \lambda_k g_k$, where $w_k \in \partial_\mu h(x_k)$, for sufficiently large k, i.e., $k \geq K_2 \geq K_1$.

Then, from (7.15) we have $(w_k, x-x_k) < 0$ and, therefore, (7.16) is satisfied in this case, too.

Finally, let $-c_2 \le h(x_k) \le c_2$. Then, according to (7.11),

$$g_k = -(\alpha_k v_k + \beta_k w_k) , \qquad \alpha_k \ge 0, \quad \beta_k \ge 0, \quad ||g_k|| = 1 .$$

Substitute v_k into (7.14) and w_k into (7.15) and multiply (7.14) by α_k and (7.15) by β_k. Summing the obtained inequalities, we prove (7.16).

Thus, (7.16) holds for all $k \ge K_2$.

Let $\tilde{x}_k = \tilde{x} - rg_k$. It is clear that $\tilde{x}_k \in S_r(\bar{x})$. Then, from (7.16) we have $(g_k, x_k - \tilde{x}_k) < 0$. Hence,

$$(g_k, x_k - \tilde{x}) < -r \qquad \forall k \ge K_2 . \tag{7.17}$$

By virtue of (7.5) and (7.11),

$$||x_{k+1} - \tilde{x}||^2 = ||x_k - \tilde{x}||^2 + 2\lambda_k (g_k, x_k - \tilde{x}) + \lambda_k^2 .$$

From this, considering (7.17), we obtain

$$||x_{k+1} - \tilde{x}||^2 < ||x_k - \tilde{x}||^2 - (2r - \lambda_k)\lambda_k .$$

Now it follows from (7.2) that $||x_k - \tilde{x}|| \xrightarrow[k\to\infty]{} -\infty$, which is impossible. ∎

<u>COROLLARY</u>. If the set $D = \{x \in \Omega \mid f(x) \le f^*_\mu + \varepsilon\}$ is bounded, then there exists an accumulation point of the sequence $\{x_k\}$, which belongs to the set D.

To prove this, we note that in this case the set D_{c_1,c_2} is bounded for any $c_1 \ge 0$, $c_2 \in (0, d_2]$. Then, by virtue of Theorem 7.1, there exist accumulation points of the sequence $\{x_k\}$ and the assertion that they belong to the set D follows from the upper semicontinuity of the mapping D_{c_1,c_2} with respect to $c_1 \ge 0$, $c_2 \in (0, d_2]$.

PROBLEM 7.1. Construct an (ε,μ)-subgradient-projection method for the minimization of the function $f(x)$ on the set $G \cap \Omega$, where $G \subset E_n$ is an arbitrary compact set such that $\operatorname{int}\{x \in G \mid h(x) \le -\gamma < 0\}$ becomes nonempty. Prove a convergence theorem.

8. THE NONSMOOTH PENALTY-FUNCTION METHOD

All the methods for the minimization of a convex function on a convex set which were considered in the preceding sections of this chapter solve the original problem directly. However, it is possible to reduce the original constrained minimization problem to some unconstrained minimization problem, or to a minimization problem on a set with a simple structure. The penalty-function method is one of those methods. In this section, we shall show how the problem of minimizing a convex function $f(x)$ on a set $\Omega = \{x \in E_n \mid h(x) \le 0\}$ can be reduced to the problem of solving a finite number of some minimization problems on the entire space E_n.

1. Let $f(x)$ and $h(x)$ be finite convex functions on E_n. The problem consists in finding

$$f^* = \inf_{x \in \Omega} f(x) , \tag{8.1}$$

where $\Omega = \{x \in E_n \mid h(x) \le 0\}$.

Assume that problem (8.1) has a solution, i.e., there exists an $x^* \in \Omega$ such that $f(x^*) = f^*$ and, moreover, that the set $M_0(x^*) = \{x \in E_n \mid f(x) \le f^*\}$ is bounded. By virtue of the continuity of $f(x)$, the set $M_0(x^*)$ is closed.

Let

$$h^+(x) = \max\{h(x), 0\} , \qquad F(x,\lambda) = f(x) + \lambda h^+(x) ,$$

$$\Phi(\lambda) = \min_{x \in E_n} F(x,\lambda) , \qquad R(\lambda) = \{x \in E_n \mid F(x,\lambda) = \Phi(\lambda)\} .$$

Note that $\inf_{x \in E_n} F(x, \lambda)$ is attained for each fixed $\lambda \geq 0$ because the set $M_0(x^*)$ is bounded.

Let

$$\partial_{x,\varepsilon} F(x, \lambda) = \{v \in E_n \mid F(z, \lambda) - F(x, \lambda) \geq (v, z-x) - \varepsilon \ \forall z \in E_n\} .$$

$$\partial_x F(x, \lambda) = \partial_{x,0} F(x, \lambda) , \qquad \partial_\lambda F(x, \lambda) = \{h^+(x)\} .$$

By virtue of Lemma 5.2 in Chapter 1,

$$\partial_x F(x, \lambda) = \begin{cases} \partial f(x) & \text{if } h(x) < 0 , \\ \partial f(x) + \lambda \partial h^+(x) & \text{if } h(x) \geq 0 . \end{cases}$$

Here, we have used the fact that if $h(x) < 0$, then $\partial h^+(x) = \{0\}$.

We note also that

$$\partial h(x) = \partial h^+(x) \qquad \text{if } h(x) > 0 ,$$

$$\partial h(x) \subset \partial h^+(x) \qquad \text{if } h(x) = 0 .$$

The following assertions hold.

1. The function $\Phi(\lambda)$ is concave on $[0, \infty)$.

This follows from the fact that the function $F(x, \lambda)$ is linear with respect to λ and, hence, concave and that the operation of taking minimum preserves the concavity. ∎

2. $\Phi(\lambda) \leq \min_{x \in \Omega} f(x) \quad \forall \lambda \geq 0 .$ (8.2)

Indeed, for any $\lambda \geq 0$ we have

$$\Phi(\lambda) = \min_{x \in E_n} F(x, \lambda) \leq \min_{x \in \Omega} F(x, \lambda) = \min_{x \in \Omega} f(x) . \qquad ∎$$

Let $\Phi^* = \max_{\lambda \geq 0} \Phi(\lambda)$.

3. Let $x_0 \in R(\lambda_0)$, $\lambda_0 \geq 0$. If $h^+(x_0) = 0$, then

$$f(x_0) = f^* = \Phi(\lambda_0) = \Phi^* . \tag{8.3}$$

Let us show this. Since $\Phi(\lambda_0) = f(x_0) + \lambda_0 h^+(x_0) = f(x_0)$, then

$$f(x_0) = \min_{x\in E_n} F(x,\lambda_0) \le \min_{x\in\Omega} F(x,\lambda_0) = \min_{x\in\Omega} f(x) \le f(x_0) .$$

Hence, from (8.2)

$$\Phi^* = \max_{\lambda\ge 0} \Phi(\lambda) \le f^* = f(x_0) = \Phi(\lambda_0) .$$

This implies that (8.3) holds. ■

4. Let $\Phi(\lambda_0) = \Phi^*$, $\lambda_0 \ge 0$. Then, $\Phi(\lambda) = \Phi(\lambda^*)$ $\forall \lambda \ge \lambda_0$.

Indeed, if $\lambda_1 \ge \lambda_0$, then for any $x \in E_n$, by virtue of the fact that $h^+(x) \ge 0$, the inequality $F(x,\lambda_1) \ge F(x,\lambda_0)$ should be satisfied and, hence, we have

$$\Phi^* \ge \Phi(\lambda_1) \ge \Phi(\lambda_0) = \Phi^* .$$ ■

5. For the derivative of the function $\Phi(\lambda)$ with respect to an arbitrary direction g, we have

$$\frac{\partial\Phi(\lambda)}{\partial g} = \min_{x\in R(\lambda)} (h^+(x), g) . \tag{8.4}$$

Indeed (see Section 1.5), since $F'_\lambda(x,\lambda) = h^+(x)$ and

$$\min_{x\in E_n} F(x,\lambda) = -\max_{x\in E_n} [-F(x,\lambda)] ,$$

then we have

$$\frac{\partial\Phi(\lambda)}{\partial g} = -\max_{x\in R(\lambda)} (-h^+(x), g) = \min_{x\in R(\lambda)} (h^+(x), g) .$$ ■

Using (8.4), we conclude that if for each $\lambda > 0$ the set $R(\lambda)$ is a singleton $(R(\lambda) = \{x(\lambda)\})$, then

$$\frac{\partial\Phi(\lambda)}{\partial g} = (h^+(x(\lambda)), g) \qquad \forall g ,$$

and in this case the function $\Phi(\lambda)$ is continuously differentiable with respect to $\lambda > 0$. In particular, we have $R(\lambda) = \{x(\lambda)\}$ if the function $f(x)$ is strictly convex on E_n.

6. Let $\Phi(\lambda_0) = \Phi^*$, $\lambda_0 \ge 0$. Then, for some $x_0 \in R(\lambda_0)$

$$x_0 \in \Omega , \qquad f(x_0) = f^* ,$$

and for any $\lambda_1 > \lambda_0$

$$x_1 \in \Omega , \qquad f(x_1) = f^* \quad \forall x_1 \in R(\lambda_1) .$$

By virtue of property 3, it is sufficient to show that

$$\min_{x\in R(\lambda_0)} h^+(x) = 0 , \qquad \max_{x\in R(\lambda_1)} h^+(x) = 0 .$$

Since, according to property 4, $\Phi(\lambda_0+\alpha) = \Phi(\lambda_0)$ for any $\alpha \geq 0$, then

$$\Phi'_+(\lambda_0) = \lim_{\alpha\to+0} \frac{\Phi(\lambda_0+\alpha) - \Phi(\lambda_0)}{\alpha} = 0 ,$$

and according to property 5, $\Phi'_+(\lambda_0) = \min_{x\in R(\lambda_0)} h^+(x)$. Thus, $\min_{x\in R(\lambda_0)} h^+(x) = 0$.

Now let $\lambda_1 > \lambda_0$. Then, again using properties 4 and 5, we obtain

$$0 = \Phi'_-(\lambda_1) = \lim_{\alpha\to+0} \frac{\Phi(\lambda_1-\alpha) - \Phi(\lambda_1)}{\alpha} = \min_{x\in R(\lambda_1)} [-h^+(x)]$$

$$= - \max_{x\in R(\lambda_1)} h^+(x) .$$

■

Sufficient conditions for the existence of $\lambda^* \geq 0$ such that $\Phi(\lambda^*) = \Phi^*$ are formulated in the following property.

7. We assume that problem (8.1) has a solution and that the Slater constraint qualification is satisfied: i.e., for some $\bar{x} \in E_n$

$$h(\bar{x}) = -\gamma < 0 . \tag{8.5}$$

If $x^* \in \Omega$ is a minimum point of the function $f(x)$ on the set Ω and $0 = v^* + \lambda^* w^*$, where $\lambda^* \geq 0$, $v^* \in \partial f(x^*)$, $w^* \in \partial h(x^*)$ are from condition (1.16), then

$$x^* \in R(\lambda^*) , \qquad \Phi(\lambda^*) = \Phi^*$$

and for any $\lambda_1 > \lambda^*$ and $x_1 \in R(\lambda_1)$

$$x_1 \in \Omega , \qquad f(x_1) = f^* . \tag{8.6}$$

<u>Proof</u>. First of all, we note that $\lambda^* \leq \infty$ because (8.5) is satisfied and $h^+(x^*) = 0$ (it follows from (1.16)).

The equality $\Phi(\lambda^*) = \Phi^*$ is satisfied by virtue of property 3 if we can show that $x^* \in R(\lambda^*)$.

If $\lambda^* = 0$, then, on the one hand, we have $0 = v^* \in \partial f(x^*)$ and, hence, x^* is a minimum point of the function $f(x)$ on E_n and, on the other hand, $F(x,\lambda^*) = f(x)$ and, therefore, $R(\lambda^*) = \{x \in E_n \mid f(x) = \min_{z \in E_n} f(z)\}$ and $x^* \in R(\lambda^*)$.

Let $\lambda^* > 0$. Then, it should be $h(x^*) = 0$ and, hence,

$$\partial h(x^*) \subset \partial h^+(x^*) \quad . \tag{8.7}$$

The condition $0 = v^* + \lambda^* w^*$ implies that $0 \in \partial f(x^*) + \lambda^* \partial h(x^*)$.

From this and (8.7), we have

$$0 \in \partial_x F(x^*,\lambda^*) = \partial f(x^*) + \lambda^* \partial h^+(x^*) ,$$

i.e., x^* is a minimum point of the function $F(x,\lambda^*)$ on E_n and $x^* \in R(\lambda^*)$.

By virtue of property 6, (8.6) is satisfied for any $\lambda_1 > \lambda^*$ and any $x_1 \in R(\lambda_1)$. ■

8. Let $0 \in \partial_{x,\varepsilon} F(x_0,\lambda_0)$, $\varepsilon \geq 0$, $\lambda_0 \geq 0$. Then,

$$\Phi^* \leq F(x_0,\lambda_0) + (h^+(x_0), \lambda^*-\lambda_0) + \varepsilon , \tag{8.8}$$

where λ^* is such that $\Phi(\lambda^*) = \Phi^* = \max_{\lambda \geq 0} \Phi(\lambda)$.

To show this, we note that

$$\Phi(\lambda^*) \leq \Phi(\lambda_0) + (\beta,\lambda^*-\lambda_0) + \varepsilon \qquad \forall \beta \in \partial_\varepsilon \Phi(\lambda_0), \tag{8.9}$$

where $\partial_\varepsilon \Phi(\lambda_0)$ is the ε-superdifferential of the concave function $\Phi(\lambda)$ at λ_0.

On the other hand, we have

$$\text{co}\,\{h^+(x) \mid x \in R_\varepsilon(\lambda_0)\} \subset \partial_\varepsilon \Phi(\lambda_0)$$

(see Section 1.17). Since $0 \in \partial_{x,\varepsilon} F(x_0,\lambda_0)$, then

$$0 \le F(x_0,\lambda_0) - \min_{x\in\Omega} F(x,\lambda_0) \le \varepsilon .$$

This implies that $0 \le F(x_0,\lambda_0) - \Phi(\lambda_0) \le \varepsilon$, i.e., $x_0 \in R_\varepsilon(\lambda_0)$. Then, $h^+(x_0) \in \partial_\varepsilon \Phi(\lambda_0)$. Now, (8.8) follows from (8.9). ∎

<u>COROLLARY</u>. If $0 \in \partial_{x,\varepsilon} F(x_0,\lambda_0)$ and $x_0 \in \Omega$, then

$$\Phi(\lambda^*) \le F(x_0,\lambda_0) + \varepsilon .$$

9. If $\lambda_0 \ge \lambda^*$ and $0 \in \partial_{x,\varepsilon} F(x_0,\lambda_0)$, then

$$f(x_0) + \lambda_0 h^+(x_0) - \min_{x\in\Omega} f(x) \le \varepsilon ; \tag{8.10}$$

in particular, for $x_0 \in \Omega$ we have

$$0 \le f(x_0) - f^* \le \varepsilon .$$

Indeed, it follows from $\lambda_0 \ge \lambda^*$ that $\Phi(\lambda_0) = \Phi(\lambda^*)$, and since $0 \in \partial_{x,\varepsilon} F(x_0,\lambda_0)$, then we obtain

$$f(x_0) + \lambda_0 h^+(x_0) - f^* = F(x_0,\lambda_0) - \Phi(\lambda_0) \le \varepsilon .$$ ∎

2. Now let us consider the penalty-function method.

First we describe a «conceptual» algorithm.

Take an arbitrary $\lambda_0 > 0$. Using any of the methods in Chapter 3, find a point $x_0 = x(\lambda_0) \in R(\lambda_0)$, i.e., a point such that

$$F(x_0,\lambda_0) = \min_{x\in E_n} F(x,\lambda_0) .$$

If $h(x_0) = 0$, then, by virtue of property 3, the point x_0 is a minimum point of the function $f(x)$ on the set Ω. If $h(x_0) > 0$, then we let $\lambda_1 = \lambda_0 + 2\max\{\lambda_0, h^+(x_0)\}$ and find a point x_1 such that

$$F(x_1,\lambda_1) = \min_{x\in E_n} F(x,\lambda_1) .$$

We continue analogously.

If problem (8.1) has a solution and the Slater constraint qualification, (8.2), is satisfied, then in a finite number of steps we can find a number $\lambda_k > 0$ and a point $x_k \in E_n$ such that $x_k \in \Omega$ (i.e., $h(x_k) \leq 0$) and

$$\Phi(\lambda_k) = \Phi^* , \qquad f(x_k) = f^* .$$

Let us prove this assertion. Let x^* be a minimum point of $f(x)$ on Ω.

If $h(x^*) < 0$, then $0 \in \partial f(x^*) = \partial_x F(x^*,0)$ and, therefore, $x^* \in R(0)$.

By virtue of property 3, $\Phi(0) = \Phi^*$.

Hence, for the point x_0 found at the first step the relation

$$x_0 \in \Omega , \qquad f(x_0) = f^* = \min_{x \in E_n} f(x)$$

is satisfied (due to property 5).

Now let $h(x^*) = 0$. Then, by virtue of the necessary condition for a minimum, there exist $\lambda^* \geq 0$, $v^* \in \partial f(x^*)$, $w^* \in \partial h(x^*)$ such that $0 = v^* + \lambda^* w^*$. Due to property 6,

$$\Phi(\lambda^*) = \Phi^* = \max_{\lambda \geq 0} \Phi(\lambda) ,$$

and for the first λ_k greater than λ^*, we have

$$x_k \in \Omega , \qquad f(x_k) = f^* .$$

Moreover, let $\delta = \lambda^* - \lambda_0$. Since $\lambda_k \geq \lambda_0 + 2k\lambda_0$, then in order to obtain $\lambda_k \geq \lambda^*$, it is necessary to run a number of steps less than or equal to $\delta(2\lambda_0)^{-1}$. ∎

<u>REMARK</u>. All the results in this section are valid also for the case where

$$\Omega = \{x \in E_n \mid h_j(x) \leq 0 \ \forall j \in J\} .$$

Here $h_j(x)$, $j \in J \equiv 1:m$, are convex functions on E_n.

In this case, by λ we mean a vector $\lambda = (\lambda_1, \lambda_2, \ldots, \lambda_m)$, and by definition

$$\{\lambda \geq 0\} = \{\lambda \mid \lambda_j \geq 0 \quad \forall\, j \in 1:m\} ,$$

$$\{\lambda > \lambda^*\} = \{\lambda \mid \lambda_j > \lambda_j^* \quad \forall\, j \in 1:m\} ,$$

$$h(x) = \max_{j \in J} h_j(x) , \qquad \lambda h^+(x) = \sum_{j=1}^{m} \lambda_j h_j^+(x) ,$$

$$(h^+(x), g) = \sum_{j=1}^{m} h_j^+(x) g_j .$$

9. THE KELLEY METHOD FOR THE MINIMIZATION ON A CONVEX SET

Let $f(x)$ and $h(x)$ be convex functions on E_n.

The problem consists in finding

$$f^* = \inf_{x \in \Omega} f(x) ,$$

where $\Omega = \{x \in E_n \mid h(x) \leq 0\}$.

Assume that the set Ω is bounded and satisfies the Slater constraint qualification, i.e., there exists a point $\bar{x} \in E_n$ such that

$$h(\bar{x}) < 0 . \tag{9.1}$$

Let $S \subset E_n$ be an arbitrary compact set including Ω.

Fix $\varepsilon \geq 0$ and $\mu \in [0, -h(\bar{x}))$. Choose an arbitrary point $x_0 \in S$ and let $\sigma_1 = \{x_0\}$. We compute arbitrary $v(x_0) \in \partial_\varepsilon f(x_0)$ and $w(x_0) \in \partial_\mu h(x_0)$. Define

$$\phi_1(x) = \max_{z \in \sigma_1} B_f(x,z) = f(x_0) + (v(x_0), x-x_0) ,$$

$$\omega_1 = \{x \in S \mid B_h(x,z) \leq 0 \quad \forall\, z \in \sigma_1\}$$

$$= \{x \in S \mid h(x_0) + (w(x_0), x-x_0) \leq 0\} ,$$

where

$$B_f(x,z) = f(z) + (v(z), x-z) , \qquad v(z) \in \partial_\varepsilon f(z) ,$$

$$B_h(x,z) = h(z) + (w(z), x-z) , \qquad w(z) \in \partial_\mu h(z) .$$

Now, suppose that the sets σ_{k-1}, ω_{k-1} and the function

$\phi_{k-1}(x)$ have already been constructed, and let

$$x_{k-1} \in \omega_{k-1}, \qquad \phi_{k-1}(x_{k-1}) = \min_{x \in \omega_{k-1}} \phi_{k-1}(x) .$$

We compute arbitrary $v(x_{k-1}) \in \partial_\varepsilon f(x_{k-1})$ and $w(x_{k-1}) \in \partial_\mu h(x_{k-1})$.

Let

$$\sigma_k = \sigma_{k-1} \cup \{x_{k-1}\}, \qquad \phi_k(x) = \max_{z \in \sigma_k} B_f(x,z), \tag{9.2}$$

$$\omega_k = \{x \in S \mid B_h(x,z) \le 0 \quad \forall z \in \sigma_k\}$$

and find a point $x_k \in \omega_k$ such that

$$\phi(x_k) = \min_{x \in \omega_k} \phi_k(x) . \tag{9.3}$$

It follows from (9.2) that

$$\sigma_{k-1} \subset \sigma_k, \qquad \omega_k \subset \omega_{k-1}. \tag{9.4}$$

Then, from (9.3) and (9.4) we have

$$\phi_{k-1}(x_{k-1}) = \min_{x \in \omega_{k-1}} \max_{z \in \sigma_{k-1}} B_f(x,z) \le \min_{x \in \omega_k} \max_{z \in \sigma_k} B_f(x,z) = \phi_k(x_k) . \tag{9.5}$$

If $x_k \in \Omega$ and

$$\phi_k(x_k) \ge f(x_k), \tag{9.6}$$

then the process terminates. In this case, we have

$$f(x_k) \le f^*_\mu + \varepsilon , \tag{9.7}$$

where

$$f^* = \min_{x \in \Omega_\mu} f(x), \qquad \Omega_\mu = \{x \in E_n \mid h(x) \le -\mu\} .$$

Let us show this. Due to the definition of $\partial_\varepsilon f(z)$,

$$f(x) - f(z) \ge (v(z), x-z) - \varepsilon \qquad \forall x,z \in E_n, \ \forall v(z) \in \partial_\varepsilon f(z) .$$

Hence,

$$f(x) + \varepsilon \ge \sup_{x \in E_n} B_f(x,z) \ge \max_{x \in \sigma_k} B_f(x,z) = \phi_k(x) . \tag{9.8}$$

From the definition of $\partial_\mu h(z)$, we have

$$h(x) - h(z) \geq (w(z), x-z) - \mu \qquad \forall\, x,z \in E_n, \quad \forall\, w(z) \in \partial_\mu h(z) .$$

Therefore,

$$\begin{aligned}\Omega_\mu = \{x \in S \mid h(x) \leq -\mu\} &\subset \{x \in S \mid B_h(x,z) \leq 0 \;\forall\, z \in E_n\} \\ &\subset \{x \in S \mid B_h(x,z) \leq 0 \;\forall\, z \in \sigma_k\} = \omega_k .\end{aligned}$$

Consequently, we conclude from (9.8) that

$$f^* + \varepsilon = \min_{x\in\Omega_\mu} f(x) + \varepsilon \geq \min_{x\in\Omega_\mu} \phi_k(x) \geq \min_{x\in\omega_k} \phi_k(x) = \phi_k(x_k) . \qquad (9.9)$$

Now, (9.7) follows from (9.6) and (9.9).

If (9.7) does not hold, then let us construct σ_{k+1}, ω_{k+1} and $\phi_{k+1}(x)$ and continue analogously. As a result, we obtain a sequence $\{x_k\}$. If it is finite, then by construction, its final element belongs to the set Ω and satisfies (9.7). Otherwise, we get the following theorem.

<u>THEOREM 9.1</u>. If x^* is an accumulation point of the sequence $\{x_k\}$, then $x^* \in \Omega$ and

$$f(x^*) \leq f^*_\mu + \varepsilon . \qquad (9.10)$$

<u>Proof</u>. The existence of accumulation points of the sequence $\{x_k\}$ follows from the boundedness of the set S.

Let $x_{k_s} \xrightarrow[s\to\infty]{} x^*$. By virtue of (9.2) and the fact that $x_{k_{s_0}} \in \sigma_{k_s}$ for $k_s > k_{s_0}$, we have

$$h(x_{k_{s_0}}) + (w(x_{k_{s_0}}), x_{k_s} - x_{k_{s_0}}) \leq 0 \qquad \forall\, k_s > k_{s_0} . \qquad (9.11)$$

Since $\partial_\mu h(x)$ is bounded on S, then, taking the limit in (9.11) as $k_s, k_{s_0} \to \infty$ under the condition $k_s > k_{s_0}$, we obtain $h(x^*) \leq 0$, i.e., $x^* \in \Omega$.

The sequence $\{\phi_k(x_k)\}$ increases monotonically and, by virtue of (9.9), is bounded from above by $f^*_\mu + \varepsilon$. For $k_s > k_{s_0}$, we have

$$f_{\mu}^{*} + \varepsilon \geq \phi_{k_s}(x_{k_s}) \geq f(x_{k_{s_0}}) + (v_{k_{s_0}}, x_{k_s} - x_{k_{s_0}}) . \quad (9.12)$$

Since $\partial_\varepsilon f(x)$ is bounded on S, then, taking the limit in (9.12) as $k_s, k_{s_0} \to \infty$ under the condition $k_s > k_{s_0}$, we obtain (9.10). ∎

10. THE RELAXATION-SUBGRADIENT METHOD IN THE PRESENCE OF CONSTRAINTS

In this section, some of the results of Section 1.6 are extended to the problem of minimizing a convex function on a convex set.

1. The Minimization on an Arbitrary Closed Convex Set.

Let $\Omega \in E_n$ be a closed convex set and $f(x)$ be a convex function on some open convex set including Ω. The problem consists in finding $\inf_{x\in\Omega} f(x) \equiv f^*$.

Assume that the set Ω has interior points. Then (see Lemma 1.2), for the function $f(x)$ to attain the minimum value on the set Ω at a point x^*, it is necessary and sufficient that

$$0 \in L_\eta(x^*) ,$$

where $L_\eta(x) = \text{co}\,\{\partial f(x) \cup T_\eta(x)\}$, $T_\eta(x) = \{v \in -\Gamma^+(x) \mid \|v\| = \eta\}$ and $\eta > 0$ is an arbitrary fixed number.

Here, we are easily convinced that the mapping $T_\eta(x)$ is upper semicontinuous on Ω. Hence, by virtue of Lemmas 2.3 and 2.4 in Chapter 1, the mapping $L_\eta(x)$ is upper semicontinuous. Moreover, it is bounded on any bounded set in Ω because $\partial f(x)$ and $T_\eta(x)$ have the same property.

Let $d(x) = \min_{v \in L_\eta(x)} \|v\|$. Now we shall describe a relaxation-subgradient method for the case under consideration.

Let us fix positive numbers η, ε_0, δ and an integer m_0. As an initial approximation, we take an arbitrary point $x_0 \in \Omega$.

Assume that the set $D(x_0) = \{x \in \Omega \mid f(x) \leq f(x_0)\}$ is bounded.

Suppose the kth approximation has already been found. We shall show how we can construct the $(k+1)^{st}$ approximation $x_{k+1} \in \Omega$ from the kth cycle.

Let $x_{k0} = x_k$ and take an arbitrary vector $v_{k0} \in L_\eta(x_{k0})$. If $\|\bar{v}_{k,k-1}\| \le \varepsilon_0$, where $\bar{v}_{k,-1} = v_{k0}$, then we let $t_k = 0$, $x_{k+1} = x_{kt_k}$ and there the kth cycle terminates (as in Sections 3.6 and 3.7, we denote by t_k a number of steps in the kth cycle).

If $\|\bar{v}_{k,-1}\| \ge \varepsilon_0$, then on the ray $\{x_{k0}(\alpha) = x_{k0} - \alpha v_{k0} \mid \alpha \ge 0\}$ we find a point $x_{k0}(\alpha_{k0})$ such that

$$f(x_{k0}(\alpha_{k0})) = \min_{x_{k0}(\alpha)\in\Omega,\,\alpha\ge 0} f(x_{k0}(\alpha)) \tag{10.1}$$

(it is possible that $\alpha_{k0} = 0$).

Now let $x_{k1} = x_{k0}$. It is obvious that $f(x_{k1}) \le f(x_{k0})$ and $x_{k1} \in \Omega$.

If $\|x_{k0} - x_{k1}\| \ge \delta$ or $f(x_{k0}) - f(x_{k1}) \ge \delta$, then we let $t_k = 1$, $x_{k+1} = x_{kt_k}$ and the kth cycle terminates. Otherwise, we take a vector $\bar{v}_{k0} \in L_\eta(x_{k1})$ such that

$$(v_{k0}, \bar{v}_{k0}) \le 0 . \tag{10.2}$$

By virtue of (10.1), such a vector $\bar{v}_{k0} \in L_\eta(x_{k1})$ always exists.

Indeed, if (10.1) is satisfied at a point $x_{k1} = x_{k0}(\alpha_{k0})$, then this is possible because

a. either $\dfrac{\partial f(x_{k1})}{\partial(-v_{k0})} = \max_{v\in\partial f(x_{k1})} (v, -v_{k0}) \ge 0$, and in this case we can find a subgradient $\bar{v}_{k0} \in \partial f(x_{k1})$ for which (10.2) holds,

b. or $x_{k1} - \beta v_{k0} \notin \Omega$ for any $\beta > 0$, i.e., $-v_{k0} \notin \gamma(x_{k1}) = \{v = \lambda(z - x_{k1}) \mid \lambda > 0,\ z \in \Omega\}$ and there exists a vector $\bar{v}_{k0} \in -\Gamma^+(x_{k1})$, $\|\bar{v}_{k0}\| = \eta$, such that (10.2) holds.

If $\|\bar{v}_{k0}\| \le \varepsilon_0$, then we let $t_k = 1$, $x_{k+1} = x_{kt_k}$ and the kth cycle terminates. If $\|\bar{v}_{k0}\| > \varepsilon_0$, then let us find a vector v_{k1} such that

$$v_{k1} = \beta_{k0}\bar{v}_{k0} + (1-\beta_{k0})v_{k0}, \qquad \beta_{k0} \in (0,1),$$

$$\|v_{k1}\| = \min_{\beta\in[0,1]} \|\beta\bar{v}_{k0} + (1-\beta)v_{k0}\|$$

(see Subsection 1.6.1).

Now, in the intersection of the ray $\{x_{k1}(\alpha) = x_{k1} - \alpha v_{k1} \mid \alpha \ge 0\}$ and the set Ω, we find a point $x_{k2} = x_{k1}(\alpha_{k1})$ such that

$$f(x_{k2}) = \min_{x_{k1}(\alpha)\in\Omega,\,\alpha\ge 0} f(x_{k1}(\alpha)) .$$

We continue analogously until we get a number of steps $t_k \le m_0$ for which at least one of the following conditions is satisfied:

a. $t_k = m_0$; (10.3)

b. $\sum_{t=1}^{t_k} \|x_{k,t-1} - x_{kt}\| \ge \delta$; (10.4)

c. $f(x_{k0}) - f(x_{kt_k}) \ge \delta$; (10.5)

d. $\|\bar{v}_{k,t_k-1}\| \le \varepsilon_0$. (10.6)

After that, we let $x_{k+1} = x_{kt_k}$ and the kth cycle terminates. Furthermore, if (10.6) is satisfied, the whole process of successive approximations terminates.

We note that

$$(v_{kt}, \bar{v}_{kt}) \le 0 \qquad \forall t \in 0:(t_k-1) . \qquad (10.7)$$

As a result, we have constructed a sequence $\{x_k\}$. If it is finite, then, by construction, for the final point x_k we have $d(x_k) \le \varepsilon_0$. Otherwise, we have the following theorem.

<u>THEOREM 10.1</u>. Let the function $f(x)$ be strictly convex on the set Ω having interior points and let the set $\mathcal{D}(x_0)$ be bounded.

Then, if a point x^* is an accumulation point of the sequence $\{x_k\}$, $x^* \in \Omega$ and

$$d(x^*) \le \max\{\varepsilon_0, bq^{m_0-1}\} \equiv d_0 , \qquad (10.8)$$

where

$$q = (1+(\varepsilon_0 b^{-1})^2)^{-\frac{1}{2}} , \qquad b = \max\{\eta, \max_{x\in D(x_0)} \max_{v\in\partial f(x)} \|v\|\} .$$

In particular, if $m_0 > 1+(\ln q)^{-1}\ln(\varepsilon_0 b^{-1})$, then $d(x^*) \le \varepsilon_0$.

Proof is similar to that of Theorem 6.1 in Chapter 1 if we take into account the fact that the point-to-set mapping $L_\eta(x) = \text{co}\,\{\partial f(x) \cup T_\eta(x)\}$ is upper semicontinuous on Ω and inequalities (10.7) are satisfied. ∎

Let D be the diameter of the set $D(x_0)$ and let r be the radius of the maximum sphere included in Ω. Then, by virtue of (1.15) and (10.8), the estimate

$$f(x^*) - f^* \le D(1+2bD(\eta r)^{-1})d_0$$

holds for any accumulation point x^* of the sequence $\{x_k\}$.

This method can be used to find minimum points of the function $f(x)$ on the set Ω. In that case, the scheme of this method remains unchanged, except that one needs to verify the conditions $t_k = m_k$ and $\|v_{kt_k}\| \le \varepsilon_k$, where $\varepsilon_k \to 0$, $m_k \to 0$, instead of conditions (10.3) and (10.6), respectively.

<u>THEOREM 10.2</u>. Let a function $f(x)$ be strictly convex on the set Ω having interior points, the set $D(x_0)$ be bounded and let $\varepsilon_k \to 0$, $m_k \to 0$.

Then, any accumulation point of the sequence $\{x_k\}$ is a minimum point of the function $f(x)$ on the set Ω.

<u>PROBLEM 10.1</u>. Prove Theorems 10.1 and 10.2.

2. Let a set Ω be described by

$$\Omega = \{x \in E_n \mid h(x) \le 0\} ,$$

where $h(x)$ is a convex function on E_n.

Assume that the Slater constraint qualification is satisfied: there exist an $\gamma > 0$ and an $\bar{x} \in E_n$ such that

$$h(\bar{x}) = -\gamma < 0 .$$

Then, the set Ω has interior points.

By the definition of the set Ω, it is possible to describe the method in Subsection 10.1 more precisely.

Let

$$L(x) = \begin{cases} \partial h(x) & \text{if } h(x) < 0 , \\ \text{co } \{\partial f(x) \cup \partial h(x)\} & \text{if } h(x) = 0 , \\ \partial h(x) & \text{if } h(x) > 0 . \end{cases}$$

For the function $f(x)$ to attain its minimum value on the set Ω at a point $x^* \in \Omega$, it is necessary and sufficient that

$$0 \in L(x^*)$$

(see Theorem 1.2).

Successive approximations will be constructed analogously, as done in Subsection 10.1, if we take vectors $\bar{v}_{kt} \in L(x_{k,t+1})$ instead of vectors $\bar{v}_{kt} \in L_\eta(x_{k,t+1}) = \text{co } \{\partial f(x_{k,t+1}) \cup T_\eta(x_{k,t+1})\}$.

Notes and Comments

Chapter 1.

Sections 1 and 3. For a more detailed discussion, see V.F. Dem'yanov and V.N. Malozemov [36]; I.I. Eremin and N.N. Astaf'ev [49]; A.D. Ioffe and V.M. Tikhomirov [57]; S. Karlin [60]; V.L. Makarov and A.M. Rubinov [78]; B.N. Pshenichnyj (Psenicnyi) [103]; R.T. Rockafellar [110].

Section 2. This description is based on Chapter I of [78]. The properties of point-to-set mappings are studied in [57], [99], [111], [137], [144], [146], [162], [165].

Sections 4 and 5. The properties of convex functions and their subgradients can be found in [56], [57], [67], [68], [69], [70], [71], [77], [110], [111], [119], [166].

Section 6. Lemma 6.4 is given, essentially, in [80]. Necessary conditions for an extremum have been investigated by many authors: [6], [22], [24], [26], [32], [39], [45], [57], [73], [74], [82], [102], [147], [158], [159], [163].

Section 7. Lemmas 7.1 and 7.2 are stated in [43], [44]. Theorem 7.3 is a special case of the corresponding theorem in S.S. Kutateladze [66]. Example 4 has been suggested by V.M. Tikhomirov.

Section 8. Relation (8.2) is established in [110], where the concept of an ε-subdifferential is also introduced. E.A. Nurminskij (Nurminskii) [90] was the first who drew attention to the continuity of an ε-subdifferential with respect to x for any fixed $\varepsilon > 0$. Theorem 8.2 is stated in [43], [44]. Lemma 8.1 and estimate (8.19) have been obtained in [132].

Section 9. Estimates (9.21) and (9.22) generalize the corresponding estimates in [37].

Sections 10 and 11. Most of the results can be found in [43], [44].

Section 13. Subsections 13.1 and 13.2 follow the lines of [100].

Section 14. The results of this section have been presented in [42]. Similar problems are examined in [5], [27], [62], [116], [143].

Section 15. Our discussion is close to that in [110]. The differentiability of a convex function of one variable has been treated in [8], [121]. The multi-variable case is studied in [109]. The almost-everywhere-existence of the second differential of a convex function is established in [2].

Section 16. Our discussion follows the lines of S.S. Kutateladze and A.M. Rubinov [68]. The proof of Lemma 16.3 is, apparently, new. Conjugate convex functions and other aspects of a duality property have been investigated in [57], [110], [128], [138], [156], [157], [169].
Section 17. The results are new, except for Lemma 17.2. For $\mu=0$, inclusion (17.10) has been observed in [92].

Chapter 2

The material of this chapter is new. Quasidifferentiable functions have been introduced in [38], [41]. Therein and in [40], [101], the basic properties of these functions have been formulated. B.N. Pshenichnyj (Pshenichnyi) [104] introduced the concept of upper convex approximation in order to study minimum problems for a large class of functions (in particular, Lipschitzian functions). The quasidifferential and the upper convex approximation are related in the following way: if $\mathcal{D}f(x_0) = [\underline{\partial} f(x_0), \overline{\partial} f(x_0)]$, then for any $w \in \overline{\partial} f(x_0)$ the set $[w + \underline{\partial} f(x_0)]$ is an upper convex approximation function $f(x)$ at the point x_0.

L.N. Polyakova has proved Theorems 5.1-5.4, 7.2, 7.3 and Lemma 6.3. She has also obtained the necessary conditions for an extremum of quasidifferentiable functions on a set given by a quasidifferentiable equality.
Sections 6 and 7 follow the lines of [40]. Lemma 8.1 has been established by B.N. Pshenichnyj.

Using a result obtained by A.Ya. Zaslavskij, A.M. Rubinov has recently proved a theorem on the quasidifferentiability of a composition of quasidifferentiable functions and obtained the formula for its quasidifferential.

It is not hard to establish a relation between quasidifferentiable functions and "tents" that have been studied by V.G. Boltyanškij (Bolt'yanskii) in [6].

Chapter 3

Various problems of mathematical programming have been treated in [16], [22], [49], [55], [61], [84], [93], [105], [112], [114], [129], [145], [164], [172].
Section 1. Lemma 1.3 is new.
Section 2. Continuous analogues of iterative processes were apparently considered for the first time in [23]. A detailed discussion of the methods for minimizing smooth functions can be found in M. Aoki [3], F.P. Vasil'ev [16],

L.V. Kantorovich and G.P. Akilov [59], V.G. Karmanov [61], E. Polak [96], B.M. Pshenichnyj (Pshenichnyi) and Yu.M. Danilin [105].

Section 3. Methods of steepest descent for minimizing a maximum (not necessarily convex) function are discussed in [36]. It has been shown therein that for $\varepsilon = 0$ the method of steepest descent can diverge. In [132], D.P. Bertsekas and S.K. Mitter observed a relative arbitrariness in choosing a direction of descent while using ε-subdifferentials. Subsection 3.5 follows the lines of [132].

Section 4. The subgradient method (the initial variant involved a constant non-normed step-size) is developed in [125]. A similar idea is used in [47]. The convergence to a set of minimum points (if the step (4.1) is chosen) is proved in [50]. Corollary of Theorem 4.1 has been observed in [124]. The ε-subgradient method for a sufficiently wide class of nonsmooth functions (including convex functions) is suggested in [92]--this has been, apparently, one of the first attempts to devise constructive minimization methods. The idea used in Section 4 is due to M.Z. Khenkin.

A generalization of the subgradient method to stochastic programming problems is systematically studied in Yu.M. Ermol'ev [51]; also, see [30], [91], [127]. In [30], A.M. Gupal has considered an application of the subgradient method to minimizing locally Lipschitzian functions, using a smoothing operation. The subgradient method has been further developed and improved in [4], [28], [58], [124], [127].

Section 5. The discussion follows the lines of [34].

Section 6. The method considered in this section has been suggested in somewhat different formulations in [149], [150] and [170]. In [171], P. Wolfe shows that in the case of an exact line-search the method suggested in [170] is reduced to the conjugate-gradient method if the objective function is quadratic. In [11], L.V. Vasil'ev has established this relationship for an arbitrary smooth function.

Section 7. The results of this section are new.

Section 8. The Kelley method has been suggested for the first time in [148], in order to minimize a linear form on a convex set. V.P. Bulatov [7] investigates thoroughly various modifications of the Kelley method. Our discussion follows the lines of N.K. Khramova [122].

Section 9. The method described in this section has been developed in [1] (see also [39]).

Section 10. The discussion follows the lines of [33]. The idea of the method is closely related to [1], [108], [148].

Chapter 4

Section 1. Lemma 1.2 is proved in [12] and Lemma 1.3 in [20]. Theorem 1.2 is an obvious generalization of Theorem 3.1 of Chapter V of [36].

Section 2. Lemma 2.3 has been obtained in [13].

Section 3. The method described is a generalization of the method used in §6 of Chapter IV of [36].

Section 4. The results of Subsections 3.1 and 3.2 are due to V.K. Shomesova.

Section 5. For $\varepsilon = \mu = 0$, the method described has been taken from [97]. In [52], the method used in [97] is generalized to the case of quasiconvex functions. Our version of the method, for $G = E_n$, has first been published in [13].

Section 7. Our discussion follows the lines of [13].

Section 8. Property 6 which has been established in [48] is most essential in this section. The penalty-function method has been treated in [25], [106] and [120].

Section 10. The method described in Subsection 10.1 has been elaborated in [12]--it is actually a generalization of the methods considered in [149]-[151] and [170] to constrained problems. In [155], a similar method is suggested for minimizing semismooth [154] functions.

Sections 6 and 9. This material is new.

References

[1] Akilov, G.P., and Rubinov, A.M. "The Method of Successive Approximations for Determining the Polynomials of Best Approximation." *Soviet Math. Doklady*, vol.5, no.4 (1964): 951-954. (English transl.)

[2] Aleksandrov, A.D. "The Almost-everywhere Existence of the Second Derivative of a Convex Function and Some Related Properties of Convex Surfaces" (in Russian). *Uchenye zapiski LGU.* Ser. matem., 1936, no.6: 3-35.

[3] Aoki, Masanao. *Introduction to Optimization Techniques; Fundamentals and Applications of Nonlinear Programming.* New York: Macmillan; London: Collier-Macmillan, 1971.

[4] Bazhenov, L.G. "Convergence Conditions of a Minimization Method of Almost-Differentiable Functions." *Cybernetics*, vol.8, no.4 (1972): 607-609. (English transl.)

[5] Beresnev, V.V. "Necessary Conditions for an Extremum in Convex Maximin Problems of Related Sets." *Cybernetics*, vol.8, no.2 (1972): 275-281. (English transl.)

[6] Bolt'yanskii, V.G. "The Method of Tents in the Theory of Extremal Problems." *Russian Math. Surveys* 30: 3 (1975): 1-54. (English transl.)

[7] Bulatov, V.P. *Metody pogruzheniya v zadachakh optimizatsii* (Imbedding Methods in Optimization Problems). Novosibirsk: Nauka, 1977.

[8] Bourbaki, Nicolas, pseud. *Fonctions d'une variable réele.* Livre 5 of *Éléments de mathématique.* Paris: Herman, 1958. Russian transl.: *Funktsii dejstvitel'nogo peremennogo.* Moscow: Nauka, 1965.

[9] Warga, J. *Optimal Control of Differential and Functional Equations.* New York London: Academic Press, 1972.

[10] Vasil'ev, L.V. Equalizing maxima. In *Voprosy teorii i elementy programmnogo obespecheniya minimaksnykh zadach* (Problems in theory and software for minimax problems), ed. V.F. Dem'yanov and V.N. Malozemov, 34-40. Leningrad: Izdatel'stvo LGU, 1977.

[11] Vasil'ev, L.V. 1979. On the relation between the relaxation generalized-gradient method and the conjugate-gradient method. In *Chislennye metody nelinejnogo programmirovaniya.* Tezisy III Vsesoyuznogo Seminara (Numerical methods of nonlinear programming. Abstracts of the Third All-Union Seminar), 45-49. Kharkov (USSR).

[12] Vasil'ev, L.V., and Dem'yanov, V.F. 1976. A relaxation method for the optimization of a convex function on a convex set. In *Chislennye metody nelinejnogo programmirovaniya.* Tezisy II Vsesoyuznogo Seminara (Numerical methods of nonlinear programming. Abstracts of the Second All-Union Seminar), 100-102. Kharkov (USSR).

[13] Vasil'ev, L.V., and Dem'yanov, V.F. "The Method of (ε,μ,τ)-generalized Descent in the Presence of Constraints" (in Russian). *Vestnik Leningradskogo universiteta* 19 (1979): 19-23.

[14] Vasil'ev, L.V., Dem'yanov, V.F., and Lisina, S.A. The minimization of a convex function using ε-subgradients. In *Upravlenie dinamicheskimi sistemami* (Dynamic system control), 3-22. Leningrad: Izdatel'stvo LGU, 1978.

[15] Vasil'ev, L.V., and Tarasov, V.N. "On a Method for Solving Minimax Problems" (In Russian). *Optimizatsiya* 9, no.36 (1977): 53-57.

[16] Vasil'ev, F.P. *Lektsii po metodam resheniya ekstremal'nykh zadach* (Lectures on methods for solving extremal problems). Moscow: Izdatel'stvo MGU, 1974.
[17] Vojton, E.F. On methods for solving some extremal problems of the synthesis of electric circuits. In *Issledovanie operatsij (modeli, sistemy, resheniya)* (Operations research (models, systems, solutions)), 16-23. Computing Center of the USSR Academy of Sciences, Moscow, 1976.
[18] Vojton, E.F. A problem with linked constraints. In *Voprosy teorii i elementy programmnogo obespecheniya minimaksnykh zadach* (Problems in theory and software for minimax problems), ed. V.F. Dem'yanov and V.N. Malozemov, 106-114. Leningrad: Izdatel'stvo LGU, 1977.
[19] Vojton, E.F., and Polyakova, L.N. "On a Class of Problems with Linked Constraints" (in Russian). *Optimizatsiya* 10 (1973): 41-46.
[20] Dem'yanov, V.F., and Malozemov, V.N., eds. *Voprosy teorii i elementy programmnogo obespecheniya minimaksnykh zadach* (Problems in theory and software for minimax problems). Leningrad: Izdatel'stvo LGU, 1977.
[21] Vorob'ev, N.N. "The present State of the Theory of Games." *Russian Math. Surveys*, vol.25, no.2 (1970): 77-136. (English transl.)
[22] Gabasov, R., and Kirillova, F.M. *Metody optimizatsii* (Optimization Techniques). Minsk: Izdatel'stvo BGU, 1975.
[23] Gavurin, M.K. "Nonlinear Functional Equations and Continuous Analogs of Iterative Methods" (in Russian). *Izvestiya VUZov. Matematika*, 1958, no.5: 18-31.
[24] Gamkrelidze, R.V. 1971. First-order necessary conditions and axiomatics for extremal problems. In *Proceedings of the Steklov Institute of Math.*, no.112: 156-186. American Math. Soc., Providence, Rhode Island. (English transl.)
[25] Germejer, Yu.B. *Vvedenie v teoriyu issledovaniya operatsij* (Introduction to operations research theory). Moscow: Nauka, 1971.
[26] Girsanov, I.V. *Lectures on Mathematical Theory of Extremum Problems.* Berlin Heidelberg New York Tokyo: Springer-Verlag, 1972. (English transl.)
[27] Gol'shtejn, E.G. *Teoriya dvojstvennosti v matematicheskom programmirovanii i ee prilozheniya* (Duality theory in mathematical programming and its applications). Moscow: Nauka, 1971. German transl.: *Dualitatstheorie in der nichtlinearen optimierung und ihrer anwendung.* Berlin: Akademie-Verlag, 1975.
[28] Gol'shtejn, E.G. "A Generalized Method for Finding Saddle Points" (in Russian). *Ekonomika i matematicheskie metody*, vol.8, no.4 (1972): 569-579.
[29] Gorelik, V.A. "Approximate Search for a Maximum Under Constraints Connecting the Variables." *USSR Comput. Maths. Math. Phys.* 12 , 2 (1972): 294-305. (English transl.)
[30] Gupal, A.M. *Stokhasticheskie metody resheniya negladkikh ekstremal'nykh zadach* (Stochastic methods for solving nonsmooth extremal problems). Kiev: Naukova Dumka, 1979.
[31] Danskin, John M. *The Theory of Max-min and Its Application to Weapons Allocation Problems.* New York Berlin Heidelberg: Springer-Verlag, 1967.
[32] Dem'yanov, V.F. *Minimaks: Differentsiruemost' po napravleniyam* (Minimax: Directional Differentiability). Leningrad: Izdatel'stvo LGU, 1974.
[33] Dem'janov, V.F. "On the Extremal Basis Method." *Soviet Math. Doklady*, vol.17, no.4 (1974): 991-995. (English transl.)

[34] Dem'yanov, V.F. "A multi-step Method of Generalized Gradient Descent." *USSR Comput. Maths. Math. Phys.*, vol.18, no.2 (1978): 45-55. (English transl.)
[35] Dem'yanov, V.F. "A Modified Generalized-Gradient Method in the Presence of Constraints" (in Russian). *Vestnik Leningradskogo universiteta* 19 (1978): 25-29.
[36] Dem'janov, V.F., and Malozemov, V.N. *Introduction to Minimax*. New York: Wiley, 1974. (English transl.)
[37] Dem'yanov, V.F., and Pevnyi, A.B. "Some Estimates in Minimax Problems." *Cybernetics*, vol.8, no.1 (1972): 116-123. (English transl.)
[38] Dem'yanov, V.F., Polyakova, L.N., and Rubinov, A.M. 1979. On a generalization of the concept of a subdifferential. In *Vsesoyuznaya konferentsiya "Dinamicheskoe upravlenie". Tezisy dokladov* (The All-Union conference on dynamic control. Abstracts of Reports), 79-84. Sverdlovsk (USSR).
[39] Dem'yanov, V.F., and Rubinov, A.M. *Approximate Methods of Solving Extremal Problems*. New York: American Elsevier, 1970. (English transl.)
[40] Dem'yanov, V.F., and Polyakova, L.N. "Minimization of a Quasidifferentiable Function on a Quasidifferentiable Set." *USSR Comput. Maths. Math. Phys.*, vol.20, no.4 (1980): 34-43. (English transl.)
[41] Dem'yanov, V.F., and Rubinov, A.M. "On Quasidifferentiable Functionals." *Soviet Math.Doklady*, vol.21, no.1 (1980): 14-17. (English transl.)
[42] Dem'yanov, V.F., and Shomesova, V.K. "Directional Differentiability of a Supremum Function" (in Russian). *Vestnik Leningradskogo universiteta*, no.7 (1978): 15-20.
[43] Dem'yanov, V.F., and Shomesova, V.K. (Dem'janov, Somesova) "Conditional Subdifferentials of Convex Functions." *Soviet Math. Doklady*, vol.19, no.5 (1978): 1185. (English transl.)
[44] Dem'yanov, V.F., and Shomesova, V.K. Conditional subgradients and conditional subdifferentials of convex functions. In *Sovremennoe sostoyanie teorii issledovaniya operatsij* (The current status of the operations research), 311-335. Edited by N.N. Moiseev. Moscow: Nauka, 1979.
[45] Dubovitskii, A.Ya., and Milyutin, A.A. "Extremum Problems in the Presence of Restrictions." *USSR Comput. Maths. Math. Phys.*, vol.5, no.3 (1965): 1-80. (English transl.)
[46] Evtushenko, Yu.G. "A Numerical Method for Finding Best Guaranteed Estimates." *USSR Comput. Maths. Math. Phys.*, vol.12, no.1 (1972): 109-128. (English transl.)
[47] Eremin, I.I. "Iteration Method for Cebysev Approximations for Sets of Incomplete Linear Inequalities." *Soviet Math. Doklady* 3 (1962): 570-572. (English transl.)
[48] Eremin, I.I. "On the Penalty Method in Convex Programming" (in Russian). *Kibernetika* 4 (1967): 63-67.
[49] Eremin, I.I., and Astaf'ev, N.N. *Vvedenie v teoriyu linejnogo i vypuklogo programmirovaniya* (Introduction to the theory of linear and convex programming). Moscow: Nauka, 1976.
[50] Ermol'ev, Yu.M. "Methods for Solving Nonlinear Extremal Problems" (in Russian). *Kibernetika* 4 (1966): 1-17.
[51] Ermol'ev, Yu.M. *Metody stokhasticheskogo programmirovaniya* (Stochastic programming methods). Moscow: Nauka, 1976.
[52] Zabotin, Ya.I., Korablev, A.I., and Khabibullin, R.F. "On the Minimization of Quasiconvex Functionals" (in Russian). *Izvestiya VUZov, Matematika* 10 (1972): 27-33.

[53] Zabotin, Ya.I., and Kreinin, M.I. "On the Convergence of Methods of Determining a Minimax." *Soviet Math. Iz. VUZ*, vol.21, no.10 (1977): 45-52. (English transl.)

[54] Zangwill, W.I. *Nonlinear Programming: A Unified Approach.* Englewood Cliffs, N.J.: Prentice-Hall, 1969.

[55] Zoutendijk, G. *Methods of Feasible Directions; a Study in Linear and Nonlinear Programming.* Amsterdam: Elsevier, 1960.

[56] Ioffe, A.D., and Levin, V.L. "Subdifferentials of Convex Functions." *Trans. Moscow Math. Soc.* 26 (1972): 1-72. (English transl.)

[57] Ioffe, A.D., and Tikhomirov, V.M. *Theory of Extremal Problems.* Amsterdam New York: North-Holland, 1979. (English transl.)

[58] Belen'kij, V.Z., and Volkonskij, B.A., eds. *Iterativnye metody v teorii igr i programmirovanii* (Iterative methods in the theory of games and programming). Moscow: Nauka, 1974.

[59] Kantorovich, L.V., and Akilov, G.P. *Functional Analysis in Normed Spaces.* New York: Macmillan, 1964. (English transl.)

[60] Karlin, Samuel. *Mathematical Methods and Theory in Games, Programming, and Economics.* Reading, Mass.: Addison-Wesley Pub. Co., 1959.

[61] Karmanov, V.G. *Matematicheskoe programmirovanie* (Mathematical Programming). Moscow: Nauka, 1975.

[62] Kirin, N.E. *Metody posledovatel'nykh otsenok v zadachakh optimizatsii upravlyaemykh sistem* (Successive estimation methods in optimization problems of controlled systems). Leningrad: Izdatel'stvo LGU, 1975.

[63] Krassovskij, N.N., and Subbotin, A.I. *Pozitsionnye differentsial'nye igry* (Closed-loop differential games). Moscow: Nauka, 1974.

[64] Kurzhanskij, A.B. *Upravlenie i nablyudenie v usloviyakh neopredelennosti* (Control and observation under uncertainty conditions). Moscow: Nauka, 1977.

[65] Kusraev, A.G. "On Necessary Conditions for an Extremum of Nonsmooth Vector-valued Mappings." *Soviet Math. Doklady*, vol.19, no.5 (1978): 1057-1060. (English transl.)

[66] Kutateladze, S.S. "Convex ε-programming." *Soviet Math. Doklady*, vol.20, no.2 (1979): 391-393. (English transl.)

[67] Kutateladze, S.S. "Convex Operators." *Russian Math. Surveys* 34, 1 (1979): 181-214. (English transl.)

[68] Kutateladze, S.S., and Rubinov, A.M. *Dvojstvennost' Minkovskogo i ee prilozheniya* (The Minkowski duality and applications). Novosibirsk: Nauka, 1976.

[69] Lebedev, V.N., and Tynjanskii, N.T. "Duality Theory for Concave-Convex Games." *Soviet Math. Doklady*, vol.8, no.3 (1967): 752-756. (English transl.)

[70] Levin, V.L. "Subdifferentials of Convex Functionals" (in Russian). *Uspekhi matemat. nauk* XXV, no.6 (1970): 183-184.

[71] Levin, V.L. "Subdifferentials of Convex Mappings and of Compositions of Functions." *Siberian Math. J.*, vol.13, no.6 (1972): 903-909. (English transl.)

[72] Levitin, E.S. "A General Minimization Method for Unsmooth Extremal Problems." *USSR Comput. Maths. Math. Phys.*, vol.9, no.4 (1969): 63-93. (English transl.)

[73] Levitin, E.S., Milyutin, A.A., and Osmolovskij, N.P. 1974. Conditions for a local minimum in a constrained problem. In *Matematicheskaya ekonomika i funktsional'nyj analiz* (Mathematical economics and functional analysis), 139-202. Moscow: Nauka.

[74] Levitin, E.S., Milyutin, A.A., and Osmolovskii, N.P. "Conditions of High Order for a Local Minimum in Problems with Contraints." *Russian Math. Surveys*, vol.33, no.6 (1978): 97-168. (English transl.)
[75] Lisina, S.A. 1979. The minimization of an ε-Convex Function. In *Chislennye metody nelinejnogo programmirovaniya. Tezisy III Vsesoyuznogo Seminara* (Numerical methods of nonlinear programming. Abstracts of the Third All-Union seminar), 122-124. Kharkov (USSR).
[76] Laurent, Pierre-Jean. *Approximation et optimisation*. Paris: Hermann, 1972.
[77] Luderer, B. "Minimax Problems and Convex Analysis" (in Russian). *Vestnik Leningradskogo universiteta* 6 (1975): 25-31.
[78] Makarov, V.L., and Rubinov, A.M. *Mathematical Theory of Economic Dynamics and Equilibria*. Berlin Heidelberg New York: Springer Verlag, 1977. (English transl.)
[79] Malozemov, V.N. "Equalization of Maxima." *USSR Comput. Maths. Math. Phys.*, vol.16, no.3 (1976): 232-236. (English transl.)
[80] Malozemov, V.N. "On Sufficient Conditions for a Local Minimax" (in Russian). *Vestnik Leningradskogo universiteta* 7 (1976): 55-59.
[81] Moiseev, V.N., and Pevnyj, A.B. Equalizing maxima. In *Voprosy teorii i elementy programmnogo obespecheniya minimaksnykh zadach* (Problems in theory and software for minimax problems), ed. V.F. Dem'yanov and V.N. Malozemov, 27-34. Leningrad: Izdatel'stvo LGU, 1977.
[82] Moiseev, N.N. *Elementy teorii optimal'nykh sistem* (The theory of optimal systems). Moscow: Nauka, 1975.
[83] Mockus, J.B. *Mnogoekstremal'nye zadachi v proektirovanii* (Multiextrema Problems in engineering design). Moscow: Nauka, 1975.
[84] Mukhacheva, Eh.A., and Rubinshtejn, G.Sh. *Matematicheskoe programmirovanie* (Mathematical programming). Novosibirsk: Nauka, 1977.
[85] Natanson, I.P. *Theory of Functions of a Real Variable*. New York: F. Ungar Publ. Co., 1955-60.
[86] Nikaido, Hukukane. *Convex Structures and Economic Theory*. New York: Academic Press, 1968.
[87] Norkin, V.I. "Nonlocal Minimization Algorithms of Nondifferentiable Functions." *Cybernetics*, vol.14, no.5 (1978): 704-707. (English transl.)
[88] Nurminskii, E.A. "Differentiability of Multivalued Mappings." *Cybernetics*, vol.14, no.5 (1978): 692-695. (English transl.)
[89] Nurminskii, E.A. "The Quasigradient Method for Solving Nonlinear Programming Problems." *Cybernetics*, vol.9, no.1 (1973): 145-150. (English transl.)
[90] Nurminskii, E.A. "Continuity of ε-subgradient Mappings." *Cybernetics*, vol.13, no.5 (1977): 790-791. (English transl.)
[91] Nurminskij, E.A. *Chislennye metody resheniya determinirovannykh i stokhasticheskikh minimaksnykh zadach* (Numerical methods for solving deterministic and stochastic minimax problems). Kiev: Nauka, 1979.
[92] Nurminskii, E.A., and Zhelikhovskii, A.A. "ε-quasigradient Method for Solving Nonsmooth Extremal Problems." *Cybernetics*, vol.13, no.1 (1977): 109-114. (English transl.)
[93] Ortega, J.M., and Rheinboldt, W.C. *Iterative Solution of Nonlinear Equations in Several Variables*. New York: Academic Press, 1970.
[94] Pevnyi, A.B. "Differentiation of a Maximin Function." *USSR Comput. Maths. Math. Phys.*, vol.11, no.2 (1971): 253-259. (English transl.)
[95] Podinovskij, V.V. On the relative importance of criteria in multicriteria decision-making. In *Mnogokriterial'nye zadachi prinyatiya reshenij* (Multicriteria decision-making problems), 48-92. Moscow: Mashinostroenie, 1978.

[96] Polak, E. *Computational Methods in Optimization; a Unified Approach.* New York: Academic Press, 1971.
[97] Polyak, B.T. (Poljak) "A General Method of Solving Extremum Problems." *Soviet Math. Doklady*, vol.8, no.3 (1967): 593-597. (English transl.)
[98] Polayk, B.T. "Minimization of Unsmooth Functionals." *Comput. Maths. Math. Phys.*, vol.9, no.3 (1969): 14-29. (English transl.)
[99] Polayk, B.T. (Poljak) "On Finding a Fixed Point of a Class of Set-Valued Mappings." *Soviet Math. Doklady*, vol.19, no.5 (1978): 1284-1288. (English transl.)
[100] Polyakova, L.N. "The Continuity of Normed Cones" (in Russian). *Vestnik Leningradskogo universiteta* 7 (1979): 112-113.
[101] Polyakova, L.N. 1979. Minimization of a class of nondifferentiable functions. In *Chislennye metody nelinejnogo programmirovaniya. Tezisy III Vsesoyuznogo Seminara* (Numerical methods of nonlinear programming. Abstracts of the Third All-Union Seminar), 145-148. Kharkov (USSR).
[102] Pontryagin, L.S., Boltyanskij, V.G., Gamkrelidze, R.V., and Mishchenko, E.F. *Matematicheskaya teoriya optimal'nykh protsessov* (Mathematical theory of optimal processes). Moscow: Fizmatgiz, 1961.
[103] Pshenichnyj, B.N. (Psenicnyi) *Necessary Conditions for an Extremum.* New York: Marcel Dekker, 1971. (English transl.)
[104] Pshenichnyi, B.N. "On Necessary Conditions for an Extremum for Non-differentiable Functions." *Cybernetics*, vol.13, no.6 (1977): 886-891. (English transl.)
[105] Pshenichnyi, B.M., and Danilin, Yu.M. *Numerical Methods in Extremal Problems*. Moscow: Mir, 1978. (English transl.)
[106] Razumikhin, B.S. *Fizicheskie modeli i metody teorii ravnovesiya v programmirovanii i ekonomike* (Physical models and methods of equilibrium theory in programming and economics). Moscow: Nauka, 1975.
[107] Rastrigin, L.A. *Statisticheskie metody poiska ekstremuma* (Statistical methods for finding an extremum). Moscow: Nauka, 1968.
[108] Remez, E.Ya. *Osnovy chislennykh metodov chebyshevskogo priblizheniya* (Fundamentals of computational methods of Chebyshev approximation). Kiev: Naukova Dumka, 1969.
[109] Reshetnyak, Yu.G. (Resetnjak, Ju.G.) "Generalized Derivatives and Differentiability Almost Everywhere." *Soviet Math. Doklady*, vol.4, no.3 (1968): 293-302. (English transl.)
[110] Rockafellar, R.T. *Convex Analysis*. Princeton Math. Series 28 (1970).
[111] Rubinov, A.M. "Sublinear Operators and Their Applications." *Russian Math. Surveys* 32: 4 (1977): 115-175. (English transl.)
[112] Rubinshtejn, G.Sh. *Konechnomernye modeli optimizatsii* (Finite-dimensional models of optimization). Novosibirsk: Izdatel'stvo Novosibirskogo Gosudarstvennogo universiteta, 1970.
[113] Rudin, Walter. *Principles of Mathematical Analysis*. 3d ed. New York: McGraw-Hill, 1976.
[114] Céa, Jean. *Optimisation--théorie et algorithmes*. Paris: Dunod, 1971. Russian transl.: *Optimizatsiya. Teoriya i Algoritmy*. Moscow: Mir, 1973.
[115] Smirnov, V.I. *A Course of Higher Mathematics*. Reading, Mass.: Addison-Wesley, 1964. (English transl.)
[116] Sotskov, A.I. "Necessary Conditions for a Minimum for a Type of Non-smooth Problem." *Soviet Math. Doklady*, vol.10, no.6 (1969): 1410-13. (English transl.)
[117] Strongin, R.G. *Chislennye metody v mnogoekstremal'nykh zadachakh* (Numerical methods in multiextrema problems). Moscow: Nauka, 1978.

[118] Tarasov, V.N. Acceleration of convergence in solving nonlinear minimax problems. In *Voprosy teorii i elementy programmnogo obespecheniya minimaksnykh zadach* (Problems in theory and software for minimax problems), ed. V.F. Dem'yanov and V.N. Malozemov, 76-81. Leningrad: Izdatel'stvo LGU, 1977.

[119] Tynyanskij, N.T. (Tynjanskii) "Conjugate Concave-Convex Functions in Linear Topological Spaces and Their Saddle Points." *Math. USSR Sbornik*, vol.7, no.4 (1969): 503-31. (English transl.)

[120] Fedorov, V.V. *Chislennye metody maksimina* (Numerical methods of maximin). Moscow: Nauka, 1979.

[121] Hardy, G.H., Littlewood, J.E., and Pólya, G. *Inequalities.* 2nd ed. Cambridge, England: Cambridge University Press, 1959.

[122] Khramova, N.K. "Some Generalizations of the Cutting-Plane Method" (in Russian). *Vestnik Leningradskogo universiteta* 7 (1979): 116-117.

[123] Dem'yanov, V.F., and Malozemov, V.N., eds. *Chislennye metody nelinejnogo programmirovaniya. Tesizy III Vsesoyuznogo Seminara* (Numerical method of nonlinear programming. Abstracts of the Second All-Union Seminar). Kharkov (USSR), 1979.

[124] Shepilov, M.A. "Method of the Generalized Gradient for Finding the Absolute Minimum of a Convex Function." *Cybernetics*, vol.12, no.4 (1976): 547-66. (English transl.)

[125] Shor, N.Z. 1962. An application of the gradient descent method for solving a transportation network problem. In *Materialy nauchnogo seminara po teoreticheskim i prikladnym voprosam kibernetiki i issledovaniya operatsij* (Proceedings of the scientific seminar on theoretical and application problems of cybernetics and operations research), 9-17. Vyp.1. Institut kibernetiki Akademii Nauk Ukrainskoj SSR, Kiev.

[126] Shor, N.Z. "A Class of Almost-Differentiable Functions and a Minimization Method for This Class." *Cybernetics*, vol.8, no.4 (1972): 599-606. (English transl.)

[127] Shor, N.Z. *Minimization Methods for Non-differentiable Functions.* Berlin Heidelberg New York: Springer-Verlag, 1985.

[128] Auslender, A. "Méthodes et théorèmes de dualité." *C.R. Ac. Sci. Paris*, vol.267 (1968): 1-4.

[129] Auslender, A. *Optimisation. Méthodes numériques.* Paris: Masson, 1976.

[130] Auslender, A. "Differentiable Stability in Nonconvex and Nondifferentiable Programming." *Math. Progr. Study*, vol.10 (1979): 29-41.

[131] Bertsekas, D.P. Nondifferentiable optimization via approximation. In *Mathematical Programming Study* 3. Nondifferentiable Optimization, ed. M.L. Balinski and P. Wolfe, 1-25. Amsterdam: North-Holland, 1975.

[132] Bertsekas, D.P., and Mitter, S.K. "A Descent Numerical Method for Optimization Problems with Nondifferentiable Cost Functionals." *SIAM J. Control*, vol.11, no.4 (1973): 637-52.

[133] Clarke, F.H. "Generalized Gradients and Applications." *Trans. Amer. Math. Soc.*vol.205 (1975): 247-62.

[134] Clarke, F.H. "A New Approach to Lagrange Multipliers." *Mathematics of Operations Research*, vol.1, no.2 (1976): 165-174.

[135] Cullum, Jane, Donath, W.E., and Wolfe, P. The minimization of certain nondifferentiable sums of eigenvalues of symmetric matrices. In *Mathematical Programming Study* 3. Nondifferentiable Optimization, ed. M.L. Balinski and P. Wolfe, 35-55. Amsterdam: North-Holland, 1975.

[136] Ekeland, Ivar. "Nonconvex Minimization Problems." *Bull. Amer. Math. Soc.*, vol.1, no.3 (1979): 443-74.

[137] Ekeland, I., and Valadier, M. "Representation of Set-valued Functions." *J. Math. Anal. Applications*, vol.35 (1971): 621-29.

[138] Fenchel, W. "On Conjugate Convex Functions." *Canad. J. Math.*, vol.1 (1949): 73-77.

[139] Goldstein, A.A. Optimization with corners. In *Nonlinear Programming*. Vol.2, 215-230. New York: Academic Press, Inc., 1975.

[140] Goldstein, A.A. "Optimization of Lipschitz Continuous Functions." *Math. Programming*, vol.13 (1977): 14-22.

[141] Halkin, H. Mathematical programming without differentiability. In *Calculus of Variations and Control Theory*, ed. D.L. Russel, 279-88. New York: Academic Press, Inc., 1976.

[142] Hiriart-Urruty, J.B. "On optimality Conditions in Nondifferentiable Programming." *Math. Programming*, vol.14 (1978): 73-86.

[143] Hogan, W.W. 1971. Directional derivatives for extremal values functions. West. Manag. Sci. Inst., Working paper, no.177. Los Angeles.

[144] Hogan, W.W. "Point-to-set Maps in Mathematical Programming." *SIAM Review*, vol.15 (1973): 591-603.

[145] Huard, P. Resolution of mathematical programming problems with nonlinear constraints by the method of centers. In *Nonlinear Programming*, ed. J. Abadie, 206-19. Amsterdam: North-Holland, 1967.

[146] Huard, P. "Optimization Algorithms and Point-to-set Maps." *Math. Programming*, vol.8 (1975): 308-331.

[147] Ioffe, A.D. "Necessary and Sufficient Conditions for a Local Minimum. 1-3." *SIAM J. Control and Optimization*, vol.17, no.2 (1979): 245-88.

[148] Kelley, J.F. "The Cutting-plane Method for Solving Convex Problems." *SIAM J. Appl. Math.*, vol.8, no.4 (1960): 703-12.

[149] Lemarechal, C. "Note on an Extension of Davidon Methods to Nondifferentiable Functions." *Math. Programming*, vol.7, no.3 (1974): 384-87.

[150] Lemarechal, C. 1974. An algorithm for minimizing convex functions. In *Proceedings of the IFIP Congress-74*, 552-56. Amsterdam.

[151] Lemarechal, C. On extension of Davidon methods to nondifferentiable problems. In *Mathematical Programming Study* 3. Nondifferentiable Optimization, ed. M.L. Balinski and P. Wolfe, 95-109. Amsterdam: North-Holland, 1975.

[154] Mifflin, R. "Semismooth and Semiconvex Functions in Constrained Optimization." *SIAM J. Control and Optimization*, vol.15, no.6 (1977): 959-72.

[155] Mifflin, R. 1977. An algorithm for constrained optimization with semismooth functions. RR-77-3, IIASA. Laxenburg, Austria.

[156] Moreau, J.-J. "Fonctionelles sous-différentiables." *C.R. Ac. Sci. Paris*, vol.257 (1963): 4117-19.

[157] Moreau, J.-J. Convexity and duality. In *Functional Analysis and Optimization*, ed. E.R. Caiancello, 145-69. New York: Academic Press, 1966.

[158] Neustadt, L.W. "A General Theory of Extremals." *J. Comput. System Sci.*, vol.3 (1969): 57-92.

[159] Neustadt, L.W. *Optimization: A Theory of Necessary Conditions*. Princeton, N.J.: Princeton University Press, 1976.

[160] *Nonsmooth Optimization*. Proceedings of a IIASA Workshop, 28 March-8 April, 1977. Ed. C. Lemarechal and R. Mifflin. Oxford: Pergamon Press, 1978.

[161] Osborne, M.R., and Watson, G.A. "An Algorithm for Minimax Approximation in the Non-linear Case." *Comput. J.*, vol.12, no.1 (1969): 63-68.

[162] Robinson, S.M., and Meyer, R.R. "Lower Semicontinuity of Multivalued Linearization Mappings." *SIAM J. on Control*, vol.11, no.3 (1973): 525-33.

[163] Robinson, S.M. "First-Order Conditions for General Nonlinear Optimization." *SIAM J. on Appl. Math.*, vol.30 (1976): 597-603.
[164] Rockafellar, R.T. "Augmented Lagrangian Multiplier Rule and Duality in Nonconvex Programming." *SIAM J. Control and Optimization*, vol.12, no.2 (1974): 268-85.
[165] Rockafellar, R.T. *The Theory of Subgradients and its Applications to Problems of Optimization*. Lecture Notes, University of Montreal, Montreal, 1978.
[166] Valadier, M. "Sous-différentiels d'une borne supérieure et d'une somme continue de fonctions convexes." *C.R. Ac. Sci. Paris*, vol.268 (1969): 39-42.
[167] Warga, J. "Necessary Conditions Without Differentiability Assumptions in Optimal Control." *J. Different. Equations* 18 (1975): 41-62.
[168] Warga, J. Derivative containers, inverse functions and controllability. In Symposium on *Calculus of Variations and Control Theory*, 13-45. University of Wisconsin-Madison, 1975. New York, 1976.
[169] Weiss, E.A. "Konjugierte Funktionen." *Arch. Math.*, Bd.XX (1969): 538-45.
[170] Wolfe, P. "Note on a Method of Conjugate Subgradients for Minimizing Nondifferentiable Functions." *Math. Programming*, vol.7, no.3 (1974): 380-83.
[171] Wolfe, P. A method of conjugate subgradients for minimizing nondifferentiable functions. In *Mathematical Programming Study* 3. Nondifferentiable Optimization, ed. M.L. Balinski and P. Wolfe, 145-73. Amsterdam: North-Holland, 1975.
[172] Wolfe, P. Convergence theory in nonlinear programming. In *Integral and Nonlinear Programming*, ed. J. Abadie, 1-36. Amsterdam: North-Holland, 1970.

Appendix 1

BIBLIOGRAPHY AND GUIDE TO PUBLICATIONS ON QUASIDIFFERENTIAL CALCULUS

This book (the Russian edition: Nedifferentsiruemaya Optimizatsiya) was submitted for publication in the summer of 1979, several months after the discovery of quasidifferentiable functions. At that time, the class of quasidifferentiable functions was viewed simply as a very rich class of functions with good properties which seem promising from the point of view of applications. Since then many new results have been obtained. But the main point, which is now recognized, is that *the study of the first-order properties of any directionally differentiable Lipschitzian function* can be limited to quasidifferentiable functions.

This is because the directional derivative of any Lipschitzian function is a positively homogeneous continuous function of a direction and can therefore be approximated to within any given accuracy by the difference of two positively homogeneous convex functions. Since any positively homogeneous convex function can be represented in the form $\max_{v \in A}(v,g)$, where $A \in E_n$ is a convex compact set, the directional derivative can be approximated by a function of the form

$$\max_{v \in A} (v,g) + \min_{w \in B} (w,g) ,$$

where A and B are convex compact sets in E_n. But this is the directional derivative of a quasidifferentiable function, and we can use the techniques of Quasidifferential Calculus to investigate its properties.

We shall now briefly mention only some of the properties of quasidifferentiable functions discovered during the last five years.

1. Quasidifferentiable functions form a linear space closed with respect to all "differentiable" operations and, more importantly, with respect to the operations of taking pointwise maximum and minimum. This has led to the development of Quasidifferential Calculus (which is a natural generalization of classical Differential Calculus (see [3*], [12*]-[15*]).

2. Necessary conditions for both unconstrained and constrained minima and maxima have been discovered (these are expressed in terms of quasidifferentials). The problem of optimizing a quasidifferentiable function on a quasidifferentiable set (i.e., a set which is described by a quasidifferentiable function) can now be treated in terms of the quasidifferentials of the functions involved (i.e., the function to be optimized and the functions describing the set over which optimization is to be performed) (see [5*], [8*], [29*]-[31*], [35*], [36*]).

3. If a point does not satisfy the necessary condition (say for a minimum), then steepest descent directions can be found (in the nonsmooth case, a steepest descent direction is not necessarily unique--this property is very important since it has been responsible for the failure of many approaches to nonsmooth problems) (see [5*], [6*], [8*], [16*]).

This opens the way to the development of numerical methods (see [6*], [20*], [21*], [25*], [26*], [32*], [33*], [37*]).

4. Many important properties of quasidifferentiable functions have been established: a chain rule (see [12*]-[14*]), implicit and inverse function theorems (see [1*], Chapter 2 of this book, [4*]), and a relation between the Clarke subdifferential and a quasidifferential (see [2*], [4*]).

Many practical problems can now be treated as quasidifferentiable (see, e.g. [38*]) and therefore it is important to develop and implement numerical methods for solving such problems. Software originally designed for smooth and convex problems can easily be incorporated into new software currently under development.

Note also that parallel computers are particularly suitable for treating quasidifferentiable problems.

Nondifferentiable functions and mappings can be extended to the infinite-dimensional case ([4*], [11*], [13*], [14*]), to control problems ([7*], [23*], [24*]), and to game theory ([28*]). Some interesting generalizations of quasidifferentiability are suggested in [17*]-[19*] and [34*].

A number of special topics in Quasidifferential Calculus are discussed in [22*], [27*], [29*].

To conclude, we can say that although a lot of problems can now be solved, many more important, interesting and challenging problems have arisen.

Appendix 2

BIBLIOGRAPHY
on
QUASIDIFFERENTIAL CALCULUS
as of January 1, 1985

[1*] Demidova, V.A., and Dem'yanov, V.F. 1983. A directional implicit function theorem for quasidifferentiable functions. Working paper WP-83-125. International Institute for Applied Systems Analysis, Laxenburg, Austria.

[2*] Dem'yanov, V.F. "On a Relation Between the Clarke Subdifferential and the Quasidifferential." *Vestnik Leningrad University Math.* 13 (1981): 183-89. (English transl.)

[3*] Dem'yanov, V.F. "Problems of Nonsmooth Optimization and Quasidifferentials" (in Russian). *Technical Cybernetics* 1 (1983): 9-19.

[4*] Dem'yanov, V.F. *Negladkie zadachi teorii upravleniya i optimizatsii* (Nonsmooth problems of control theory and optimization). Leningrad: Izdatel'stvo Leningradskogo universiteta, 1982.

[5*] Dem'yanov, V.F. 1983. Quasidifferentiable functions: necessary conditions and descent directions. Working Paper WP-83-64. International Institute for Applied Systems Analysis, Laxenburg, Austria.

[6*] Dem'yanov, V.F., Gamidov, S., and Sivelina, T.I. 1983. An algorithm for minimizing a certain class of quasidifferentiable functions. Working Paper WP-83-122. International Institute for Applied Systems Analysis, Laxenburg, Austria.

[7*] Dem'yanov, V.F., Nikulina, V.N., and Shablinskaya, I.R. 1984. Quasidifferentiable problems in optimal control. Working Paper WP-84-2. International Institute for Applied Systems Analysis, Laxenburg, Austria.

[8*] Dem'yanov, V.F., and Polyakova, L.N. "Minimization of a Quasidifferentiable Function on a Quasidifferentiable Set." *USSR Comput. Maths. Math. Phys.*, vol.20, no.4 (1981): 34-43. (English transl.)

[9*] Dem'yanov, V.F., Polyakova, L.N., and Rubinov, A.M. 1979. On a generalization of the concept of a subdifferential. In *Vsesoyuznaya konferentsiya "Dinamicheskoe upravlenie". Tezisy dokladov*. (The All-Union conference "Dynamic Control." Abstracts of reports), 79-84. Sverdlovsk (USSR).

[10*] Dem'yanov, V.F., Polyakova, L.N., and Rubinov, A.M. 1984. Nonsmoothness and quasidifferentiability. Working Paper WP-84-22. International Institute for Applied Systems Analysis, Laxenburg, Austria.

[11*] Dem'yanov, V.F., and Rubinov, A.M. "On Quasidifferentiable Functionals." *Soviet Math. Doklady* 21 (1980): 14-17. (English transl.)

[12*] Dem'yanov, V.F., and Rubinov, A.M. "On Some Approaches to Nonsmooth Optimization Problems" (in Russian). *Ekonomika i matematicheskie metody* 17 (1981): 1153-74.

[13*] Dem'yanov, V.F., and Rubinov, A.M. Elements of quasidifferential calculus (in Russian). In [4*], pp.5-127.

[14*] Dem'yanov, V.F., and Rubinov, A.M. "On Quasidifferentiable Mappings." *Mathematische Operationsforschung und Statistik, Series Optimization* 14, 1 (1983): 3-21.

[15*] Dem'yanov, V.F., and Vasil'ev, L.V. *Nondifferentiable Optimization*. New York Los Angeles: Optimization Software, Inc., 1985.

[16*] Dem'yanov, V.F., and Zabrodin, I.S. 1983. Directional differentiability of a continual maximum function of quasidifferentiable functions. Working Paper WP-83-58. International Institute for Applied Systems Analysis, Laxenburg, Austria.
[17*] Gorokhovik, V.V. "On Quasidifferentiability of Real-Valued Functions." *Soviet Math. Doklady* 26 (1982): 491-94. (English transl.)
[18*] Gorokhovik, V.V. "Quasidifferentiability of Real-Valued Functions and Local Extremum Conditions." *Siberian Math. J.*, vol.25, no.3 (1984): 388-395. (English transl.)
[19*] Gorokhovik, V.V. "ε-quasidifferentiability of Real-Valued Functions and Optimality Conditions in Extremal Problems." *Math. Programming Study.* Forthcoming.
[20*] Kiwiel, K.C. "A Linearization Method for Minimizing Certain Quasidifferentiable Functions." *Math. Programming Study.* Forthcoming.
[21*] Kiwiel, K.C. 1984. Randomized search directions in descent methods for minimizing certain quasidifferentiable functions. Collaborative Paper CP-84-56. International Institute for Applied Systems Analysis, Laxenburg, Austria.
[22*] Melzer, D. "On the Expressibility of Piecewise-Linear Continuous Functions as the Difference of Two Piecewise-Linear Convex Functions." *Math. Programming Study.* Forthcoming.
[23*] Nikulina, V.N., and Shablinskaya, I.R. Optimality conditions in control problems with a quasidifferentiable functional (in Russian). In [4], pp. 175-204.
[24*] Nikulina, V.N., and Shablinskaya, I.R. 1984. Quasidifferentiable terminal problems of optimal control. In *Abstracts of the IIASA Workshop on Nondifferentiable Optimization: Motivations and Applications*. September 17-22, 1984, Sopron, Hungary, 121-27. International Institute for Applied Systems Analysis, Laxenburg, Austria.
[25*] Pallaschke, D. 1984. On numerical experiences with a quasidifferentiable optimization algorithm. Ibid., 138-40.
[26*] Pallaschke, and Recht, P. 1984. On the steepest descent method for a class of quasidifferentiable optimization problems. Collaborative Paper CP-84-57. International Institute for Applied Systems Analysis, Laxenburg, Austria.
[27*] Pecherskaya, N.A. "Quasidifferentiable Mappings and the Differentiability of Maximum Functions." *Math. Programming Study.* Forthcoming.
[28*] Pechersky, S.L. 1984. Positively homogeneous quasidifferentiable functions and their applications in cooperative game theory. Collaborative Paper CP-84-26. International Institute for Applied Systems Analysis, Laxenburg, Austria.
[29*] Polyakova, L.N. "Necessary Conditions for an Extremum of Quasidifferentiable Functions." *Vestnik Leningrad University Math.* 13 (1981): 241-47.
[30*] Polyakova, L.N. "On One Nonsmooth Optimization Problem." *Cybernetics* 3 (1982):
[31*] Polyakova, L.N. 1984. On the minimization of a quasidifferentiable function subject to equality-type quasidifferentiable constraints. Collaborative Paper CP-84-27. International Institute for Applied Systems Analysis, Laxenburg, Austria.
[32*] Polyakova, L.N. 1984. On minimizing the sum of a convex function and a concave function. Collaborative Paper CP-84-28. International Institute for Applied Systems Analysis, Laxenburg, Austria.

[33*] Polyakova, L.N. 1984. On the minimization of a quasidifferentiable function subject to an equality-type constraint. In *Abstracts of the IIASA Workshop on Nondifferentiable Optimization: Motivations and Applications.* September 17-22, 1984, Sopron, Hungary, 141-44. International Institute for Applied Systems Analysis, Laxenburg, Austria.
[34*] Rubinov, A.M., and Yagubov, A.A. 1984. The space of star-shaped sets and its application in nonsmooth optimization. Collaborative Paper CP-84-29. International Institute for Applied Systems Analysis, Laxenburg, Austria.
[35*] Shapiro, A. "On Optimality Conditions in Quasidifferentiable Optimization." *SIAM J. Control and Optimization* 23, 4 (1984): 610-17.
[36*] Shapiro, A. "Quasidifferential Calculus and First-Order Optimality Conditions in Nonsmooth Optimization." *Math. Programming Study*. Forthcoming.
[37*] Silevina, T.I. "On the Minimization of One Class of Quasidifferentiable Functions." *Vestnik Leningrad University Math.* 7 (1983):
[38*] Vojton, E.F. Quasidifferentiable functions in problems of optimal synthesis of electric circuits (in Russian). In [4], 291-307.
[39*] Xia, Z.Q. 1984. On mean value theorems in quasidifferential calculus. In *Abstracts of the IIASA Workshop on Nondifferentiable Optimization: Motives and Applications*. September 17-22, 1984, Sopron, Hungary, 186-89. International Institute for Applied Systems Analysis, Laxenburg, Austria.

Index

TRANSLATIONS SERIES IN MATHEMATICS AND ENGINEERING

V.F. Dem'yanov, and L.V. Vasil'ev
Nondifferentiable Optimization
1985, xvii + 455 pp.
ISBN 0-911575-09-X Optimization Software, Inc.
ISBN 0-387-90951-6 Springer-Verlag New York Berlin Heidelberg Tokyo
ISBN 3-540-90951-6 Springer-Verlag Berlin Heidelberg New York Tokyo

V.P. Chistyakov, B.A. Sevast'yanov, and V.K. Zakharov
Probability Theory For Engineers
1985, approx. 200 pp.
ISBN 0-911575-13-8 Optimization Software, Inc.
ISBN 0-387-96167-4 Springer-Verlag New York Berlin Heidelberg Tokyo
ISBN 3-540-96167-4 Springer-Verlag Berlin Heidelberg New York Tokyo

B.T. Polyak
Introduction To Optimization
1985, approx. 450 pp.
ISBN 0-911575-14-6 Optimization Software, Inc.
ISBN 0-387-96169-0 Springer-Verlag New York Berlin Heidelberg Tokyo
ISBN 3-540-96169-0 Springer-Verlag Berlin Heidelberg New York Tokyo

V.A. Vasilenko
Spline Functions: Theory, Algorithms, Programs
1985, approx. 280 pp.
ISBN 0-911575-12-X Optimization Software, Inc.
ISBN 0-387-96168-2 Springer-Verlag New York Berlin Heidelberg Tokyo
ISBN 3-540-96168-2 Springer-Verlag Berlin Heidelberg New York Tokyo

V.F. Kolchin
Random Mappings
1985, approx. 250 pp.
ISBN 0-911575-16-2 Optimization Software, Inc.
ISBN 0-387-96154-2 Springer-Verlag New York Berlin Heidelberg Tokyo
ISBN 3-540-96154-2 Springer-Verlag Berlin Heidelberg New York Tokyo

A.A. Borovkov, Ed.
Advances In Probability Theory:
Limit Theorems For Sums of Random Variables
1985, approx. 400 pp.
ISBN 0-911575-17-0 Optimization Software, Inc.
ISBN 0-387-96100-3 Springer-Verlag New York Berlin Heidelberg Tokyo
ISBN 3-540-96100-3 Springer-Verlag Berlin Heidelberg New York Tokyo

Continued on page 454

TRANSLATIONS SERIES IN MATHEMATICS AND ENGINEERING

V.V. Ivanishchev, and A.D. Krasnoshchekov
Control of Variable Structure Networks

1985, approx. 200 pp.
ISBN 0-911575-05-7 Optimization Software, Inc.
ISBN 0-387-90947-8 Springer-Verlag New York Berlin Heidelberg Tokyo
ISBN 3-540-90947-8 Springer-Verlag Berlin Heidelberg New York Tokyo

A.N. Tikhonov, Ed.
Problems In Modern Mathematical Physics and Computational Mathematics

1985, approx. 500 pp.
ISBN 0-911575-10-3 Optimization Software, Inc.
ISBN 0-387-90952-4 Springer-Verlag New York Berlin Heidelberg Tokyo
ISBN 3-540-90952-4 Springer-Verlag Berlin Heidelberg New York Tokyo

N.I. Nisevich, G.I. Marchuk, I.I. Zubikova, and I.B. Pogozhev
Mathematical Modeling of Viral Diseases

1985, approx. 400 pp.
ISBN 0-911575-06-5 Optimization Software, Inc.
ISBN 0-387-90948-6 Springer-Verlag New York Berlin Heidelberg Tokyo
ISBN 0-387-90948-6 Springer-Verlag Berlin Heidelberg New York Tokyo

V.G. Lazarev, Ed.
Processes and Systems In Communication Networks

1985, approx. 250 pp.
ISBN 0-911575-08-1 Optimization Software, Inc.
ISBN 0-387-90950-8 Springer-Verlag New York Berlin Heidelberg Tokyo
ISBN 3-540-90950-8 Springer-Verlag Berlin Heidelberg New York Tokyo

TRANSLITERATION TABLE

R	E	R	E
а А	a	р Р	r
б Б	b	с С	s
в В	v	т Т	t
г Г	g	у У	u
д Д	d	ф Ф	f
е Е	e	х Х	kh
ё Ё	e	ц Ц	ts
ж Ж	zh	ч Ч	ch
з З	z	ш Ш	sh
и И	i	щ Щ	shch
й Й	j	ъ Ъ	"
к К	k	ы Ы	y
л Л	l	ь Ь	'
м М	m	э Э	eh
н Н	n	ю Ю	yu
о О	o	я Я	ya
п П	p		